Developments in Soil Science 28B

SOIL MINERAL–ORGANIC MATTER–MICROORGANISM INTERACTIONS AND ECOSYSTEM HEALTH

Ecological Significance of the Interactions Among Clay Minerals, Organic Matter and Soil Biota

Volume 28B

Developments in Soil Science
Series Editors: A.E. Hartemink and A.B. McBratney

Titles currently available in this Series

11A PEDOGENESIS AND SOIL TAXONOMY. I. Concepts and Interactions
L.P. Wilding, N.E. Smeck and G.F. Hall (Editors)
ISBN: 0-444-42100-9

13 ELEMENTS OF SOIL PHYSICS
P. Koorevaar, G. Menelik and C. Dirksen
ISBN: 0-444-42242-0

21 VOLCANIC ASH SOILS: Genesis, Properties and Utilization
S. Shoji, M. Nanzyo and R.A. Dahlgren
ISBN: 0-444-89799-2

23 SOIL CONSERVATION AND SILVICULTURE
J. Dvorák and L. Novák (Editors)
ISBN: 0-444-98792-4

24 VERTISOLS AND TECHNOLOGIES FOR THEIR MANAGEMENT
N. Ahmad and A. Mermut (Editors)
ISBN: 0-444-88789-X

25 SOIL QUALITY FOR CROP PRODUCTION AND ECOSYSTEM HEALTH
E.G. Gregorich and M.R. Carter (Editors)
ISBN: 0-444-81661-5

26 ENZYMOLOGY OF DISTURBED SOILS
S. Kiss, D. Paşca and M. Drăgan-Bularda
ISBN: 0-444-50057-X

27 FRACTALS IN SOILS SCIENCE
Y. Pachepsky, J.W. Crawford and W.J. Rawls (Editors)
ISBN: 0-444-50530-X

28A SOIL MINERAL–ORGANIC MATTER–MICROORGANISM INTERACTIONS AND ECOSYSTEM HEALTH: Dynamics, Mobility and Transformation of Pollutants and Nutrients
A. Violante, P.M. Huang, J.-M. Bollag and L. Gianfreda
ISBN: 0-444-51038-9

28B SOIL MINERAL–ORGANIC MATTER–MICROORGANISM INTERACTIONS AND ECOSYSTEM HEALTH: Ecological Significance of the Interactions Among Clay Minerals, Organic Matter and Soil Biota
A. Violante, P.M. Huang, J.-M. Bollag and L. Gianfreda
ISBN: 0-444-51039-7

Developments in Soil Science 28B

SOIL MINERAL-ORGANIC MATTER-MICROORGANISM INTERACTIONS AND ECOSYSTEM HEALTH

Ecological Significance of the Interactions Among Clay Minerals, Organic Matter and Soil Biota

Volume 28B

Edited by

A. Violante
Dipartimento di Scienze Chimico-Agrarie
Università di Napoli Federico II, Portici (Napoli)
Italy

P.M. Huang
Department of Soil Science
University of Saskatchewan, Saskatoon
Canada

J.-M. Bollag
Laboratory of Soil Biochemistry
Center for Bioremediation and Detoxification
The Pennsylvania State University
University Park, PA
USA

L. Gianfreda
Dipartimento di Scienze Chimico-Agrarie
Università di Napoli Federico II, Portici (Napoli)
Italy

2002
ELSEVIER
Amsterdam – Boston – London – New York – Oxford – Paris
San Diego – San Francisco – Singapore – Sydney – Tokyo

ELSEVIER SCIENCE B.V.
Sara Burgerhartstraat 25
P.O. Box 211, 1000 AE Amsterdam, The Netherlands

First edition 2002

Library of Congress Cataloging in Publication Data
A catalog record from the Library of Congress has been applied for.

British Library Cataloguing in Publication Data
A catalogue record from the British Library has been applied for.

ISBN: 0-444-51038-9 (Volume 28A)
ISBN: 0-444-51039-7 (Volume 28B)
ISBN: 0-444-51040-0 (Set comprising volumes 28A and 28B)

ISSN: 0166-2481 (Series)

Transferred to Digital Printing 2006

PREFACE

The Working Group M.O. (*Interactions of Soil Minerals with Organic Components and Microorganisms*) of the International Union of Soil Sciences (IUSS) was founded in 1990 at the 14th World Congress of Soil Sciences (Kyoto, Japan). It organized two International Symposia in Edmonton (Canada) in 1992 and in Nancy (France) in 1996. Specialized and shorter co-sponsored meetings were also held in Acapulco (Mexico) in 1994, in St Louis (USA) in 1995, in Charlottetown (Canada) in 1999 and in Guelph (Canada) in 2001. As a result of these Symposia and Meetings, seven volumes or special books were published in the last 6 years.

The present book consists of two volumes presenting 59 of a total of 175 invited and volunteered papers presented at the 3rd Symposium on "Soil Mineral-Organic Matter-Microorganism Interactions and Ecosystem Health". Organized by the Working Group MO it was held in Naples-Capri (Italy) from May 22 to 26, 2000. All papers accepted for publication in this book have been subjected to critical peer review. Volume A deals with the dynamics, mobility and transformation of pollutants and nutrients and Volume B covers the ecological significance of the interactions among clay minerals, organic matter and soil biota.

The aim of the 3rd Symposium was to provide a forum for the interaction of soil chemists, soil mineralogists, soil microbiologists, soil biochemists and environmental scientists with the intention of promoting discussions and exchanging information on many topics of mutual interest in this important area of science. It was also meant to stimulate research leading to an integration of knowledge on "soil minerals-natural organics-microorganisms" and their impact on agricultural production and environmental protection.

Soil is a dynamic system in which soil minerals constantly interact with organic matter and microorganisms. The close association and interactions between abiotic and biotic entities in soil environments govern (1) mineral weathering reactions, aggregate formation, and surface reactivity of soil minerals with respect to nutrients and environmental pollutants, (2) the dynamics and transformation of metals, metalloids, and natural and anthropogenic organics, and metabolic processes, growth and ecology of microbes, and (3) has an impact on soil development, agricultural production, environmental protection, and ecosystem integrity. Among soil processes, chemical and biogeochemical reactions have an important role in the speciation, bioavailability, toxicity, transformations and transport of metals and anthropogenic organics.

The 3rd Symposium was sponsored by the Commissions II (Soil Chemistry), III (Soil Biology), VII (Soil Mineralogy) and VIII (Soils and the Environment) of the IUSS, by the University of Naples Federico II and by the Working Group "NAMOX" of the Società Italiana di Scienza del Suolo (SISS). More than 220 scientists of 32 different countries (Australia, Austria, Bangladesh, Canada, Chile, China, Columbia, Czech Republic, Denmark, Egypt, Finland, France, Germany, Hungary, Israel, Italy, Japan, New Zealand, Poland, Spain, The Netherlands, Rumania, Russia, USA, UK, South Africa, Korea, Sri Lanka, Switzerland, Taiwan, Sweden, and Zimbabwe) participated at this scientific event. One hundred and seventy five papers were presented at the meeting. The participants of the Symposium represented several subdisciplines of Soil Sciences as well as Ecology, Environmental Science, Toxicology and Health Science.

We are highly appreciative of the response of the authors to our request for the preparation of updated and original manuscripts and are grateful to the external referees for their expert critiques and inputs to maintain the quality of this book. Sincere appreciation is extended to the Dipartimento di Scienze Chimico-Agrarie, University of Naples Federico II for support during the preparation of these volumes. Mrs. I. Crovella and Mr. M. Clumez deserve a special mention for their help in organizing the Symposium.

The Editors acknowledge with deep gratitude Dr. M. A. Rao for her active collaboration in the preparation of the programme and during the Symposium and mainly for her tireless effort and qualified help in editing this book.

The Editors are also grateful to Mrs Joy Drohan for her excellent technical help in editing the papers in respect to the English style and the typeset format used in these volumes.

Finally, the Editors express their gratitude to the Ministero delle Politiche Agricole e Forestali (Rome), to the University of Naples Federico II, to the Società Italiana di Scienza del Suolo (SISS), to the Gruppo Italiano AIPEA, to the Banco di Napoli, to Shimadzu, Perkin Elmer, Dionex for financial support for organizing the Symposium and for publishing these volumes.

A. Violante
P.M. Huang
J.-M. Bollag
L. Gianfreda

ABOUT THE EDITORS

Antonio Violante is Professor of Agricultural Chemistry at the University of Naples (Italy). He took his Ph.D. in Chemistry at the University of Naples in 1969. He was awarded postdoctoral fellowships from the University of Wisconsin, USA (1976-1977) and the University of Saskatchewan, Canada (1981-1982) and was invited professor at the Department of Soil Science, University of Saskatchewan, Canada in 1985 and 1992.

Dr. Violante was head of the Dipartimento di Scienze Chimico-Agrarie and is Coordinator of the *Doctoral School in Agricultural Chemistry* of the University of Naples Federico II. He has served on many committees of the Italian Society of Soil Science (President of the Session Soil Chemistry), Italian Society of Agricultural Chemistry. He is vice-president and liaisons officer of Gruppo Italiano AIPEA. He was the scientific chairman and chief organizer of International and National Congresses.

Dr. Violante has contributed to promote research on the interface between soil chemistry and mineralogy and soil biology. The areas of research include the formation mechanisms of Al-hydroxides and oxyhydroxides, the surface chemistry and reactivities of short-range ordered precipitation products of Al and Fe, the influence of biomolecules on the sorption/desorption of nutrients and xenobiotics on/from variable charge minerals and soils and on the factors which influence the sorption and residual activity of enzymes on phyllosilicates, variable charge minerals, organo-mineral complexes, and soils.

Dr. Violante is the author or co-author of 135 refereed research articles and book chapters and over 70 additional scientific contributions. He presented papers at many scientific Congresses and Symposia and gave invited lectures at Universities and Research Institutes worldwide. Dr. Violante has international research/teaching experience in Canada, USA, Europe, China and Chile. He has trained students for Master Degree, Ph.D., and postdoctoral fellows and received visiting scientists worldwide. He serves on the editorial board of three international journals.

Pan Ming Huang received his Ph.D. degree in Soil Science at the University of Wisconsin, Madison, in 1966. He is currently Professor of Soil Science at the University of Saskatchewan, Saskatoon, Canada. His research work has significantly advanced the frontiers of knowledge on the nature and surface reactivity of mineral colloids and organomineral complexes of soils and sediments and their role in the dynamics, transformations, and fate of nutrients, toxic metals, and xenobiotics in terrestrial and aquatic environments. His research findings, embodied in over 270 refereed scientific publications, including research papers, book chapters, and 10 books, are fundamental to the development of sound strategies for managing land and water resources.

He has developed and taught courses in soil physical chemistry and mineralogy, soil analytical chemistry, and ecological toxicology. He has successfully trained and inspired M.Sc. and Ph.D. students and postdoctoral fellows, and received visiting scientists world-wide. He has served on numerous national and international scientific and academic committees. He has served as a member of many editorial boards such as the *Soil Science Society of America Journal, Geoderma, Chemosphere, and Advances in Environmental*

Science. He is currently a titular member of the Commission of Fundamental Environmental Chemistry of the International Union of Pure and Applied Chemistry and is the founding and current Chairman of the Working Group MO "*Interactions of Soil Minerals with Organic Components and Microorganisms*" of the International Union of Soil Sciences. He received the distinguished Researcher Award from the University of Saskatchewan and the Soil Science Research Award from the Soil Science Society of America, the American Society of Agronomy, and the American Association for the Advancement of Science.

Jean-Marc Bollag is Professor of Soil Biochemistry and Director of the Center of Bioremediation and Detoxification, Environmental Resources Research Institute, at The Pennsylvania State University. He is the author or coauthor of more than 220 professional papers and serves on the editorial board of five international journals. He is a frequent lecturer at conferences and seminars throughout the world. A recipient of the Julius Baer Fellowship, the Gamma Sigma Delta Research Award, and the Badge of Merit from the Polish Ministry of Agriculture. Dr. Bollag is a Fellow of the American Academy of Microbiology, the Soil Science Society of America, and the American Society of Agronomy. He is also recipient of the "Environmental Quality Research Award" from the American Society of Agronomy.

Dr. Bollag received the Ph.D. degree in Plant Physiology from the University of Basel, Switzerland, and conducted postdoctoral work at the Weitzmann Institute of Science, Rehovoth, Israel, and at Cornell University, Ithaca, New York. He was also a Visiting Scientist in the Biochemistry Section of Agrochemicals at Ciba-Geigy, Basel, Switzerland. Most of his research is related to the fate of pollutants in the environment and to bioremediation problems (incorporation of pollutants into soil organic matter as a detoxification method and application of enzymes for pollution control).

Liliana Gianfreda is Professor of Agricultural Biochemistry and Soil Biochemistry at University of Naples Federico II. She is author or coauthor or more than 150 professional papers and has participated as invited lecturer to several National and International conferences and symposiums.

Dr. Gianfreda received her Ph.D. in Chemistry from the University of Naples, Italy and performed most of her post-doctoral work at the University of Naples. She was Visiting Scientist in the Laboratory of Soil Biochemistry of the Environmental Resources Research Institute of the Pennsylvania State University, USA, and in the Institute of Chemistry, Universidada Estadual de Campinas, Sao Paolo, Brasil.

She is President of the Commission Soil Biology of the Italian Soil Science Society. She is also member of several National and International Scientific Societies (Italian Society of Chemistry, Italian Society of Biochemistry, Italian Society of Soil Sciences, International Union of Soil Sciences, and Italian Society of Agricultural Chemistry

Dr. Gianfreda was Coordinator of several National and International Research projects. She has a large experience in soil enzymology and interactions between enzymatic and non-enzymatic proteins and soil organic and inorganic components. She is also expert of the use of biological agents for the restoration and bioremediation of polluted sites.

REFEREES

The following scientists gave their time and talent as technical referees of manuscripts submitted for publication in this book . Their devotion is greatly appreciated.

Alexander M.	(U.S.A)	**Farini A.**	(Italy)
Andreux F.	(France)	**Fent G.**	(Germany)
Arshad M.A.	(Canada)	**Filip Z.**	(Germany)
Barton C.	(U.S.A.)	**Gadd G.M.**	(U.K.)
Baveye P.	(U.S.A.)	**Gaillard J-F.**	(France)
Berthelin J.	(France)	**Gennari M**	(Italy)
Bespalova A.	(Russia)	**Germida J.**	(Canada)
Bollag J.-M.	(U.S.A.)	**Gerzabek M.H.**	(Austria)
Braun J.	(U.S.A.)	**Gianfreda L.**	(Italy)
Brookes P.C.	(U.K.)	**Gigliotti C.**	(Italy)
Burns R.G.	(U. K.)	**Hsu Pa Ho**	(U.S.A.)
Buurman P.	(The Netherlands)	**Insam H.**	(Austria)
Chenu C.	(France)	**Jackson T.**	(Canada)
Chiu C.Y.	(Taiwan)	**Kandeler E.**	(Germany)
Chorover J.	(U.S.A.)	**Kirchmann H.**	(Sweden)
Colin P.	(France)	**Knackmuss H-J.**	(Germany)
Curtin D.	(Canada)	**Krishnamurti G.**	(U.S.A.)
Davies G.	(U.S.A.)	**Kubicki J.D.**	(U.S.A.)
De Freiteis R. J.	(Canada)	**Kurek E.**	(Poland)
de Kimpe C.	(Canada)	**Leifeld J.**	(Germany)
Dec J.	(U.S.A.)	**Leyval C.**	(France)
Dick R.	(U.S.A.)	**Markkola A.M.**	(Finland)
Djurhuus J.	(Denmark)	**Matschonat G.**	(Germany)
Dubbin W.	(U.K.)	**Mbagwu J.**	(Nigeria)
Ehrlich E.	(Canada)	**McGrath S.P.**	(U. K.)
Eriksson J.	(Sweden)	**Naidja A.**	(Canada)
Etana A.	(Sweden)	**Nannipieri P.**	(Italy)

Olsen J.E.	(Denmark)
Page A.L.	(U.S.A.)
Pampura T.	(Russia)
Piccolo A.	(Italy)
Quiquampoix H.	(France)
Reinhold-Hurek B.	(Germany)
Ristori G.	(Italy)
Robert M.	(France)
Ruggiero P.	(Italy)
Sakurai K.	(Japan)
Schaeffer A.	(Germany)
Schnitzer M.	(Canada)
Schulten H.-R.	(Germany)
Segat A.	(Argentina)
Senesi N.	(Italy)
Sequi P.	(Italy)
Shindo H.	(Japan)
Speir T.	(New Zealand)
Staunton S.	(France)
Stotzky G.	(U.S.A.)
Tabatabai M.A.	(U.S.A)
Tani M.	(Japan)
Torrent J.	(Spain)
Trasar-Cepeda C.	(Spain)
Violante A.	(Italy)
Violante P.	(Italy)
Walia S.	(India)
Wang M.C.	(Taiwan)
Wilson J.	(U. K.)
Xing G.X.	(China)
Yang J.	(Korea)
Yuan G.	(New Zealand)

CONTRIBUTORS

Adamo P. Dipartimento di Scienze Chimico-Agrarie, Università di Napoli Federico II, Via Università 100, 80055 Portici (NA), Italy

Ahn M.-Y. Laboratory of Soil Biochemistry, The Pennsylvania State University, University Park, PA 16802, U.S.A.

Andreoni V. Dipartimento di Scienze e Tecnologie Alimentari/Microbiologiche, Università degli Studi di Milano, Via Celoria 2, 20133 Milano, Italy

Benedetti A. Istituto Sperimentale per la Nutrizione delle Piante di Roma, Via della Navicella 2/4, 00184 Roma, Italy

Bespalova A.Y. Moscow State University, Faculty of Soil Science, 119899 Vorobiovy Hills, Moscow, Russia

Bollag J.-M. Laboratory of Soil Biochemistry, The Pennsylvania state University, University Park, PA 16802, U.S.A.

Brandi G. Istituto Sperimentale per lo Studio e la Difesa del Suolo, Piazza D'Azeglio 30, 50121 Firenze, Italy

Brookes P.C. Soil Science Department, IACR-Rothamsted, Harpenden, AL5 2JQ, Hertfordshire, U. K.

Burgos P. Institute for Natural Resources and Agrobiology, Sevilla (IRNAS) (CSIC), Avenida Reina Mercedes, 10 P.O. Box 1052, 41080-Sevilla, Spain.

Burns R.G. Research School of Biosciences Biological Laboratory, University of Kent, Canterbury, Kent U. K.

Cabrera F. Institute for Natural Resources and Agrobiology, Sevilla (IRNAS) (CSIC). Avenida Reina Mercedes, 10 P.O. Box 1052, 41080-Sevilla, Spain

Calamai L. Istituto di Genesi e di Ecologia del Suolo, CNR, P.le delle Cascine 28, 50144 Firenze, Italy

Castaldini M. Istituto Sperimentale per lo Studio e la Difesa del Suolo, Piazza D'Azeglio 30, Firenze, Italy

Cavalca L. Dipartimento di Scienze e Tecnologie Alimentari/Microbiologiche, Università degli Studi di Milano, Via Celoria 2, 20133 Milano, Italy

Chang A.C. Department of Environmental Sciences, University of California, Riverside, California 92521, U.S.A.

Cheng H.H. Department of Soil, Water & Climate, University of Minnesota, St Paul, Minnesota 55108, U.S.A.

Chenu C. Unitè de Science du Sol, INRA Versailles, France

Civilini M. Dipartimento Scienze degli Alimenti Università degli Studi di Udine, Via Marangoni 97, 33100 Udine, Italy

Crecchio C. Dipartimento di Biologia e Chimica Agroforestale ed Ambientale, Università degli Studi di Bari, Via Amendola 165a, 70126 Bari, Italy

Curcio E. Dipartimento di Scienze della Vita, Seconda Università di Napoli, Via Vivaldi 45, 81100 Caserta, Italy

D'Oriano P. Dipartimento di Scienze della Vita, Seconda Università di Napoli, Via Vivaldi 45, 81100 Caserta, Italy

Dazzi C. Dipartimento di Agronomia, Coltivazioni Erbacee e Pedologia, Università di Palermo, Viale delle Scienze, 90128 Palermo, Italy

Dec J. Laboratory of Soil Biochemistry, The Pennsylvania State University, University Park, PA 16802, U.S.A.

De Marco A. Dipartimento di Biologia Vegetale, Università di Napoli Federico II, Via Foria 223, 80139 Napoli, Italy

De Kimpe C. Agricolture and Agri-Food, Ottawa, Canada

De Pascale R.A. Dipartimento di Scienze Ambientali, Seconda Università di Napoli, Via Vivaldi 43, 81100 Caserta, Italy

Dell'Abate M.T. Istituto Sperimentale per la Nutrizione delle Piante, Via della Navicella 2/4, 00184 Roma, Italy

Dell'Amico E. Dipartimento di Scienze e Tecnologie Alimentari e Microbiologiche. Università degli Studi di Milano, Via Celoria 2, 20133 Milano, Italy

El-Ghamry A.M. College of Environmental and Resource Sciences, Zhejiang University, Hanghou 310029 China

Fabiani A. Istituto Sperimentale per lo Studio e la Difesa del Suolo, Piazza D'Azeglio 30, Firenze, Italy

Farini A. Department of Crop Production, University of Milan, Via Celoria 2, 20133 Milano, Italy

Farrell R.E. Department of Soil Science, University of Saskatchewan, 51 Campus Drive, Saskatoon, Saskatchewan S7N 5A8, Canada

Fereres A. Consejo Superior de Investigaciones Cientificas, Centro de Ciencias Medio Ambientales Serrano, 115 dpdo, 28006 Madrid, Spain

Fierro A.R. Dipartimento di Biologia Vegetale, Università di Napoli Federico II, Via Foria 223, 80139 Napoli, Italy

Fioretto A. Dipartimento di Scienze della Vita, Seconda Università di Napoli, Via Vivaldi 45, 81100 Caserta, Italy

Fornasier F. Istituto Sperimentale per la Nutrizione delle Piante, S.O.P. Gorizia, Via Trieste 23, 34170 Gorizia, Italy

Franchi M. Department of Animal Biology and Genetics, University of Florence, Via Romana 17, 50125 Florence, Italy

Frick C.M. Department of Soil Science, University of Saskatchewan, 51 Campus Drive, Saskatoon, Saskatchewan S7N 5A8, Canada

Gallori E. Department of Animal Biology and Genetics, University of Florence, Via Romana 17, 50125 Florence, Italy

Gamba C. Istituto Sperimentale per lo Studio e la Difesa del Suolo, Piazza D'Azeglio 30, 50121 Firenze, Italy

Germida J.J. Department of Soil Science, University of Saskatchewan, 51 Campus Drive, Saskatoon, Saskatchewan S7N 5A8, Canada

Gerzabek M.H. Austrian Research Centers, Division of Life Sciences, A-2444 Seibersdorf, Austria

Gianfreda L. Dipartimento di Scienze Chimico-Agrarie, Università di Napoli Federico II, Via Università 100, 80055 Portici (NA), Italy

Gigliotti C. Department of Crop Production, University of Milan, Via Celoria 2, 20133 Milano, Italy

González P. Consejo Superior de Investigaciones Cientificas, Centro de Ciencias Medioambientales Serrano, 115 dpdo, 28006 Madrid, Spain

Haberhauer G. Austrian Research Centers, Division of Life Sciences, A-2444 Seibersdorf, Austria

Huang C.Y. College of Environmental and Resource Sciences, Zhejiang University, Hanghou 310029 China

Kandeler E. University Hohenheim, Institute of Soil Science, Emil-Wolff-Straße, D-70599 Stuttgart, Germany

Kirchmann H. Swedish University of Agricultural Sciences, Department of Soil Sciences, Box 7014, S-75007 Uppsala, Sweden

Koo Bon-Jun Department of Environmental Sciences, University of California, Riverside, California 92521, U.S.A.

Koskinen W.C. USDA-Agriculture Research Service, St Paul, Minnesota 55108, U.S.A.

Kreslavski V.D. Institute of Basic Biological Problems, RAS, Pushchino, Moscow Region, 142290 Russia

Lombardi A. Dipartimento di Scienze della Vita, Seconda Università di Napoli, Via Vivaldi 45, 81100 Caserta, Italy

Madejón E. Institute for Natural Resources and Agrobiology, Sevilla (IRNAS) (CSIC). Avenida Reina Merceds, 10 P.O. Box 1052, 41080-Sevilla Spain

Marfenina O.E. Moscow State University, Faculty of Soil Science, 119899 Vorobiovy Hills, Moscow, Russia

Miclaus N. Istituto Sperimentale per lo Studio e la Difesa del Suolo, Piazza D'Azeglio 30, Firenze, Italy

Mirabella A. Istituto Sperimentale per lo Studio e la Difesa del Suolo, Piazza D'Azeglio 30, Firenze, Italy

Motuzova G.V. Moscow State University, Faculty of Soil Science, 119899 Vorobiovy Hills, Moscow, Russia

Mpepereki S. Department of Soil Science and Agric. Eng., University of Zimbabwe, Harare, Zimbabwe

Muchaonyerwa P. Department of Soil Science and Agric. Eng,, University of Zimbabwe, Harare, Zimbabwe

Nannipieri P. Department of Soil Science and Plant Nutrition, University of Florence, P.le delle Cascine 28, 50144 Firenze, Italy

Nyamugafata P. Department of Soil Science and Agric. Eng., University of Zimbabwe, Harare, Zimbabwe

Page A.L. Department of Environmental Sciences, University of California, Riverside, California 92521, U.S.A.

Pantani O.L. Istituto per la Genesi e l'Ecologia del Suolo (IGES), CNR Firenze, P.le delle Cascine 28, 50144 Firenze, Italy

Papa S. Dipartimento di Scienze della Vita, Seconda Università di Napoli, Via Vivaldi 45, 81100 Caserta, Italy

Papini R. Istituto Sperimentale per lo Studio e la Difesa del Suolo, Piazza D'Azeglio 30, 50121 Firenze, Italy

Petronio B.M. Dipartimento di Chimica, Università di Roma, La Sapienza, Piazzale A. Moro 5, 00185 Roma, Italy

Pinzari F. Istituto Sperimentale per la Nutrizione delle Piante di Roma, Via della Navicella 2/4, 00184 Roma, Italy

Piovanelli C. Istituto Sperimentale per lo Studio e la Difesa del Suolo, Piazza D'Azeglio 30, 50121 Firenze, Italy

Pizzigallo M.D.R. Dipartimento di Biologia e Chimica Agroforestale ed Ambientale, Università degli Studi di Bari, Via Amendola 165a, 70126 Bari, Italy

Rao M.A. Dipartimento di Scienze Chimico-Agrarie, Università di Napoli Federico II, Via Università 100, 80055 Portici (NA), Italy

Renella G. Department of Soil Science and Plant Nutrition, University of Florence, P.le delle Cascine 28, 50144 Firenze, Italy

Rinaldi L. Department of Animal Biology and Genetics, University of Florence, Via Romana 17, 50125 Florence, Italy

Ristori G.G. Istituto per la Genesi e l'Ecologia del Suolo (IGES) CNR Firenze, P.le delle Cascine 28, 50144 Firenze, Italy

Rogers B.F. Department of Environmental Sciences, Rutgers University, New Brunswick, NJ 08901, U.S.A.

Ruggiero P. Dipartimento di Biologia e Chimica Agroforestale ed Ambientale, Università degli Studi di Bari, Via Amendola 165a, 70126 Bari, Italy

Rutigliano F.A. Dipartimento di Scienze Ambientali, Seconda Università di Napoli, Via Vivaldi 43, 81100 Caserta, Italy

Saccomandi F. Dipartimento di Scienze Chimico-Agrarie, Università di Napoli Federico II Via Università 100, 80055 Portici (NA), Italy

Sannino F. Dipartimento di Scienze Chimico-Agrarie, Università di Napoli Federico II, Via Università 100, 80055 Portici (NA), Italy

Santomassimo F. Istituto Sperimentale per lo Studio e la Difesa del Suolo, Piazza D'Azeglio 30, Firenze, Italy

Sartori G. Museo Tridentino di Scienze Naturali, Via Calepina 14, Trento, Italy

Sessitsch A. Austrian Research Centers, Division of Life Sciences, A-2444 Seibersdorf, Austria

Shea P.J. Institute of Agriculture and Natural Resources, University of Nebraska, Lincoln, NE 68583, U.S.A.

Stach J.E.M. Research School of Biosciences Biological Laboratory, University of Kent, Canterbury, Kent U. K.

Stotzky G. Laboratory of Microbial Ecology, Department of Biology, New York University, New York NY 10003, U.S.A.

Strynar M. Laboratory of Soil Biochemistry, The Pennsylvania State University, University Park, PA 16802, U.S.A.

Tate III R.L. Department of Environmental Sciences, Rutgers University, New Brunswick, NJ 08901, U.S.A

Vasilyeva G.K. Institute of Physicochemical and Biological Problems in Soil Science, RAS, Pushchino, Moscow Region, 142290 Russia

Vingiani S. Dipartimento di Scienze Chimico-Agrarie, Università di Napoli Federico II, Via Università 100, 80055 Portici (NA), Italy

Violante A. Dipartimento di Scienze Chimico-Agrarie, Università di Napoli Federico II, Via Università 100, 80055 Portici (NA), Italy

Violante P. Dipartimento di Scienze Chimico-Agrarie, Università di Napoli Federico II, Via Università 100, 80055 Portici (NA), Italy

Virzo De Santo A. Dipartimento di Biologia Vegetale, Università di Napoli Federico II, Via Foria 223, 80139 Napoli, Italy.

Xu J.M. College of Environmental and Resource Sciences, Zhejiang University, Hanghou 310029 China.

CONTENTS

Clays and Humic Acids Affect Persistence and Biological Activity of Insecticidal Proteins from *Bacillus Thuringiensis* in Soil
G. Stotzky 1

Microbial Ecology of Soil Biofilms: Substrate Bioavailability, Bioremediation and Complexity
R.G. Burns and J.E.M. Stach 17

Biogeochemistry of Phosphorus, Iron, and Trace Elements in Soils as Influenced by Soil-Plant-Microbial Interactions
A.C. Chang, A.L. Page and Bon-Jun Koo 43

Adsorption of the Insecticidal Toxin from *Bacillus Thuringiensis* Subspecies *Tenebrionis* to Clay Fractions of Tropical Soils
P. Muchaonyerwa, C. Chenu, O.L. Pantani, L. Calamai, P. Nyamugafata and S. Mpepereki 59

Toxins from *Bacillus Thuringiensis* Subsp. *Israelensis* Sorbed on Clays: Persistence and Activity Against the Mosquito *Culex Pipiens*
P. González, O.L. Pantani, G.G. Ristori and A. Fereres 69

Soil Ecosystem Properties, Microbial Diversity, and Ecosystem Assessments
R.L. Tate III and B.F. Rogers 79

Effects of Soil Abiotic Processes on the Bioavailability of Anthropogenic Organic Residues
P. Ruggiero, M.D.R. Pizzigallo and C. Crecchio 95

Interactions of Minerals-Organic Matter-Living Organisms on the Fate of Allelochemicals and Xenobiotics in Soil: A Methodological Evaluation
H.H. Cheng and W.C. Koskinen 135

Environmental Effect of Metsulfuron-Methyl on the Size of Microbial Biomass and N-Mineralization in a Loamy Sand Soil
A.M. El-Ghamry, C.Y. Huang and J.M. Xu 147

Transformation and Binding of 2,4,6-Trinitrotoluene in Soil Amended with Activated Carbon
G.K. Vasilyeva, V.D. Kreslavski, P.J. Shea and J.-M. Bollag 157

Phytoremediation of Oil-Contaminated Soils
J.J. Germida, C.M. Frick and R.E. Farrell 169

Secondary Mobilization of Heavy Metals in Polluted Soils under Microbial Influence (Model Experiment)
A.Yu. Bespalova, G.V. Motuzova and O.E. Marfenina 187

Microbial Biomass Response to Heavy Metals in the Field
C. Gigliotti and A. Farini 195

Role of Fire on Soil Organic Matter Turnover and Microbial Activity in a Mediterranean Burned Area
F.A. Rutigliano, A.R. Fierro, R.A. De Pascale, A. De Marco and A. Virzo De Santo 205

Soil Development and Microbial Community along an Altitudinal Transect in Trentino Mountains
M. Castaldini, A. Mirabella, G. Sartori, A. Fabiani, F. Santomassimo and N. Miclaus 217

Soil Microbial Activity in Three Evergreen Oak (*Quercus ilex*) Woods in a Mediterranean Area
S. Papa, E. Curcio, A. Lombardi, P. D'Oriano and A. Fioretto 229

A Method to Recover *Salmonella* from Compost by IMS-PCR
M. Civilini 239

Study of Temporal Variability in the Microbial Community of Two Forest Soils: Biomass and Activity
C. Gigliotti and A. Farini 247

Effect of Cu^{2+} and Hg^{2+} on Gene Transfer by Transformation and Transduction in *Bacillus subtilis*
M. Franchi, L. Rinaldi and E. Gallori 257

Cadmium and Zinc Toxicity to Soil Microbial Biomass and Activity
G. Renella, P.C. Brookes and P. Nannipieri 267

Impact of Tillage Practices on Seasonal Nitrate Dynamics, Leaching and Microbial Activity in a Crop Rotation
R. Papini, C. Gamba, C. Piovanelli and G. Brandi 275

Characterization of Enzymatic or Abiotic Immobilization of Xenobiotics in Soil
J.-M. Bollag, M. Strynar, M.-Y. Ahn and J. Dec 289

Enzymes in Soil: Properties, Behavior and Potential Applications
L. Gianfreda, M.A. Rao, F. Sannino, F. Saccomandi and A. Violante 301

Response of Organic Matter Pools and Enzyme Activities in Particle Size Fractions to Organic Amendments in a Long-Term Field Experiment
M.H. Gerzabek, G. Haberhauer, E. Kandeler, A. Sessitsch and H. Kirchmann 329

Indirect Approaches for Assessing Intracellular Arylsulfatase Activity in Soil
F. Fornasier 345

Changes in Soil Organic Matter, Enzymatic Activities and Heavy Metal Availability Induced by Application of Organic Residues 353
P. Burgos, E. Madejón and F. Cabrera

Oxygenase Systems in an Oligotrophic Bacterial Community of a Subsurface Water Polluted by BTEX
L. Cavalca, E. Dell'Amico and V. Andreoni 363

Lichen-Rock Interactions and Bioformation of Minerals
P. Adamo, S. Vingiani and P. Violante 377

Soil Humic Acids Formation and Characteristics in a Xeric Mollisol Reforested with two Tree Species
M.T. Dell'Abate, F. Pinzari, A. Benedetti, B.M. Petronio and C. Dazzi 393

Energy Use in the A and B Horizons of the Soil under a Pine and a Cedar Stand
F. Pinzari, M.T. Dell'Abate, A. Benedetti and C. Dazzi 405

Future Perspectives on the Environment and Human Health Interactions
C. De Kimpe 415

Index 433

Developments in Soil Science, Volume 28B
Editors: A. Violante, P.M. Huang, J.-M. Bollag and L. Gianfreda

CLAYS AND HUMIC ACIDS AFFECT THE PERSISTENCE AND BIOLOGICAL ACTIVITY OF INSECTICIDAL PROTEINS FROM *BACILLUS THURINGIENSIS* IN SOIL

G. Stotzky

Laboratory of Microbial Ecology, Department of Biology, New York University, New York, NY 10003, USA

Insecticidal proteins produced by various subspecies of *Bacillus thuringiensis* bound rapidly and tightly on clays, both pure mined clay minerals and soil clays, and on humic acids extracted from soil. This binding reduced the susceptibility of these proteins to microbial degradation, and the bound proteins retained their insecticidal activity. Both purified proteins and proteins released from the biomass of transgenic *Bt* corn and in root exudates of growing *Bt* corn exhibited binding and persistence in soil. Insecticidal protein was also released in the root exudates of *Bt* potato but not in root exudates of *Bt* cotton, canola, and tobacco. The vertical movement of the proteins, either purified or in root exudates and biomass of *Bt* corn, decreased as the concentration of kaolinite or montmorillonite in the soil increased. The biomass of transgenic *Bt* corn decomposed less in soil than did the biomass of isogenic non-*Bt* corn, possibly because the biomass of *Bt* corn had a significantly higher content of lignin than did the biomass of non-*Bt* corn. The toxins did not appear to have any consistent effects on organisms (earthworms, nematodes, protozoa, bacteria, fungi) in soil or *in vitro*. The toxins were not taken up from soil by non-*Bt* corn, carrot, radish, or turnip grown in soil in which *Bt* corn had previously been grown or into which purified toxin or biomass of *Bt* corn had been incorporated.

1. INTERACTIONS OF PURIFIED INSECTICIDAL PROTEINS WITH CLAYS AND HUMIC ACIDS

Bacillus thuringiensis (*Bt*) is a gram-positive, aerobic, spore-forming, rod-shaped bacterium that produces a parasporal, proteinaceous, crystalline inclusion during sporulation. This inclusion, which may contain more than one type of insecticidal crystal protein (ICP), is solubilized and hydrolyzed in the midgut of larvae of susceptible insects when ingested, releasing polypeptide toxins that eventually cause death of the larvae [*see* 1, 2]. More than 3000 isolates of *Bt* from 50 countries have been collected [*see* 2-4]. The ICPs have been classified on the bases of their structure, encoding genes, and host range and on the flagellar H-antigens of the bacteria that produce them [1, 4]. Numerous distinct crystal protein (*cry*) genes have been identified that code for insecticidal proteins (Cry proteins): CryI and CryIIB proteins are specifically toxic to Lepidoptera; CryIIA proteins to Lepidoptera and Diptera; CryIII proteins to Coleoptera; and four CryIV proteins to Diptera. In addition, two genes (*cytA, cytB*) that code for cytolytic proteins (CytA, CytB) are present with the CryIV proteins. This nomenclature has been revised [4] but will be partially retained here, as many of the published studies discussed were done while the old nomenclature was used. Some ICPs also exhibit activity against other orders of insects (e.g., Homoptera, Hymenoptera, Orthoptera, Mallophaga), as well as against nematodes, mites, *Collembola*, protozoa, and other organisms [2-5].

Preparations of *Bt*, usually as sprays that contain a mixture of cells, spores, and parasporal crystals, have been used as insecticides for more than 30 years. Until recently, no unexpected toxicities from such sprays have been recorded, probably because *Bt* does not survive or grow well in soil [e.g., 6-12], and its spores are rapidly inactivated by UV radiation [13, 14]. Consequently, there is probably little production of the toxins in soil, and the persistence of introduced toxins is a function primarily of the: 1) amount added; 2) rate of consumption and inactivation by insect larvae; 3) rate of degradation by microorganisms; and 4) rate of abiotic inactivation. However, when the genes that code for these toxins are genetically engineered into plants, the toxins continue to be synthesized during growth of the plants. If production exceeds consumption, inactivation, and degradation, the toxins could accumulate to concentrations that may enhance the control of target pests or constitute a hazard to nontarget organisms, such as the soil microbiota, beneficial insects (e.g., pollinators, predators and parasites of insect pests) [e.g., 5, 15-23], and other animal classes. The accumulation and persistence of the toxins could also result in the selection and enrichment of toxin-resistant target insects [e.g., 24-30]. Persistence is enhanced when the toxins are bound on surface-active particles in the environment (e.g., clays and humic substances) and, thereby, rendered less accessible for microbial degradation but are still retentive of toxic activity [31].

These potential hazards and benefits are affected by modifications (e.g., truncation and rearrangement of codons) [*see* 2] of the introduced toxin genes to code only for the synthesis of "active" toxins, or a portion of the toxins, rather than of nontoxic crystalline protoxins. Consequently, it will not be necessary for an organism that ingests the active toxins to have a high midgut pH (*ca.* 10.5) for solubilization of the IPCs and specific proteolytic enzymes to cleave the protoxins into toxic subunits. Therefore, nontarget insects and organisms in higher and lower trophic levels could be susceptible to the toxins, even if they do not have an alkaline gut pH and appropriate proteolytic enzymes. This leaves only the third of the three barriers that appear to be responsible for the host specificity of the ICPs: i.e., specific receptors for the toxins on the midgut epithelium that are often, but not always, present in larger numbers in susceptible larvae [24, 32, 33].

We have studied the equilibrium adsorption and binding of the purified toxins and protoxins produced by *B. thuringiensis* subsp. *kurstaki* (*Btk*; 66 kDa; active against Lepidoptera), subsp. *tenebrionis* (*Btt*; 68 kDa; active against Coleoptera), and subsp. *israelensis* (*Bti*; 25 to 130 kDa; active against Diptera) on the clay minerals, montmorillonite and kaolinite, and on the clay-, silt-, and sand-size fractions of soil [34-41], as well as the adsorption and binding of the toxin from *Btk* on humic acids from different soils and on complexes of clay-humic acids-Al hydroxypolymers [42, 43]. Montmorillonite and kaolinite are the predominant clay minerals in many soils, and these clays differ in structure and numerous physicochemical characteristics (e.g., cation-exchange capacity, specific surface area), and in their effects on biological activity in soil [*see* 44].

The toxins and protoxins adsorbed rapidly (in <30 min, the shortest time studied) on montmorillonite and kaolinite and on the clay-size fraction of soil, indicating that toxins released upon disintegration of transgenic plant cells and in root exudates would be only briefly in a free state susceptible to rapid biodegradation in soil. The greater adsorption of the toxin than of the protoxin from *Btk* was probably a result of differences in their molecular mass (M_r) (toxin = 66

kDa; protoxin = 132 kDa) and, possibly, in their conformation [31, 45]. Inasmuch as toxins, rather than protoxins, are expressed in most transgenic plants, essentially only the results of studies with toxins will be discussed.

Adsorption of the toxins increased with their concentration and then plateaued, indicating that the clays became saturated with the toxins, as observed with other proteins and other organic compounds [e.g., 31, 44]; was maximal between pH 6 and 8; was affected by the type of cations on the exchange complex of the clays (adsorption generally decreased as the valency of the cations increased, as has also been observed with other proteins) [*see* 44]; and was significantly lower on kaolinite than on montmorillonite. Only about 10 and 30% of the toxin from *Btk* and *Btt*, respectively, adsorbed at equilibrium were desorbed by one or two washes with distilled water. Additional washings desorbed no more toxin, indicating that the toxins were tightly bound on the clays and suggesting that they would not be easily desorbed *in situ* and leached into groundwater by rain, irrigation, snow melts, etc. Interaction of the toxins with the clays did not alter significantly their structure, as indicated by sodium dodecyl sulfate polyacrylamide gel electrophoresis (SDS-PAGE) [40] and enzyme-linked immunosorbent assays (ELISA) [36] of the equilibrium supernatants and desorption washes and by Fourier-transform infrared analyses and insect bioassays of the bound toxins. The toxins partially intercalated montmorillonite, as determined by X-ray diffraction analyses; there was more intercalation by the toxin from *Btt* than by the toxin from *Btk*, but the entire protein from either *Btk* or *Btt* did not penetrate. There was no intercalation of kaolinite, a nonexpanding clay mineral. Even though the M_r of the toxins from *Btk* and *Btt* was similar, the toxin from *Btt* intercalated montmorillonite more but adsorbed and bound less on montmorillonite and kaolinite than did the toxin from *Btk*, indicating that the structure of these proteins differs sufficiently to result in different characteristics of adsorption on clays.

A dot-blot ELISA method was developed that detects the toxins from *Btk* and *Btt* in soil, with a lower limit of detection of *ca.* 3 ng/g of soil [36]. Toxins added to nonsterile soil, unamended or amended with montmorillonite or kaolinite, were detected by this method on the clay-size fraction but not on the silt- and sand-size fractions on which the toxins do not appear to bind. The use of flow cytometry as a rapid and sensitive method with which to detect the toxins in soil was also developed [38]. Currently, Lateral Flow Quickstix (EnviroLogix, Portland, ME), which are rapid (<10 min) Western blot detection systems with a lower detection limit of *ca.* 10 parts per 10^9, are being used. The immunological results are confirmed by larvicidal assays [e.g., 46, 47].

The bound toxins retained insecticidal activity. The toxin from *Btk* or *Btt* bound on montmorillonite, kaolinite, or the clay-size fraction of soil was insecticidal to the larvae of the tobacco hornworm (*Manduca sexta*) and the Colorado potato beetle (*Leptinotarsa decemlineata*), respectively, in standard larvicidal assays where comparable amounts of purified adsorbed, bound, or free toxins were added to appropriate larval media [37]. Bound toxin from *Btk*, but not *Btt*, had a higher toxicity (i.e., had lower LC_{50} values - concentration necessary to kill 50% of the larvae) than free toxin, possibly as the result of the toxin from *Btk* being concentrated on the clays, and the larvae ingested more clay-bound toxin than free toxin spread over the surface of the larval growth medium.

When free toxin from *Btk* was added to nonsterile soils maintained at the --33-kPa water tension and 24 ± 2°C (i.e., under optimal conditions for microbial activity), insecticidal activity was detected after 234 days, the longest time evaluated, albeit with a reduction in activity during the incubation period [39]. This persistence was considerably longer than persistences reported in the literature, which ranged in "half-life" from *ca.* 8 to 17 days for purified toxins and from *ca.* 2 to 41 days for biomass of transgenic corn, cotton, and potato [e.g., 48-51]. Insecticidal activity was greater and persisted longer in soil naturally containing or amended with kaolinite than in soils naturally containing or amended with montmorillonite, possibly because montmorillonite-containing soils had a higher pH (5.8 to 7.3) and, therefore, more bacterial activity [44], which resulted in greater biodegradation of the toxin. A decrease in the persistence of the toxin in kaolinite-containing soils adjusted to *ca.* pH 7 from pH 4.9 with $CaCO_3$ further indicated that pH is a factor in the persistence of this toxin in soil. The importance of pH and other physicochemical characteristics, as well as of biological characteristics, of soil in the persistence of the toxins in soil needs to be determined, especially to explain the reported differences in the persistence of the toxins.

The binding of the toxins from *Btk* and *Btt* on clays reduced their availability to microbes, which is probably responsible for their persistence in soil [41]. The free toxins were readily utilized as sources of carbon and/or nitrogen by pure and mixed cultures of microbes, including soil suspensions, especially after the microbes had been "conditioned" (i.e., induction of proteases) on the toxins or other proteins. In contrast, the bound toxins were not utilized as a source of carbon, utilized slightly as a source of nitrogen, and did not support growth in the absence of exogenous sources of both available carbon and nitrogen. After exposure of the bound toxins to microbes, *in vitro* and in soil, the toxins retained insecticidal activity, even after alternately freezing and thawing or wetting and drying the soil for 40 days (albeit wetting and drying somewhat reduced the activity). It was anticipated that such treatments might have released toxins from surface-active particles and rendered them more susceptible to biodegradation.

Preliminary studies with the toxins from *Bti* indicated that they also bind rapidly and tightly on montmorillonite and kaolinite and that the bound toxins retain activity against the larvae of mosquito (*Culex pipiens*) [Lee, Saxena, and Stotzky, unpublished data]. Further studies with the toxins from *Bti* are in progress, as the larvae of mosquitoes, as well as of some other dipterans, are filter-feeders. Hence, toxins bound on clays and humic substances may be more effective for the control of mosquitoes and other undesirable dipterans and pose less risk to the environment than the use of transgenic *Bt* cyanobacteria, which could transfer the toxin genes to other bacteria [*see* 26, 52].

Results similar to those obtained with clays were observed when the toxin from *Btk* was reacted with humic acids prepared from four different soils and differing in total acidity (5.0 to 7.4 cmol/kg), in the content of carboxyl and phenolic groups (3.2 to 4.2 and 1.0 to 3.4 cmol/kg, respectively), and in the degree of polymerization, as evaluated by the ratio of absorbance at 464 and 665 nm (E4:E6; 5.5 to 7.0): 75 to 80% of the toxin adsorbed at equilibrium was strongly bound, with the exception of humic acids from soil on which sugar beets had been grown and which bound only *ca.* 43% of adsorbed toxin; the bound toxin was toxic to the larvae of *M. sexta*, with a LC_{50} lower than that of the free toxin; and the bound toxin did not support the growth of a mixed microbial culture from soil, although the free toxin was readily utilized as a source of carbon [42]. When the toxin was bound on complexes of montmorillonite-humic acids-Al hydroxypolymers, which more closely approximate the presumed composition of surface-active particles in soil, similar results were obtained [43].

The toxins from *Btk*, *Btt*, and *Bti*, free or bound on clays, had no effect on the growth *in vitro* of pure and mixed cultures of bacteria [both gram-positive (*Bacillus subtilis, B. cereus, Btk, Bti, Arthrobacter globiformis*) and gram-negative (*Agrobacterium radiobacter, Pseudomonas aeruginosa, Proteus vulgaris, P. mirabilis, Escherichia coli, Enterobacter aerogenes, E. cloacae, Oscillatoria* sp.)], fungi [both yeast (*Saccharomyces cerevisiae, Candida albicans*) and filamentous forms (*Rhizopus nigricans, Cunninghamella elegans, Aspergillus niger, Fusarium solani, Penicillium* sp.)], and algae [primarily green (*Chlamydomonas* sp., *Oedogonium* sp., *Euglena* sp.) and diatoms] [53]. Although these results were not unexpected, they differed from those indicated by Yudina and coworkers [54-57], who suggested that the toxins from *Bt* have antibiotic properties.

2. INTERACTIONS IN SOIL OF INSECTICIDAL PROTEINS IN TRANSGENIC *BT* PLANTS

The addition of biomass from transgenic corn (*Zea mays L.*) (NK6800Bt) expressing the Cry1Ab toxin from *Btk* resulted in a significantly lower gross metabolic activity (i.e., CO_2 evolution) of soil than did the addition of nontransgenic biomass (i.e., same variety but without the toxin gene; NK6800) [Flores, Saxena, and Stotzky, unpublished data; 31, 45]. The amounts of C evolved as CO_2 increased in proportion to the amounts of biomass added when compared with the amounts evolved from the unamended control soil. However, the amounts evolved were statistically significantly lower from soil amended with biomass of *Bt* corn than from soil amended with biomass of non-*Bt* corn. This difference occurred with stems and leaves from two separate batches of corn tissue, even when glucose was added with the tissue. Changes in the C:N ratio of the soil-biomass systems by the addition of glucose, NH_4NO_3, or glucose plus NH_4NO_3 did not alter the relative differences in CO_2 evolution between soil amended with biomass of *Bt* corn and soil amended with biomass of non-*Bt* corn.

The activities of enzymes (i.e., proteases, dehydrogenases, alkaline and acid phosphatases, and arylsulfatases) and the numbers of culturable bacteria and fungi fluctuated throughout the incubations and sometimes differed with the various treatments, but there were no consistent statistically significant differences in activities and numbers between soil amended with biomass of *Bt* corn and soil amended with biomass of non-*Bt* corn. All soil samples amended with biomass of *Bt* corn were lethal to larvae of *M. sexta* (LC_{50} values ranged from 0.27 to 0.59 mg of biomass, with confidence intervals of 0.144 to 0.495 and 0.351 to 1.070 mg of biomass, respectively), whereas there was no mortality with soil amended with biomass of non-*Bt* corn and with soil that was not amended.

The reasons for the lower biodegradation of the biomass of *Bt* corn than of the biomass of non-*Bt* corn are not known. It was not the result of differences in the C:N ratios of the biomass, as leaf and stem tissue of one batch of both *Bt* corn and non-*Bt* corn had similar C:N ratios, and changes in the ratios of the soil-biomass systems, as well as the addition of an available carbon and energy source in the form of glucose, did not significantly alter the relative differences in biodegradation between biomasses. The lower biodegradation was apparently not the result of the inhibition of the activity of the soil microbiota by the biomass of *Bt* corn, as the numbers of culturable bacteria and fungi and the activity of enzymes representative of those involved in the degradation of plant biomass were not significantly different between soil amended with biomass of *Bt* corn and soil amended with biomass of non-*Bt* corn. These results confirmed *in vitro* observations that the toxins from *Btk*, *Btt*, and *Bti* were not toxic to pure and mixed cultures of microbes [53] and *in situ* observations with transgenic plants that showed no consistent and lasting effects of the biomass of these plants on the soil microbiota [58, 59]. Although it is tempting to suggest that the insertion of the *cry1Ab* gene into the plant genome affected the susceptibility of *Bt* corn to biodegradation, there are no data to support this other than the observation that tissues of *Bt* corn show greater resistance to breakage and maceration

[Flores, Saxena, and Stotzky, unpublished data] and anecdotal reports that *Bt* corn has greater standability (i.e., less lodging) and is preferred less as feed by cattle than is non-*Bt* corn [Pat Porter, personal communication].

Because these observations suggested some differences in the chemical composition between *Bt* and non-*Bt* corn, the lignin content of 10 *Bt* corn hybrids, representing three transformation events (Bt11, MON810, and 176), and of their respective non-*Bt* isolines, grown in both a plant-growth room and the field, was evaluated. Uniform free-hand sections of fresh stems of corn, harvested after tasseling and ear production and of the same age, from between the third and fourth node were examined for lignin by fluorescence microscopy at 400 nm [60]. A higher content of lignin was observed in the vascular bundle sheath and in the sclerenchyma cells surrounding the vascular bundle of all *Bt* corn hybrids than of their respective non-*Bt* isolines, which was confirmed by staining the sections with toluidine blue [61]. The average diameter of the vascular bundle and surrounding lignified cells in *Bt* corn was 21.5 ± 0.84 μm, whereas that of non-*Bt* corn was 12.4 ± 1.14 μm [62].

The content of lignin of the same portion of the stems was determined chemically by the acetyl bromide method [63]. The lignin content of all hybrids of *Bt* corn, whether grown in the plant-growth room or in the field, was significantly higher (33 to 97% higher) than that of their respective non-*Bt* isolines [62]. There was a significantly higher lignin content ($P < 0.002$) in plants transformed by event Bt11 than by event MON810. There were no significant differences in the lignin content among isogenic non-*Bt* plants. The lignin content of the only available hybrid transformed by event 176 was lower than that of hybrids transformed by events Bt11 and MON810. These results differed from those reported by Faust [64], which indicated no significant differences in lignin content between dried biomass of whole plants of *Bt* (event MON810) and non-*Bt* corn but indicated that *Bt* corn had a higher moisture content and a lower level of ammonia than non-*Bt* corn ($P < 0.05$). However, Masoero *et al.* [65] reported a 16% higher lignin content in *Bt* than in non-*Bt* corn.

Lignin is a major structural component of plant cells that confers strength, rigidity, and impermeability to water. Any modifications in lignin content could result in effects that may have ecological implications [66]. For example, the increase in lignin content in *Bt* corn may be beneficial, as it can provide greater resistance to attack by second-generation European corn borer [67], reduce susceptibility to molds [65], and retard litter degradation and decomposition by microbes [68, 69], as also indicated in our studies by the lower evolution of CO_2 from soils amended with biomass of *Bt* corn than with biomass of non-*Bt* corn [Flores, Saxena, and Stotzky, unpublished data].

The lower degradation of the biomass of *Bt* corn may be beneficial, as the organic matter derived from *Bt* corn may persist and accumulate longer and at higher levels in soil, thereby improving soil structure and reducing erosion. By contrast, the longer persistence of the biomass of *Bt* corn may extend the time that the toxin is present in soil and, thereby, enhance the hazard to nontarget organisms and the selection of toxin-resistant target insects. Additional studies are necessary to clarify the impacts of the lower degradation of the biomass of *Bt* corn, especially as about 8 million hectares of *Bt* corn, approximately 26% of the total acreage of corn, were planted in the United States alone in 1999 [70].

The Cry1Ab protein was present in root exudates from transgenic *Bt* corn (NK4640Bt) grown in sterile hydroponic culture and in sterile and nonsterile soil in a plant-growth room [46]. The presence of the toxin was indicated by a major band migrating on SDS-PAGE to a position corresponding to a M_r of 66 kDa, the same as that of the Cry1Ab protein, and confirmed by immunological and larvicidal assays. After 25 days, when the hydroponic culture was no longer sterile, the band at 66 kDa was not detected (there were several new protein bands of smaller M_r) and the immunological and larvicidal assays were negative, indicating that microbial proteases had hydrolyzed the toxin. By contrast, the toxin was detected after 25 days in both sterile and nonsterile soil, indicating that the released toxin was bound on surface-active particles in rhizosphere soil, which protected the toxin from hydrolysis, similar to results observed with purified toxins (*see* above).

To verify these results and to estimate the importance of the clay mineralogy and other physicochemical characteristics, which influence the activity and ecology of microbes in soil [44, 71, 72], on the persistence of the toxin released in root exudates from *Bt* corn, studies were done in soil amended with various concentrations (3 to 12%) of montmorillonite or kaolinite in a plant-growth room. All samples of rhizosphere soil from plants of *Bt* corn were positive 10, 20, 30, and 40 days after germination for the presence of the toxin when assayed immunologically with Lateral Flow Quickstix [47]. No toxin was detected in any soil with plants of non-*Bt* corn or without plants. All samples of soil in which *Bt* corn was grown were toxic to the larvae of *M. sexta*, with mortality ranging from 25 to 100% on day 10 and increasing to 88 to 100% on day 40, whereas there was no significant mortality with any soil from non-*Bt* plants and with soil without plants. In addition, the size and weight of surviving larvae exposed to soils from *Bt* corn were significantly lower (*ca.* 50 to 92% lower) than those exposed to soil from non-*Bt* corn or to soil without plants, and these larvae usually died 2 to 3 days later. The larvicidal activity was generally higher initially in soil amended with montmorillonite than with kaolinite, probably because montmorillonite, a swelling 2:1, Si:Al, clay mineral with a significantly higher cation-exchange capacity and specific surface area than kaolinite, a nonswelling 1:1, Si:Al, clay, bound more toxin in the root exudates than did kaolinite, as has been observed with pure toxin [e.g., 35, 40] (*see* above). However, mortality in soil amended with montmorillonite or kaolinite was essentially the same after 40 days, indicating that over a longer time, the persistence of larvicidal activity appeared to be independent of the clay mineralogy and other physicochemical characteristics of the soils. The increase in larvicidal activity between day 10 and day 40 indicated that the toxin in the root exudates was concentrated when adsorbed on surface-active components of the soils.

The immunological and larvicidal assays of soil from the rhizosphere of *Bt* corn grown in the field were also positive, even in soil collected after frost from plants that had been dead for several months, whereas they were negative for rhizosphere soil from non-*Bt* corn [47]. Although larval mortality in rhizosphere soil from plants of field-grown *Bt* corn ranged from 38 to 100%, and the coefficients of variation were large, the size and weight of the surviving larvae were reduced by 40 to 50% when compared with soil from non-*Bt* corn or from soil without plants. Moreover, most larvae died a few days later.

To determine whether the release of the Cry1Ab protein is a common phenomenon with transgenic *Bt* corn and is not restricted to the NK4640Bt hybrid studied originally, the release of the protein in the exudates of 12 additional *Bt* hybrids, representing three different transformation events (Bt11, MON810, and 176), and of their isogenic nontransgenic counterparts was studied with plants grown in the plant-growth room and in the field. All samples of rhizosphere soil from the 12 hybrids grown in the plant-growth room were positive 40 days after germination for the presence of the toxin when assayed immunologically with Quickstix, and all samples were toxic to the larvae of *M. sexta*, with mortality ranging from 38 to 100%. No toxin was detected immunologically or by larvicidal assay in any soil in which plants of non-*Bt* corn or no plants had been grown [73]. In addition, the weight of surviving larvae exposed to soils from *Bt* corn was significantly lower (80 to 90%) than that of larvae exposed to soils from non-*Bt* corn or without plants, and these larvae usually died after an additional 2 to 3 days.

The immunological and larvicidal assays of soil from the rhizosphere of all *Bt* hybrids grown in the field and harvested after the production of ears of corn were also positive, whereas they were negative for all non-*Bt* corn isolines. The larval mortality in rhizosphere soil from plants of field-grown *Bt* corn ranged from 37 to 100%, and the weight of the surviving larvae was reduced by 85 to 98% when compared with soil from non-*Bt* corn or without plants, and most of these larvae died after a few more days. There were no discernable or consistent differences in the exudation of the toxin (as evaluated by mortality, weight of surviving larvae, or immunologically) between plants derived from different transformation events, regardless of whether they were grown in the plant-growth room or in the field.

These results indicated that the toxin released in exudates from roots of *Bt* corn could accumulate in soil and retain insecticidal activity, especially when the toxin is bound on surface-active soil particles and, thus, becomes resistant to degradation by microorganisms. Although some toxin was probably released from sloughed and damaged root cells, the major portion was derived from exudates, as there was no discernable root debris after centrifugation of the Hoagland's solution when plants were grown for 25 days in hydroponic culture.

In addition to the large amount of toxin that will be introduced into soil in plant biomass after harvest and some that will be introduced in pollen released during tasseling [22, 23], these results indicate that the toxin will also be released to soil from roots during the entire growth of a *Bt* corn crop. The presence of the toxin in soil could improve the control of insect pests, or the persistence of the toxin in soil could enhance the selection of toxin-resistant target insects and constitute a hazard to nontarget organisms. Because *Bt* corn contains truncated genes that encode toxins rather than the nontoxic crystalline protoxins produced by *Bt*, potential hazards are exacerbated, as it is not necessary for an organism ingesting the toxins to have a high gut pH (*ca.* 10.5) for solubilization of the protoxins and for specific proteases to cleave the protoxins into toxins. Moreover, receptors for the toxins are present in both target and nontarget insects [e.g., 1]. Consequently, nontarget insects and organisms in higher and, perhaps also, in lower trophic levels could be susceptible to the toxins.

The CrylAc protein was not released in hydroponic culture or in soil in the root exudates of *Bt* cotton, canola, and tobacco containing the *crylAc* gene, whereas the Cry3A protein was released in the root exudates of *Bt* potato containing the *cry3A* gene. Immunological assays of soil from the rhizosphere and of the hydroponic solution of *Bt* cotton, canola, and tobacco were negative, and the soil and solution were not toxic to the larvae of *M. sexta*. However, the immunological assays of the soil and solution of *Bt* potato were positive for the Cry3A protein, and they were toxic to the larvae of the Colorado potato beetle (*L. decemlineata*). The percent mortality of soil and hydroponic solution from *Bt* potato was 35 and 38%, respectively, and the size of the surviving larvae was reduced by 25 to 40% when compared with soil or solution from non-*Bt* potato or from soil and solution without plants [Saxena and Stotzky, unpublished data]. These observations raise some interesting questions. For example, how is a 66-kDa protein, such as the Cry1Ab protein, released intact from roots? The release of molecules with such high M_r from roots usually requires the presence of a "signal peptide" [e.g., 74]. Apparently, the endoplasmic reticulum is close to or associated with the plasma membrane in the roots of corn and, apparently, of potato, but this does not appear to be the situation in cotton, canola, and tobacco. Studies are in progress to answer this question.

There were no significant differences in the percent mortality and weight of earthworms (*Lumbricus terrestris*) after 40 days in soil planted with *Bt* or non-*Bt* corn or not planted or after

45 days in soil amended with biomass of *Bt* or non-*Bt* corn or not amended [75]. However, the toxin was present in both casts and guts of worms in soil planted with *Bt* corn or amended with biomass of *Bt* corn, whereas it was absent in casts and guts of worms in soil planted with non-*Bt* corn or not planted and in soil amended with biomass of non-*Bt* corn or not amended. When worms from soils with plants of *Bt* corn or amended with biomass of *Bt* corn were transferred to fresh soil, the toxin was cleared from the guts in 1 to 2 days. All samples of soil amended with biomass of *Bt* corn and from the rhizosphere of *Bt* corn were positive for the presence of the toxin and were lethal to the larvae of *M. sexta* after 45 and 40 days, respectively, whereas there was no mortality in soil amended with biomass of non-*Bt* corn, in rhizosphere soil of non-*Bt* corn, or in soil with no plants or not amended, which were also negative in the immunological assays. There were also no statistically significant differences ($P>0.5$) in the numbers of nematodes and culturable protozoa, bacteria (including actinomycetes), and fungi between rhizosphere soil of *Bt* and non-*Bt* corn or between soil amended with *Bt* or non-*Bt* biomass.

These results suggested that the toxin released in root exudates of *Bt* corn or from the degradation of the biomass of *Bt* corn is not toxic to a variety of organisms in soil. However, the presence of the toxin in the guts and casts of earthworms grown with *Bt* corn and in soil amended with biomass of *Bt* corn indicated again that the released toxin bound on surface-active particles in soil, which protected the toxin from biodegradation, similar to what has been observed with purified toxins. Because only one species of earthworms and only culturable microorganisms and nematodes were evaluated, more detailed studies, including the use of techniques of molecular biology (e.g., denaturing gradient and temperature gradient gel electrophoresis, single-strand conformation polymorphism), on the composition and diversity of these groups of organisms are necessary to confirm the absence of effects of the Cry1Ab toxin on biodiversity in soil.

Because of concern that plants will take up *Bt* toxins from soil, non-*Bt* corn, carrot, radish, and turnip were grown in soil in which *Bt* corn had previously been planted or which was amended with purified toxin or with ground biomass of *Bt* corn. No Cry1Ab protein was detected in the tissues of any of the plants, even after 120 or 180 days of growth, whereas the protein was present in all samples of soil [76, 77]. To determine whether the Cry1Ab protein is taken up in hydroponic culture, *Bt* corn was grown aseptically in Hoagland's solution for 15 days in the plant-growth room, and two-day old seedlings derived from surface-sterilized seeds of non-*Bt* corn that had been germinated on agar were transferred aseptically and grown in the same solution for 7 and 15 days. No Cry1Ab protein was detected in the tissues of non-*Bt* corn, whereas the toxin was easily detected in the hydroponic solution. Consequently, the apparent lack of uptake of the toxin from soil was not the result of the binding of the protein on surface-active particles, as no such surfaces were present in the hydroponic solution. These results were not unexpected, as it is doubtful that plants can take up molecules as large as 66 kDa. However, the possibility that the toxin was taken up and then degraded within the plants cannot be excluded based on these studies. As importantly, these results demonstrated that the toxin released in root exudates and from biomass of *Bt* corn persists in soil for at least 120 to 180 days, probably because it is bound on surface-active particles.

When 0.8, 1.6, or 3.2 $\mu g\ g^{-1}$ of soil, oven-dry weight equivalent, of purified CrylAb protein was added to the top of plastic columns containing 50 g of soil, the toxin was detected in leachates from all columns 1 and 3 h after addition, with the largest amount (*ca.* 75%) detected from columns containing soil not amended with clay and the lowest amount (*ca.* 16%) detected from columns containing soil amended to 12% with montmorillonite or kaolinite; intermediate amounts of protein were leached from soils amended to 3, 6, or 9% with the clays [78].

Larvicidal activity against *M. sexta* was higher with leachates from soil not amended or amended to 3 or 6% with montmorillonite or kaolinite (mortality ranged from 12 ± 6.3 to 68 ± 11.9%, and the weight of a single larva ranged from 0.03 ± 0.01 to 0.3 ± 0.03 g) than from soil amended to 9 or 12% with the clays (mortality ranged from 12 ± 6.3 to 37 ± 12.5%, and the weight of a single larva ranged from 0.05 ± 0.01 to 0.6 ± 0.03 g), indicating that the protein moved less through soil as the clay concentration was increased. However, after 12 and 24 h, no protein was detected in any of the leachates, even by the immunological assay with a lower detection limit of 10 parts per 10^9, indicating that the protein bound on the soils and that its desorption was reduced. The presence of the CrylAb protein in the leachates was confirmed by SDS-PAGE.

The vertical distribution in the columns of protein that was not recovered in the leachates confirmed that the protein moved less through soil amended with the higher concentrations of the clays. For example, when soil at 2 to 4, 70 to 80, and 140 to 150 mm from the top of randomly-selected columns of soil unamended or amended to 3 or 12% with montmorillonite or kaolinite and to which 0.8 or 3.2 μg of protein g^{-1} of soil had been added was analyzed after periodic leaching over a 24-h period for the presence of the protein, larvicidal activity at 2 to 4 mm of soil amended to 12% with montmorillonite or kaolinite was higher (mortality ranged from 37 ± 7.3 to 100%, and the weight of a single larva ranged from 0.04 ± 0.01 to 0.06 ± 0.03 g) than that of soil not amended or amended to 3% with the clays (mortality ranged from 12 ± 7.3 to 81 ± 11.9%, and the weight of a single larva ranged from 0.05 ± 0.03 to 0.4 ± 0.03 g). Mortality generally decreased with depth in the columns as the clay concentration was increased.

The CrylAb protein was present in the leachates from soil columns in which hybrids of *Bt* corn were grown, whereas it was absent in the leachates from columns in which their respective non-*Bt* isolines were grown. Hybrids of three transformation events (Bt11: NK4640Bt, N7590Bt; MON810: Dk647Bty; and 176: Maximizer) and their isolines (NK4640, N7590, and DK647; isoline for Maximizer 176 was not available) were grown in glass tubes for 40 days in a plant-growth room (26 ± 2°C, 12 h light-dark cycle) when the tubes were leached with 10 ml of water, the leachates centrifuged, and the supernatants analyzed for the presence of the protein. Although the mortality of the leachates was only 12 ± 7.3 to 20 ± 10.5%, there was a *ca.* 62% reduction in larval weight (0.2 ± 0.06 to 0.5 ± 0.07 g) compared with larvae exposed to leachates from columns in which non-*Bt* corn or no plants were grown (0.8 ± 0.04 to 1.1 ± 0.07 g), indicating vertical movement of the protein from the rhizosphere.

The protein was also present in leachates from soil columns containing biomass of *Bt* corn, whereas it was absent in leachates from columns amended with biomass of non-*Bt* corn. Mortality was 43 ± 6.3%, and the weight of a single larva was 0.08 ± 0.02 g, indicating that as the biomass degraded, the protein was released and some bound on soil particles and some dissolved in soil water and moved down with excess water.

The movement of the CrylAb protein through soil was influenced by its tendency to stick to surface-active particles, particularly clay and organic matter [31, 35, 39, 40, 42, 43]. The protein exhibited both strong binding and high persistence in soils that contained the higher clay concentrations (e.g., 9 and 12% montmorillonite or kaolinite), and it remained near the soil surface, increasing its probability of being transported to surface waters via erosion and runoff. In contrast, in soils with lower clay concentrations (e.g., unamended or amended to 3 and 6% with the clays), the protein was leached more through the soils, and it may more likely contaminate groundwater. The possibility of contamination of surface or groundwater, which depends greatly on the desorption of the protein and on the amount of water impacting on the soil as rain, irrigation, snow melts, etc., may pose a hazard to nontarget aquatic Lepidoptera, which are more plentiful in waters (e.g., 279 species in North America alone) than in soil [79.

80]. Without an adequate input of water, the protein is more likely to remain within the biologically-active root zone, where some protein, especially that not bound on surface-active particles, will be mineralized.

3. CONCLUSIONS

These studies on the interaction of insecticidal proteins with two types of surface-active particles that differ greatly in composition and structure demonstrate further the importance of surface-active particles to the biology of natural habitats. These studies also confirm and extend previous observations on the influence of clays and other surface-active particles on the activity, ecology, and population dynamics of microbes and viruses in soil and other natural habitats, as well as on the transfer of genetic information among bacteria by conjugation, transduction, and transformation [e.g., 31, 44, 52, 81, 82].

Moreover, the results obtained with these proteins indicate their potential environmental importance when bound on surfaces in soil. For example, the persistence of the bound toxins from *Bt* could pose a potential hazard to nontarget organisms and result in the selection of toxin-resistant target insects and, thereby, negate the benefits of using a biological, rather than a synthetic chemical, insecticide. However, the persistence of the bound toxins could also enhance the control of target pests. These aspects require more study, especially a case-by-case evaluation of each toxin.

In addition to suggesting potential hazards and benefits of bound toxins from *Bt*, the results of these studies indicate that caution must be exercised before transgenic plants and animals genetically modified to function as "factories" ("pharms") for the production of vaccines, hormones, antibodies, blood substitutes, toxins, pharmaceuticals, and other bioactive compounds are released to the environment. Because of the large differences in the chemical composition and structure of proteins and other organic molecules studied [e.g., 31, 44], as well as between clays and humic acids, these studies can serve as models for the potential fate and effects of other biomolecules, which are also chemically and structurally diverse, that will be introduced to soil from such factories. As with *Bt* plants, where only a portion of the plants is harvested (e.g., ears of corn, bolls of cotton, kernels of rice, tubers of potato) and the remainder of the biomass is incorporated into soil wherein the toxins released from disintegrating biomass are rapidly bound on surface-active particles, some of the biomass of these plant factories will also be incorporated into soil. With transgenic animal factories, feces, urine, and subsequently even carcasses containing bioactive compounds will eventually reach soil and other natural habitats (e.g., surface and groundwaters). If these bioactive compounds (including prions from contaminated animal carcasses) bind on clays and humic substances - and as many of these compounds are proteinaceous, they most likely will - they may also persist in natural habitats. If they retain their bioactivity, they could affect the biology of these habitats. Consequently, before the use of such plant and animal factories (and, probably, also microbial factories), the persistence of their products and the potential effects of the products on the inhabitants of soil and other habitats must be thoroughly evaluated.

ACKNOWLEDGMENTS

Most of the studies from this laboratory discussed in this paper were supported, in part, by grants from the US Environmental Protection Agency (most recently R826107-01), the US National Science Foundation, and the NYU Research Challenge Fund. The opinions expressed herein are not necessarily those of the Agency, the Foundation, or the Fund. Sincere appreciation is expressed to colleagues who contributed to these studies.

REFERENCES

1. Höfte, H., Whiteley, H.R., 1989. Insecticidal crystal proteins of *Bacillus thuringiensis*. Microbiol. Rev. 53, 242-255.
2. Schnepf, E., Crickmore, N., Van Rie, J., Lereclus, D., Baum, J., Feitelson, J., Zeigler, D.R., Dean, D.H., 1998. *Bacillus thuringiensis* and its pesticidal crystal proteins. Microbiol. Mol. Biol. Rev. 62, 775-806.
3. Feitelson, J.S., Payne, J., Kim, L., 1992. *Bacillus thuringiensis*: insects and beyond. Bio/Technology 10, 271-275.
4. Crickmore, N., Zeigler, D.R., Feitelson, J., Schnepf, E., Van Rie, J., Lereclus, D., Baum, J., Dean, D.H., 1998. Revision of the nomenclature for the *Bacillus thuringiensis* pesticidal crystal proteins. Microbiol. Mol. Biol. Rev. 62, 807-813
5. Addison, J.A., 1993. Persistence and nontarget effects of *Bacillus thuringiensis* in soil: a review. Can. J. For. Res. 23, 2329-2342.
6. Saleh, S.M, Harris, R.F., Allen, O.N., 1970. Recovery of *Bacillus thuringiensis* from field soils. J. Invertebr. Pathol. 15, 55-59.
7. West, A.W., 1984. Fate of the insecticidal, proteinaceous, parasporal crystal of *Bacillus thuringiensis* in soil. Soil Biol. Biochem. 16, 351-360.
8. West, A.W., Burges, H.D., 1985. Persistence of *Bacillus thuringiensis* and *Bacillus cereus* in soil supplemented with grass or manure. Plant Soil 83, 389-398.
9. West, A.W., Burges, H.D., Wyborn, C.H., 1984. Effect of incubation in natural and autoclaved soil upon potency and viability of *Bacillus thuringiensis*. J. Invertebr. Pathol. 44, 121-127.
10. West, A.W., Burges, H.D., White, R.J., Wyborn, C.H., 1984. Persistence of *Bacillus thuringiensis* parasporal crystal insecticidal activity in soil. J. Invertebr. Pathol. 44, 128-133.
11. West, A.H., Burges, H.D., Dixon, T.J., Wyborn, C.H., 1985. Survival of *Bacillus thuringiensis* and *Bacillus cereus* spore inocula in soil: effects of pH, moisture, nutrient availability and indigenous microorganisms. Soil Biol. Biochem. 17, 657-665.
12. Petras, S.F., Casida, L.E., 1985. Survival of *Bacillus thuringiensis* spores in soil. Appl. Environ. Microbiol. 50, 1496-1501.
13. Griego, V.M., Spence, K.D., 1978. Inactivation of *Bacillus thuringiensis* spores by ultraviolet and visible light. Appl. Environ. Microbiol. 35, 906-910.
14. Ignoffo, C.M., Garcia, C., 1978. UV-photoinactivation of cells and spores of *Bacillus thuringiensis* and effects of peroxide on inactivation. Environ. Entomol. 7, 270-272.
15. Flexner, J.L., Lighthart, B., Croft, B.A., 1986. The effects of microbial pesticides on non-target, beneficial arthropods. Agric. Ecosys. Environ. 16, 203-254.
16. Goldburg, R.J., Tjaden, G., 1990. Are B.t.k. plants really safe to eat? Bio/Technology 8, 1011-1015.
17. James, R.R., Miller, J.C., Lighthart, B., 1993. *Bacillus thuringiensis* var. *kurstaki* affects a beneficial insect, the Cinnabar moth (Lepidoptera: Arctiidae). J. Econ. Entomol. 86, 334-

339.
18. Johnson, K.S., Scriber, J.M., Nitas, J.K., Smitley, D.R., 1995. Toxicity of *Bacillus thuringiensis* var. *kurstaki* to three nontarget lepidoptera in field studies. Environ. Entomol. 24, 288-297.
19. Hilbeck, A., Baumgartner, M., Fried, P.M., Bigler, F., 1998. Effects of transgenic *Bacillus thuringiensis* corn-fed prey on mortality and development time of immature *Chrysoperla carnea* (Neuroptera: Chrysopidae). Environ. Entomol. 27, 480-487.
20. Hilbeck, A., Moar, W.J., Pusztai-Carey, M., Filippini, A., Bigler, F., 1998. Toxicity of *Bacillus thuringiensis* Cry1Ab toxin to the predator *Chrysoperla carnea* (Neuroptera: Chrysopidae). Environ. Entomol. 27, 1255-1265.
21. Hilbeck, A., Moar, W.J., Pusztai-Carey, M., Filippini A., Bigler, F., 1999. Prey-mediated effects of Cry1Ab toxin and protoxin and Cry2A protoxin on the predator *Chrysoperla carnea*. Entomol. Exp. Appl. 91, 305-316.
22. Losey, J.E., Rayor, L.S., Carter, M.E., 1999. Transgenic pollen harms monarch larvae. Nature 399, 214-214.
23. Obrycki, J.J., Losey, J.E., Taylor, O.R., Jesse, L.C.H., 2001. Transgenic insecticidal corn: beyond insecticidal toxicity to ecological complexity. Bioscience 51, 353-361.
24. Van Rie, J., McGaughey, W.H., Johnson, D.E., Barnett, B.D., Van Mellaert, H., 1990. Mechanism of insect resistance to the microbial insecticide of *Bacillus thuringiensis*. Science 247, 72-74.
25. McGaughey, W.H., Whalon, M.E., 1992. Managing insect resistance to *Bacillus thuringiensis* toxins. Science 258, 1451-1455.
26. Entwistle, P.F., Cory, J.S., Bailey, M.J., Higgs, S. (Eds.), 1993. *Bacillus thuringiensis*, an Environmental Biopesticide: Theory and Practice. John Wiley & Sons, Chichester.
27. Tabashnik, B.E., 1994. Evolution of resistance to *Bacillus thuringiensis*, Ann. Rev. Entomol. 39, 47-79.
28. Tabashnik, B.E., Liu, Y.-B., Malvar, T., Heckel, D.G., Masson, L., Ballester, V., Granero, F., Ménsua, J.L., Ferré, J., 1997. Global variation in the genetic and biochemical basis of diamondback moth resistance to *Bacillus thuringiensis*. Proc. Nat. Acad. Sci. USA 94, 12780-12785.
29. Bauer, L.S., 1995. Resistance: a threat to the insecticidal crystal proteins of *Bacillus thuringiensis*. Florida Entomol. 78, 414-443.
30. Ferré, J., Escriche, B., Bel, Y., Van Rie, J., 1995. Biochemistry and genetics of insect resistance to *Bacillus thuringiensis* insecticidal crystal proteins. FEMS Microbiol. Lett. 132, 1-7.
31. Stotzky, G., 2000. Persistence and biological activity in soil of insecticidal proteins from *Bacillus thuringiensis* and of bacterial DNA bound on clays and humic acids. J. Environ. Qual. 29, 691-705.
32. Wolfersberger, M.G., 1990. The toxicity of two *Bacillus thuringiensis* δ-endotoxins to gypsy moth larvae is inversely related to the affinity of binding sites on midgut brush border membranes for the toxins. Experientia 46, 475-477.
33. Garczynski, S.F., Crim, J.W., Adang, M.J., 1991. Identification of putative brush border membrane-binding protein specific to *Bacillus thuringiensis* δ-endotoxins by protein blot analysis. Appl. Environ. Microbiol. 57, 2816-2820.
34. Venkateswerlu, G., Stotzky, G., 1990. A simple method for the isolation of the antilepidopteran toxin from *Bacillus thuringiensis* subsp. *kurstaki*. Biotech. Appl. Biochem. 12, 245-251.

35. Venkateswerlu, G., Stotzky, G., 1992. Binding of the protoxin and toxin proteins of *Bacillus thuringiensis* subsp. *kurstaki* on clay minerals. Curr. Microbiol. 25, 225-233.
36. Tapp, H., Stotzky, G., 1995. Dot blot enzyme-linked immunosorbent assay for monitoring the fate of insecticidal toxins from *Bacillus thuringiensis* in soil. Appl. Environ. Microbiol. 61, 602-609.
37. Tapp, H., Stotzky, G., 1995. Insecticidal activity of the toxins from *Bacillus thuringiensis* subspecies *kurstaki* and *tenebrionis* adsorbed and bound on pure and soil clays. Appl. Environ. Microbiol. 61, 1786-1790.
38. Tapp, H., Stotzky, G., 1997. Monitoring the fate of insecticidal toxins from *Bacillus thuringiensis* in soil with flow cytometry. Can. J. Microbiol. 43, 1074-1078.
39. Tapp, H., Stotzky, G., 1998. Persistence of the insecticidal toxins from *Bacillus thuringiensis* subsp. *kurstaki* in soil. Soil Biol. Biochem. 30, 471-476.
40. Tapp, H., Calamai, L., Stotzky, G., 1994. Adsorption and binding of the insecticidal proteins from *Bacillus thuringiensis* subsp. *kurstaki* and subsp. *tenebrionis* on clay minerals. Soil Biol. Biochem. 26, 663-669.
41. Koskella, J., Stotzky, G., 1997. Microbial utilization of free and clay-bound insecticidal toxins from *Bacillus thuringiensis* and their retention of insecticidal activity after incubation with microbes. Appl. Environ. Microbiol. 63, 3561-3568.
42. Crecchio, C., Stotzky G., 1998. Insecticidal activity and biodegradation of the toxins from *Bacillus thuringiensis* subsp. *kurstaki* bound to humic acids from soil. Soil Biol. Biochem. 30, 463-470.
43. Crecchio, C., Stotzky, G., 2000. Biodegradation and insecticidal activity of the toxin from *Bacillus thuringiensis* subsp. *kurstaki* bound on complexes of montmorillonite-humic acids-Al hydroxypolymers. Soil Biol. Biochem. 33, 573-581.
44. Stotzky, G., 1986. Influence of soil mineral colloids on metabolic processes, growth, adhesion, and ecology of microbes and viruses. *In*: Huang, P.M., Schnitzer, M. (Eds.), Interactions of Soil Minerals with Natural Organics and Microbes. Soil Sci. Soc. Am., Madison, pp. 305-428.
45. Stotzky, G., 2001. Release, persistence, and biological activity in soil of insecticidal proteins from *Bacillus thuringiensis*. *In*: Letourneau, D., Burrows, B. (Eds.), Genetically Engineered Organisms: Assessing Environmental and Human Health Effects. CRC Press, Boca Raton, 187-222.
46. Saxena, D., Flores, S., Stotzky, G., 1999. Insecticidal toxin from *Bacillus thuringiensis* in root exudates of transgenic corn. Nature 402, 480.
47. Saxena, D., Stotzky, G., 2000. Insecticidal toxin from *Bacillus thuringiensis* is released from roots of transgenic *Bt* corn *in vitro* and *in situ*. FEMS Microbiol. Ecol. 33, 35-39.
48. Palm, C.J., Donegan, K.K., Harris, D., Seidler, R.J., 1994. Quantification in soil of *Bacillus thuringiensis* var. *kurstaki* δ-endotoxin from transgenic plants. Mol. Ecol. 3, 145-151.
49. Palm, C.J., Schaller, D.L., Donegan, K.K., Seidler, R.J., 1996. Persistence in soil of transgenic plant produced *Bacillus thuringiensis* var *kurstaki* δ-endotoxin. Can. J. Microbiol. 42, 1258-1262.
50. Sims, S.R., Holden, L.R., 1996. Insect bioassay for determining soil degradation of *Bacillus thuringiensis* subsp. *kurstaki* CryIA(b) protein in corn tissues. Environ. Entomol. 25, 659-664.
51. Sims, S.R., Ream, J.E., 1997. Soil inactivation of the *Bacillus thuringiensis* subsp. *kurstaki* CryIIA insecticidal protein within transgenic cotton tissue: laboratory microcosms and field studies. J. Agric. Fd. Chem. 45, 1502-1505.

52. Stotzky, G., 1989. Gene transfer among bacteria in soil. In: Levy, S.B., Miller, R.V. (Eds.), Gene Transfer in the Environment. McGraw-Hill, New York, pp. 165-222.
53. Koskella, J., Stotzky, G., 2001. Insecticidal toxins from *Bacillus thuringiensis* subspp. *kurstaki, tenebrionis,* and *israelensis* have no microbiostatic or microbicidal activity. Can. J. Microbiol. (submitted).
54. Egorov, N.S., Yudina, T.C., Baranov, A.Y., 1990. Correlation between the insecticidal and antibiotic activities of the parasporal crystals of *Bacillus thuringiensis*. Mikrobiologiya 59, 448-452.
55. Yudina, T.G., 1996. Comparison of antibacterial activity of the parasporal bodies from various bacilli. Izvestiya Akademii Nauk. Series Biology 5, 451-452.
56. Yudina, T.G., Mil'ko, E.S., Egorov, N.S., 1996. Sensitivity of *Micrococcus luteus* dissociation variants to δ-endotoxins of *Bacillus thuringiensis*. Mikrobiologiya 63, 365-369.
57. Yudina, T.G., Burtseva, L.I., 1997. Activity of δ-endotoxins of four *Bacillus thuringiensis* subspecies against prokaryotes. Mikrobiologiya 66, 25-31.
58. Donegan, K.K., Palm, C.J., Fieland, V.J., Porteous, L.A., Ganio, L.M., Schaller, D.L., Bucao, L.Q., Seidler, R.J., 1995. Changes in levels, species and DNA-fingerprints of soil microorganisms associated with cotton expressing the *Bacillus thuringiensis* var. *kurstaki* δ-endotoxin. Appl. Soil Ecol. 2, 111-124.
59. Donegan, K.K., Shaller, D.L., Stone, J.K., Ganio, L.M., Reed, G., Hamm, P.B., Seidler, R.J., 1996. Microbial populations, fungal species diversity and plant pathogen levels in field plots of potato plants expressing the *Bacillus thuringiensis* var. *tenebrionis* endotoxin. Transgen. Res. 5, 25-35.
60. Hu, W., Harding, S.A., Lung, J., Popko, J.L., Ralph, J., Stokke, D.D., Tsai, C., Chiang, V.L., 1999. Repression of lignin biosynthesis promotes cellulose accumulation and growth in transgenic trees. Nature Biotech. 17, 808-812.
61. Sylvester, A.W., Ruzin, S.E., 1994. Light microscopy I: dissection and microtechnique. *In*: Freeling, M., Walbot, B. (Eds.), The Maize Handbook. Springer-Verlag, NY, pp. 83-95.
62. Saxena, D., Stotzky, G., 2001. *Bt* corn has a higher lignin content than non-*Bt* corn, Am. J. Bot. 88, 1704-1706.
63. Hatfield, R.D., Grabber, J., Ralph, J., Brei, K., 1999. Using the acetyl bromide assay to determine lignin concentration in herbaceous plants. Some cautionary notes. J. Agric. Fd. Chem. 47, 628-632.
64. Faust, M.A., 1999. Research update on *Bt* corn silage. Proc. Four-State Applied Nutrition & Mgmt. Conf., Iowa State Univ., Ames.
65. Masoero, F., Moschini, M., Rossi, F., Prandini, A., Pietri, A., 1999. Nutritive value, mycotoxin contamination and *in vitro* rumen fermentation of normal and genetically modified corn (Cry1A(B)) grown in northern Italy. Maydica 44, 205-209.
66. Halpin, C., Knight, M.E., Foxon, G.A., Campbell, M.M., Bondet, A.M., Boon, J.J., Chabbert, B., Tollier, M., Schuch, W., 1994. Manipulation of lignin quality by down regulation of cinnamyl alcohol dehydrogenase. Plant J. 6, 339-350.
67. Ostrander, B.N., Coors, J.G., 1997. Relationship between plant composition and European corn borer resistance in three maize populations. Crop Sci. 37, 1741-1745.
68. Reddy, C.A., 1984. Physiology and biochemistry of lignin degradation. *In*: Klug, M.J., Reddy, C.A. (Eds.), Current Perspectives in Microbial Ecology. Am. Soc. Microbiol., Washington, D.C., pp. 558-571.
69. Tovar-Gomez, M.R., Emile, J.C., Michalet-Doreau, B., Barriere, Y., 1997. *In situ*

degradation kinetics of maize hybrid stalks. Anim. Fd. Sci. Technol. 68, 77-88.
70. U.S. Environmental Protection Agency, 2000. Preliminary Risk and Benefit Assessments for *Bt* Plant-Pesticide. US EPA, Washington, D.C.
71. Stotzky, G., 1974. Activity, ecology, and population dynamics of microorganisms in soil. *In*: Laskin, A.I., Lechevalier, H. (Eds.), Microbial Ecology. CRC Press, Cleveland, pp. 57-135.
72. Stotzky, G., 1997. Soil as an environment for microbial life. *In*: van Elsas, J.D., Wellington, E.M.H., Trevors, J.T. (Eds.), Modern Soil Microbiology. Dekker, New York, pp. 1-20.
73. Saxena, D., Flores, S., Stotzky, G., 2001. *Bt* toxin is released in root exudates from 12 transgenic corn hybrids representing three transformation events. Soil Biol. Biochem. (in press).
74. Borisjuk, N.V., Borisjuk, L.G., Logendra, S., Peterson, F., Gleba, Y., Raskin, I., 1999. Production of recombinant proteins in plant root exudates. Nature Biotech. 17, 466-469.
75. Saxena, D., Stotzky, G., 2001. *Bacillus thuringiensis* (*Bt*) toxin released from root exudates and biomass of *Bt* corn has no apparent effect on earthworms, nematodes, protozoa, bacteria, and fungi in soil. Soil Biol. Biochem. 33, 1225-1230.
76. Saxena, D., Stotzky, G., 2001. *Bt* toxin is not taken up from soil by plants. Nature Biotech. 19, 199.
77. Saxena, D., Stotzky, G., 2001. *Bt* toxin is not taken up from soil or hydroponic culture by corn, carrot, radish, or turnip. Plant Soil (in press).
78. Saxena, D., Flores, S., Stotzky, G., 2001. Vertical movement in soil of insecticidal Cry1Ab protein from *Bacillus thuringiensis*. Soil Biol Biochem. (in press).
79. Lange, W.H., 1984. Aquatic and semiaquatic Lepidoptera. *In*: Merritt, R.W., Cummins, K.W. (Eds.), An Introduction to the Aquatic Insects of North America. Kendall/Hunt, Dubuque, pp. 348-360.
80. Williams, D.D., Feltmate, B.W., 1994. Aquatic Insects. CAB International, Oxford, p. 358.
81. Yin, X., Stotzky, G., 1997. Gene transfer among bacteria in natural environments. Adv. Appl. Microbiol. 45, 153-212.
82. Vettori, C., Stotzky, G., Yoder, M., Gallori, E., 1999. Interaction between bacteriophage PBS1 and clay minerals and transduction of *Bacillus subtilis* by clay-phage complexes. Environ. Microbiol. 1, 347-355.

Developments in Soil Science, Volume 28B
Editors: A. Violante, P.M. Huang, J.-M. Bollag and L. Gianfreda

MICROBIAL ECOLOGY OF SOIL BIOFILMS: SUBSTRATE BIOAVAILABILITY, BIOREMEDIATION AND COMPLEXITY

R. G. Burns and J. E. M. Stach

Research School of Biosciences, University of Kent, Canterbury, Kent CT2 7NJ, United Kingdom

Microorganisms accumulate at soil surfaces where many xenobiotic carbon and energy sources are concentrated. At this soil-water interface interactive microbial communities develop and are often contained within extracellular polysaccharide structures known as biofilms. These microenvironments are sites of comparatively high microbial activity and the many complex interactions that take place within them need to be better understood before rational approaches to soil bioremediation can be developed. Xenobiotic bioavailability at soil surfaces and within biofilms is a dynamic property and determined by the properties of the soil, the organic in question, and the composition and activities of the microbial community. In addition, the period to which the soil has been exposed to the contaminant and the horizon in which the xenobiotic or its metabolites are located are important factors to consider in any clean-up programme. Traditional methods for accelerating pollutant transformations in soil have had only mixed success and both stimulation of the indigenous population and augmentation with novel bacteria or fungi often fail. However, newer molecular techniques for assessing the genotypic and phenotypic potential of a polluted soil are providing information that will permit a more rational approach to remediation. Some of these methods, which involve DNA and RNA extraction and analysis, are evaluated and novel ways of selecting soil- and biofilm-competent inoculants are presented. Finally, the problems associated with studying and manipulating such a diverse environment as soil are discussed both in the narrow context of the needs of bioremediation and the broader context of understanding infinitely complex systems.

1. INTRODUCTION

Microorganisms in soil function within communities composed of a number of different species and are influenced by the biological, physical and chemical properties of their immediate surroundings. This environment has dimensions and gradients measured in nanometres and micrometres (microenvironments) and millimetres and centimetres (macroenvironments). Understanding relationships within and between these different scales is a big challenge to soil biologists.

Some soil microenvironment communities are organised within three-dimensional biofilms composed predominantly of extracellular polysaccharides. This entrapped state may enable the microbes to communicate with each other and facilitate genetic exchange, culminating in the development of a number of biochemical and physiological properties that would not be displayed by individual and spatially isolated species. The result of this microenvironment association is that microbial interactions in biofilms are dynamic and reciprocal and change in time and space in ways that are not easily understood using conventional techniques. Fortunately, in recent years, it has become possible to study complex microbial interactions using combinations of molecular, microscopic, analytical and immunological techniques which offer a high degree of sensitivity and the opportunity for real time, non-destructive measurements. A clearer comprehension of how microbes function in soil is a real prospect.

Microorganisms, for the most part, reside on or in the very near vicinity of soil (and root) surfaces and have a variety of mechanisms for ensuring that they are attracted to and retained within that comparatively nutritious zone. It is at the surfaces of silts, clays and humates that many natural and anthropogenic organic and inorganic ions accumulate offering microenvironment concentrations of energy sources and nutrients in excess of those prevailing in the aqueous phase. Thus, microbes at surfaces usually survive longer, grow faster and respire more rapidly than those located away from surfaces. Not surprisingly, selection pressures in generally oligitrophic soil environments favour microorganisms that can migrate towards surfaces, adhere to those surfaces and then, in combination with other species, form structured and mutually supportive communities. Notwithstanding these advantages, the soil surface microenvironment is not all good news for the resident population: some adsorbed (and absorbed within the biofilm) compounds are toxic to microorganisms and the community has to find ways of avoiding or eliminating that problem. Furthermore, once the surface becomes nutrient depleted, microorganisms must either revert to a resting 'survival' state or detach themselves [1-3] and search for a more favourable environment.

Understanding microbe, substrate, enzyme and product associations with soil surfaces and within biofilm structures presents exciting challenges to those who wish to describe and modify the microbial ecology of the soil. Important and economically-significant manipulations include increasing microbial activities that lead to the accelerated degradation of potentially harmful organic compounds, and the topic of bioremediation has grown in importance. This is because contaminated soils account for tens of thousand of hectares of land throughout Europe and the rest of the world and have a limited value for development for housing, recreation, agriculture and forestry until they are cleaned up. Furthermore, polluted soil will often be a source of ground water contamination which, in turn, becomes a problem for the water industry as it strives to purify abstracted water to the high standards demanded of potable supplies.

Many of these 'brownfield' sites are a legacy from our industrial past or arose during a time when neither the chemical industry nor the general public were aware of the dangers of uncontrolled waste disposal. This means that contaminated sites may contain, not only a mixture of organic and inorganic pollutants, but also will be chronically polluted with residues that have accumulated over decades. These are very significant factors to consider when analysing a soil prior to determining an appropriate remediation strategy. These strategies are very much in their infancy but, at the most fundamental level, depend on full details of pollutant composition, concentration and distribution, as well as an understanding of the microbial and process diversity already present in the polluted soil, and the relationships between substrate sorption kinetics and potential microbial response (i.e. bioavailability) within the microenvironment of the biofilm. This knowledge will allow the development and testing of predictive and experimental soil models or microcosms.

2. SOIL BIOFILMS

When microbes become deposited at surfaces they give rise to, and become part of, complex and often multispecies structures delineated by extracellular polysaccharides (EPS). The attraction of planktonic bacteria (often the first surface colonizers), the adhesion phase and the subsequent community development, have been well documented [4-7] and the ecology and evolutionary implications of such assemblages (which in soil may be composed of fungi, yeasts, algae and protozoa, as well as bacteria) is much discussed [8]. These diverse microbial communities are interactive within their biofilm microenvironment and 'communicate' with each other by producing a number of molecules which presumably diffuse limited distances within the biofilm and induce responses in neighbouring cells. Prominent among these are the N-acylhomoserine lactones of Gram-negative bacteria and various peptides and lactone-like molecules of Gram-positive bacteria. These signals may regulate community composition and density and are referred to, collectively, as quorum sensing molecules.

In the past ten years the spatial organisation and phenotypic and genotypic diversity of EPS biofilms has been interrogated using a number of molecular, immunological and microscopic techniques; prominent amongst the last mentioned are confocal laser scanning microscopy [9] and, more recently, atomic force microscopy [10, 11]. Biofilm structure and the 3-dimensional organisation (i.e. architecture) of the microorganisms are influenced by : substratum properties (e.g. roughness, electrochemistry, hydrophobicity); the hydrodynamics of the soil (e.g. porosity, mass transfer, flow rates); concentrations and types of nutrients and energy sources; the production and composition of exopolysaccharides; and, of course, the indigenous microbial population. This means that no single biofilm model can be appropriate for all soils under all conditions. Consequentially, studies have involved a number of surfaces ranging from glass slides [12], through various membranes and 'inert' supports, to soil particles [13] and root surfaces [14]. Examining these biofilms usually reveals that they are non-uniform and composed of dense EPS structures containing microcolonies and voids (Figure 1) which increase the surface area of the biofilm and, no doubt, enhance nutrient and carbon flux and improve oxygen diffusion.

The most studied biofilm systems are aquatic [7, 15, 16] and the knowledge derived from their study has found application in wastewater and drinking water treatment where adhered microorganisms absorb and degrade various organic and inorganic contaminants [17, 18]. However, soil and its inorganic and organic components, as well as the root/soil interface (i.e. the rhizosphere), are included increasingly in biofilm studies. Soil biofilms, unlike their aquatic counterparts, need to survive under conditions of dehydration and there is some evidence to suggest that greater quantities of protective EPS are generated under low water conditions [19]. Studies of communities degrading 2,4,6-trichlorobenzoic acid and diclofop have revealed that biofilm architecture is directly related to the complexity of the carbon source utilised [20-22]. Furthermore, the EPS components were shown to play a role in the uptake by, and accumulation of xenobiotics within, degradative biofilms [23].

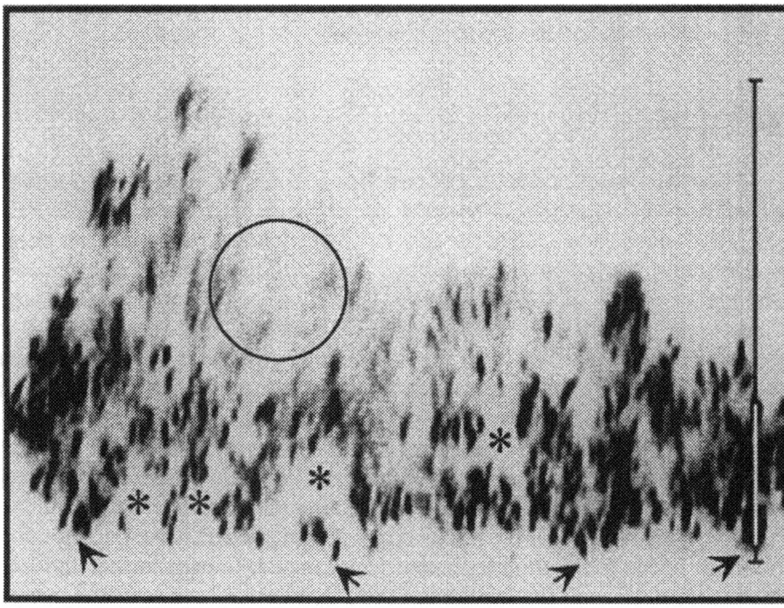

Figure 1. An x-z sagital section of a biofilm growing on naphthalene and phenanthrene as sole carbon sources [120]. Arrows indicate 'mushroom-stacks' of bacteria growing from a single point of attachment. Asterisks indicate water channels. Encircled is a large region of extracellular polysaccharide void of bacteria. Scale-bar = 30 μm.

The use of fluorescent lectins has revealed that specific EPS components are required for accumulation of diclofop, and that the components changed from a nonuniform to a uniform distribution when a more labile carbon source (e.g. tryptone soya broth) replaced the herbicide [24]. However, it has not been determined if the typical biofilm structures associated with the degradation of xenobiotic compounds described above [23], are due to the chemical complexity or uniqueness of the substrate or to the higher oxygen requirement needed to catabolise, detoxify or transform the xenobiotic [25].

It is essential to have a clear understanding of the three-dimensional microbial structures associated with substrata to which natural and xenobiotic substrates are attracted. This is because the bioavailability of the incoming pollutant chemical to the biofilm community will be strongly influenced by surface topology and physico-chemistry as well as the spatial arrangements of the indigenous microbes.

3. BIOAVAILABILITY

The term bioavailability has a number of theoretical and experimentally-determined definitions. Depending on the objectives of the experiment, xenobiotics may be bioavailable: as toxicants to microbes, plants and invertebrates; for microbial transformation involving catabolism, co-metabolism or some partial transformation or modification of the molecule; or (following leaching or run-off) to inhabitants and consumers of receiving waters. With regard to inducing and monitoring bioremediation, it is of paramount importance to understand the determinants and dynamics of all three of these types of bioavailability.

It is tempting to provide a general definition of bioavailability, one that describes it as "that proportion of the organic that is in the liquid phase and therefore, by definition, capable of uptake and transformation by suitable microorganisms in the vicinity". However, the utilisation of substrates and nutrients concentrated at surfaces is not merely defined by sorption kinetics and aqueous phase concentrations.

3.1. Bioavailability of 2,4-D and simazine

The degradation of the herbicide 2,4-dichlorophenoxy acetic acid (2,4-D) associated with the highly adsorptive clay mineral chlorite is a good example of the dynamic and complex nature of substrate bioavailability [26]. At the beginning of the experiment, 254μg of 2,4-D were strongly adsorbed to each gram of clay. In the control, non-inoculated flask, after shaking in sterile minimal salts medium for 28 days, 114μg/g (45%) of the initially strongly adsorbed 2,4-D was desorbed and 140 μg/g (55%) remained adsorbed. This 114μg/g in the solution phase is the fraction that the simple definition of bioavailabilty (given above) would identify as the maximum available for degradation. However, in flasks inoculated at day zero with a known 2,4-D degrading bacterium, 148μg/g (58%) of the original 2,4-D was mineralised (i.e converted to CO_2, water and microbial biomass). Furthermore, at the end of the 28-d experiment, 55μg/g of 2,4-D was detected in solution (and probably would have been degraded also had the experiment been prolonged) and only 51 μg/g remained strongly adsorbed. Thus, it appears from these data that the inoculated *Pseudomonas* species 'accessed' 34μg/g (148-114) of the strongly adsorbed 'non-bioavailable' 2,4-D and this may have risen to at least 89μg/g (134+55) with further incubation.

A second example of how microbes utilise apparently immobilised xenobiotics comes from our studies of s-triazine herbicides adsorbed, not to clays, but to granular activated carbon (GAC) [27-30]. A minimal salts medium containing GAC saturated with strongly adsorbed simazine and simazine in solution (as sole C and N sources) was inoculated with a known triazine degrader, *Rhodococcus rhodochrous* SL1. After 21d at 20°C, 96% of the soluble simazine had been degraded but, in addition, 28% of the strongly adsorbed simazine had been degraded. Once again, a microbial species appears to be degrading a xenobiotic that might be considered as non-bioavailable. A recent paper [31] describes the mineralization of atrazine within microbial biofilms and highlights the importance of this microenvironment in the dynamics of pollutant degradation.

3.2. Substrate accessibility

How do microbes access adsorbed substrates? There are a number of obvious and less obvious possibilities and four are suggested here.

A. As the soluble substrate is metabolised, adsorbed substrate is desorbed in an attempt to maintain equilibrium concentration.

B. During the metabolism of soluble compounds, microbes externalise H^+ ions or produce acidic metabolites which become concentrated at the anionic clay and humic surfaces and within biofilms; this reduces the microenvironment pH. As a consequence, adsorbants and adsorbates that depend on ionogenic properties for mutual association may (depending on their pKa values) undergo a change in the distribution, intensity or even sign of their exposed functional groups. Thus, what was once an attraction due to oppositely charged interactants (and a constant condition in a sterile environment), in the presence of metabolically-active microbes becomes one of repulsion - at which point substrates pass into the solution phase.

C. Microbes are known to produce surfactants such as glycolipids, lipopeptides, lipoproteins and heteropolysaccharides [32, 33] that will render hydrophobic organics associated with surfaces more available by creating emulsions that can be transported into the cell and metabolised.

D. Microbes produce extracellular enzymes that will be attracted to soil clay and humic surfaces with which they form complexes [34, 35]. This means that soil particles carry a persistent enzymatic catalytic capacity capable of responding to arriving substrates. Two very powerful techniques, Fourier Transform Infrared Spectroscopy [36] and Atomic Force Microscopy [37], have the potential to reveal the orientation of an enzyme (or a protein) at a negatively-charged clay surface and to show that it is dependent on a number of factors, including the isoelectric point of the enzyme and the pH. This means that at one pH value steric hinderence will reduce the exposure of an enzymes to its soluble substrate (e.g. a soil applied pesticide) whereas at another pH the active site is exposed to the substrate [38]. The degradation of the substrate will be much less rapid in the former situation that in the latter.

If one considers a condition where *both* substrate and enzyme are adsorbed, then the likelihood of interaction is density dependent. In other words, a substrate adsorbed to saturation on a clay surface is more likely to interact with immobilized enzymes than a surface with a few spatially separated substrate and enzyme molecules.

A third possible case (and one that is pertinent to bioremediation) is one in which the xenobiotic is already adsorbed (or otherwise associated) on or within soil colloids and aggregates, and the existing microbes and/or their enzymes are being stimulated (biostimulation) or new ones added (bioaugmentation), in an attempt to accelerate degradation. During these processes, the newly generated or directly applied enzymes may interact with all or part of the xenobiotic depending upon its configuration in the adsorbed state.

This model of bioavailability awaits detailed investigation, but, no doubt, organics can be associated with surfaces such that rings, substituents and/or side chains are accessible to the active site of the appropriate enzyme, even though the molecule is anchored. Under these circumstances, it is possible that parts of a xenobiotic are cleaved by extracellular or cell wall bound enzymes and enter the aqueous phase or pass directly into the adjacent producer cell whilst a 'footprint' of the parent compound is left behind. Recent developments in atomic force microscopy and the use of differential radio-labelling of complex molecules may enable us to visualise and even quantify this form of bioavailability. The point being made here is that ascertaining the spatial location and configuration of a pollutant and its potential degraders within the soil microenvironment may prove to be important in predicting bioavailability and, consequently, deciding on bioremediation strategy.

3.3. Ageing of pollutants

Further changes in bioavailability may occur the longer the organic remains in the soil. This is because a process known as 'ageing' usually increases the recalcitrance of potential substrates. In other words, chronic pollutants are less bioavailable (and significantly less toxic and mobile – at least in the short-term) than acute pollutants [39-41]. Once this property of long-term pollution is acknowledged, then the dangers of extrapolating data derived from laboratory experiments involving recently spiked and often pristine soils (the usual methodology described in the literature) to the field situation becomes readily apparent. Ageing is a consequence of the many different mechanisms whereby substrates become complexed with soil surfaces and the variety of biotic and abiotic transformations that can take place with the passage of time. One of the best known examples of ageing involves the comparatively rapid process that occurs when highly cationic compounds (such as the bipyridylium herbicide, paraquat) are added to soils [42, 43]. These persistent compounds are adsorbed on the first negatively charged surface (usually a clay) with which they come in contact. Weak van der Waals forces may allow limited desorption and migration within the alumino-silicate lattices where hydrogen bonding will occur rendering the compound non-bioavailable. If the first point of contact is humic matter rather than a clay colloid some greater mobility may take place prior to irreversible association with the clay fraction [44].

A more complex and longer-term ageing process affects many xenobiotics. A good example of this is afforded by acylanilide herbicides, such as propanil. This pesticide is degraded in days by microbes and, as such, the parent compound can be described as non-persistent. However, its rapidly generated metabolite, 3,4-dichloroaniline [45, 46], can condense abiotically and be converted by peroxidases to form the recalcitrant genotoxin tetrachloroazobenzene [47] or be copolymerised and oxidatively coupled to form a humic-aniline complex – one of the many examples of so-called 'bound residue' [e.g. 48-52]. Bound residues are persistent and presumably in states that are not chemically-or physically-accessible to microbes or their enzymes. But the dynamic state of bioavailability, described above, tells us that we must be vigilant with regard to regarding bound residues as inert. Especially significant in this context, but outside the purview of this article, is the detection of pesticides associated with humic and fulvic acid fractions that have leached through soils and into groundwater [53].

The various types of ageing tell us that bioavailabilities change with time and that bioremediation strategies must take account of whether the pollution is acute or chronic as well as considering the fate of likely metabolites. The complex relationships between ageing and bioavailability have been reported in much detail in a series of papers by Alexander and his colleagues [54-60] and the reader is referred to these for more detail.

3.4. Bioavailability and soil depth

Another obvious but often ignored property of soil is that chemical, physical and biological characteristics change, not only over small distances (i.e. colloids, microaggregates, macroaggregates) but also with depth [61]. This may have a significant impact on the rates of degradation [62-69] and transport [70-72] of pesticides and other potential pollutants and is well illustrated by two examples from our recent work.

The first involves the herbicide 2,4-D (Figure 2) which was shown to be degraded much more rapidly in surface soils than in subsurface soils [73, 74]. This was certainly due to the comparatively initial low numbers of 2,4-D degraders in the sub-soil but differences in soil physico-chemical properties with depth also may have influenced bioavailability. A related property of soil depth, which requires investigation, is that in soils to which herbicide has been applied over a period of years, the surface layers may contain high numbers of the parent compound degraders whilst the lower horizons (which have been enriched by the breakdown products rather than the parent compound) will have higher levels of metabolite degraders. For example, in the case of 2,4-D these sub-soil microbes could be 2,4-DCP (2,4- dichlorophenol) degraders.

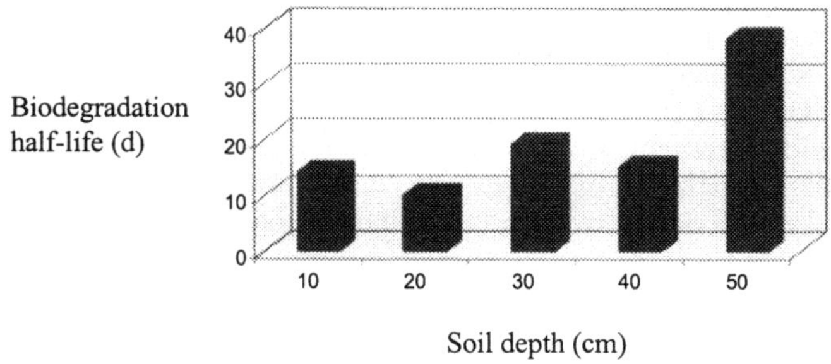

Figure 2. Biodegradation half-lives of 2,4-D in soil collected from different depths.

The second example involves dichlorprop, an enantiomeric herbicide with two stereoisomers of the same compound (Shaw and Burns, unpublished). When the effect of soil depth on the degradation of this compound was studied it was revealed that the surface layers contained only R-enantiomer degraders (Figure 3) while the sub-surface contained both R- and S-enantiomer degraders. The enantioselectivity of microbial degraders is of interest not only as a guide to bioaugmentation but also because analysis of ground water will reveal the ratio of one form of the herbicide to the other. Then, if the chemical, physical and microbiological characteristics of the soil are defined, the ratio of enantiomers in the original application could be calculated. This means that if manufacturers were required to register the ratios in their formulations and, if enough were known about the properties of the soils to which they were applied and the impact of rainfall and other climatic factors, then the origins of a pollution event could be identified and liability assessed.

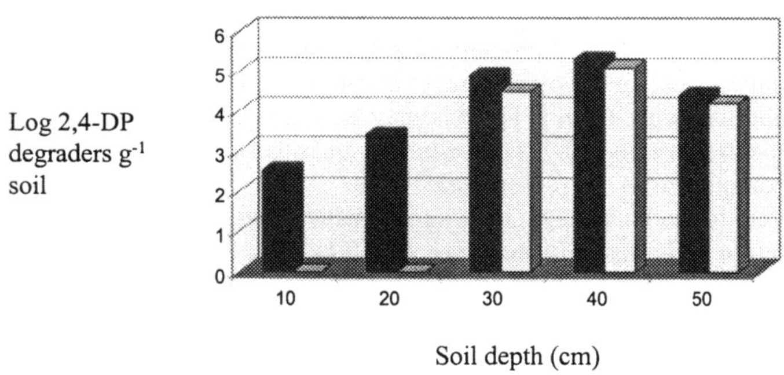

Figure 3. Microorganisms degrading R (■) and S (□) enantiomers of dichlorprop (2,4-DP) in soil collected from different depths.

In summary, these properties of soils and the xenobiotics that pass through them have implications for the development of bioremediation technology. It must be remembered that soils change with depth (effectively a deep horizon may be a totally different soil from a surface horizon in the same profile), substrate concentrations and metabolites and their bioavailabilities change with depth and that, although microbes in sub-surface horizons and deep soils are far more abundant than previously thought, species and process diversity may be very different to those detected and expressed at the surface. In other words, bioremediation (whether by biostimulation or bioaugmentation) may require or target different species and different processes and require different strategies at different depths within the same soil.

4. MICROBIAL AND MOLECULAR POTENTIAL FOR BIOREMEDIATION

4.1. Traditional methods

One of the first and most important questions to ask when considering if bioremediation is a possible solution to a contaminated site is - do the appropriate microbes already exist in the soil? If the soil has been contaminated for a period of time it is likely that selection pressures have given rise to a population of pollutant-resistant bacteria and fungi and it is probable that some of these strains have the capacity to transform at least some of the xenobiotics. This process of microbial acclimation involves the enrichment of existing populations, the induction of uptake mechanisms and degradative enzymes, or changes to the existing gene pool as a result of mutations. If time is not of importance then these processes may eventually degrade a significant proportion of the contaminants. This process is described as *natural attenuation*. However, if the site has been designated as contaminated (that is why it is in need of remediation) it is evident that, even if competent microbes exist, they are not degrading all the problem compounds to harmless products (i.e. carrying out mineralisation) at the desired rate.

Attempts to answer the question about the presence of competent microorganisms posed above have, traditionally, involved isolating bacteria on selective media (often after further directed and prolonged enrichment *in vitro* using the target pollutant or its metabolites) and then attempting to cultivate them in pure cultures (the viable plate method) with the target pollutant(s) as the sole carbon (and/or nitrogen) source. This will sometimes provide a partial answer and will often generate bacteria that will allow the biochemical study of xenobiotic metabolism. However, many bacteria are difficult if not impossible to isolate using conventional media and growth conditions, and most pollutants are mineralised in the environment by a community of microorganisms rather than a single species. Furthermore, the high degree of selectivity and bias of traditional isolation techniques means that only about 5000 species of soil bacteria have been accurately described [75], with the proportion of culturable cells seldom reaching 5% of the total number measured by direct counting procedures [76] or the use of phospholipid fatty acid analysis [77]. In other words, a soil may contain all the species necessary to degrade the pollutants but you would never discover this using traditional methods.

4.2. Biostimulation

Another approach is to assume that all the right species are already in place (whether you can isolate them or not) and to manipulate the soil to encourage these putative pre-existing microbes to reorganise themselves and form degradative communities (with an appropriate number of cells and the optimum species composition) and to express their degradative potential at a predictable and rapid rate. This is known as enhanced natural attenuation or biostimulation, and may be initiated by additions of nitrogen, potassium, phosphate and organic carbon, shifting the pH with lime or sulphur, aeration and irrigation [78]. Biostimulation assumes that the rate limiting factors are due to the composition, numbers and activities of the existing competent degraders. Equally likely is that all this adjustment of the indigenous population will have only a small effect because it is the physical and chemical association of the pollutant substrates with the soil components or their uneven distribution within the soil matrix that renders them recalcitrant or non-bioavailable (see above). Under these circumstances, bioavailability may be increased by adding surfactants [33]. Of course, even if one or a combination of these modifications give rise to accelerated degradation, the fundamental microbiology behind the success of the clean-up is unknown and any knowledge generated may be of little value when dealing with the next contaminated site. Bioremediationists need rapid and accurate identification of a soil's potential for pollutant degradation and clear advice with regard to its manipulation.

4.3. Molecular site assessment

In the last few years the shortcomings of the hit-or-miss approaches to bioremediation has been questioned and a more logical analysis of a soil's potential has been advocated. This involves what has been termed 'molecular site assessment' [79]. Molecular site assessment (MSA) involves the application of molecular diagnostic and monitoring techniques for assessing environments contaminated with xenobiotic compounds and involves DNA and mRNA extraction and quantification. The MSA approach prevents biases imposed by selective isolation of microorganisms or sampling from experiments conducted under laboratory imposed conditions [80] and over-rides the problem of non-culturable species.

A prerequisite of MSA is the extraction of nucleic acid from the contaminated soil. This provides a genetic library which can be interrogated, and will yield information with regard to the existing catabolic potential of a soil to degrade specific xenobiotics. On this basis, a rational choice can be made whether to augment or stimulate. Of course, the quality of the extracted nucleic acid, in terms of how accurately it reflects the composition of the soil microbial community, is absolutely critical. Genomic DNA extracted from soil is derived from all extant organisms (whether viable or not) and also DNA bound to soil components [81, 82]. Thus, a genetic fingerprint based on total DNA may overestimate the number of active species in the soil at the time of sampling. In contrast the RNA content is higher in active bacteria than in inactive cells and may provide a more accurate measure of the current activities of the microbial community.

So, using nucleic acids extracted directly from contaminated environments is a promising technology in environmental remediation [83]. The detailed characterisation of genotypic and phenotypic diversity of a soil should allow predictions about bioremediation potential as well as the enabling microbial changes that take place during biostimulation and bioaugmentation in the field to be monitored.

4.4. DNA extraction from soil

Many methods have been published for the extraction and purification of total DNA from soils and sediments for the culture-independent investigation of microbial communities [85-90]. The extraction methods are varied (and the outcomes not necessarily identical) but all rely on a number of physical treatments such as bead-milling [86, 91, 92], freeze-thawing [93], lysis using detergents, or a combination of these. However, one oversight is that evaluations of these techniques are based often on the quantity and purity of the isolated DNA [89, 94], comparatively few studies compare the relative composition of the DNA in terms of the microbial sequence diversity [88, 90, 95]. Evaluation of the yield of extracted DNA represents the lysis efficiency of the extraction method, i.e. the percentage of cells lysed (estimated using direct microscopic observation of cell numbers pre- and post-extraction). Sequence diversity in the extract is required to assess the proportion of distinct phylogenetic groups sampled by the extraction method. Cloning and sequencing or fingerprint analysis of the DNA extract can be used to optimise the extraction method for the detection of specific species and/or functional genes [90]. Fingerprint analysis of DNA extracted from soil may be conducted using Terminal-Restriction fragment length polymorphism (T-RFLP) [96], denaturing gradient gel electrophoresis (DGGE) or temperature gradient gel electrophoresis (TGGE) [95]. We have recently evaluated a single-strand conformation polymorphism (SSCP) approach for evaluating sequence diversity in soil extracted DNA. Results using five different extraction techniques revealed that a greater quantity of extracted DNA is not necessarily synonymous with greater sequence representation [90].

4.5. RNA extraction from soil

Genomic DNA extracted from soil is derived from all extant organisms (whether viable or not) and may also contain 'old' DNA bound to soil components [97, 98]. Thus a genetic fingerprint based on total DNA may overestimate the number and variety of active species in the soil at the time of sampling. Furthermore, DNA extracted from soil only represents the genotypic potential of the soil to produce active xenobiotic-degrading enzymes; it is not possible to infer biodegradative activity from the mere presence of degradative genes. Sayler and colleagues [79] state that "The issue of activity, not just the presence of degradative genes or bacteria, is critical to understanding the dynamics of degradation processes and the effectiveness of engineering practices in promoting bioremediation." The transcription of DNA to mRNA is the prelude to production of active proteins. Therefore, mRNA-based gene detection records the activity of genes at the moment of sampling (i.e. in real-time). This is because prokaryotic mRNA transcripts are short-lived, with half-lives of minutes *in vivo* [99] and are, therefore, markers of current metabolism *in situ* [80]. A number of techniques have been reported for the extraction of RNA from environmental samples [100-105]. Many of these co-extract DNA and RNA [89, 90, 104] and, therefore, both the presence and activity of species/genes can be monitored from a single nucleic acid extract.

Extraction of DNA and RNA from environmental samples can be accomplished directly (e.g. washing cells from soil) or indirectly (e.g. filtering cells from aqueous environments prior to extraction). Direct extraction of DNA and RNA from environmental samples is favoured over indirect extraction because there is a greater yield and a less biased sample of microbial community diversity is provided [89, 106-108].

4.6. Examples of molecular site assessment

Nucleic acids extracted from xenobiotic-contaminated soil can be used to quantify species and genes involved in biodegradation and to assess the diversity of those species and genes. As previously mentioned, the activity of specific species and genes in contaminated environments is critical to our understanding of biodegradation.

Extraction of mRNA has been used to study microbial processes including bacterial transformation of mercury [109], manganese peroxidase activity during polycyclic aromatic hydrocarbon bioremediation [110], *in situ* carbon fixation aquatic in environments [111, 84], degradation of 1,2,4-trichlorobenzene in river sediments [112], biodegradation potential in deep subsurfaces sediments [113] and expression of *nif*H genes in microbial assemblages [114].

Probe detection of *nahA* (naphthalene dioxygenase) mRNA transcripts, in a soil contaminated with polycyclic aromatic hydrocarbons, indicated that biodegradation was occurring in soil at the time of sampling. Furthermore, quantification of the transcripts indicated that mRNA levels were directly correlated to naphthalene concentration [79, 115, 116]. However, many individual bacterial species will contribute to the overall *nah*A mRNA detected. This means that when the polymerase chain reaction (PCR) or reverse transcriptase-PCR (RT-PCR) is used for detection and quantification of genes, the primers or probes employed may (whether designed to or not) detect a number of genetically homologous but (and this is critical) not functionally identical or equally active genotypes. Therefore, in order to determine the activity of individual catabolic genes and their relevance to bioremediation (or biogeochemical processes in general), amplified genes must be separated, in order that individual genotypes can be evaluated.

Wilson and colleagues [80] analysed the diversity of individual biodegradative genotypes expressed in the field. In their study mRNA, extracted from groundwater microorganisms catabolising naphthalene at a site contaminated with coal-tar, was amplified (using degenerate primers for *nahAc*-type dioxygenases), cloned and sequenced. The results showed that a variety of dioxygenases were being simultaneously transcribed at the time of sampling. This contrasted with a culture-based assessment of the same site, which revealed near identical *nahAc* sequences from a diverse group of bacterial species [117]. Whilst cloning and sequencing may represent the 'gold standard' for sequence diversity investigations, the high-cost and extended turn-around time make it impractical for impact studies that depend on the rapid analysis of large numbers of samples. Improvements in the molecular site assessment of contaminated sites will be provided by techniques which enable the rapid collection of data, *in situ* and in real-time, and which can be applied at low-cost in scaled-down laboratories on site.

In our laboratory, methods have been designed to compare direct DNA- with RNA- based molecular site assessment, without the need of laborious and costly cloning and sequencing techniques. This was achieved by: (i) techniques that provided PCR/RT-PCR ready DNA/mRNA directly from PAH-contaminated soil in less than one hour; (ii) the construction of a culture library of phylogenetically distinct species representing the known diversity of polycyclic aromatic hydrocarbon degrading isolates; (iii) the use of existing primers (and novel primers designed specifically for this study) to amplify all currently known naphthalene dioxygenase genes from both Gram-positive and Gram-negative bacteria; and (iv) the use of single-strand conformation polymorphism (SSCP) to detect and identify diverse naphthalene dioxygenase genes recovered directly from soil. Whole community and cloned naphthalene dioxygenase (NDO) cDNA was subjected to single-strand conformation polymorphism (SSCP) and in-gel phylogenetic analysis to uncover the diversity of NDO genes present in a soil contaminated with polycyclic aromatic hydrocarbons. This study showed that: *cnd*A-, *nah*Ac-, and *phn*Ac- like NDO genes were being actively transcribed at the time of sampling; there was significant intra- specific diversity in the NDO genes; and that it was possible to rapidly identify novel genes (without the need for cloning and sequencing) using culture libraries and SSCP. Use of both DNA and mRNA highlighted differences in both the presence and activity of specific NDOs. To our knowledge, these results are the first to report the diversity of NDO genes in contaminated soil expressed in real-time. The study shows that it was possible to employ rapid, cost effective techniques in molecular site assessment. Comparisons between DNA- and RNA-based molecular site assessment highlighted *nah*Ac genotypes that were present but not active at the time of sampling, and *cnd*A, *nah*Ac, and *phn*Ac genotypes that were being expressed at the time of sampling. Results of this kind are directly applicable to the design of bioremediation strategies, i.e. biostimulation of species harboring PAH-degradative genotypes that are currently inactive, or the bioaugmentation of contaminated soils with species that are already actively catabolising PAHs.

4.7. Bioaugmentation

If the pre-existing catabolic capacity or genotypic potential of a soil is inadequate or cannot be exploited then remediation will depend on the addition of specialised microbes to the contaminated soil. Microbial inocula will either contain the missing biodegradative pathways or serve as donor strains providing a source of catabolic genes for the indigenous microflora. In the former instance, the microbes will need to establish themselves as part of the indigenous population and to proliferate and express their desired metabolic properties. However, establishing inoculants in complex soil environments has proved difficult because new additions must be compatible with the established highly selected microbial community as well as tolerant of the prevailing chemical, physical and climatic conditions. Nonetheless, introduced strains do have one major advantage: the capacity to utilize a previously-unexploited resource (i.e. the persistant pollutant). It is just such a opportunity that improves the success rate of rhizobial (colonizing introduced legume seeds) and mycorrhizal (colonizing seedlings) inoculants. However, the distribution of the target pollutants is likely to be heterogenous and associated with surfaces and the inoculant will not only have to survive and proliferate but also will need to display chemotaxis, access its surface located substrate, and compete with established biofilm communities. Some of the desired properties of a successful bioremediation species or a community of microbes are presented in Table 1.

Table 1

Some properties of an ideal bioremediation inoculant species or microbial community

- has soil, biofilm and rhizosphere ‘competence’ genes
- does not disrupt beneficial microbial processes
- can degrade a range of target xenobiotics at both high and low concentrations
- is resistant to heavy metal pollutants
- displays chemotaxis – locates and moves towards pollutant ‘hot spots’
- increases bioavailability by producing surfactants
- functions as a donor in horizontal transfer of catabolic plasmids
- can be detected and tracked once released
- serves as a reporter of contaminant concentrations and their location

The second possibility, introducing donor strains that are compatible (i.e. are recombinants and form transconjugants) with the already well-adapted indigenous population, is an attractive idea because this means that the prolonged survival of the inoculant is not a prerequisite for successful degradation. The transfer of 2,4-D degrading plasmids and catabolic genes from inoculants to indigenous bacteria, and the consequent acceleration in 2,4-D breakdown, has been demonstrated [118, 119]

Table 2 summarises the molecular approach to bioremediation assessment, the various decision points with regard to the choice of biostimulation and bioaugmentation, and a number of ways in which the accelerated degradation of target compounds may be monitored and achieved.

Table 2

Soil bioremediation – the molecular site assessment approach

Does the soil contain indigenous microbes with the potential to transform or metabolise the xenobiotic(s)?

- if yes -

Can this potential be transcribed?

- if yes -

Is the transcribed protein active?

- if yes -

- Stimulate full expression: O_2, H_2O, C, N, P, K, pH, etc
- Improve bioavailability: surfactants, chelators, pH, etc

- if no -

- Inoculate with one or more microbial species
- Stimulate and improve as before

ν Use microbial biosensors (e.g. *lux*, *gfp*) to report on bioavailability, distribution, concentrations and progress with degradation

4.8. Generating appropriate inoculants

Recently, we have developed a continuous culture flow cell [120] which increases the likelihood of enriching biofilm-competent biodegraders that will not only transform the target pollutant and its metabolites but (and equally important) will be compatible with and compliment the activities of the indigenous community. In addition, these 'customised' communities should be able to establish themselves at soil surfaces, resist shear forces and predation, and access sorbed xenobiotics.

It is apparent that biofilm communities are not only far more diverse than those isolated from conventionally enrichments but also degrade xenobiotics more rapidly. For example, we identified 38 bacterial and 2 fungal species in a PAH-degrading biofilm whereas batch enrichments generated only 12 bacterial isolates. Furthermore, the biofilm community degraded naphthalene four times faster than the batch community. We believe that rational bioremediation depends on a complete knowledge of the relevant microbial processes (i.e genotypic potential and phenotypic response) already present in the contaminated soil and that conventional enrichment and isolation methods give a incomplete and probably misleading picture. Research into the removal of s-triazine herbicides, such as atrazine, from drinking water using established and defined biofilms in activated carbon filters has illustrated the potential application of biofilm bioremediation technology to cleaning up the environment [30]).

With this in mind we have investigated the degradation of PAHs in both conventional shake flask enrichments and in continuous-flow slide culture (CFSC), used for the production of biofilm communities. Direct comparisons of culturable species diversity, active species diversity (detected by RT-PCR amplification of RNA extracted from both systems) and intra- and inter-specific PAH-catabolic genes diversity were made. As stated above, the biofilm culture system produced three times the culturable bacteria diversity than that of the enrichment culture, fungal species implicated in the degradation of aromatic compounds in contaminated soils were also isolated from the biofilm community. Naphthalene dioxygenase mRNA from *cnd*-, *nah*Ac, and *phn*Ac-like gene classes was detected in the biofilm culture only *nah*Ac mRNA transcripts were detected in the enrichment culture. Finally, bacterial species that were active during PAH-degradation in bulk soil (used as the source of inoculants in both culture systems) were identified by extracting RNA from soil and RT-PCR amplification of 16S rRNA genes. The biofilm and soil active species comprised bacteria from the *Cytophaga/Flavobacterium*, High G+C, and α-, β-, γ-Proteobacterial divisions; bacteria active in the enrichment all belonged to β- and γ-Proteobacteria. Diversity within the *nah*Ac gene class was also greater in the biofilm culture system which expressed up to five *nah*Ac genotypes whilst the enrichment culture system expressed only one. However, plating of enrichment and biofilm culture onto solid media (in order to isolate pure culture PAH-degraders) resulted in the same bias for Proteobacteria. This indicates that growth on solid media and physical separation of the PAH-degrading communities may be the primary mechanism influencing the recovery of phylogenetically distinct bacteria harbouring novel PAH-degradative genes.

5. SOIL COMPLEXITY

Soil can be described as "a heterogenous matrix in which microbial communities and processes are discontinuous, dynamic and usually chaotic in time and space." This is both a daunting definition and a research and conceptual challenge. The fact is that soils are complex systems, yet, for many, the received wisdom is that if we can dissect and describe this complexity, we will be able to predict and manipulate the environment for our benefit [121]. The targets for manipulation include the obvious, such as solubilising nutrients for plant uptake, inducing beneficial microbe-plant symbioses (especially those involving rhizobia and legumes and mycorrhizal associations), inducing or improving 'natural' biological control of plant pests and pathogens, and the bioremediation of contaminated soils. There is also potential for improving soil aggregation and reducing erosion, increasing the production of numerous plant growth promoting compounds, and even manipulating carbon sequestration to reduce factors contributing to global warming. However, soil biologists are well aware that these processes are not performed by individual species or genera and are not even the sum of the activities of a large collection of independent organisms. What they are is the product of closely integrated soil chemical and physical properties as well as a myriad of biological factors. It, therefore, follows that the functions of these systems cannot be fully understood by investigating the component parts, and such all-embracing properties as soil fertility, quality or sustainability can only be studied in an entire system. Without doubt, the traditional reductionist approach to scientific research will generate reproducible data concerning optimal degradation rates, metabolic pathways, enzyme induction, repression and activities, microbial growth rates etc., but extrapolation to the complex holistic world in which soil microorganisms function is fraught with danger The oft-discussed poor relationship of laboratory experiments (even those incorporating sophisticated microcosms) to events in the real environment is a clear example of the difficulties faced by soil microbiology researchers.

One important decision for researchers who are looking for the solution to the contaminated land problem concerns the amount of detailed knowledge that is necessary to enable the reliable remediation of a high proportion of the sites [122]. Put another way - do we need a complete understanding of all components of the soil system before it can be manipulated? Confidence in our ability to attain a complete understanding of soil (or any complex system) assumes a Newtonian deterministic probability that is build on order and rules (i.e. basic laws). It also accepts the principal that the component parts of a complex system should be not only individually definable and measurable but also give rise to some sort of stability and homeostasis. But the study of soil is not just the study physical and chemical phenomena, it isn't just hydrology and biology and it certainly isn't just ion exchange and enzyme kinetics. What it is is microbial ecology, and ecology has an irreducible complexity that may not be rule-based or even predictable. In other words, the sum is not necessarily defined by the properties of the component parts. In ecology the relationships between laws and behaviour is not straightforward – even if it exists at all. In a complex ecological system, experimental or theoretical reductionism tells you a lot about the properties of the component parts but little about the synergistic interactions between those parts. In fact, a reductionist approach may render the whole system unworkable and may even (somewhat heretically) suggest that complex interdependent systems can arise from functionally independent (or even functionless) component parts [123].

Some have argued that the measure of a system's complexity is algorithmic – that is its complexity is as great as the shortest possible comprehensible message describing it. As a system becomes more complex (and the message gets longer) the closer it approaches randomness. Of course, messages may be shorter and better understood if mathematical principles and its language are applied but there is a limited value to that, in as much as the mathematicians ability to visualise, manipulate and extrapolate from a complex equation is an ability shared by only a small number of biologists. Inevitably at some point complexity becomes chaos – although this is used here to define irregular yet patterned behaviour governed by non-random laws. However, there is a real danger that intense examination of a chaotic system with a great depth of complexity will result in a description no less complicated than that used to define the original system. In other words, there has been no real advance in understanding.

5.1. An artistic metaphor

Multi-component, interactive systems develop from comparably simple states and, therefore, the more complex a system is the longer or more involved must have been its evolution. Thus, we could try to define the complexity of a soil process or a soil/water interface microenvironment containing a biofilm in terms of the amount of time and free energy that has gone into its development. This period of development has been termed the *depth of complexity*. The Nobel physicist Murray Gell-Mann, based at the Sante Fe Institute in New Mexico, has used pictorial art metaphors to describe complexity [124]. He regards a minimalist painting (by such as Mark Rothko or Barnett Newman) as having a low algorithmic complexity because it doesn't take many words or very long to describe its visual components with a high degree of accuracy. The coloured stripes and shapes also have a low depth of complexity because it appears that it wouldn't take much time or effort for a novice to reproduce the picture. An 'action painting' by Jackson Pollock, on the other hand, has such a high degree of algorithmic complexity, one that is close to randomness and chaos (but see [125, 126] for a fractal analysis of Pollock's art), so it would be virtually impossible to describe a composition such as Summertime or one of the 'black paintings' with any degree of accuracy so the listener (at the other end of a telephone, for example) would learn enough to produce a copy with any degree of accuracy. But, like the Rothko's, Jack the Dripper's paintings have a low depth of complexity. Gell-Mann compares these 20th century paintings with those of the early 16th century Flemish artist Hieronymous Bosch. These fantastical and mysterious works of art have a very high algorithmic complexity (but are certainly not random) and would require a very long descriptive message. Furthermore, the often multi-component pictures (the famous The Garden of Earthly Delights has four panels) have a great depth of complexity because of the time, thought and effort (no doubt many sketches, months of planning, overpaintings, etc.) that prefaced their production. Of course, an art historian (or the artist himself) could argue that a minimalist painting has a great depth of complexity because its evolutionary development includes the entire history of 20th century art! Or that Bosch was completely mad and the only logic and organisation in his clusters of symbols and shapes and distorted flora and fauna are imposed on them by a viewer desperate to create order and meaning out of chaos. In other words, everything from modern art to a soil surface smothered in microbes, enzymes and inorganic and organic reactants, can be simple or complex depending on your point of view and your needs. Gathering more information about the Bosch paintings, and learning what little is known of the artist, may prove to be an enjoyable, if time-consuming, exercise but it may not help you resolve the mysteries and meanings of the The Garden of Earthly Delights.

The more we learn about soil systems the more it appears that, like the Bosch paintings, they have a high algorithmic complexity; after all we use whole books and volumes of journals in our attempt to describe them. Soil also has a great depth of complexity in terms of its morphogenesis and the evolution of its microbial community - and our long history of trying to understand it. However, some of its individual components, such as a highly weathered, homoionic clay may, at first sight, have low algorithmic complexity. But even this begins to look unsustainable because recent atomic force micrographs and infra red analyses show us how heterogenous a clay surface really is with variable topology, charge density and distribution, configuration of sorbed organics and inorganics, and so on. In other words, a clay surface presents many different nanosites and the possible interactions of microbes, substrates, enzymes, toxicants and nutrients are almost endless.

However, other than as a challenging intellectual exercise and a test of experimental design and expensive pieces of equipment, is an understanding of the microenvironment complexities of clays going to help us quantify, predict and manipulate pollutant breakdown? Or is it probable that many of the features of a complex system have a negligible impact on the overall processes and may be redundant under many circumstances. Modellers might define the soil system and its bioremediation capacity according to such properties as microbial biomass, organic carbon content, average temperature, rainfall and pH and might suggest that these 'big' factors explain and predict 90% of the events that take place in soil. Scientists often assume that complex systems require complicated explanations and will spend a lifetime (or as many research grants as they can get their hands on) dissecting and reconstructing the system under study. However, understanding, for example, substrate adsorption and desorption kinetics or microbial chemotaxis towards remote pollutant 'hot spots', may only increase the predictability to 91% - and may not be a top research priority.

5.2. How much do we need to know?

There are some controversial questions posed by the previous sections. Firstly, how do we decide what soil components and how much detailed knowledge of each are relevant to our attempt to quantify, predict, manipulate and monitor pollutant degradation? Secondly, will a complete knowledge of soil components and processes and where these are located at the microenvironment level help us to resolve and design a rationale and successful bioremediation strategy? And then there is a third disturbing question arising from these two: is there a possibility that when we know all about the individual components of a process (be they biological, chemical or physical) they will turn out to be no less complex or easier to interpret than the process itself?

We believe these are important issues because we are in a conservative research era when funding is increasingly determined by the needs (or at least the perceived needs) of the end users and curiosity-lead research is expensive and seen by the funding agencies to embody a high risk of failure. The black box of soil processes, once so dark and so deep that one could only hypothesise as to its contents, is now having its lid prised open and strong lights shone into its corners. This is an exciting period in soil biology and the rapid progress that we hope for will be made as long as researchers are aware of the need to present their research (and research proposals) in a realistic and applied context. Relating effort and the acquisition of knowledge to a set of defined objectives is an issue for funders, programme managers and researchers alike.

6. CONCLUSIONS

More than 100 years of soil microbiology research has helped us to appreciate the essential contribution that bacteria and fungi make to the continuation of life on Earth. Nonetheless, we are still struggling to understand the complex functions of the millions of microbes per gram of soil and how their activities can be predicted and manipulated. In the context of this article the manipulation is targeted towards degrading various toxic organics and returning the soil to something approaching its pristine non-contaminated state. Given that there are estimated to be as much as 360,000 hectares of contaminated land in the UK alone, the importance of developing rational and successful remediation processes is evident. In situ bioremediation is a cost-effective alternative to chemical and physical ex situ processes and a significant proportion of brown-field sites could be cleaned up to a standard appropriate for use. We are on the verge of major breakthroughs in our understanding of the fundamental processes involved in soil decontamination and the long and sometimes frustrating gestation period of bioremediation science is coming to an end.

REFERENCES

1. Boyd, A., Chakrabarty, A.M., 1994. Role of alginate lyase in cell detachment of *Pseudomonas aeruginosa*. Appl. Environ. Microb. 60, 2355-2359.
2. Burns, R.G., 1995. Enumeration, survival and beneficial activities of microorganisms introduced into soil. *In*: Berthelin, J., Bollag, J-M., McGill, W.B., Page, A.L., (Eds.), Environmental Impacts of Soil Component Interactions. Lewis CRC, Boca Raton, Florida, pp.141-160.
3. Lee, S.F., Li, Y.H., Bowden, G.H., 1996. Detachment of *Streptococcus mutans* biofilm cells by an endogenous enzymatic activity. Infect Immun 64, 1035-1038.
4. Characklis, W.G., Marshall, K.C., 1990. Biofilms: A basis for an interdisciplinary approach. *In*: Characklis, W.G., Marshall, K.C., (Eds.), Biofilms. Wiley, N.Y. pp. 3-15.
5. Costerton, J.W., Lewandowski, Z., Caldwell, D.E., Korber, D.R., Lappin-Scott, H.M., 1995. Microbial biofilms. Annu. Rev. Microbiol. 49, 711-745.
6. Wimpenny, J., Gilbert, P., Walker, J., Brading, M., Bayston, R., (Eds.), 1999. Biofilms: The Good, the Bad and the Ugly. Cardiff: BioLine, pp 263-268.
7. Jass, J., Roberts, S.K., Lappin-Scott, H.M., 2002. Microbes and enzymes in biofilms. *In*: Burns, R.G., Dick, R.P. (Eds.), Enzymes in the Environment. Marcel Dekker, NY. pp.307-326.
8. Caldwell, D.E., Wolfaardt, G.M., Korber, D.R., Lawrence, J.R., Brannan, D.K., 1996. Cultivation of microbial consortia and communities. *In*: Knudsen, G.R., McInerney, M.J., Stetzenbach, L.D., Walter, M.V. (Eds.), Manual of Environmental Microbiology. ASM Press, Washington, D.C. pp. 79-90.
9. Caldwell, D.E., Korber, D.R., Lawrence, J.R., 1992. Confocal laser microscopy and digital image-analysis in microbial ecology. Adv. Microb. Ecol. 12, 1-67.
10. Forsythe, J., Maurice, P., Hersman, L., 1998. Attachment features between an aerobic *Pseudomonas sp.* bacteria and hematite observed with atomic-force microscopy. *In*: Arehart S.V. Hulston, A (Eds.), Water-Rock Interaction International IX, , Balkema, Rotterdam, pp. 927-930.
11. Lower, S.K., Maurice, P.A., 2001. Use of atomic force microscopy to study fundamental interactions between biomolecules, microorganisms, and mineral surfaces. Earth-Sci. Rev. (in press)
12. Christensen, B,B,, Sternberg, C., Andersen, J.B., Palmer, R.J., Nielsen, A.T., Givskov, M., Molin, S., 1999. Molecular tools for study of biofilm physiology. Method Enzymol. 310, 20-42.
13. Aakra, A., Hesselsoe, M., Bakken, L.R., 2000. Surface attachment of ammonia-oxidizing bacteria in soil. Microbial Ecol. 39, 222-235.
14. Bloemberg, G.V., Wijfjes, A.H.M., Lamers, G.E.M., Stuurman, N., Lugtenberg, B.J.J., 2000. Simultaneous imaging of *Pseudomonas fluorescens* WCS365 populations expressing three different autofluorescent proteins in the rhizosphere: New perspectives for studying microbial communities. Mol. Plant. Microbe. In. 13, 1170-1176.
15. Battin, T.J., Sengschmitt, D. 1999. Linking sediment biofilms, hydrodynamics, and river bed clogging: evidence from a large river. Microbial. Ecol. 37, 185-196.

16. Yallop, M.L., Paterson, D.M., Wellsbury, P., 2000. Interrelationships between rates of microbial production, exopolymer production, microbial biomass, and sediment stability in biofilms of intertidal sediments. Microbial. Ecol. 39, 116-127.
17. Janning, K.F., Mesterton, K., Harremoes, P., 1997. Hydrolysis and degradation of filtrated organic particulates in a biofilm reactor under anoxic and aerobic conditions. Water Sci. Technol. 36, 279-286.
18. Peys, K., Diels, L., Leysen, R., Vandecasteele, C., 1997. Development of a membrane biofilm reactor for the degradation of chlorinated aromatics. Water Sci. Technol. 36, 205-214.
19. Roberson, E.B., Firestone, M.K., 1992. Relationship between desiccation and exopolysaccaride production in a soil *Pseudomonas* sp. Appl. Environ. Microb. 58, 1284-1291.
20. Wolfaardt, G.M., Lawrence, J.R., Robarts, R.D., Caldwell, S.J., Caldwell, D.E., 1994. Multicellular Organization in a Degradative Biofilm Community. Appl. Environ. Microb. 60, 434-446.
21. Moller, S., Korber, D.R., Wolfaardt, G.M., Molin, S., Caldwell, D.E., 1997. Impact of nutrient composition on a degradative biofilm community. Appl. Environ. Microb. 63, 2432-2438.
22. Karthikeyan, S., Wolfaardt, G.M., Korber, D.R., Caldwell, D.E., 1999. Functional and structural responses of a degradative microbial community to substrates with varying degrees of complexity in chemical structure. Microbial. Ecol. 38, 215-224.
23. Wolfaardt, G.M., Lawrence, J.R., Headley, J.V., Robarts, R.D., Caldwell, D.E., 1994. Microbial Exopolymers Provide a Mechanism for Bioaccumulation of Contaminants. Microbial. Ecol. 27, 279-291.
24. Wolfaardt, G.M., Lawrence, J.R., Robarts, R.D., Caldwell, D.E., 1998. In situ characterization of biofilm exopolymers involved in the accumulation of chlorinated organics. Microbial. Ecol. 35, 213-223.
25. Molin, S., 2000. Probing complex biofilms. Abstracts Society for General Microbiology, 147th Ordinary Meeting, Exeter, UK.
26. McGhee, I., Sannino, F., Gianfreda, L., Burns, R.G., 1999. Bioavailability of 2,4-D sorbed to a chlorite-like complex. Chemosphere 39, 285-291.
27. Feakin, S.J., Blackburn, E., Burns, R.G., 1994. Biodegradation of s-triazine herbicides at low concentrations in surface waters. Water Res. 28, 2289-2296.
28. Feakin, S.J., Blackburn, E., Burns, R.G., 1995. Inoculation of granular activated carbon in a fixed bed with s-triazine-degrading bacteria as a water treatment process. Water Res. 29, 819-825.
29. Feakin, S.J., Gubbins, B., McGhee, I., Shaw, L., Burns, R.G., 1995. Inoculation of granular activated carbon with s-triazine-degrading bacteria for water treatment at pilot scale. Water Res. 29, 1681-1688.
30. Jones, L.R., Owen, S.A., Horrell, P., Burns, R.G., 1998. Bacterial inoculation of granular activated carbon filters for the removal of atrazine from surface waters. Water Res. 32, 2542-2549.
31. Lawrence, J.R., Kopf, G., Headley, J.V., Neu, T.R., 2001. Sorption and metabolism of selected herbicides in river biofilm communities. Can. J. Microbiol. 47, 634-641.
32. Desai, J.D., Banat, I.M., 1997. Microbial production of surfactants and their commercial potential. Microbiol. Mol. Biol. R. 61, 47-70.

33. Ron, E.Z., Rosenberg, E., 2001. Natural roles of biosurfactants. Environ. Microbiol. 3, 229-236.
34. Burns, R.G., 1982. Enzyme activity in soil: location and a possible role in microbial ecology. Soil Biol. Biochem. 14, 423-427.
35. Burns, R.G., 1989. Microbial and enzymic activites in soil. *In*: Characklis, W.G., Wilderer, P.A., (Eds.), Structure and Function of Biofilms. John Wiley & Sons, New York, pp 333-349.
36. Baron, M.H., Revault, M., Servagent-Noinville, S., Abadie, J., Quiquampoix, H., 1999. Chymotrypsin adsorption on montmorillonite: Enzymatic activity and kinetics FTIR structural analysis. J. Colloid Interf. Sci. 214, 319-332.
37. Willemsen, O.H., Snel, M.M.E., van Noort, S.J.T., van der Werf, K.O., de Grooth, G.B., Figdor, C.G., Greve, J., 1999. A physical approach to reduce nonspecific adhesion in molecular recognition atomic force microscopy. Biophys. J. 76, 716-724.
38. Quiquampoix, H., Servagent-Noinville, S., Baron, M-H., 2002. Enzyme adsorption on soil mineral surfaces and consequences for the catalytic activity. *In*: Burns, R.G., Dick, R.P., (Eds.), Enzymes in the Environment, Marcel Dekker, New York.
39. Smith, M., Lethbridge, G., Burns, R.G., 1997. Bioavailability and biodegradation of polycyclic aromatic hydrocarbons in soil. FEMS Microbiol. Lett. 152, 141-147.
40. Smith, M.J., Lethbridge, G., Burns, R.G., 1999. Fate of phenanthrene, pyrene and benzo(a)pyrene during biodegradation of crude oil added to soils. FEMS Microbiol. Lett. 173, 445-452.
41. Reid, B.J., Jones, K.C., Semple, K.T., 2000. Bioavailability of persistent organic pollutants in soils and sediments - a perspective on mechanisms, consequences and assessment. Environ. Pollut. 108, 103-112.
42. Weber, J.B., Weed, S.B., 1974. Effects of Soil on the Biological Activity of Pesticides. *In*: Guenzi, W.D. (Ed.) Pesticides in Soil and Water. Soil Sci. Soc. Am., Madison, WI. pp. 223-250.
43. Tarasevich, Y.I., Zhukova, A.L. Bondarenko, S.V., 1996. Study of diquat and paraquat sorption by layered silicates with expanding structural cells. Colloid J. 58, 522-527.
44. Burns, R.G., Audus, L.J., 1970. Distribution and breakdown of paraquat in soil. Weed Res., 10, 49-58.
45. Bartha, R., Pramer, D., 1967. Pesticide transformation to aniline and azo compounds in soil. Science 156, 1617-1618.
46. Hoalgland, R.E., Zablotowicz, R.M., Locke, M.A., 1994. Bioremediation through rhizosphere technology. ACS Symposium Series 563, 160-183.
47. Pothuluri, J.V., Hinson, J.A., Cerniglia, C.E., 1991. Propanil – Toxicological characteristics, metabolism and biodegradation potential in soil. J. Environ. Qual. 20, 330-347.
48. Benoit, P., Preston, C.M., 2000. Transformation and binding of C-13 and C-14 labelled atrazine in relation to straw decomposition in soil. Eur. J. Soil Sci. 51, 43-54.
49. Burauel, P., Fuhr F, 2000. Formation and long-term fate of non-extractable residues in outdoor lysimeter studies. Environ. Pollut. 108, 45-52.
50. Northcott, G.L., Jones, K.C., 2001. Partitioning, extractability, and formation of nonextractable PAH residues in soil. 1. Compound differences in aging and sequestration. Environ. Sci. Technol. 35, 1103-1110.

51. Kopinke, F.D., Georgi, A., Mackenzie, K., 2001. Sorption and chemical reactions of PAHs with dissolved humic substances and related model polymers. Acta Hydrochim. Hydrobiol. 28, 385-399.
52. Reuter, S., Ilim, M., Munch, J.C., Andreux, F., Scheunert, I., 1999. A model for the formation and degradation of bound residues of the herbicide C-14-isoproturon in soil. Chemosphere 39, 627-639.
53. Zwiener, C., Kumke, M.U., Abbt-Braun, G., Frimmel, F.H., 1999. Adsorbed and bound residues in fulvic acid fractions of a contaminated ground water – Isolation, chromatographic and spectroscopic characterization. Acta Hydrochim. Hydrobiol. 27, 208-213.
54. Chung, N.H., Alexander, M., 1998. Differences in sequestration and bioavailability of organic compounds aged in dissimilar soils. Environ. Sci. Technol. 32, 855-860.
55. Tang, W.C., White, J.C., Alexander, M., 1998. Utilization of sorbed compounds by microorganisms specifically isolated for that purpose. Appl. Microbiol. Biot. 49, 117-121.
56. White, J.C., Alexander, M., Pignatello, J.J., 1999. Enhancing the bioavailability of organic compounds sequestered in soil and aquifer solids. Environ. Toxicol. Chem. 18, 182-187.
57. Alexander, R.R., Alexander, M., 1999. Genotoxicity of two polycyclic aromatic hydrocarbons declines as they age in soil. Environ. Toxicol. Chem. 18, 1140-1143.
58. Chung, N., Alexander, M., 1999. Effect of concentration on sequestration and bioavailability of two polycyclic aromatic hydrocarbons. Environ. Sci. Technol. 33, 3605-3608.
59. Alexander, M., 2000. Aging, bioavailability, and overestimation of risk from environmental pollutants. Environ. Sci. Technol. 34, 4259-4265.
60. Nam, K., Alexander, M., 2001. Relationship between biodegradation rate and percentage of a compound that becomes sequestered in soil. Soil Biol. Biochem. 33, 787-792.
61. Taylor, J., Wilson, B., Mills, M.S., Burns, R.G., 2002. Comparison of microbial numbers and enzymatic activities in surface soils and subsoils using different techniques. Soil Biol. Biochem. 34 (in press)
62. Karpouzas, D.G., Walker, A., Drennan, D.S.H., Froud-Williams, R.J., 2001. The effect of initial concentration of carbofuran on the development and stability of its enhanced biodegradation in top-soil and sub-soil. Pestic. Man. Sci. 57, 72-81.
63. Reffstrup, T.K., Sorensen, H. Helweg, A., 1998. Degradation of mecoprop at different concentrations in surface and sub-surface soil. Pestic. Sci. 52, 126-132.
64. Accinelli, C., Dinelli, G., Vicari, A., Catizone, P., 2001. Atrazine and metolachlor degradation in subsoils. Biol. Fertil. Soils 33, 495-500.
65. Mills, M.S., Hill, I.R., Newcombe, A.C., Simmonds, N.D., Vaughan, P.C., Verity, A.A., 2001. Quantification of acetochlor degradation in the unsaturated zone using two novel in situ field techniques: comparisons with laboratory-generated data and implications for groundwater risk assessments. Pest Manage. Sci. 57, 351-359.
66. Mbuya, O.S., Nkedi-Kizza, P., Boote, K.J., 2001. Fate of atrazine in sandy soil cropped with sorghum. J. Environ. Qual. 30, 71-77.
67. Di, H.J., Sparking, G.P., Lee, R., Magesan, G.N., 2001. The effect of mineralisation rates of atrazine in surface and subsurface soils on its groundwater contamination potential. Aus. J. Soil Res. 39, 175-183.
68. Nicholls, P., Campbell, A.T., Williams, R.H., 2000. Factors influencing rates of degradation of an arylamide and a benzoic acid in subsoils. Pest Manage. Sci. 56, 607-614.

69. Shelton, D.R., Sadeghi, A.M., Isensee, A.R., 1998. Effect of tillage on atrazine bioavailability. Soil Sci. 163, 891-896.
70. Pivetz, B.E., Steenhuis, T.S., 1995. Biodegradation and bioremediation – soil matrix and macropore biodegradation of 2,4-D. J. Environ. Qual. 24, 564-570.
71. Zehe, E., Fluhler, H., 2001. Preferential transport of isoproturon at a plot scale and a field scale tile-drained site. J. Hydrol. 247, 100-115.
72. Pang, L.P., Close, M.E., Watt, J.P.C., Vincent, K.W., 2000. Simulation of picloram, atrazine, and simazine leaching through two New Zealand soils and into groundwater using HYDRUS-2D. J. Contam. Hyrol. 44, 19-46.
73. Shaw, L.J., Burns, R.G., 1998a. Biodegradation of 2,4-D in a noncontaminated grassland soil profile. J. Environ. Qual. 27, 1464-1471.
74. Shaw, L.J., Burns, R.G., 1998b. Biodegradation-transport interactions of pulse applied 2,4-D in repacked soil column microcosms. J. Environ. Qual. 27, 1472-1478.
75. Amann, R.I., Ludwig, W., Schleifer, K.H, 1994. Identification of uncultured bacteria: a challenging task for molecular taxonomists. ASM News 60, 360-365.
76. Bakken, L.R., 1995. Separation and purification of bacteria from soil. Appl. Environ. Microb. 49, 143-169.
77. Zeller, V., Bardgett, R.D., Tappeiner, U., 2001. Site and management effects of soil microbial properties of subalpine meadows: a study of land abandonment along a north-south gradient in the European Alps. Soil Biol. Biochem. 33, 639-649.
78. McGhee, I., Burns, R.G., 1995. Bioremediation of soil contaminated with the herbicides 2,4-dichlorophenoxyacetic acid (2,4-D) and 2-methyl-4-chlorophenoxyacetic acid (MCPA). Appl. Soil Ecol. 2, 143-154.
79. Sayler, G. S.,Layton, A., Lajoie, C., Bowman, J., Tschantz, M., Fleming, J. T., 1995. Molecular site assessment and process monitoring in bioremediation and natural attenuation. Appl. Biochem. Biotechnol. 54, 277-290.
80. Wilson, M.S., Bakermans, C., Madsen, E.L., 1999. In situ, real-time catabolic gene expression: Extraction and characterization of naphthalene dioxygenase mRNA transcripts from groundwater. Appl. Environ. Microb. 65, 80-87.
81. Widmer, F., Seidler, R.J., Watrud, L.S., 1996. Sensitive detection of transgenic plant marker gene persistence in soil microcosms. Mol. Ecol. 5, 603-613.
82. Paget, E., Lebrun, M., Freyssinet, G., Simonet, P., 1998. The fate of recombinant plant DNA in soil. Eur. J. Soil Biol. 34, 81-88.
83. Pieper, D.H., Reineke, W., 2000. Engineering bacteria for bioremediation. Curr. Opin. Biotech. 11, 262-270
84. Xu, H.H., Tabita, F.R., 1996. Ribulose-1,5-bisphosphate carboxylase/oxygenase gene expression and diversity of Lake Erie planktonic microorganisms. Appl. Environ. Microb. 62, 1913-1921.
85. Cullen, D.W., Hirsch, P.R., 1998. Simple and rapid method for direct extraction of microbial DNA from soil for PCR. Soil Biol. Biochem. 30, 983-993.
86. Kowalchuk, G.A., Bodelier, P.L.E., Heilig, G.H.J., Stephen, J.R., Laanbroek, H.J., 1998. Community analysis of ammonia-oxidising bacteria, in relation to oxygen availability in soils and root-oxygenated sediments, using PCR, DGGE and oligonucleotide probe hybridisation. FEMS Microbiol. Ecol. 27, 339-350.
87. Kuske, C.R., Banton, K.L., Adorada, D.L., Stark, P.C., Hill, K.K., Jackson, P.J., 1998. Small-scale DNA sample preparation method for field PCR detection of microbial cells and spores in soil. Appl. Environ. Microb. 64, 2463-2472.

88. Krsek, M., Wellington, E.M.H., 1999. Comparison of different methods for the isolation and purification of total community DNA from soil. J. Microbiol. Meth. 39, 1-16.
89. Miller, D.N., Bryant, J.E., Madsen, E.L., Ghiorse, W.C., 1999. Evaluation and optimization of DNA extraction and purification procedures for soil and sediment samples. Appl. Environ. Microb. 65, 4715-4724.
90. Stach, J.E.M., Bathe, S., Clapp, J.P., Burns, R.G., 2001. PCR-SSCP comparison of 16S rDNA sequence diversity in soil DNA obtained using different isolation and purification methods. FEMS Microbiol. Ecol. 36, 139-151.
91. More, M.I., Herrick, J.B., Silva, M.C., Ghiorse, W.C., Madsen, E.L., 1994. Quantitative Cell-Lysis of Indigenous Microorganisms and Rapid Extraction of Microbial DNA from Sediment. Appl. Environ. Microb. 60, 1572-1580.
92. Yeates, C., Gillings, M.R., 1998. Rapid purification of DNA from soil for molecular biodiversity analysis. Lett. Appl. Microbiol. 27, 49-53.
93. Picard, C., Di Cello, F., Ventura, M., Fani, R., Guckert, A., 2000. Frequency and biodiversity of 2,4-diacetylphloroglucinol- producing bacteria isolated from the maize rhizosphere at different stages of plant growth. Appl. Environ. Microb. 66, 948-955.
94. Frostegard, A., Courtois, S., Ramisse, V., Clerc, S., Bernillon, D., Le Gall, F., Jeannin, P., Nesme, X., Simonet, P., 1999. Quantification of bias related to the extraction of DNA directly from soils. Appl. Environ. Microb. 65, 5409-5420.
95. Heuer, H., Smalla, K., 1997. Application of denaturing gradient gel electrophoresis and temperature gradient electrophoresis for studying soil microbial communities. *In*: van Elsas, J.D., Trevors, J.T. and Wellington, E.M.H., (Eds.), Modern Soil Microbiology. Marcel Dekker, New York, pp. 353-373.
96. Liu, W.T., Marsh, T.L., Cheng, H., Forney, L.J., 1997. Characterization of microbial diversity by determining terminal restriction fragment length polymorphisms of genes encoding 16S rRNA. Appl. Environ. Microb. 63, 4516-4522.
97. Alvarez, A.J., Khanna, M., Toranzos, G.A., Stotzky, G., 1998. Amplification of DNA bound on clay minerals. Mol. Ecol. 7, 775-778.
98. Stotzky, G., 2000. Persistence and biological activity in soil of insecticidal proteins from *Bacillus thuringiensis* and of bacterial DNA bound on clays and humic acids. J. Environ. Qual. 29, 691-705.
99. Rauhut, R., Klug, G., 1999. mRNA degradation in bacteria. FEMS Microbiol. Rev. 23, 353-370.
100. Borneman, J., Triplett, E.W., 1997. Rapid and direct method for extraction of RNA from soil. Soil Biol. Biochem. 29, 1621-1624.
101. Mendum, T.A., Sockett, R.E., Hirsch, P.R., 1998. The detection of Gram-negative bacterial mRNA from soil by RT- PCR. FEMS Microbiol. Lett. 164, 369-373.
102. Miskin, I.P., Farrimond, P., Head, I.M., 1999. Identification of novel bacterial lineages as active members of microbial populations in a freshwater sediment using a rapid RNA extraction procedure and RT-PCR. Microbiol.-UK 145, 1977-1987.
103. Chandler, D.P., Stults, J.R., Cebula, S., Schuck, B.L., Weaver, D.W., Anderson, K.K., Egholm, M., Brockman, F.J., 2000. Affinity purification of DNA and RNA from environmental samples with peptide nucleic acid clamps. Appl. Environ. Microb. 66, 3438-3445.
104. Griffiths, R.I., Whiteley, A.S., O'Donnell, A.G., Bailey, M.J., 2000. Rapid method for coextraction of DNA and RNA from natural environments for analysis of ribosomal DNA-

and rRNA-based microbial community composition. Appl. Environ. Microb. 66, 5488-5491.

105. Kozdroj, J., van Elsas, J.D., 2000. Application of polymerase chain reaction-denaturing gradient gel electrophoresis for comparison of direct and indirect extraction methods of soil DNA used for microbial community fingerprinting. Biol. Fert. Soils 31, 372-378.

106. Holben, W.E., Jansson, J.K., Chelm, B.K., Tiedje, J.M., 1988. DNA probe method for the detection of specific microorganisms in the soil bacterial community. Appl. Environ. Microbiol. 54, 703-711.

107. Steffan, R.J., Goksoyr, J., Bej, A.K., Atlas, R.M., 1988. Recovery of DNA from soils and sediments. Appl. Environ. Microb. 54, 2908-2915.

108. Leff, L.G., Dana, J.R., McArthur, J.V., Shimkets, L.J., 1995. Comparison of methods of DNA extraction from stream sediments. Appl. Environ. Microb. 61, 1141-1143.

109. Jeffrey, W.H., Nazaret, S., Barkay, T., 1996. Detection of the *mer*A gene and its expression in the environment. Microbial. Ecol. 32, 293-303.

110. Bogan, B.W., Schoenike, B., Lamar, R.T., Cullen, D., 1996. Manganese peroxidase mRNA and enzyme activity levels during bioremediation of polycyclic aromatic hydrocarbon-contaminated soil with *Phanerochaete chrysosporium*. Appl. Environ. Microb. 62, 2381-2386.

111. Paul, J.H., 1996. Carbon cycling: Molecular regulation of photosynthetic carbon fixation. Microbial. Ecol. 32, 231-245.

112. Meckenstock, R., Steinle, P., van der Meer, J.R., Snozzi, M., 1998. Quantification of bacterial mRNA involved in degradation of 1,2,4-trichlorobenzene by *Pseudomonas* sp. strain P51 from liquid culture and from river sediment by reverse transcriptase PCR (RT/PCR). FEMS Microbiol. Lett. 167, 123-129.

113. Ogram, A., Sun, W.H., Brockman, F.J., Fredrickson, J.K., 1995. Isolation and Characterization of Rna from Low-Biomass Deep- Subsurface Sediments. Appl. Environ. Microb. 61, 763-768.

114. Zani, S., Mellon, M.T., Collier, J.L., Zehr, J.P., 2000. Expression of *nif*H genes in natural microbial assemblages in Lake George, New York, detected by reverse transcriptase PCR. Appl. Environ. Microb. 66, 3119-3124.

115. Fleming, J.T., Sanseverino, J., Sayler, G.S., 1993. Quantitative relationship between naphthalene catabolic gene- frequency and expression in predicting PAH degradation in soils at town gas manufacturing sites. Environ. Sci. Technol. 27, 2614-2614.

116. Sanseverino, J., Werner, C., Fleming, J., Applegate, B., King, J.M.H., Sayler, G.S., 1993. Molecular diagnostics of polyaromatic hydrocarbon biodegradation in manufactured gas plant soils. Biodegradation 4, 303-321.

117. Herrick, J.B., StuartKeil, K.G., Ghiorse, W.C., Madsen, E.L., 1997. Natural horizontal transfer of a naphthalene dioxygenase gene between bacteria native to a coal tar-contaminated field site. Appl. Environ. Microb. 63, 2330-2337.

118. Dejonghe, W., Goris, J., El Fantroussi, S., Hofte, M., De Vos, P., Verstraete, W., Top, E.M., 2000. Effect of dissemination of 2,4-dichlorophenoxyacetic acid (2,4-D) degradation of plasmids on 2,4-D degradation and on bacterial community structure in two different soil horizons. Appl. Environ. Microb. 66, 3297-3304.

119. de Lipathy, J.R., Barkay, T., Sorensen, S.J., 2001. Enhanced degradation of phenoxyacetic acid in soil by horizontal transfer of the tfdA gene encoding a 2,4-dichlorophenoxyacetic acid dioxygenase. FEMS Microbiol. Ecol. 35, 75-84.

120. Stach, J.E.M., Burns, R.G., 1999. Microbial community development in polycyclic aromatic hydrocarbon degrading biofilms. Society for General Microbiology, 143rd Meeting, Exeter, UK.
121. Burns, R.G., 2001. Bioavailability, bioremediation and biofilms in soil: how complex is complex? Mitteilung der Duetschen Bodenkundlichen Gesellschaft 95, 19-22.
122. Burns, R.G., 2001. Degradation of xenobiotics: Soil surfaces and biofilms and their relevance to bioavailability and bioremediation. Royal Society of Chemistry. Newsletter, 14, 3-8.
123. Behe, M.J., 1996 Darwin's Black Box: The Biochemical Challenge to Evolution. The Free Press, pp 307.
124. http://www.santefe.edu
125. Taylor R.P., Micolich, A.P., Jones, D., 1999. Fractal analysis of Pollock's drip paintings. Nature 39, 422.
126. http://www.sciencenews.org/sn_arc99/9_18_99/mathland.htm

Developments in Soil Science, Volume 28B
Editors: A. Violante, P.M. Huang, J.-M. Bollag and L. Gianfreda

BIOGEOCHEMISTRY OF PHOSPHORUS, IRON, AND TRACE ELEMENTS IN SOILS AS INFLUENCED BY SOIL-PLANT-MICROBIAL INTERACTIONS

A. C. Chang, A. L. Page and Bon-Jun Koo

Department of Environmental Sciences, University of California, Riverside, California 92521 USA

Interactions among soil, plant, and microorganisms are most prevalent in rhizosphere, the region adjacent to plant roots where soil-chemical, -physical, and -biological properties differs from that of the bulk soils as a result of the interactive processes. The boundary where the rhizosphere ends and the bulk soil begins, however is not easy to define and varies with plant species and age and morphology of roots. At any given time, the rhizosphere accounts for only a very small fraction of the soils in the root zone. Yet, all of the surface soils, at one time or another, may be under the influence of roots. Its chemical composition is dynamic and complex and its microbial community is dense and diverse and with presently available methodology chemical and microbiological properties of the rhizosphere are impossible to determine *in situ*. Present knowledge of the rhizosphere biogeochemistry is based largely on *ex situ* experiments employing test tube cultures and hydroponics, mostly under aseptic conditions. The rhizosphere is enriched with organic substances of plant and microbial origins that include organic acids, sugars, amino acids, lipids, coumarins, flavonoids, proteins, enzymes, aliphatics, aromatics, carbohydrates and other substances. Among them, the organic acids are the most significant as substrates for microbial metabolism and precursors of chemical reactions. In the rhizosphere, organic acids of microbial origin are virtually of the same composition as those synthesized by plants and enter the rhizosphere as cell lysis, root leakage and secretion of root exudates. The most commonly found organic acids in the rhizosphere are acetic, butyric, citric, fumaric, lactic, malic, malonic, oxalic, propionic, and succinic acids. Chemical processes involving organic acids release not readily available plant nutrients such as iron, phosphorus, and trace elements and induce weathering of soil minerals through reduction, dissolution, and complexation reactions.

1. INTRODUCTION

Soil is an entry point from which nutrients and toxicants may access the human food chain. The fact that elemental profiles of biota often resemble the inherent geochemical composition of their habitats reflects the processes by which elements are routinely transferred from lithosphere to the biosphere. Under the natural setting, trace elements are frequently present in the soil in low concentrations and tend to form sparingly soluble minerals or be adsorbed by other solid phases. The extent they are transferred from the terrestrial to the biotic environment is dependent on the biogeochemical interactions among soil, plants, and microorganisms at the region immediately adjacent to the plant roots, referred as the rhizosphere. At this interface, substances produced by plants may affect the chemical reactions and biological activities in the soil and vice versa (Figure 1).

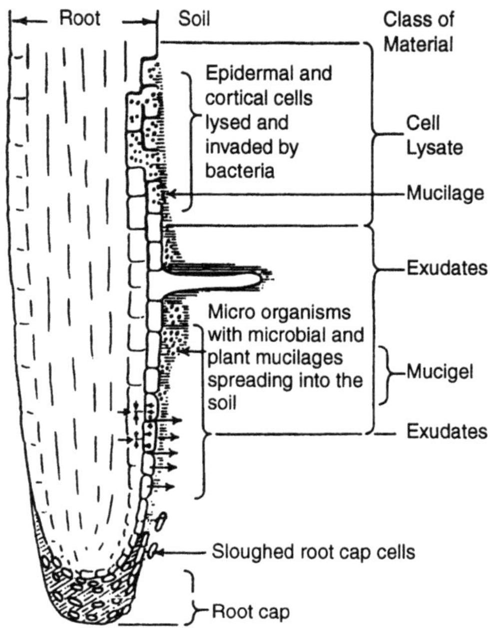

Figure 1. Origin of organic material in the rhizosphere (modeled after reference [6])

Jenny and Overstreet [1] postulated that plants absorbed mineral elements through contact and ion exchange between soil colloids and roots. Eaton [2] demonstrated that plants absorbed P and Fe from insoluble minerals across particle-root interfaces. Merckx et al. [3] reported that the dissolved metal concentrations in fallow soils were considerably less than that in adjacent vegetated soils. It has also been illustrated that the solubility of metals in soils underwent seasonal changes, reflecting the effects induced by the absence and presence of root and the gradual expansion of the root mass over a growing season [4] (Figure 2). The flow of chemicals in soil may be further improved through enhanced soil aggregation in the near root environment [5].

2. RHIZOSPHERE AND ROOT EXUDATES

The rhizosphere is the portion of soil that is next to the roots of living plants and is subject to the influence of chemicals excreted by roots of living plants and the microbial community supported by the root excretion. Its domain varies with the plant species and age and morphology of roots. The soil in the rhizosphere supports a diverse and densely populated microbial community and is subject to chemical transformations caused by presence of root exudates and metabolites of microbial degradation. Because of the association with plant roots and microorganisms, chemical, physical and biological properties of the soil in the rhizosphere differ from that of the bulk soil [6].

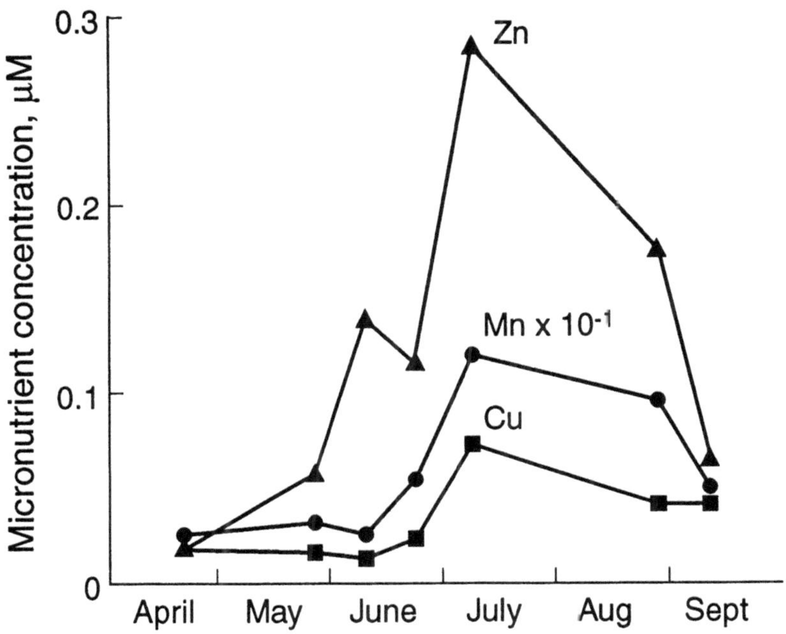

Figure 2. Seasonal effects on the micronutrient concentration of soil solution in the rhizosphere of spring barley (derived from reference [4]).

The chemical and biological activities of the rhizosphere, ideally, should be examined *in situ*. None of the methods devised so far, however, are sufficient to tract the dynamic changes in soils surrounding the roots of any living plant. More often, the chemistry of the rhizosphere is deduced from *ex situ* experimental data derived from plants cultured in growth chambers under, relatively speaking, simple growing media and sterile environment. For examples, seedlings may be grown aseptically on nutrient solution-soaked filter paper placed in Petri dishes, on nutrient agar filled glass tubes, or with static hydroponics culture in sealed containers. These approaches allowed investigators to collect substances excreted by roots of living plants free from the influence of microorganisms. The substances collected in this manner may not be representative of those found in the rhizosphere of plants grown in field conditions. *In vivo*, plants are grown for a rather short duration due to constraints on space and roots propagate without experiencing the impedance created by soils and without the presence of microorganisms. The chemistry of rhizosphere may also be examined by growing plants to maturity in large-scale sand culture hydroponics under septic conditions or by sampling of soil adjacent to roots of growing plants in fields. Although these techniques allow for the culture of plants through their life cycle, they do not permit the distinction between products of plant and microbial origins.

The root exudates are composed of complex mixtures of compounds which include sugars, amino acids, organic acids, lipids, coumarins, flavonoids, proteins, enzymes, aliphatics, aromatics, carbohydrates and other substances (Table 1). The rhizosphere also contains cell

Table 1
Substances found in plant root exudates

Category	Substances Identified
Sugar	Glucose, fructose, maltose, galactose, ribose, xylose, rhamnose, arabinose, raffinose, sucrose, fucose, oligosaccharides
Amino acid and amide	Leucine/isoleucine, valine, γ-amino butyric acid, glutamine, α-alanine, β-alanine, asparagine, arginine, serine/homoserine, glutamic acid, aspartic acid, glycine, tryptophane, cystine/cysteine, phenylalanine, threonine, tyrosine, lysine, proline, methionine, cystathionine, histidine
Organic acid	Oxalic, malic, acetic, propionic, butyric, valeric, citric, succinic, fumaric, glycolic, tartaric, malonic, galacturonic, hydroxyglutaric, adipic, glucuronic
Growth factor	Choline, inositol, pyridoxine, ρ-amino benzoic acid, η-methyl nicotinic acid, biotin, thiamine, niacin, pantothenate
Phenolic acids and coumarin	Caffeic acid, cinnamic acid, protocatechuic acid, ρ-hydroxybenzoic acid, vanillic acid, syringic acid, ρ-coumaric acid, scopoletin, salycilic acid, ferulic acid, coumarin, o-coumaric acid
Enzyme	Invertase, amylase, protease, phosphatase, polygalacturonase
Nucleotide and flavonone	Flavonone, adenine, guanine, uridine, cytidine
Fatty acid	Stearic, oleic, linoleic, linolenic acids, palmitic
Protein	Green fluorescent protein, human placental alkaline phosphatase, bacterial xylanase
Sterol	Cholesterol, campesterol, stigmasterol, sitosterol
Others	Auxins, fluorescent substances, hydrocyanic acids, organic phosphorus compounds, vitamins, photosynthesis inhibitors, germination stimulators, bacterial stimulants and inhibitors, fungal growth stimulants, hydroxamic acids, sorgoleone (mitotic inhibitor), polypeptide

lysates that are released when root cortex cells rupture either by mechanical damage, grazing of soil organisms, or the emergence of lateral roots that break through the root epidermis. Organic acids and sugars make up the bulk of the soluble rhizosphere excretions with lesser amounts of amino acids and others [7, 8]. Measurements in relation to distance from the root epidermis indicate that the rhizosphere extends to a distance approximately 1-3 millimeters outward from the root surface [9, 10].

The most commonly found organic acids in rhizosphere include acetic, butyric, citric, fumaric, lactic, malic, malonic, oxalic, propionic, and succinic [8, 11-14]. They may enter the rhizosphere through either the active excretion by roots, leakage at damaged root surface, or cell lysis. They may also simply be metabolites derived from microbial degradation of organic substances originated from roots. In the rhizosphere, the density and diversity of microorganisms far exceed other parts of the soil. Because of the continued supply of substrates by roots, the rhizosphere is expected to contain more organic acids than the bulk soils surrounding it. The mucilage region surrounding the root contains organic substances that possess functional groups capable of complexing inorganic ions and thereby creating a barrier selectively controlling their entry into plants and microorganisms [15]. A number of plant and environmental factors

(Table 2) affect the kind and amount of organic acids present in the rhizosphere. These include plant species and developmental stage, plant nutrition, plant injury, temperature and light intensity, soil water content, the ambient soil atmosphere and the nature and extent of the microbial population [26].

Table 2
Effects of environmental factors on root exudates

Factor	Effects	Reference
Soil pH	Acidification of rhizosphere change composition of root exudates.	16, 17, 18
Soil type	Roots of plants growing in sand produced greater amounts of exudates than those grown in clay or nutrient solution because of abrasive injury caused by the sand particles.	19, 20
Oxygen status	Plants produced root exudates of different composition under aerobic or anaerobic condition.	21
Light intensity	Light intensity affected rates of photosynthesis of plants that in turn affected nutrient translocation at the rhizosphere and the secretion of root exudates. High light intensity increased root exudates.	17, 22- 24
Soil temperature	Temperature affected rates of photosynthesis, translocation of photosynthates to the root, and enzymatic reactions. High temperature stimulated root exudation.	23
Soil moisture content	When a root system experiencing water stress was irrigated, it caused an increase in root exudation.	22
Nutrient availability	P, Fe, Zn, and Cu deficiencies increased excretion of root exudates.	7, 13, 18, 22, 24-28
Plant species and cultivars	Root exudates of different plants and cultivars have different chemical compositions.	13, 18, 24, 26, 29-31
Root density, root growth patterns, and rhizosphere physical properties	Different physical structures in rhizosphere could form different microhabitats for different microorganisms and induce different root exudates. Cluster roots have been associated with localized increased concentration of root exudates.	32
Age of the root	Exudates were secreted more frequently by roots during the initial stages than the later stages of growth. Mother roots exuded more root exudates than mycorrhizal roots. Root exudates come primarily from the emerging lateral root tips.	27
Microorganism	Increases root exudation and they use exudates as growth substrates. Bacterial population lowers exudation of mycorrhizal roots and changes in exudate compositions.	16, 18, 20, 33, 34-38
Stress condition	Stresses caused by nutrient deficiency, phytotoxicity, water stress, O_2 and CO_2 concentration, temperature extremes, pH and salt concentration can change the composition of root exudates.	23, 24, 30

3. PLANT NUTRITION – P, Fe, AND TRACE ELEMENTS

Small amounts of low molecular organic acids, phenols, amino acids and reduced sugars are always found in the rhizosphere. These compounds and the metabolites of microorganisms supported by such compounds may influence the solubility of sparingly soluble elements in the rhizosphere [39, 40]. In soils, trace metals are tied up in sparingly soluble chemical forms or are specifically adsorbed onto surfaces of oxides of iron and manganese. The organic acids present in the rhizosphere are by far the most significant chemicals in dissolving and complexing trace metal elements. Amino acids and phenols, that are also present in the rhizosphere, have only a limited role in reactions with the trace elements [41]. The mobilization and transport of trace elements in the rhizosphere are further influenced by processes such as oxidation/reduction, acidification/alkalinization induced by differential plant uptake of cations/anions, and mass action associated with the uptake of nutrients by plants and microorganisms (Figure 3).

3.1. Phosphorus

Some plants excrete greater amounts of organic acids when they are cultured in phosphorus- and/or iron-deficient growing conditions. The release of reducing substances by roots and microbial metabolism serves to mobilize the insoluble phosphorus and iron in the plant growing media. Likewise, protons released by plant roots in response to the preferential uptake of cations over anions also encourage the mobilization of iron and phosphorus in the rhizosphere (Figure 3).

Grierson [42], in short term experiments, observed the excretion of succinic, malic, tartaric and citric acids by radish and rape plants subjected to phosphorus-deficient and -sufficient conditions. The sum of organic acids excreted by radish and rape plants cultured under phosphorus-deficient conditions was 643 and 268 μg^{-1} pot^{-1} 4 hr^{-1}, respectively, compared to 15 and 7.1 μg^{-1} pot^{-1} 4 hr^{-1}, respectively for radish and rape cultured under phosphorus-sufficient conditions [42]. Similar observations have been made for alfalfa [43], and lupin [10]. Microorganisms also secrete organic acids that dissolve iron phosphates. Young et al. [12] identified bacteria that dissolved ferric phosphate through secretion of citric and malonic acids. The phosphorus deficiencies in plants were avoided when the root excretions induced:

- Reduction of ferric iron from iron phosphate resulting in the release of phosphate and its transport to the root surface;
- Production of protons resulting in dissolution of iron, aluminum or calcium phosphate and transport of the phosphate to the root surface; and/or
- Release of phosphatase by plants or microorganisms enabling the hydrolysis of organic phosphates [10].

3.2. Iron

When iron present in the growing media was not readily available, plants excreted organic substances that mobilize the insoluble iron [37, 44, 45, 46, 47]. For example, plants subjected to iron deficiency stress acquired iron through phytosiderophores that were produced by *Graminaceous* species present in the rhizosphere [28]. Phytosiderophores are non-proteinogenic amino acids which have a unique capacity to complex ferric iron and transport it across the plasma membrane [28]. Other processes that contribute to the dissolution of iron in the growing media include dissolution in response to increase acidity, complexation by

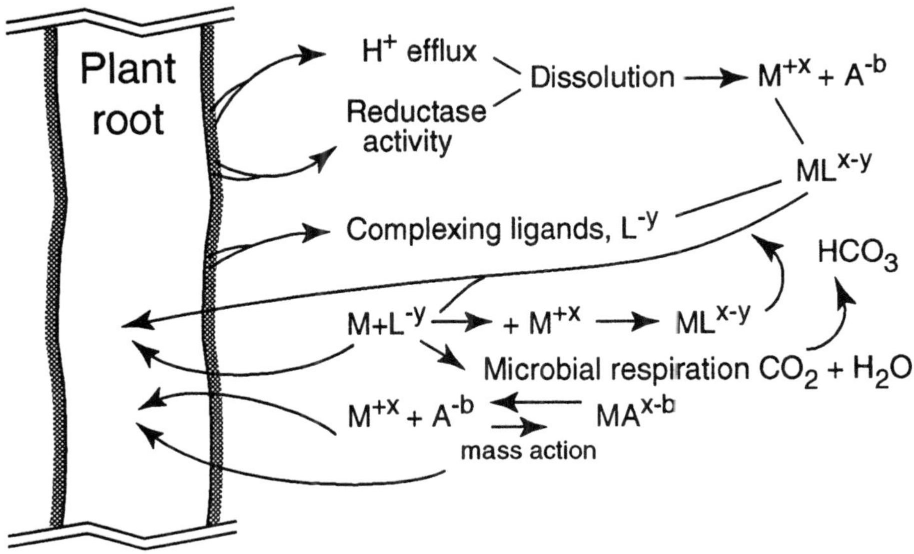

Figure 3. Rhizosphere processes involving the mobilization of metals in which M, A, and L represent metal ions, anions, metal complexing ligands, respectively and x, y, and b are integers.

organic acid ligands, and reduction of Fe^{+++} to Fe^{++} by reducing substances excreted by roots. It has been postulated that the reduction of Fe in the rhizosphere may involve the following processes in sequence [48, 49]:

1. Precursor complex formation in which surface adsorbed Fe^{3+} ($>Fe^{III}OH$) reacts with a phenoxy reductant (ArOH)

$$>Fe^{III}OH + ArOH = >Fe^{III}OAr + H_2O \quad (1)$$

2. Electron transfer resulting in the formation of phenoxy radical (•OAr)

$$>Fe^{III}OAr = >Fe^{III}\bullet OAr \quad (2)$$

3. Release of phenoxy radical and reduction of surface adsorbed Fe^{3+} to Fe^{2+}

$$>Fe^{III}\bullet OAr + H_2O = >Fe^{II}OH_2 + \bullet OAr \quad (3)$$

4. Release of reduced Fe (Fe^{2+}) and regeneration of trivalent iron on the surface of the solid phase ($>Fe^{III}OH$)

$$>Fe^{II}OH_2 = Fe^{2+} + >Fe^{III}OH \quad (4)$$

Vempati et al. [49] showed that root exudates of soybean reduced the iron in smectite and ferric oxide. White lupine excreted citrate and protons that dissolved iron and aluminum phosphates and increased the capacity to reduce ferric iron in the plant-growing medium [50, 51]. The exudates isolated from alfalfa cultured in hydroponics under iron-deficient conditions dissolved substantially greater amounts of iron phosphate than exudates from alfalfa grown in hydroponics under iron-sufficient conditions. A number of plants under iron stress have been shown to excrete caffeic acid or similar compounds capable of mobilizing iron by the reductive process [52-54]. Vempati et al. [49] demonstrated that exudates of soybean seedlings grown in an *in vitro* geoponic system reduced the structural iron in iron bearing minerals.

3.3. Other elements

Organic acids released from plants/microorganisms in response to stresses due to nutrient deficiencies are also capable of reacting with other trace elements. Marschner et al. [55] and Takagi, et al. [56] demonstrated that the root exudates of wheat plants cultured under zinc- and iron-deficient media mobilized zinc and iron at a comparable rate. Root exudates of 11 day old barley plants grown under iron deficiency stress increased the steady state concentrations of Cu, Fe, Zn, and Mn in a calcareous soil by a factor 20 over the root exudates of similar plants that received sufficient amounts of iron [44]. Merckx et al. [3] reported that the water-soluble organic material of soil from roots of maize and wheat possessed greater capacity to complex Co, Zn and Mn than similarly extracted soluble organic matter from the soil of a not cultivated adjacent field. Uren and Reisenauer, [38] and Godo and Reisenauer [57] demonstrated the mobilization of manganese through reduction of manganic oxides by soluble soil organic matter and root exudates. Compared to the solubility in distilled water, root exudates of *Zea mays* and 2 varieties of tobacco increased the solubility of Fe, Mn and Cu in the soil, but exhibited little effect on the solubility of Ni and Zn [58].

Substances excreted by plants may protect them against phytotoxic levels of substances in the soil environment. Delhaize et al. [59] observed that roots of aluminum tolerant wheat plants excreted 5-10 times more malic acid than genotypes of aluminum-sensitive plants (Figure 4). Presumably the malic acid forms a complex with aluminum thereby lowering the concentration of soluble inorganic aluminum to non-toxic levels.

4. SOIL WEATHERING

Plant absorption and microbial uptake of potassium contributed to the weathering of mica and other phyllosilicate primary minerals. In general, potassium in the interlayer region of micaceous minerals is in equilibrium with potassium ion in the soil solution and it is absorbed into and released from the interlayer region in response to the activity of potassium in solution phase. Studies have shown that the uptake of potassium by plants and microbes can reduce the concentration of potassium in the soil solution to levels where, to maintain equilibrium, potassium is progressively released to the soil solution from the interlayer region. The removal of potassium from the interlayer region and its replacement by basic cations lead to the formation of partially potassium depleted minerals, such as illite and hydrobiotite, and a completely potassium depleted mineral, vermiculite. Chemically, this process may be represented by the following equilibrium reaction:

$$\underset{\text{(Biotite)}}{K_2[Si_6Al_2][Mg_{6-x}Fe_x]O_{20}(OH)_{4(s)}} + M^{2+}{}_{(aq)} = \underset{\text{(Vermiculite)}}{M[Si_6Al_2][Mg_{6-x}Fe_x]O_{20}(OH)_{4(s)}} + 2K^{+}{}_{(aq)} \quad (5)$$

where M denotes Ca^{2+} and/or Mg^{2+} at the soil/root interface. As the K^+ in soil solution is depleted by plant uptake or microbial synthesis, the equilibrium shifts toward the right resulting in the formation of vermiculite, a secondary mineral. Under humid climates where the annual precipitation is considerably greater than the annual evapotranspiration loss, the leaching of potassium and other basic cations results in the weathering of micaceous minerals and leads to the formation of illite, vermiculite, montmorilinite, kaolinite, and eventually oxides of iron and aluminum [60].

Mortland et al. [61] grew successive crops of wheat over a one-year period in sand culture with biotite as the sole source of potassium. They observed the continuous uptake of potassium by plants and alteration of biotite to vermiculite. Potassium ions in the interlayer region of micas are not readily exchangeable. Its replacement by other basic cations resulted in increases in the cation exchange capacity of the minerals. In the system containing 0.5% biotite, the cation exchange capacity increased from 14 $cmol^{-1}$ kg for the original biotite to 54 cmol kg^{-1} for the weathered end products. Spyidakis et al. [62] planted seven tree seedlings in sand media that contained 0.2% biotite as the sole source of potassium. When the clay minerals were recovered 13 months later, their particle sizes were significantly reduced and biotite was converted to vermiculite and kaolinite.

The synergism of plants and microorganisms in the rhizosphere often accelerates the weathering of clay minerals. Leyval and Berthelin [63] planted pine seedlings that were inoculated with bacteria, mycorrizal fungi, and both bacteria and mycorriza fungi and pine seedlings that were not inoculated in sand with 1% phlogopite (a ferromagnesium phyllosilicate similar to biotite) as the sole source of potassium, magnesium, and iron. The

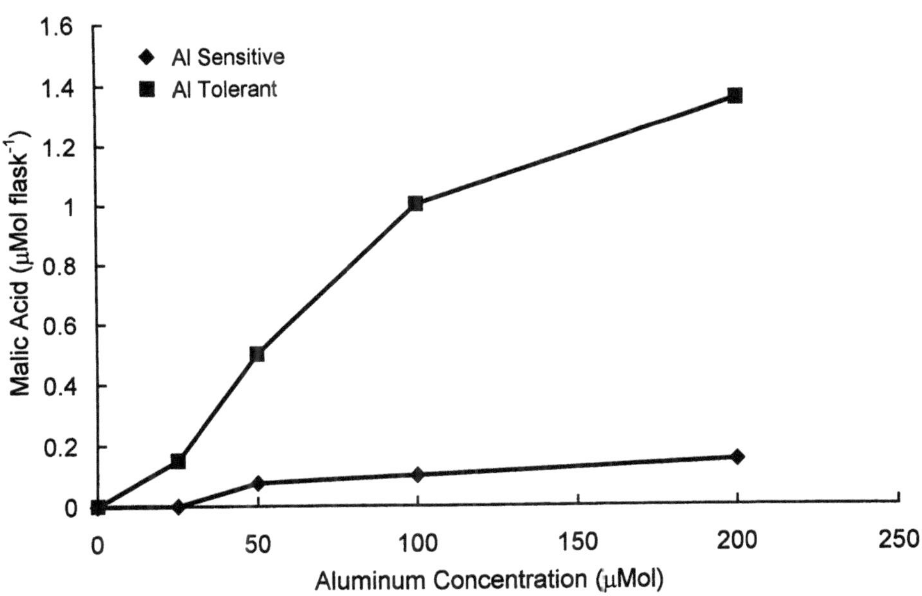

Figure 4. Excretion of malic acid by Al-sensitive and Al-tolerant plants in relation to Al concentration of growing media (derived from reference [59]).

phlogopite was weathered to vermiculite in one year. The extent of weathering by the inoculated tree seedlings was more intense than the non-inoculated tree seedlings suggesting that the products of microbial activities in the rhizosphere hastened the weathering process.

Weed et al. [64] exposed micas to 7 different fungi (*Ascomycetes*, *Basidiomycetes*, *Phycomycetes* and 4 *Imperfecti ssp.*) and demonstrated that biotite, phlogopite and muscovite were converted to vermiculite with the liberation of potassium from the mineral structures. Figure 5 shows the conversion of biotite and muscovite (10 Å diffraction spacing) to intermediate K-depleted micas (10 Å and 14.2 Å diffraction spacing) and vermiculite (14.2 Å diffraction spacing) following exposure to fungi over time. The strength of the mineral's binding for potassium determined the extent of the weathering. The potassium on muscovite was more difficult to be removed than potassium on biotite and phlogopite, which exhibited comparable extent of weathering. Potassium, rubidium, and cesium are selectively adsorbed on edge sites of micas. Root exudates of plants may affect their adsorption. Delvaux et al. [65] exposed root mats of rye seedlings to cesium-137 treated forest and meadow soils from 47 pedons located in Austria, Belgium, Germany, Ireland, Italy and Switzerland and found that the selectivity of the soils for cesium is proportional to their vermiculite content. The rhizosphere-enhanced uptake of cesium is greatest in soils low in vermiculite and potassium.

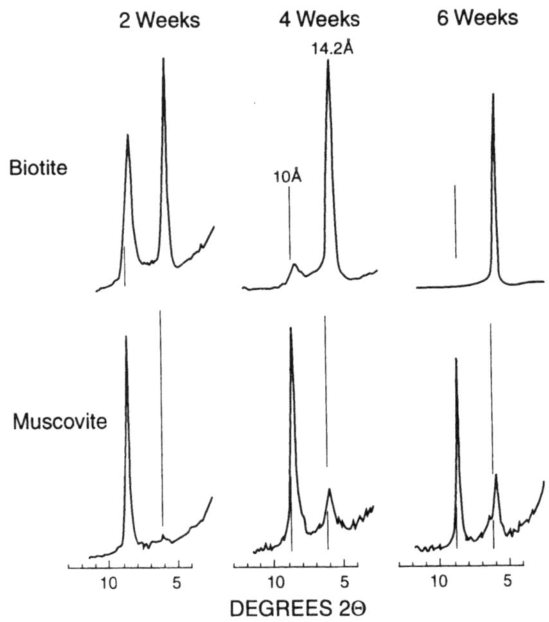

Figure 5. Weathering of biotite and muscovite to vermiculite after exposed to fungi for 2, 4 and 6 weeks (according to reference [64]).

Root excretion and microbial metabolism induced organic substances in the rhizosphere that contributed to the further weathering of micas and other soil minerals. Kosta et al. [66] showed that Fe^{3+} atoms in the mineral structure of phyllosilicates may be reduced and released as Fe^{2+} when the clay minerals are incubated with the culture of *Shewanella putrefaciens* - an iron reducing bacterium (Figures 6). Reduction of iron present in the crystal lattice caused dissolution of the clay. Rates of dissolution were proportional to the surface area of the layer silicates. Boyle et al. [67] attributed the removal of tetrahedral and octahedral cations from mica following equilibration with oxalic, malonic, citric, malic, propionic and lactic acids to the organic acid chelation of octahedral and tetrahedral cations. The pH of the systems ranged from 0.65 and 2.4, however, and the dissolution of structural cations created by the acidity should not be ruled out all together.

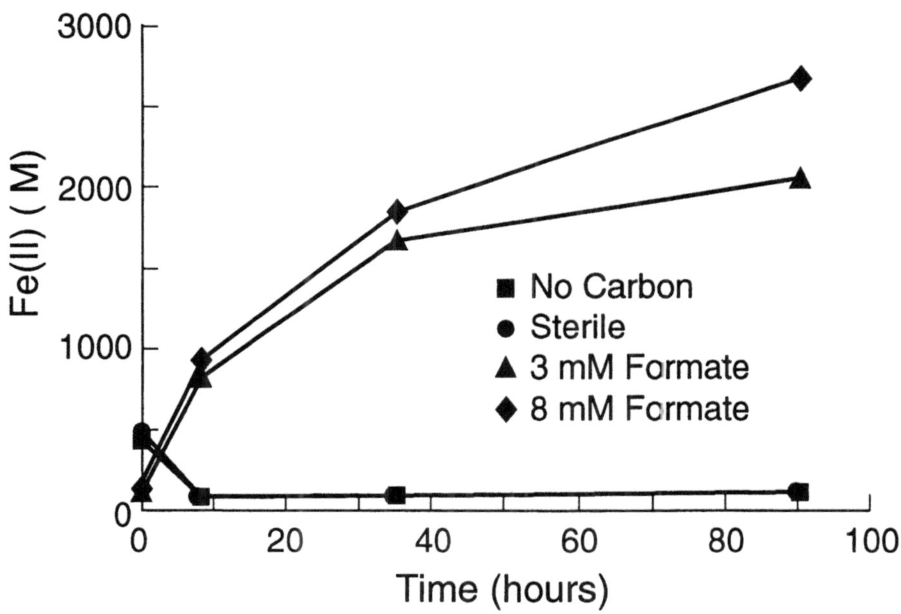

Figure 6. Bacterial reduction of structural Fe^{3+} in smectite suspension (derived from reference [66]).

5. CONCLUDING REMARKS

The rhizosphere, a thin layer of soil adjacent to roots of living plants, is a chemically complex and microbiologically dynamic segment of the soil. At any given time, it accounts for only a very small fraction of the bulk soil in the root zone. Yet, all of the surface soils, at one time or another, may be under the influenced of roots. The rhizosphere receives exudates from roots, supports a dense and diverse population of microorganisms, and its chemistry is affected by the organic substances exuded by plant roots and by metabolites produced from microbial degradation of the organic substrates. The extraction of nutrients such as potassium ions by plants often initiates the weathering of primary minerals (micas) to secondary minerals (vermiculite). The biogeochemical activities in the rhizosphere have profound influence on evolution of soils. They induced dissolution of not readily soluble plant nutrients and potentially toxic elements and accelerated the weathering of clay minerals further affecting the soil ability to accommodate plant nutrients and to attenuate toxic elements.

REFERENCES

1. Jenny, H., Overstreet, R., 1938. Contact effects between plant roots and soil colloids. Proc. Nat. Academ. Sci. 24, 384-392.
2. Eaton, F. M., 1936. Automatically operated sand-culture equipment. J. Agri. Res. 53, 433-444.
3. Merckx, R., Ginkel, J.H., Sinnaeve, J., Cremers, A., 1986. Plant-induced changes in the rhizosphere of maize and wheat. II. Complexation of cobalt, zinc and manganese in the rhizosphere of maize and wheat. Plant Soil. 96, 95-107.
4. Linehan, D.J., Sinclair, A.H., Mitchell, M.C., 1989. Seasonal changes in Cu, Mn, Zn and Co concentrations in soil in the root-zone of barley (*Hordeum vulgare* L.). J. Soil Sci. 40, 103-115.
5. Habib, L., Chenu, C., Morel, J.L., Guckert, A., 1990. Adsorption of maize root mucilages on homoionic clays - consequences on the microstructure of the complexes. Comptes Rendus de L Academie des Sciences Serie II - Mecanique Physique Chimie Sciences de L univers Sciences de La Terre. 310, 1541-1546.
6. Rovira, A.D., Foster, R.C., Martin, J.K., 1979. Origin, nature, and nomenclature of organic materials in the rhizosphere. *In:* Harley, J.L., Scott-Russel, R. (Eds.), The soil-root interface. Academic Press Inc., New York, pp.1-4.
7. Kraffczyk, I., Trolldenier, G., Beringer, H., 1984. Soluble root exudates of maize: influence of potassium supply and rhizosphere microorganisms. Soil Biol. Bioch. 16, 315-322.
8. Mench, M., Morel, J.L., Guckert, A., 1988. Influence of metal [Cd (II), Cu (II), Pb (II), Zn (II)] ions on soluble root exudates of maize (*Zea mays* L.). Agronomie. 8, 237-241.
9. Hoffland, E., Findenegg, G., Nelemans, J., 1989. Solubilization of rock phosphate by rape. II. Local root exudation of organic acids as a response to P-starvation. Plant Soil. 113, 161-165.
10. Li, M.G., Shinano, T., Tadano, T., 1997. Distribution of exudates of lupin roots in the rhizosphere under phosphorus deficient conditions. Soil Sci. Plant Nutr. 43, 237-245.
11. Otani, T., Ae, N., Tanaka, H., 1996. Phosphorus (P) uptake mechanisms of crops grown in soils with low P status. Soil Sci. Plant Nutr. 42, 553-560.
12. Young, C.C., Chang, C.H., Chen, L.F., Chao, C.C., 1998. Characterization of the nitrogen fixation and ferric phosphate solubilizing bacteria isolated from a Taiwan soil. J. Chinese Agr. Chem. Soc. 36, 201-210.
13. Laheurte, F., Leyval, C., Berthelin, J., 1990. Root exudates of maize, pine and beech seedlings influenced by mycorrhizal and bacterial inoculation. Symbiosis. 9, 111-116.
14. Kothari, S.K., Marschner, H., Romheld, V., 1991. Effect of a vesicular-arbuscular mycorrhizal fungus and rhizosphere micro-organisms on manganese reduction in the rhizosphere and manganese concentrations in maize (*Zea mays* L.). New Phytol. 117, 649-655.
15. Morel, J.L., Mench, M., Guckert, A., 1987. Heavy metal dynamics in the rhizosphere: role of root exudates. Revue d'Ecologie et de Biologie du Sol. 24, 485-492.

16. Meharg, A.A., Killham, K., 1990. The effect of soil pH on rhizosphere carbon flow of *Lolium perenne*. Plant Soil. 123, 1-7.
17. Misra, A., Ramani, S., 1992. Identification of caffeic acid in the root exudate of Fe- efficient Japanese mint subjected to Fe stress: its action on Fe- chlorosis recovery. Agrochimica. 36, 93-100.
18. Szmigielska, A.M., Vanrees, K.C.J., Cieslinski, G., Huang, P.M., Knott, D.R., 1995. Determination of low molecular weight dicarboxylic acids in root exudates by gas chromatography. J. Agr. Food Chem. 43, 956-959.
19. Van Veen, J.A., Ladd, J.N., Amatom, M., 1985. Turnover of carbon and nitrogen through the microbial biomass in sandy loam and clay soil incubated with [^{14}C(U)]glucose and [^{15}N]$(NH_4)_2SO_4$ under different moisture regimes. Soil Biol. Biochem. 17, 747-756.
20. Burken, J.G., Schnoor, J.L., 1996. Phytoremediation: plant uptake of atrazine and role of root exudates. J. Environ. Eng. 122, 958-963.
21. Wiedenroth, E., Poskuta, J., 1981. The influence of oxygen deficiency in roots on CO_2 carbon dioxide exchange rates of shoots and distribution of ^{14}C carbon isotope-photoassimilates of wheat seedlings. Z. Pflanzenphysiol. 103, 459-467.
22. Nye, P.H., Vaughan, D., Edwards, A.C., 1977. Movement and uptake of solutes in field conditions. In Solute movement on the soil root system. Studies in Ecology. Blackwell, Oxford. 243-323.
23. Graham, J.H., Leonard, R.T., Menge, J.A., 1982. Interaction of light intensity and soil temperature with phosphorus inhibition of vesicular-arbuscular mycorrhizal formation. New Phytol. 91, 683-690.
24. Kato-Noguchi, H., Mizutani, J., Hasegawa, K., 1994. Allelopathy of oats. II. Allelochemical effect of *L*-tryptophan and its concentration in oat root exudates. J. Chem. Ecol. 20, 315-319.
25. Pojasok, T., Kay, B.D., 1990. Effect of root exudates from corn and bromegrass on soil structural stability. Can. J. Soil Sci. 70, 351-362.
26. Anderson, T.A., Guthrie, E.A., Walton, B.T., 1993. Bioremediation in the rhizosphere. Environ. Sci. Tech. 27, 2630-2636.
27. Schilling, G., Gransee, A., Deubel, A., Lezovic, G., Ruppel, S., 1998. Phosphorus availability, root exudates, and microbial activity in the rhizosphere. Z. Pflanz. Bodenk. 161, 465-478.
28. Marschner, H., 1998. Role of root growth, arbuscular mycorrhiza, and root exudates for the efficiency in nutrient acquisition. Field. Crop Res. 56, 203-207.
29. Smith, W.H., 1970. Root exudates of seedling and mature sugar maple. Phytopathology. 60, 701-703.
30. Zhang, F., Romheld, V., Marschner, H., 1989. Effect of zinc deficiency in wheat on the release of zinc and iron mobilizing root exudates. Z. Pflanz. Bodenk. 152, 205-210.
31. Bozhkov, A.I., Menzyanova, N.G., Leontovich, V.P., 1996. Lipid composition and antibacterial activity of root exudates secreted by wheat seedlings. Russian J. Plant Physiol. 43, 795-799.
32. Chaboud, A., Rougier, M., 1991. Effect of root density in incubation medium on root exudate composition of axenic maize seedlings. J. Plant Physiol. 137, 602-606.
33. Barber, D.A., Martin, J.K., 1976. The release of organic substances by cereal roots into soil. New Phytol. 76, 69-80.
34. Martin, J.K., 1977. Factors influencing the loss of organic carbon from wheat roots [Soils]. Soil Biol. Biochem. 9, 1-7.

35. Prikryl, Z., Vancura, V., 1980. TI: Root exudates of plants. VI. Wheat root exudation as dependent on growth, concentration gradient of exudates and the presence of bacteria *Pseudomonas putida*. Plant Soil. 57, 69-83.
36. Schonwitz, R., Ziegler, H., 1982. Exudation of water-soluble vitamins and of some carbohydrates by intact roots of maize seedlings (*Zea mays* L.) into a mineral nutrient solution. Z. Pflanzenphysiol. Int. J. Plant Physiol. 107, 7-14.
37. Gardner, W.K., Barber, D.A., Parbery, D.G., 1983. Non-infecting rhizosphere micro-organisms and the mineral nutrition of temperate cereals. J. Plant Nutr. 6, 185-199.
38. Uren, N.C., Reisenauer, H.M., 1988. The role of root exudates in nutrient acquisition. Advances in Plant Nutrition [edited by Tinker, B.]. 3, 79-114.
39. Welch, R.M., 1995. Micronutrient nutrition of plants. Critical Reviews in Plant Sciences. 14, 49-82.
40. Krishnamurti, G.S.R., Cieslinski, G., Huang, P.M., Van Rees, K.C.J., 1997. Kinetics of cadmium release from soils as influenced by organic acids: Implication in cadmium availability. J. Environ. Qual. 26, 271-277.
41. Jones, D.L., Edwards, A.C., Donachie, K., Darrah, P.R., 1994. Role of proteinaceous amino acids released in root exudates in nutrient acquisition from the rhizosphere. Plant Soil. 158, 183-192.
42. Grierson, P.F., 1992. Organic acids in the rhizosphere of *Banksia integrifolia* L. Plant Soil. 144, 259-265.
43. Lipton, D.S., Blanchar, R.W., Blevins, D.G., 1987. Citrate, malate, and succinate concentration in exudates from P- sufficient and P-stressed *Medicago sativa* L. seedlings. Plant Physiol. 85, 315-317.
44. Treeby, M., Marschner, H., Romheld, V., 1989. Mobilization of iron and other micronutrient cations from a calcareous soil by plant-borne, microbial, and synthetic metal chelators. Plant Soil. 114, 217-226.
45. Awad, F., Romheld, V., Marschner, H., 1988. Mobilization of ferric iron from a calcareous soil by plant-borne chelators (phytosiderophores). J. Plant Nutr. 11, 701-713.
46. Awad, F., Romheld, V., Marschner, H., 1994. Effect of root exudates on mobilization in the rhizosphere and uptake of iron by wheat plants. Plant Soil. 165, 213-218.
47. Awad, F., Mostafa, M.A., El-Sebaay, A.S., El-Ashry, M.S., 1999. Effect of wheat root exudates on mobilization of iron and zinc from calcareous soils. Com. Soil Sci. Plant Anal. 30, 1897-1907.
48. LaKind, J.S., Stone, A.T., 1989. Reductive dissolution of goethite by phenolic reductants. Geochim. Acta. 53, 961-971.
49. Vempati, R.K., Kollipara, K.P., Stucki, J.W., Wilkinson, H., 1995. Reduction of structural iron in selected iron-bearing minerals by soybean root exudates grown in an *in vitro* geoponic system. J. Plant Nutr. 18, 343-353.
50. Gerke, J., Romer, W., Jungk, A., 1994. The excretion of citric and malic acid by proteoid roots of *Lupinus albus* L.; Effects on soil solution concentrations of phosphate, iron, and aluminum in the proteoid rhizosphere in samples of an oxisol and a luvisol. Z. Pflanz. Bodenk. 157, 289-294.
51. Dinkelaker, B., Romheld, V., Marschner, H., 1989. Citric acid excretion and precipitation of calcium citrate in the rhizosphere of white lupin (*Lupinus albus* L.). Plant Cell Environ. 12, 285-292.
52. Olsen, R.A., Bennett, J.H., Blume, D., Brown, J.C., 1981. Chemical aspects of Fe iron stress response mechanism in tomatoes. J. Plant Nutr. 3, 905-921.

53. Hether, N.H., Olsen, R.A., Jackson, L.L., 1984. Chemical identification of iron reductants exuded by plant roots. J. Plant Nutr. 7, 667-676.
54. Romheld, V., Marschner, H., 1983. Mechanism of iron uptake by peanut plants. I. Fe(III) reduction, chelate splitting, and release of phenolics. Plant Physiol. 71, 949-954.
55. Marschner, H., Romheld, V., Kissel, M., 1986. Different strategies in higher plants in mobilization and uptake of iron. J. Plant Nutr. 9, 695-713.
56. Takagi, S., Kamei, S., Yu, M.H., 1988. Efficiency of iron extraction from soil by mugineic acid family phytosiderophores. J. Plant Nutr. 11, 643-651.
57. Godo, G.H., Reisenauer, H.M., 1980. Plant effects on soil manganese availability. Soil Sci. Soc. Am. J. 44, 993-995.
58. Mench, M., Martin, E., 1991. Mobilization of cadmium and other metals from two soils by root exudates of *Zea mays* L., *Nicotiana tabacum* L. and *Nicotiana rustica* L. Plant Soil. 132, 187-196.
59. Delhaize, E., Ryan, P.R., Randall, P.J., 1993. Aluminum tolerance in wheat (*Triticum aestivum* L.). II. Aluminum- stimulated excretion of malic acid from root apices. Plant Physiol. 103, 695-702.
60. Jackson, M.L., 1964. Chemical composition of soils. *In:* Bear, F. E. (Ed.), Chemistry of the Soil, 2nd Ed. Renhold Publishing Corp. New York, pp. 71–141.
61. Mortland, M.M., Lawton, K., Uehara, G., 1956. Alteration of biotite to vermiculite by plant growth. Soil Sci. 82, 477-481.
62. Spyridakis, D.E., Chester, S.G., Wilde, S.A., 1967. Kaolinization of biotite as a result of coniferous and deciduous seedling growth. Soil Sci. Soc. Am. Proc. 31, 203-210.
63. Leyval, C., Berthelin, J., 1991. Weathering of a mica by roots and rhizospheric microorganisms of pine. Soil Sci. Soc. Am. J. 55, 1009-1016.
64. Weed, S.B., Davey, C.B., Cook, M.G., 1969. Weathering of mica by fungi. Soil Sci. Soc. Am. Proc. 33, 702-706.
65. Delvaux, B., Kruyts, N., Cremers, A., 2000. Rhizospheric mobilization of radiocesium in soils. Environ. Sci. Tech. 34, 1489-1493.
66. Kostka, J.E., Haefele, E., Viehweger, R., Stucki, J.W., 1999. Respiration and dissolution of Fe(III)-containing clay minerals by bacteria. Environ. Sci.Technol. 33, 3127 – 3133.
67. Boyle, J.R., Voigt, G.K., Sawhney, B.L., 1974. Chemical weathering of biotite by organic acids. Soil Sci. 117, 42-45.

Developments in Soil Science, Volume 28B
Editors: A. Violante, P.M. Huang, J.-M. Bollag and L. Gianfreda

ADSORPTION OF THE INSECTICIDAL TOXIN FROM *BACILLUS THURINGIENSIS* SUBSPECIES *TENEBRIONIS* TO CLAY FRACTIONS OF TROPICAL SOILS

P. Muchaonyerwa[a, b], C. Chenu[a*], O.L. Pantani[c], L. Calamai[c], P. Nyamugafata[b] and S. Mpepereki[b]

[a]Unité de Science du Sol, INRA, 78026 Versailles, France

[b]Department of Soil Science and Agric. Eng., University of Zimbabwe, Harare, Zimbabwe

[c]Istituto Genesi Ecologia Suolo, CNR Firenze, Italy

[*]Corresponding author

Genetically modified plants, which produce *Bacillus thuringiensis* toxins, release the toxins into the soil through root exudates and upon decomposition of residues. There is need to assess the fate of these toxins in the soil. Our aim was to analyze the adsorption of the toxin from *Bacillus thuringiensis* subsp. *tenebrionis* (Btt) to surface active soil constituents. We separated clay-sized fractions from a vertisol, an alfisol and an oxisol after complete dispersion of the soils by mechanical means and purified the toxin from a Btt commercial preparation. Adsorption isotherms were established by the difference method in deionized water or in phosphate buffer (pH 6.8). Adsorption of Btt toxin was in the order: vertisol > alfisol > oxisol, and it was mostly irreversible, except for the alfisol. Adsorption could be explained by physicochemical conditions (pH and ionic strength) and the nature of clay constituents, i.e., the dominant clay mineralogy: montmorillonite (vertisol) and kaolinite (alfisol, oxisol). The results show that Bt toxins released into the soil may be retained by soil colloids. The impact of such adsorption on biodegradation and pesticidal activity should be studied.

1. INTRODUCTION

Massive doses of chemical pesticides used in agriculture and mosquito control may pollute the environment [1, 2]. The need for environmentally friendly biopesticides has enhanced interests in proteins from sub-species of *Bacillus thuringiensis* (Bt), a gram-positive, rod-shaped aerobic bacterium, which are selectively pesticidal to larvae of various insect orders, including Lepidoptera, Coleoptera and Diptera [3, 4]. Commercial preparations of these proteins used in agricultural applications degrade in a few days [5]. Plants, such as cotton, maize and potato, have been genetically modified to express the genes that code for production of the active toxins [6]. Such transgenic plants are being introduced into tropical agriculture. For example, Bt-maize is now commercially grown in South Africa, while cotton farmers in Zimbabwe are fighting hard to get clearance from the government to grow Bt-cotton. Bt-maize releases the toxin into the soil through root exudates [7] during the growth stages and through decomposition of residues incorporated into the soil after harvesting. There are rising concerns worldwide about the environmental safety of transgenic plants on non-target insects and soil microorganisms and the transfer of the modified genetic material across related organisms.

Bacillus thuringiensis toxins were found to adsorb onto pure clay minerals and onto humic acids extracted from soils [8–10], and the complexes formed from these interactions were less available for microbial degradation and they retained their pesticidal activity longer than free forms [9, 11]. These interactions were found to be dependent on clay mineralogy, pH and the charge-compensating cation on the clay [8, 10]. The toxins from transgenic plants may persist in the environment when they complex with clays, oxihydroxides of Fe and Al, and organic matter.

Very little information is available concerning the interactions between Bt toxins and the naturally occurring soil clay particles with more complex and heterogeneous mineralogy and coated with oxyhydroxides and organic matter [12, 13]. Studies by Sims and Ream [14] have shown that the activity of Bt sub-species *kurstaki* (Btk) toxin from transgenic cotton was detectable and stable (< 25% of initial bioactivity) in the soil for over 120 days. Tapp and Stotzky [15] found that toxins from Bt sub-species added to non-sterile soil could still be detected in the clay fraction of soil by insect bioassays after 40 days. The aim of the present work was to analyze the adsorption of the toxin from *Bt* sub-species *tenebrionis* (Btt) onto naturally occurring tropical soil clay particles.

2. MATERIALS AND METHODS

2.1. Soils

We used soil samples from ploughed layers (0–20 cm) of the following soils: vertisol from Zimbabwe (Chisumbanje), alfisol from Zimbabwe (Chiredzi) and oxisol from Tanzania (Morogoro). Particle-size fractionation was performed according to the method of Balesdent et al. [16], which does not destroy organic matter and sesquioxides. Dispersion of the soil was achieved by mechanical means, i.e., agitation of the soil suspension with glass beads and dispersion of micro-aggregates by ultrasoounds. The size fractions less than 2 μm were separated by centrifugation and freeze-dried.

2.2. Clay-sized fractions

Masses of the clay fractions were recorded, their C contents were measured by combustion and mineralogy was determined with x-ray diffractometry after destruction of organic matter. Cation exchange capacity (CEC) was determined by cobaltihexamine, and crystalline iron was determined by subtracting iron measured by oxalate extraction from dithionite-citrate extractable iron [17]. The properties of the clay fractions are listed in Table 1.

2.3. Btt toxin

The toxin of Btt was purified from a commercial preparation, Novodor (Novo Nordisk) using MOPS buffer at pH 7 [8] modified by addition of β-mercaptoethanol, instead of dithiothreitol, as a reducing agent. The commercial preparation was mixed with MOPS buffer at a ratio of 1:9, stirred overnight and centrifuged at 15 000 x *g* for 25 minutes. The supernatant was sequentially filtered through 0.8 μm, 0.45 μm and 0.2 μm filters and dialyzed (molecular cutoff of 12 kDa) against distilled water (48 h). The toxin precipitate was centrifuged at

Table 1
Characteristics of clay fractions from selected tropical soils

Property	clay fraction		
	vertisol	alfisol	oxisol
mass (g kg^{-1} of oven-dry soil)	739 ± 9	250 ± 6	536 ± 15
C content (g kg^{-1})	13.9 ± 1.1	16.4 ±0.2	17.8 ± 0.0
pH (1.0 mg ml^{-1} suspension in water)	7.1 ± 0.1	6.9 ± 0.1	6.2 ± 0.1
CEC cmol(+) kg^{-1}	92.6	30.8	11.5
Total surface area (EGME*)(m^2 g^{-1})	434	166	103
Fe_2O_3 (g kg^{-1})	10.7	103.9	122.8
mineralogy	smectite	kaolinite	kaolinite
	kaolinite	illite	illite
	illite	smectite	smectite
	quartz	quartz	goethite
		feldspar	
		hematite	

*EGME = Ethylen glycol monoether

15 000 x *g* and the pellet was stored in a refrigerator at 4 °C. Purity tests with sodium dodecyl sulfate-poly-acrylamide gel electrophoresis (SDS-PAGE) showed an intense band around 66 kDa molecular weight with minor bands at 38, 40 and 56 kDa, which were probably residues of the toxin. The carbon content of the lyophilized toxin was determined by combustion. Solutions of toxin ranging from 0–2.54 mg/ml were prepared by mixing the toxin with deionized water or with phosphate buffer at pH 6.8. The toxin solutions were centrifuged at 27 200 x *g* for 20 minutes to eliminate any precipitate. Toxin concentrations were determined with the Lowry method [18] using a micro-Lowry kit from Sigma Co. Bovine serum albumin (BSA) was used as the standard. Carbon contents of the toxin solutions were also determined and a regression analysis was performed to determine the relationship between BSA equivalents and the toxin concentration. This relationship was used to convert all the data from BSA equivalents to actual toxin concentrations.

2.4. Clay suspension

Clay suspensions (1 g/L) were prepared either with deionized water or in 0.2 M phosphate buffer at pH 6.8 (in the case of vertisol). To be representative of the physicochemical conditions prevailing in the different soils, most of the adsorption measurements were performed in water, i.e., uncontrolled pH and no added salts. The use of phosphate buffer could alter the surface properties of oxyhydroxides in alfisol and oxisol as a result of fixation of phosphate and therefore affect adsorption of the toxin.

2.5. Adsorption and desorption measurements

Equilibrium adsorptions were achieved by mixing the clay suspensions with toxin solutions at a ratio of 1:1 by volume to a total volume of 5 ml in 40 ml centrifuge tubes with 3 replicate tubes at each toxin concentration. The tubes were shaken on an end-over-end shaker (32 rpm) for 3 h at 25°C, and the mixtures were centrifuged at 27 200 x *g* for 20 minutes. Four ml of the supernatant were removed, and the pellet was washed by resuspending in 4 ml of either distilled water or phosphate buffer (vertisol only) and shaken on a motorized wheel for 30 minutes before centrifugation. The pellet was repeatedly washed until no more toxin could be detected in the washes. This normally occurred after 3 washes. Protein contents of the equilibrium supernatants and all the washings were determined by the Lowry method. Amounts of toxin adsorbed at equilibrium were determined by subtracting the amount determined in the equilibrium supernatant from the amount added into the system. Quantities of protein strongly bound to the clay fractions were also determined by subtracting the amounts of toxins removed at equilibrium and in all the washes from amounts of toxin added into the system. Adsorption isotherms were plotted from the means of equilibrium concentrations and quantities of protein adsorbed per mg of clay.

In another experiment, carried out to determine the effect of pH on adsorption of the toxin, suspensions of vertisol and alfisol clay fractions (1 mg/ml) and toxin solutions were adjusted to pH 5.5, 7.0 and 7.5 with either 0.01 M HCl or 0.01 M NaOH. The toxin solutions were centrifuged (27 200 x g) for 20 minutes to remove any precipitate, and the supernatants were adjusted to a concentration of 0.4 mg/ml with water at the appropriate pH. Adsorption experiments were performed by mixing clay suspensions and toxin solutions, at the same pH, for 3 h at 25°C with end-over-end shaking. The mixtures were centrifuged (27 200 x *g*) and the supernatants removed before washing with water at the appropriate pH until no toxin could be detected in the washes.

3. RESULTS

3.1. Adsorption of Btt toxin on the different natural clay fractions in unbuffered conditions

Adsorption of Btt toxin on clay fractions of the different tropical soils increased with increasing toxin concentration (Figure 1); the adsorption isotherms were linear (C-type) [19]. A plateau was reached only with the oxisol clay fraction. For all clay fractions, no more toxin could be desorbed after 3 washes with deionized water. More than 70% of toxin adsorbed at equilibrium was retained in vertisol and oxisol, and less than 50% was retained after 3 washes (Figure 2). The vertisol clay fraction adsorbed and bound greater amounts of toxin than clay fractions of alfisol and oxisol, which bound comparable quantities. The pH values of equilibrium supernatants were slightly increased by addition of the toxin (Table 2).

Table 2
pH values of equilibrium adsorption supernatants of natural clay fractions (pure toxin solution had pH 7.9)

Clay fraction	Initial toxin concentration (mg ml^{-1})					
	0.00	0.08	0.21	0.42	0.84	1.28
vertisol (non-buffered)	7.3 ±0.0	7.3±0.3	7.3±0.1	7.5±0.3	7.8 ± 0.3	7.8 ± 0.3
alfisol (non-buffered)	7.3 ±0.0	7.1±0.0	7.3±0.1	7.2±0.0	7.1±0.1	7.0 ± 0.1
oxisol (non-buffered)	7.0 ± 0.0	7.1±0.2	7.2±0.0	7.1±0.3	7.1±0.0	7.0 ± 0.1
vertisol (buffered)	6.8 ±0.0	6.8±0.0	6.8±0.0	6.8±0.0	6.8±0.0	6.8 ± 0.0

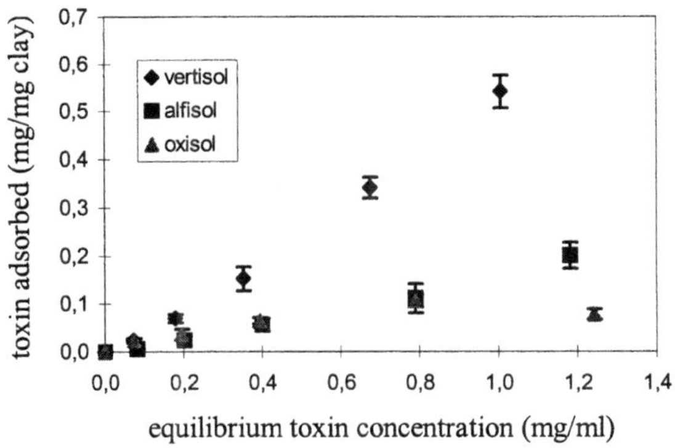

Figure 1. Adsorption of Btt toxin onto clay fractions from different tropical soils

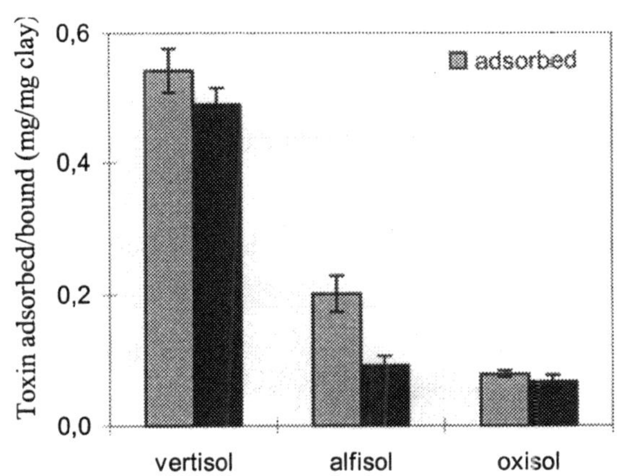

Figure 2. Desorption of Btt toxin from clay fractions of tropical soils for an initial concentration of 2.56 mg toxin/mg clay in the equilibrium mixture. Clay concentration in the equilibrium mixtures was 0.5 mg/ml.

3.2. Adsorption of Btt toxin in buffered conditions

The buffered vertisol clay fraction adsorbed all the toxin from solution until a plateau was reached (Figure 3a). The adsorption isotherm was of high affinity type (H-type) [19]. In unbuffered conditions, adsorption of the toxin on the vertisol clay fraction could be described by a linear isotherm, and no plateau was reached (Figure 3a). Bound toxin was from 85–95% of the amount adsorbed at equilibrium in both systems. However, the buffered clay bound significantly higher quantities, as shown in Figure 3b.

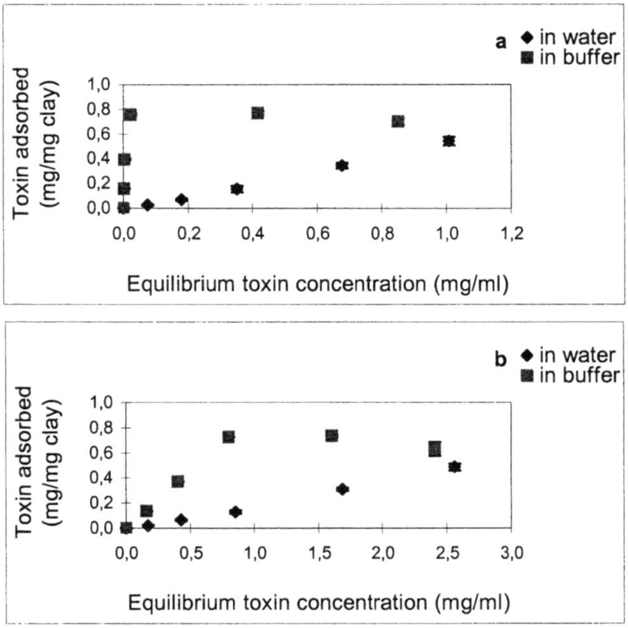

Figure 3. Adsorption (a) and retention (b) of Btt toxin on vertisol clay fraction in water and in buffered conditions

3.3. Effect of pH on adsorption of Btt toxin onto clay fractions of soil

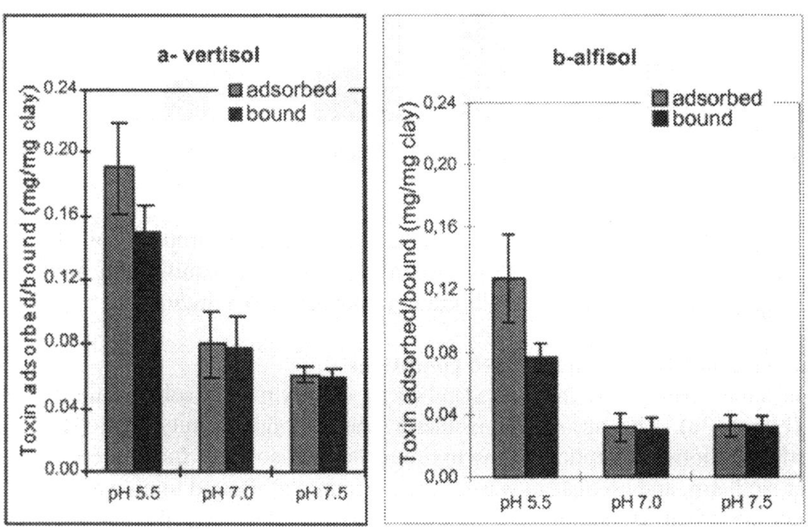

Figure 4. Effect of pH on adsorption of Btt toxin on vertisol (3a) and alfisol (3b) clay fractions

Adsorption of Btt toxin on both vertisol and alfisol clay fractions decreased with an increase in pH from pH 5.5 to pH 7.5 (Figures 4a and b). The alfisol clay fraction adsorbed less toxin than the vertisol clay fraction at each pH. Adsorption at pH 7.0 and 7.5 was irreversible (> 95% retained after washing), but at pH 5.5, 20% and 40% of adsorbed toxin could be desorbed from vertisol and alfisol clay fractions, respectively (Figures 4a and b).

4. DISCUSSION

The toxin from Btt adsorbed, in substantial amounts, to natural tropical soil clay fractions. As stated by Tapp et al. [8], Btt toxin is likely to be retained in soils when it adsorbs onto soil particles and may resist decomposition by soil microorganisms and retain its insecticidal activity [11, 12]. Tapp and Stotzky [15] observed that Btk toxin could be detected in the clay fraction of non-sterile soils even after 40 days.

4.1. Influence of clay fraction constituents

Adsorption of Btt toxin on the different soil clay fractions was in the order vertisol > alfisol > oxisol. Coefficients of distribution (Kd) were calculated by linear regression of the equilibrium toxin concentration, in mg/L, against adsorbed toxin, in mg/kg. In the case of the oxisol clay fraction, only the linear portion of the isotherm was used in the regression. Coefficients of distribution ranged from 518 L/kg (r^2=0.991) for vertisol to 159 L/kg (r^2=0.983) for alfisol and 145 L/kg (r^2=0.948) for oxisol. From the coefficients of distribution, we see that the toxin has greater affinity for the vertisol clay fraction than either the alfisol or the oxisol clay fractions. Tapp et al. [8] and Venkateswerlu and Stotzky [10] observed lower adsorption of toxins of Bt subspecies *kurstaki* and Bt subspecies *tenebrionis* on kaolinite than on montmorillonite homoionic to different cations. Montmorillonite is dominant in vertisols, while alfisols and oxisols are dominated by kaolinite (Table 1). The differences in clay mineralogy may explain the adsorption behaviors of these clay fractions. Specific surface areas and the CEC of the clay fractions (Table 1) may be the major properties that are responsible for the differences in toxin adsorption. Indeed, the ratios Kd/specific surface area were 1.19 (vertisol), 0.96 (alfisol) and 1.40 (oxisol), which are rather close values, suggesting that available surface areas control the adsorption of the toxin.

These clay fractions differ in their clay mineralogy, the presence of oxides and the amounts of organic matter (Table 1). There is little difference among the fractions in terms of organic matter, but there are substantial amounts of oxides in oxisol and alfisol clay fractions. Adsorption studies by Gianfreda and Violante [20] showed that enzymes precipitate on phenolic compounds and that coating montmorillonite with hydroxides of Al reduced enzyme adsorption. Amino compounds are known to react with phenolic and quinonoid substances to give rise to dark coloured, biologically stable complexes resembling humic compounds [21]. Condensed tannins yield more stable complexes than their hydrolyzable counterparts [21]. There are possibilities that organic matter and oxide affect adsorption of Btt toxin.

4.2. Influence of ionic strength

Physicochemical conditions (pH and ionic strength) of the medium appear to play an important role in the adsorption of the toxin. More toxin was adsorbed in higher ionic strength solutions (0.2 M phosphate buffer) than in water. The possible explanation is that the buffered clay fraction was saturated with Na^+ ions and could have been more dispersed, exposing a larger surface for adsorption, while the non-buffered clay was more flocculated because of higher concentrations of Ca^{2+} [21] than Na^+. Tapp et al. [8] observed that adsorption of both Btt and Btk toxins decreased with increase in valency of the charge-compensating cation on the homoionic clays.

4.3. Influence of pH

Higher toxin adsorption occurred at pH 5.5 than at pH 7.0 and 7.5. This trend was different from that observed by Tapp et al. [8]; they observed that maximum adsorption of Btt toxin on dirty clays (coated with oxides of Fe) occurred between pH 5 and pH 9. However, a similar trend was observed for Btk toxin [8, 10]. This can possibly be explained by the isoelectric point (pI) of the protein, which is pH 6.5 [8]. Below the pI, the toxin may interact with the clay through electrostatic attractions due to a net positive charge, but above the pI, the toxin may be repelled from the negatively charged clay surface due to its net negative charge, thereby reducing adsorption [22]. Among other mechanisms, van der Waals interactions, operative at all pH values, and multiple H-bonding are involved in adsorption of the toxin above the pI [21]. Ionic strength and pH may have interactive effects to determine adsorption of the toxin by influencing: charge characteristics of both the clay particles (pH-dependent minerals) and the toxin; solubility and conformation of the toxin; and relative accessibility of the surface [21]. Servagent-Noinville et al. [23] suggested that upon adsorption of proteins to clay surfaces, a transfer of protons may occur between the solution and the clay-protein interface to counterbalance the electrostatic repulsion between the surface of the clay and the negatively charged amino acids of the protein, such as aspartic acid or glutamic acid.

5. CONCLUSIONS

Clay fractions of tropical soils adsorbed large quantities of Btt toxin; the adsorption depended on the nature of clay fraction constituents and physicochemical conditions (pH and ionic strength) of the medium. A large proportion of agricultural soils in the tropics contains significant amounts of clay-sized particles that may adsorb Btt toxin when it is released through transgenic plants. Effects of organic matter and oxihydroxides of Fe and Al on toxin adsorption need to be studied. Further work should focus on effects of adsorption of Btt toxin onto tropical soil clays during its decomposition by soil microorganisms and on the toxin's toxicity to insect larvae.

ACKNOWLEDGMENTS

The studies were funded by grant INCO-DCIC18CT97-0135 from the European Community under the MICROBE Project. The opinions expressed in this paper are solely of the authors. We also thank the technical staff in Science du Sol at INRA Versailles for their invaluable input.

REFERENCES

1. Chikuni, O., Nhachi, C.F.B., 1996. Residues of organochlorine pesticides in human milk.

In: Nhachi, C.F.B, Kasilo, O.M.J. (Eds.), Pesticides in Zimbabwe: Toxicity and Health Implications. University of Zimbabwe Publications, Harare, pp.73-80.

2. Zaranyika, M.F., Makhubalo, J.M., 1996. Organochlorine pesticide residues in in-land waters in Zimbabwe. *In*: Nhachi, C.F.B., Kasilo, O.M.J. (Eds.), Pesticides in Zimbabwe: Toxicity and Health Implications. University of Zimbabwe Publications, Harare, pp. 89-106.
3. Hofte, H., Whiteley, H.R., 1989. Insecticidal crystal proteins of *Bacillus thuringiensis*. Microbiol. Rev. 53, 242–255.
4. Crickmore, N., Zeigler, D.R., Feitelson, J., Schepf, E., Van Rie, J., Lereclus, D., Baum, J., Dean, D.H., 1998. Revision of the nomenclature for the *Bacillus thuringiensis* pesticidal crystal proteins. Microbiol. Molec. Biol. Rev. 62, 807–813.
5. Feitelson, J.S., Quick, T.C., Gaertner, F., 1990. Alternative hosts for *Bacillus thuringiensis* delta-endotoxin gene. *In*: Baker, R.R., Dunn, P.E. (Eds.), New directions in biological control: alternatives for suppressing agricultural pests and diseases UCLA Symposia on Molecular and Cellular Biology Series, Vol. 112. Liss, New York, pp. 561-571
6. Schnepf, E., Crickmore, N., Van Rie, J., Lereclus, D., Baum, J., Feitelson, J, Zeigler, D.R., Dean, D.H.,1998. Bacillus thuringiensis and its pesticidal crystal proteins. Microbiol. Molec. Biol. Rev. 62, 775–806.
7. Saxena, D., Flores, S., Stotzky, G., 1999. Insecticidal toxin in root exudates of *Bt* corn. Nature 402, 480-483.
8. Tapp, H., Calamai, L., Stotzky, G., 1994. Adsorption and binding of the insecticidal proteins from *Bacillus thuringensis* subsp. *kurstaki* and subsp. *tenebrionis* on clay minerals. Soil Biol. Biochem. 26, 663–679.
9. Crecchio, C., Stotzky, G., 1998. Insecticidal activity and biodegradation of the toxin from *Bacillus thuringiensis* subsp. *kurstaki* bound to humic acids from soils. Soil Biol. Biochem. 30, 463–470.
10. Venkateswerlu, G., Stotzky, G., 1992. Binding of the protoxin and toxin of *Bacillus thuringiensis* subsp. *kurstaki* on clay minerals. Current Microbiol. 25, 1–9.
11. Koskella, J., Stotzky, G., 1997. Microbial utilization of free and clay-bound insecticidal toxins from *Bacillus thuringiensis* and their retention of insecticidal activity after incubation with microbes. Appl. Environ. Microbiol. 63, 3561–3568.
12. Tapp, H., Stotzky, G., 1995. Insecticidal activity of the toxins from *Bacillus thuringiensis* subspecies *kurstaki* and *tenebrionis* adsorbed and bound on pure and soil clays. Appl. Environ. Microbiol. 61, 1786–1790.
13. Sundaram, K., 1996. Sorptive interactions and binding of "delta" endotoxin protein from *Bacillus thuringiensis* subsp. *kurstaki* in forest soils. J. Environ. Sci. Health. Part B: Pesticides, Food Contaminants and Agricultural Wastes 31, 1321–1340.
14. Sims, S.R., Ream, J.E., 1997. Soil inactivation of the *Bacillus thurigiensis* subsp. *kurstaki* CryIIA insecticidal protein with transgenic cotton tissue: Laboratory microcosm and field studies. J. Agric. Food Chem. 45, 1502–1505.
15. Tapp, H., Stotzky, G., 1998. Persistence of the insecticidal protein from *Bacillus thruringiensis* subsp. *kurstaki* in soil. Soil Biol. Biochem. 30, 471–476.
16. Balesdent, J., Pétraud, J.P., Feller, C., 1991. Effet des ultrasons sur la distribution granulométrique des matières organiques des sols. Sci. Sol 29, 95–106.
17. Mehra, O.P., Jackson, M.L., 1960. Iron oxide removal from soils and clays by a dithionite-citrate system buffered with sodium bicarbonate. Clays Clay Min. 7, 317–327.
18. Lowry, O.H., Rosebrough, N. J., Farr, A.L., Randall, R.J., 1951. Protein measurement with

folin phenol reagent. J. Biol. Chem. 193, 265–275.
19. Giles, C.H., D'Silva, A.P., Easton, I.A., 1974. A general treatment and classification of solute adsorption isotherm, Part II: Experimental interpretations. J. Coll. Interf. Sci. 47, 766-778.
20. Gianfreda, L., Violante, A., 1995. Activity, stability, and kinetic properties of enzymes immobilised on clay minerals and organomineral complexes. *In*: Huang, P.M., Berthelin, J., Bollag, J.M., McGill, W.B., Page, A.L. (Eds.), Environmental Impact of Soil Component Interactions: Metals, Other Inorganic, and Microbial Activities. CRC Lewis Publishers, Boca Raton, FL, pp. 29-38.
21. Theng, B.K.W., 1979. Formation and Properties of Clay-Polymer Complexes. Elsevier Scientific Publishing Company, Amsterdam.
22. Quiquampoix, H., Staunton, S., Baron, M.H., Ratcliffe, R.G., 1993. Interpretation of the pH dependence of protein adsorption on clay mineral surfaces and its relevance to the understanding of extracellular enzyme activity in soil. Colloid Surf. A: Physicochemical and Engineering Aspects 75, 85–93.
23. Servagent-Noinville, S., Revault, M., Quiquampoix, H., Baron, M.H., 2000. Conformational changes of bovine serum albumin induced by adsorption on different clay surfaces. J. Coll. Interf. Sci. 221, 273–283.

Developments in Soil Science, Volume 28B
Editors: A. Violante, P.M. Huang, J.-M. Bollag and L. Gianfreda

TOXINS FROM *BACILLUS THURINGIENSIS* SUBSPECIES *ISRAELENSIS* SORBED ON CLAYS: PERSISTENCE AND ACTIVITY AGAINST THE MOSQUITO *CULEX PIPIENS*

P. González[a], O. L. Pantani[b], G.G. Ristori[b] and A. Fereres[a]

[a] Consejo Superior de Investigaciones Científicas
Centro de Ciencias Medioambientales Serrano, 115 dpdo, 28006, Madrid, Spain

[b] Istituto per la Genesi e l'Ecologia del Suolo, CNR Firenze,
P.le delle Cascine 28, 50144 Firenze, Italy

The parasporal crystal of the isolate of *Bacillus thuringiensis* (*Bt*) subsp. *israelensis* (*Bti*) H14 is widely used to control larvae of mosquitoes and blackflies. Its outstanding insecticidal property is the result of the synergistic activity of 4 major proteins (Cry4A, Cry4B, Cry11A and Cyt1A). However, the persistence of the insecticidal property is quite low due to a low stability under field conditions (the crystals settle rapidly). There are some examples in the literature indicating that *Bt* toxins degrade at a slower rate when they are sorbed or bound on clays. Also, the sorption of *Bti* toxins on slow-settling clays could be a way of increasing their presence near the air-water interface, where larvae of several mosquito species are usually feeding. We investigated the activity and persistence of *Bti* crystals for a period of up to 29 days against the 4th instar larvae of *Culex pipiens* (LC_{50} = 0.003 μg/ml of water) when mixed with a suspension of Na-bentonite. For comparison purposes, a similar experiment was conducted with toxins that were dissolved from the crystals and then sorbed on clay. We developed a method to improve the dissolution of the crystals by using some chemicals that were not toxic to the larvae of *C. pipiens*. On the first day, the mortality was significantly higher using crystals alone than in the clay + crystal mixtures. The efficacy of crystals declined over time in all treatments, but after 29 days, the mortality increased in some of the treatments to which clay had been added. Conversely, the mortality rate of *C. pipiens* declined over time when no clay was added to the crystals. The overall cumulative mortality was highest for crystals alone, and there were no significant differences when a low concentration of clay (0.05 mg/ml) was present. The cumulative mortality was significantly reduced when the clay concentration was increased. The LC_{50} and LC_{90} of the solubilized toxins in the absence of clays were 34.0 ng/ml and 131.6 ng/ml, respectively, indicating that their toxicity was about 10 times less than that of the intact crystals (3.3 ng/ml and 54.6 ng/ml, respectively). In all cases, the solubilized clay-sorbed toxins had lower toxicity to *C. pipiens* than intact crystals mixed with clays (at equal concentrations).

1. INTRODUCTION

The parasporal body of the isolate H14 of *Bacillus thuringiensis* subsp. *israelensis* (*Bti*) is the most toxic per unit weight of all known *Bt* isolates [1] because it combines 4 major proteins with synergistic activity: Cry4A, Cry4B, Cry11A and Cyt1A [2, 3]. It is used to control larvae of mosquito and blackfly, although its persistence is quite low due to its low stability under field conditions. This lack of residual activity is the result of photodegradation by UV light [4], heat, desiccation or changes in pH [5], together with a rapid settlement of crystals into the mud or consumption by microorganisms [6]. The efficacy of *Bti* crystals varies depending on the mosquito species, *Anopheles* spp. being more tolerant than *Culex* spp. and *Aedes* spp., probably because of the low filtration rate of the larvae of *Anopheles* [7]. It is possible to increase the efficacy of *Bti* crystals by mixing them with corn oil or wheat flour to provide a floating bait to *Anopheles* larvae [8]. It is also possible to increase the stability of *Bti* crystals to UV radiation by adding a powder of melanin pigment into a spore suspension and preparations of δ-endotoxin of *Bti* [9].

There are some examples in the literature indicating that toxins of *Bt* (obtained from subspecies *kurstaki* and *tenebrionis*) bound or sorbed to clays with a high cationic exchange capacity can delay toxin degradation and may even increase its insecticidal activity [10]. Conversely, it has also been reported that the efficacy of commercial formulations of *Bti* decreases in aqueous environments containing soil or clay particles at concentrations between 10 and 0.1 mg/ml [11, 12]. Nevertheless, Ramoska et al. [11] did not use materials that were able to form stable colloidal suspensions. A sieve was used for separating the mineral material, and particles smaller than 100 μm were considered as clay.

To prevent *Bti* crystals from settling into the mud of ponds, it would be very practical to solubilize them and attach the resulting toxins to particles with a greater capacity to remain in suspension. To adsorb *Bti* toxins to solid particles, the crystals need to be solubilized first, which results in a drastic reduction in toxicity; in fact, solubilized crystals were more than 7000 times less toxic to *Aedes aegypti* larvae than intact crystals [6], probably because mosquito larvae, as filter feeders, selectively concentrate particles of 0.5 to 10 μm in diameter while excluding water and soluble molecules from the gut [13]. Schnell et al. [6] were able to demonstrate that solubilized *Bti* crystals remain intrinsically toxic to mosquito larvae by adsorbing them to 0.8-μm latex beads; the toxicity of sorbed material was only 27 times less than intact crystals. Therefore, attachment or sorption of *Bti* toxins to particles with a greater capacity to remain in suspension in water, such as clay minerals, should enhance its persistence while retaining most of its toxicity.

The objective of our work was to assess the efficacy and persistence of *Bti* crystals when mixed with different concentrations of clays (stable colloidal suspensions) against *C. pipiens*. Another objective was to dissolve the crystals and test the bioactivity of solubilized toxins after they are sorbed on clays.

2. MATERIALS AND METHODS

2.1. Persistence of *Bti* crystals mixed with clays and bioactivity against *C. pipiens*

Crystals of *Bti* were separated from the commercial product (CP) Turbac CD (Abbott Laboratories) by isopycnic gradient centrifugation as described by Ang and Nickerson [14]. The crystals were found to float at a density of 1.33-1.35 g/cc. The procedure was improved to increase the yield of extraction. The CP was diluted with distilled water, allowed to settle in a tube (30 cm high, 1000 ml) for 24 hours and the supernatant was removed. This procedure was repeated several times (4-5) to remove most of the light material. Sodium bromide was added to the resulting suspension until the density of 1.34 g/cc was reached (approx. 37% NaBr). This suspension was centrifuged (Rotor Beckman JCF-Z at 15,000 rpm) and the supernatant discarded. The pellet (crystals) was recovered, washed and centrifuged until no NaBr was detected in the supernatant. Eventually, the dried suspension was photographed using scanning electron microscopy as described by Yamamoto et al. [15] to confirm the final result. To work with stable colloidal suspensions, the reference material (SWy-1, Crook County, Na-Montmorillonite, obtained from the Source Clay Repository, Univ. of Missouri) was allowed to settle for 1 week in a bucket whose height was 38 cm; only the material in suspension was recovered. Five different concentrations of *Bti* crystals (0.03, 0.06, 0.12, 0.15 and 0.3 μg/ml) were combined with 4 different concentrations of clays (0, 0.05, 0.5 and 5 mg/ml). After thoroughly mixing each of the clay and crystal suspensions in a beaker, the suspensions were poured into plastic cups (200 ml/cup). The mosquito larvae used for the experiments were started from an autogenic population of *C. pipiens* originally from Baix Llobregat (Barcelona, Spain), which was kindly supplied by C. Aranda and R. Eritja. Ten 4th-instar larvae of *C. pipiens* were placed into each experimental plastic cup. The mortality was assessed 24 h after exposure to the insecticidal suspension. Mosquitoes (dead and alive) were removed from the cups every time after counting, and a new set of 10 larvae was introduced at different time intervals (1, 5, 15, and 29 days after the suspension was prepared). There were twelve replicates of each clay-crystal combination. All experiments were carried out inside an environmentally controlled chamber at a constant temperature of 25 ± 1°C and a relative humidity of 75 ± 10%. The LC_{50} and LC_{90} for *C. pipiens* of the intact crystals alone were determined by using five concentrations that were determined in preliminary tests. Before the bioassays, it was determined that the clays alone were not toxic to *C. pipiens* at a concentration of 5 mg/ml under the same experimental conditions.

2.2. Persistence of solubilized clay-sorbed toxins from *Bti* crystals and bioactivity against *C. pipiens*

The procedure to dissolve the toxins in the crystals has been used in the past for identification of toxins and analytical purposes, and as far as the authors are aware, no attempt has been made to dissolve completely the crystals for evaluation of their toxicity. Insell and Fitz-James [16] studied the solubility of crystals under different conditions (pH, chemicals and so on). Our main purpose was to compare the activity of intact crystals with equivalent amounts of solubilized clay-sorbed proteins. The procedure finally adopted was able to dissolve completely the crystals. The procedure used for solubilization was similar to the one described by Ishii and Ohba [17], although we used ascorbic acid instead of β-mercaptoethanol as a reducing agent, to avoid toxic effects on mosquito larvae [6]. In summary, 3 mg of *Bti* crystals, 10 ml of 0.01M EDTA, and 0.176 g of ascorbic acid were added in a beaker with stirring. The pH was adjusted to 10 by dropwise addition of 0.1N Na_2CO_3. Finally, the volume was brought to 100 ml. The suspension was kept at 37°C for 1 h to promote the dissolution of crystals. The resultant solution was centrifuged (15,000 *g* for 15 minutes) and no pellet was detected, demonstrating the complete dissolution of the solid material. The composition of this solution was similar to intact crystals, as demonstrated by SDS-PAGE analysis (data not shown).

Appropriate amounts of the above solution and of clay were mixed, and the pH was adjusted to 7.0 with diluted HCl, to eliminate any effect on mortality due to alkaline conditions and to perform all the experiments under constant conditions. Some preliminary measurements indicated that the Crook County clay used for the sorption experiments is capable of depleting from solution around 75% of the solubilized proteins at pH 7.0 (1 ml of solution containing 0.5 mg of clay and 2 mg of solubilized crystals).

The bioassays with *C. pipiens* were carried out in the same way as described above. The number of replicates was 10, and the concentrations of crystals and clays used were the same except that an extra concentration of clay equivalent to 1 mg/ml and 0.6 μg/ml of crystals was also used. The LC_{50} and LC_{90} of the solubilized crystals for *C. pipiens* were also determined using 5 different concentrations.

2.3. Analysis of data

The corrected mortality of *C. pipiens* in all treatments was calculated using Abbott's formula [18] and the LC_{50} and LC_{90} were calculated for both intact and solubilized crystals using the POLO software for probit analysis [19]. The mean cumulative mortality (defined as the sum of mosquitoes dead after each count made on day 1, 5, 15, and 29) was calculated for each pair of clay-*Bti* concentrations. Cumulative mortality (in %) was subjected to a 2-factor analysis of variance (ANOVA) using the concentrations of toxin and clay as two independent factors. The means obtained for each pair of clay-*Bti* concentrations were compared to each other by orthogonal contrasts (planned comparisons of pair of means) at the $P = 0.05$ level. All data were transformed before running the ANOVA according to the transformation: X'=arcsin X/100. The variation over time of the insecticidal activity (calculated as corrected mortality) was plotted, and the persistence of *Bti* mixed or sorbed to clays was analyzed by a 1-factor ANOVA for each pair of clay-*Bti* concentrations (in this case, the variable was time). Mean comparisons were then made according to Fisher's unprotected LSD test (P=0.05), after transforming the mortality values by $X'=\arcsin \sqrt{X/100}$.

3. RESULTS AND DISCUSSION

3.1. Persistence of *Bti* crystals mixed with clays and bioactivity against *C. pipiens*

The LC_{50} and LC_{90} for the intact crystals against *C. pipiens* were 3.3 ng/ml and 54.6 ng/ml, respectively. The LC_{50} value is not very different from the one reported in the literature for intact crystals against *Aedes aegypti* (7.5 ng/ml) [6].

Table 1A shows the cumulative mortality (%) of *C. pipiens* over the entire experiment when exposed to different concentrations of clay-crystal. In this case, the mean comparisons were made across concentrations of clays. The data show that when the concentration of clay was increased, the toxicity of *Bti* crystals decreased. However, there were no significant differences (P 0.05) in mortality between the treatments with crystals alone and those containing 0.05 mg/ml of clay (for all of the concentrations of crystals used). Moreover, when crystal concentration was low (0.03 or 0.06 μl/ml), there were no significant differences in cumulative mortality between 0.05 and 0.5 mg/ml of clay; clay significantly reduced the insecticidal activity of *Bti* crystals at concentrations ≥ 0.5 mg/ml. These results are consistent with the ones reported by Ramoska *et al.* [11], who showed a significant reduction in toxicity with clay concentrations higher than 0.5 mg/ml. Table 1B shows the cumulative mortality of *C. pipiens* over the entire experiment when the mean comparison was made across the concentration of *Bti* crystals. In this case, there was a clear increase in toxicity with increasing concentrations of crystals. This increase was larger when low concentrations of clays were used.

Table 1
Cumulative mortality (%) (Mean ± SEM) of *C. pipiens* when using mixtures of *Bti* intact crystals and clay at different concentrations over the entire experiment

A) comparisons across [clay]

Clay mg ml^{-1}	*Bti* $\mu g\ ml^{-1}$				
	0.03	0.06	0.12	0.15	0.30
0.00	34.37 ± 2.8 a	46.87 ± 2.7 a	62.71 ± 2.3 a	74.79 ± 1.7 a	85.21 ± 2.9 a
0.05	27.50 ± 1.9 ab	43.54 ± 2.5 ab	64.58 ± 2.6 a	71.04 ± 2.8 a	89.79 ± 1.8 a
0.50	23.54 ± 2.9 b	35.21 ± 3.7 b	43.12 ± 2.4 b	49.79 ± 2.3 b	72.71 ± 2.6 b
5.00	10.83 c ± 2.2 c	21.04 ± 4.2 c	24.79 ± 2.6 c	25.83 ± 2.6 c	32.29 ± 3.7 c

B) comparisons across [*Bti* crystals]

Bti $\mu g\ ml^{-1}$	Clay mg ml^{-1}			
	0.00	0.05	0.50	5.00
0.03	34.37 ± 2.8 a	27.50 ± 2.0 a	23.54 ± 2.9 a	10.83 ± 2.2 a
0.06	46.87 ± 2.7 b	43.54 ± 2.5 b	35.21 ± 3.7 b	21.04 ± 4.2 b
0.12	62.71 ± 2.3 c	64.58 ± 2.6 c	43.12 ± 2.4 bc	24.79 ± 2.6 bc
0.15	74.79 ± 1.7 d	71.04 ± 2.8 c	49.79 ± 2.3 c	25.83 ± 2.6 bc
0.30	85.21 ± 2.8 e	89.79 ± 1.8 d	72.71 ± 2.6 d	32.30 ± 3.7 c

Different letters within columns indicate significant differences according to a 2-factor ANOVA and orthogonal contrasts (planned comparisons of pair of means) ($P<0.05$). Data were transformed before running a 2-factor ANOVA according to X'=arcsin X/100.

The activity of *Bti* crystals over time was quantified by calculating the mortality of *C. pipiens* at different periods of time (1, 5, 15 and 29 days after the clay-crystal suspensions were prepared). When the period of time after the preparation of the suspensions increased, the bioactivity of the mixtures was maintained in the presence of clays, whereas the activity of the preparations without clays declined significantly over time (Figure 1). On the first day, the insecticidal activity of the crystals alone was higher than when adding clays to the suspension. However, on day 29, the *Bti* activity was in general greater in the presence of clays. In some cases, the activity of the toxin was significantly higher on day 29 than on the first day after preparation of the suspension ([*Bti*]= 0.06 and 0.12 μg/ml data not reported combined with [clay]= 5 mg/ml).

These results indicate that clays can increase the persistence of *Bti* crystals over time. Alternatively, the clays may hamper the ingestion of crystals, which results in lower activity but higher persistence. To this point, it should be emphasized that depletion of crystals ingested by dead mosquitoes may reduce the activity of a *Bti* suspension over time [4]. Therefore, the exposure of 5 different sets of mosquito larvae to the crystal suspensions, such as the ones used in the experiments, could result in a reduction in the persistence of their insecticidal properties. In our experiments, the high mortality obtained on the first date in the non-clay treatments possibly resulted in a partial depletion of the crystals by the first set of mosquitoes that was

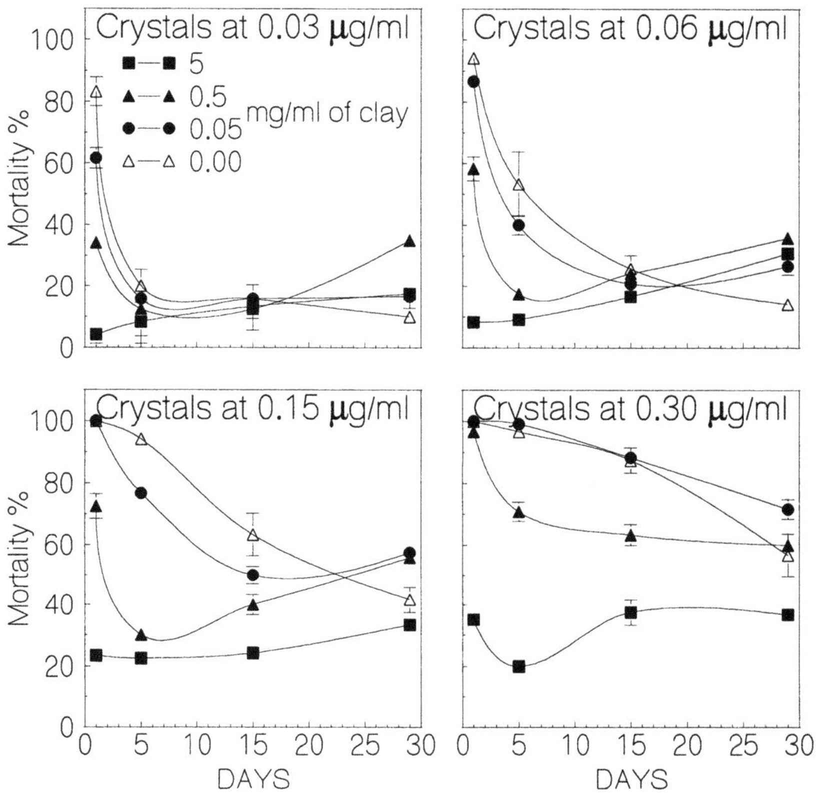

Figure 1. Variation of mortality of *C. pipiens* over time when exposed to different concentrations of *Bti* solid crystals mixed with clays (error bars = SEM).

tested. This depletion could explain the sharp reduction observed in the insecticidal activity of the non-clay suspensions over time, since their toxicity was very low when the fifth set of mosquitoes was tested on day 29.

3.2. Bioactivity and persistence of solubilized *Bti* crystals sorbed to clays against *C. pipiens*

The LC_{50} and LC_{90} of the solubilized crystals in the absence of clays were 34.0 ng/ml and 131.6 ng/ml, respectively, indicating that the toxicity was about 10 times less than that of the intact crystals. However, our solubilization protocol seems to be much more efficient than the ones previously published; the dissolution procedure decreased the activity of solid crystals only 10 times instead of 7000 times, as previously reported [6]. This difference could have been the result of the incomplete dissolution obtained by previous authors, which could result in a lack of some toxins and, consequently, in a reduction of activity because of the lack of synergistic activity of all toxins together.

Tables 2A and 2B show the cumulative mortality (%) of *C. pipiens* over the entire experiment when exposed to different concentrations of solubilized clay-sorbed toxin, across all

Table 2

Cumulative mortality (%) (Mean ± SEM) of *C. pipiens* when using solubilized (from *Bti* crystals) clay-sorbed toxins at different concentrations over the entire experiment.

A) comparisons across [clay]

	Bti µg/ml					
Clay mg/ml	0.03	0.06	0.12	0.15	0.30	0.60
0.00	15.50 ± 0.3a	24.25± 1.5a	29.00 ± 0.9a	33.25 ± 0.8a	47.25 ± 2.99a	61.50 ± 0.7a
0.05	14.00 ± 0.8ab	19.25 ± 2.1ab	27.25 ± 1.7a	33.50 ± 2.5a	46.00 ± 4.41a	59.50 ± 1.0a
0.50	17.00 ± 1.1a	14.50 ± 1.7bc	21.25 ± 1.5b	30.00 ± 3.1a	37.75 ± 4.52b	58.50 ± 0.7a
1.00	13.50 ± 0.7ac	14.50 ± 1.2bc	19.75 ± 3.4cb	17.50 ± 1.5b	30.75 ± 2.47c	49.00 ± 0.8b
5.00	10.00 ± 0.5bc	13.50± 1.3c	14.50 ± 1.6c	14.25± 1.2b	26.25 ± 2.69c	41.00 ± 1.1c

B) comparisons across [solubilized clay-sorbed *Bti*]

	Clay mg/ml				
Bti µg/ml	0.00	0.05	0.50	1.00	5.00
0.03	15.50 ± 0.3a	14.00 ± 0.8a	17.00 ± 1.1ab	13.50 ± 0.7a	10.00 ± 0.5a
0.06	24.25 ± 1.5b	19.25 ± 2.1b	14.50 ± 1.7a	14.50 ± 1.8ab	13.50 ± 1.3ab
0.12	29.00 ± 0.9cb	27.25 ± 1.7c	21.25 ± 1.5b	19.75 ± 3.4bc	14.50 ± 1.6bc
0.15	33.25 ± 0.8c	33.50 ± 2.5d	30.00 ± 3.1c	17.50 ± 1.5ac	14.25 ± 1.2ac
0.30	47.25 ± 3.0d	46.00 ± 4.4e	37.75 ± 4.5d	30.75 ± 2.5d	26.25 ± 2.7d
0.60	61.50 ± 0.7e	59.50 ± 1.0f	58.50 ± 0.7e	49.00 ± 0.8e	41.00 ± 1.1e

Different letters within columns indicate significant differences according to a 2-factor ANOVA and orthogonal contrasts (planned comparisons of pair of means) ($P<0.05$). Data were transformed before running a 2 factor ANOVA according to X'=arcsin X/100.

the concentrations of clays and toxins used, respectively. The data show that when the concentration of clay was increased, the toxicity of the toxins sorbed on clay decreased.

However, as reported for the case of the intact crystals and clay mixtures, there were no significant differences (P 0.05) in mortality when using a concentration of 0.05 mg/ml of clay and solubilized clay-sorbed toxins alone. In all cases, the toxins sorbed on clays had lower toxicity than the crystals mixed with clays (at equal concentrations), indicating that the solubilized clay-sorbed toxins were not as efficient as intact crystals mixed with clays in controlling *C. pipiens*. A similar result was reported for solubilized toxins sorbed on latex beads that were much less active against *A. aegypti* than the intact crystals mixed with latex beads [6]. The mortality of *C. pipiens* at 1, 5, 15, and 29 days after the suspensions were prepared was used to monitor over time the activity of solubilized clay-sorbed toxins. The results (Figure 2) indicate that with time, the bioactivity is maintained in a similar way as for clay-solid crystal mixtures, whereas the activity of the preparations without clays declined earlier. In some cases, the activity of the toxins on day 29 was higher for the clay-sorbed toxin than for the toxin solution alone.

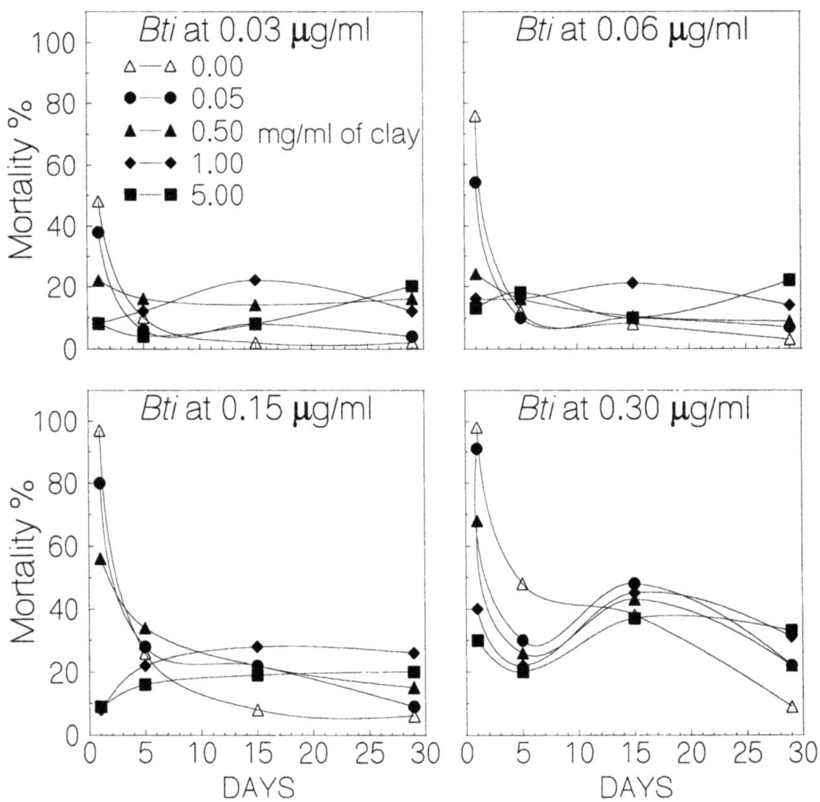

Figure 2. Variation of mortality of *C. pipiens* over time when exposed to different concentrations of solubilized clay-sorbed toxins from *Bti* (error bars = SEM).

ACKNOWLEDGMENTS

This work was funded by the European Union INCO grant nº ERBIC18CT970135.

The authors wish to thank C. Aranda and R. Eritja for providing the mosquito colonies and for their expertise in mosquito-rearing techniques and bioassays.

REFERENCES

1. Federici, B.A., 1999. *Bacillus thuringiensis* in biological control. *In*: Bellows, T.S., Fisher, T.W. (Eds.), Handbook of Biological Control. Principles and Applications of Biological Control. Academic Press, San Diego, pp. 575-593.
2. Crickmore, N., Bone, E.J., Williams, J.A., Ellar, D.J., 1995. Contribution of the individual components of the delta-endotoxin crystal to the mosquitocidal activity of *Bacillus thuringiensis* subsp. *israelensis*. FEMS Microbiol. Lett. 131, 249-254.
3. Georghiou, G.P., Wirth, M.C., 1997. Influence of exposure to single versus multiple toxins of *Bacillus thuringiensis* subsp. *israelensis* on development of resistance in the mosquito *Culex quinquefasciatus* (Diptera: Culicidae). Appl. Environ. Microbiol. 63, 1095-1101.
4. Ignoffo, C.M., Garcia, C., Kroha, M.J., Fukuda, T., Couch, T.L., 1981. Laboratory tests to evaluate the potential efficacy of *Bacillus thuringiensis* var. *israelensis* for use against mosquitoes. Mosquito News 41, 85-93.
5. Leong, K.L.H., Cano, R.J., Kubinski, A.M., 1980. Factors affecting *Bacillus thuringiensis* total field persistence. Environ. Entomol. 9, 593-599.
6. Schnell, D.J., Pfannenstiel, M.A., Nickerson, K.W., 1984. Bioassay of solubilized *Bacillus thuringiensis* var. *israelensis* crystals by attachment to latex beads. Science 223, 1191-1193.
7. Aly, C., Mulla, M.S., Xu, B.Z., Schnetter, W., 1988. Rate of ingestion by mosquito larvae (Diptera: Culicidae) as a factor in the effectiveness of a bacterial stomach toxin. J. Med. Entomol. 25, 191-196.
8. Aly, C., Mulla, M.S., Schnetter, W., Xu, B.Z., 1987. Floating bait formulations increase effectiveness of *Bacillus thuringiensis* var. *israelensis* against *Anopheles* larvae. J. Am. Mosq. Control Assoc. 3, 583-588.
9. Liu, Y.T., Sui, M.J., Ji, D.D., Wu, I.H., Chou, C.C., Chen, C.C., 1993. Protection from ultraviolet irradiation by melanin of mosquitocidal activity of *Bacillus thuringiensis* var. *israelensis*. J. Invertebr. Pathol. 62, 131-136.
10. Tapp, H., Stotzky, G., 1995. Insecticidal activity of the toxins from *Bacillus thuringiensis* subspecies *kurstaki* and *tenebrionis* adsorbed and bound on pure and soil clays. Appl. Environ. Microbiol. 61, 1786-1790.
11. Ramoska, W.A., Watts, S., Rodriguez, R.E., 1982. Influence of suspended particulates on the activity of *Bacillus thuringiensis* serotype H-14 against mosquito larvae. J. Econ. Entomol. 75, 1-4.
12. Van Essen, F.W., Hembree, S.C., 1982. Simulated field studies with four formulations of *Bacillus thuringiensis* var. *israelensis* against mosquitoes: residual activity and effect of soil constituents. Mosquito News 42, 66-72.
13. Dadd, R.H., 1971. Effects of size and concentration of particles on rates of ingestion of latex particulates by mosquito larvae. Ann. Entomol. Soc. Am. 64, 687-692.
14. Ang, B.J., Nickerson, K.W., 1978. Purification of the protein crystal from *Bacillus thuringiensis* by zonal gradient centrifugation. Appl. Environ. Microbiol. 36, 625-626.

15. Yamamoto, T., Iizuka, T., Aronson, J.N., 1983. Mosquitocidal protein of *Bacillus thuringiensis* subsp. *israelensis*: identification and partial isolation of the protein. Curr. Microbiol. 9, 279-284.
16. Insell, J.P., Fitz-James, P.C., 1985. Composition and toxicity of the inclusion of *Bacillus thuringiensis* subsp. *israelensis*. Appl. Environ. Microbiol. 50, 56-62.
17. Ishii, T., Ohba, M., 1993. Diversity of *Bacillus thuringiensis* environmental isolates showing larvicidal activity specific mosquitoes. J. Gen. Microbiol. 139, 2849-2854.
18. Abbott, W.S., 1925. A method for computing the effectiveness of an insecticide. J. Econ. Entomol. 18, 265-267.
19. Russell, R.M., Robertson, J.L., Savin, N.E., 1977. POLO: A new computer program for Probit analysis. ESA Bulletin 23, 209-213.

Developments in Soil Science, Volume 28B
Editors: A. Violante, P.M. Huang, J.-M. Bollag and L. Gianfreda

SOIL ECOSYSTEM PROPERTIES, MICROBIAL DIVERSITY, AND ECOSYSTEM ASSESSMENTS

R. L. Tate III and B. F. Rogers

Department of Environmental Sciences, Rutgers University, New Brunswick, NJ 08901, USA

The importance of a diverse, resilient soil microbial community in the development of functional, sustainable land based ecosystems has long been appreciated. An essential element for evaluating impacts of various management decisions on agricultural, urban, or reclamation sites and for studying the stability of native ecosystems is the accurate assessment of the function and diversity of the soil microbial community. Traditional methods for quantifying soil microbial processes (e.g., population density measurements, enzyme activities, and respiration measurements) are useful descriptors of microbial function, in some situations, but frequently the results do not clearly relate to site conditions. Data interpretation can be complicated, for example by the high redundancy of the microbial community (i.e., mixtures of active and inactive populations with the predominant microbes commonly being inactive or in a resting state) and the accumulation of stabilized enzymes, which may or may not contribute to total ecosystem function. More general measures of soil microbial diversity —phospholipid fatty acid analysis (PLFA) and metabolic diversity (as determined by the BIOLOG procedure)— have been shown to be useful in characterization and differentiation of ecosystem microbial populations, both in native and managed soils. These more general indicators (i.e., as compared to DNA analysis methods where specific microbial species are designated) are sensitive, reasonably easily applied techniques that are useful in assessing microbial community status and development in a wide range of soils and soil-like materials. Although reproducible "fingerprints" characteristic of specific soil systems result from these analyses, the direct relationship of specific metabolic activities or occurrence of specific fatty acids in the assay procedures to actual field situations has been questioned. Therefore primary concerns for application of metabolic diversity and phospholipid fatty acid analyses to study of native soil systems are the need to clarify the relationship between the laboratory derived data and the nature of the processes occurring in the field and the delineation of reasonable boundaries for data interpretation.

1. INTRODUCTION

As the complexity of the issues involving management and stewardship of soil systems continue to increase, the need to assess the status and function of soil microbial communities is becoming more acute. Whether the need is to evaluate the quality of agricultural soils, the impact of anthropogenic intrusion (e.g., chemical pollution), the status of nutrient cycling in native systems, or the results of reclamation management, accurate and reproducible indicators of soil microbial community sustainability or resilience are essential for achievement of project goals and development of appropriate (sound) soil stewardship plans. Commonly used procedures include the quantification of enzymic activity (e.g., proteases or phosphatases, or the more general enzyme related parameters such as dehydrogenase or microbial respiration), individual soil community functions (such as, nitrogen fixation or mineralization potentials), or variation in individual or specific microbial populations (denitrifiers, nitrifiers, etc.). Each of these methods has major advantages and limitations. Enzyme activities are useful in quantifing changes in specific soil processes, but due to complexities relating to the sample size used in enzyme quantification and the correlation of enzyme activity levels with quantities of soil organic matter, site variation can be large. Similar situations are noted when individual microbial population densities are quantified. Concerns can also be raised regarding the question of the relationship of potential enzymatic or microbial activity to the actual level of the activity expressed *in situ*.

Two recently developed methods for assessing general variation of microbial communities in soil systems are metabolic diversity (as frequently measured using the BIOLOG procedure) and phospholipid fatty acid (PLFA) diversity analysis. As will be detailed below, the data resulting from the application of these two methods relate to general changes to the soil system without some of the short term fluctuations commonly seen in enzyme activities or microbial population assessments. Thus, it may be hypothesized that seasonal or more long term alterations in soil microbial community structure are revealed by these latter two methods, thereby providing an indication of changes in general microbial community dynamics resulting from a general perturbation of system properties, as might result from ecosystem evolution, reclamation management, or chemical contamination. Thus, the topics to be addressed in this critique are a) whether metabolic or PLFA diversity analyses reflect specific ecosystem types or variation of controlling properties within an ecosystem, b) the sensitivity of metabolic or PLFA diversity to changes in system properties, and c) whether these methods reveal “real world” soil microbial properties.

2. UTILITY OF METABOLIC AND PHOSPHOLIPID FATTY ACID DIVERSITY ANALYSIS IN CHARACTERIZATION OF SOIL SYSTEMS

There are two important questions that must be addressed when considering the utility of any quantitative procedure for study of native soil systems. 1) Does the method provide results that vary between soil systems in a meaningful manner? 2) What is the relationship of the data provided by the method to *in situ* soil properties or processes? As will be documented below, both PLFA and metabolic diversity have been found to be robust tools for distinguishing soil ecosystem types and for assessing the impact of chemical contamination or system management effects on soil microbial community diversity. But, before a complete response to our questions can be provided, the assumptions and principles underlying the methods must be considered. These fundamental considerations for PLFA analysis are reasonably straightforward, but data interpretation does rely on several basic assumptions relating to the stability of the fatty acids in soil and their distribution between various microbial groupings. The metabolic diversity assessments are more problematic.

PLFA is dependent on a total or representative extraction of the phospholipid fatty acid fraction of the soil samples. (For reasons of conformity and comparison of data, fatty acids are designated in terms of total number of carbon atoms, and the number of double bonds, followed by the position of the double bond from the methyl end of the molecule. The prefixes a and i indicate anteiso and iso branching, cy indicates a cyclopropane fatty acid and methyl branching (Me) is indicated as the position of the methyl group from the carboxyl end of the chain.) A generally accepted assumption for the PLFA procedure is that the fatty acids are solely associated with living microbial cells. Once the microbial cells die the released phospholipid fatty acids must be decomposed within a short time frame. An associated assumption is that the fatty acid distribution of living microbial cells differs between microbial species—hence the distribution of phospholipid fatty acids can serve as a surrogate indicator of variations in microbial diversity. The validity of this assumption provides an additional piece of information about the identity of the microbes comprising the soil microbial community. Some phospholipid fatty acids have been determined to be indicator fatty acids for specific microbial groups. Association of the presence of specific phospholipid fatty acids as indicators of changes in community composition is exemplified by a study of microbial community structure in boreal peatlands and in the Judean desert. In the peatlands study, clusters of fatty acids characteristic of Gram negative bacteria, Gram positive and anaerobic Gram negative bacteria, sulfate-reducing bacteria and/or actinomycetes, methane-oxidizing bacteria, and eucaryotes were identified [1]. In the study of the Judean desert soil microbial community, decreases in monounsaturated fatty acids and hydroxy-substituted phospholipid fatty acids due to desiccation of soils normally having high microbial biomass during the rainy period were taken to indicate declines in Gram negative bacterial populations. Similarly, changes in the nature of the phospholipid fatty acid composition due to desiccation of soils maintaining low microbial biomass during the rainy season suggested changes in eukaryotes [2]. Also, the fatty acid patterns suggest that the bacterial community of Carex peats have a higher proportion of Gram negative bacteria and lower proportion of Gram positive bacteria compared to the Sphagnum peats [3]. Care must be taken in reaching conclusions regarding variation of specific microbial groupings based on the quantities of a specific fatty acid present in the soil extract. The assumptions are that no other microbial grouping in the soil contain appreciable amounts of the fatty acid and that the quantity of fatty acid present is proportional to the density of the population in question. The validity of the unique distribution assumption must continually be questioned as long as our understanding of the totality of the composition of the soil microbial community is limited. The precision of considerations of specific population sizes must await more research.

The scientific under-girding of metabolic diversity assessments of soil communities is more challenging. The procedure has been shown to be useful for evaluating the metabolic capabilities of a mixed microbial community extracted from soil [4, 5]. With this method, the capability of the soil microbial community to metabolize a combination of 95 separate carbon sources, based on a redox procedure using tetrazolium dye reduction, is determined. The assay is conducted on a 96 well micro-titer plate with one well serving as a no-substrate control.

An initial question relating to the utility of the BIOLOG procedure for study of soil systems relates to the concern of whether the capacity to metabolize the various carbon substances in the micro-titer plate can be expected to differ between various ecosystem types, or even soil management regimes. It appears to be somewhat illogical that the capability to metabolize the various substrates should differ with soil ecosystem properties. The illogical situation stems from the nature of the carbon substances used in the test and the great diversity of the microbial community of soil. As many as 4000-7000 different species (as indicated by DNA hybridization) have been estimated to occur in each gram of soil [6]. If the 95 individual carbon sources were unique to specific soil systems, a difference between the substances oxidized by the bacterial communities of various ecosystem types would be expected. On the contrary, the organic substances include a variety of commonly encountered carboxylic acids (e.g., succinic acid, citric acid, *cis*-aconitic acid, *p*-hydroxyphenyl-acetic acid), amino acids (e.g., L-histidine, L-asparagine, L-analylglycine), polymers (e.g., Tween-80, Tween-40), and other substances (urocanic acid, lactulose, D-raffinose, formate, glycogen, uridine, etc.), none of which are specific to soil or particularly unique from the view of metabolic process. Thus, considering the diverse metabolic capabilities associated with soil, one could easily conclude that these substances should all be readily oxidized in any soil system; yet, as will be cited for specific studies below, a definitive fingerprint, *per se*, is provided for individual soil systems by application of the BIOLOG procedure.

3. DO THE METHODS IN QUESTION PROVIDE RESULTS THAT VARY BETWEEN SOIL SYSTEMS IN A MEANINGFUL MANNER?

From even a cursory survey of the literature, it can be concluded that both the PLFA and metabolic diversity analyses provide data that allow differentiation of soil ecosystem status. A rapidly expanding array of reports are appearing in the literature demonstrating utility of both methods of soil community analyses. Thus, only a few examples of such studies (selected primarily to emphasize the breadth of the situations where such analyses are useful) will be presented here. (Several reviews are available that summarize the utility of both procedures for community structure analysis in soil [7-10].

Data from such studies must be considered within the perspective of the limitations or characteristics of the procedure used to collect them. A number of modifiers, such as "functional" and "metabolic" have been used to describe the data provided by metabolic capability analyses. The primary word modified by these descriptors is "diversity". It must be stressed that using strict definitions the methods do not provide a traditional diversity analysis *per se* of the sample, such as is provided with a traditional diversity assessment, as results from a Shannon-Weaver analysis and related procedures; but, the data from the metabolic analysis (and indeed the PLFA analysis also) do represent a "population" of metabolic capabilities and fatty acids. Thus, it could reasonably be stated that metabolic or PLFA diversities, except with limited instances in the PLFA analysis, do not represent separate species or functional guilds as traditionally studied in diversity analysis but that the assessments do provide a surrogate for more detailed taxonomic or species analyses [10]. To overcome this limitation in part, frequently although both PLFA and metabolic diversity analyses have been applied individually, it is becoming more common to use both techniques concurrently to describe soil microbial community variability, thereby strengthening the linkage of the results to changes in the microbial populations in situ. Since, as noted above, the two techniques assess variation in distinct and independent fundamental properties of the microbial communities, this is a logical and useful extension of our analytical capabilities [9].

Both assessors of general changes in microbial community diversity have been shown to vary in polluted, managed, and native soil systems. Shifts in microbial community structure due to increases in levels of toxic substances, both inorganic and organic, are revealed by both PLFA and metabolic diversity analyses. For example, the major changes that result from high metal loading from mining and smelter activities and the results of remediation of such soils are readily demonstrated [11-15]. Many of the changes in the fingerprints provided for these soils can be documented in the more controlled environment of soil microcosms amended with varying levels of specific metals [16-19]. For example, a difference noted in data from field and microcosms studies using the PLFA procedure involves elucidation of the effect of the metals on the fungal populations. The indicator fatty acid for the fungal population was not detected in the field soils with high metal loading but was found in the metal amended soils incubated in the laboratory. This apparently contradictory observation is explained in part by considering the complexity of the interactions of the fungi in the field soils, as is exemplified by a study of a forest ecosystem downwind from a Zn smelter site [11]. In the original forest ecosystem, much of the soil fungal population existed in mycorrhizal associations. The combination of sulfur dioxide and metal contamination of the ecosystem through the many decades of existence of the smelter resulted in total die-off of the aboveground population, thus, logically resulting in a loss of both the fungal and plant symbionts. Therefore, it can be postulated that in the field soils, the indicator fatty acids associated with the primarily mycorrhizal fungal population were lost from the soils. This situation contrasts to the parallel study of a laboratory incubated soil collected from an agricultural field [20]. In that soil, the free-living fungal populations established in the soil prior to Zn amendment in the laboratory were resistant to the amended metals. The methods have also been shown to provide indications of the effect of low levels of metal contamination as exemplified by that associated with the incorporation of acceptable levels of sewage sludges into soil [20-22]. Not only are the diversities of the microbial communities altered as demonstrated through a principal component analysis of the phospholipid fatty acid or metabolic diversity data but the rate of color development in the BIOLOG procedure reflects that characteristic of metal contaminated soils [11, 20], thereby providing evidence of the role of both the soil metal loading and the organic matter augmentation resulting from the sludge amendment.

A demonstration of the sensitivity of both assessors of soil microbial diversity is the indirect effect of elevated atmospheric carbon dioxide on rhizosphere populations. It is reasonable to assume that long term increases in soil temperature, as are anticipated to result from global warming or as are the simple result of seasonal climatic variation (albeit on a much shorter timeframe), could alter soil microbial diversity by selectively stimulating populations with differing optimum temperatures for growth and activity. For example, PLFA diversity was noted to change in soils incubated in at temperatures varying from 5 to 25°C [22]. Similarly, variations occurred in the metabolic diversity of a New Jersey Pinelands soil under different temperatures [23]. Due to the dynamics of air exchange between soil pores and the atmosphere and the production of carbon dioxide by microbial, plant and animal respiration, soil microbes reside in a carbon dioxide enriched environment. Thus, it could be concluded *a priori* that elevated atmospheric carbon dioxide concentrations would have minimal, if any, affect on soil microbial metabolism. But, considering that a primary controller of soil community dynamics is the nature of the fixed carbon and energy source, any change in plant chemistry or root exudate composition could alter the composition of the soil microbial community. Both metabolic and phospholipid fatty acid diversity analyses of rye grass (*Lolium perenne*) [24], white oak (*Quercus alba* L.) [25], and *Gutierrezia sarothae* [26] plant rhizosphere populations demonstrate measurable shifts in the community composition due to elevated atmospheric carbon dioxide levels.

Examples of other toxicants that have been shown to result in an at least short term modification of the PLFA and/or metabolic diversities include simulated acid rain [27], the soil fumigant metam sodium [28], trichloroethylene and toluene [29], phenyl urea herbicides [30], and 2,4,6, trinitrotoluene [31]. In all of these situations, varying portions of the overall microbial community is inhibited or killed by the toxicant resulting in a shift in the detectable metabolic capability or phospholipid fatty acid composition of the community.

One could almost conclude *a priori* that results of metabolic and PLFA diversity analyses would suggest community level changes in the more severely modified systems, such as those exemplified by the metal and chemically contaminated sites evaluated in the above cited studies. A second consideration involves the need to consider the effect of low level system changes on the metabolic and PLFA diversity patterns, such as the varying soil properties resulting from alteration of cropping practices, and even in slowly evolving native systems. As documented by individual studies cited below, considerable effort has been expended in showing varying diversity patterns in a variety of native and cropped systems. It would be useful to determine if the specific patterns detected relate to alteration of identifiable soil properties.

It could be hypothesized that a primary control of microbial diversity in soil is the nature of the carbon and energy source. Thus, changes in soil physical and chemical properties affecting the fixed carbon inputs would necessarily alter the PLFA or metabolic diversity patterns. This hypothesis is based on the well established principles that a) the vast majority of the microbes in soils are heterotrophic, b) soil microbial population densities generally correlate with the quantities of soil organic matter (humus) in the system, c) soil microbial activity is enhanced by amendments with fixed carbon (i.e., the primary limitation to soil microbial community activity is carbon and energy). (A more in depth presentation of these points is presented elsewhere [9]). Thus, it is reasonable to hypothesize that changes in carbon inputs result in an alteration of microbial community dynamics—presumably including size, composition, and activity.

Based on the hypothesis that the diversity of the heterotrophic microbial community is controlled by the nature of the fixed carbon inputs, it can be postulated that differences in metabolic diversity and PLFA diversity assessments relate to the plant species comprising the above-ground community, the productivity of the plant community—e.g., for agricultural soils, the fertilization regime (which indirectly augments carbon input through stimulation of biomass production)—and location of the input (e.g., litter vs. rhizosphere carbon inputs). The literature is replete with examples of both the metabolic and fatty acid analysis procedures being used to distinguish specific soil systems, based on type of ecosystem (swamp, forest, grassland) as well as the type of aboveground plant community (especially as exemplified by agricultural communities). A common conclusion of such studies is that both methods provide a fingerprint representative of the communities with suggestions that PLFA analyses may be more sensitive for community analysis than is the metabolism based procedure. In a variety of studies ranging from agricultural field and greenhouse studies with a differing crops and cropping systems [32-34], flooded and drained rice fields [35, 36], native and constructed salt marshes [37] reliable and reproducible separations of soil microbial communities were achieved, with the fatty acid based technique providing an indication of changes in major biological groups within the community. Application of the techniques to the analysis of forest soil systems revealed the capacity to distinguish soils impacted by varying degrees of management intervention (e.g., clearcutting, scarification and prescribed burning [38] and amendment with lime and wood-ash [39]).

Studies of grassland communities exemplify the effect of altering plant productivity on PLFA and metabolic diversity fingerprints. Seasonal changes in quality and quantity of grass biomass has been shown to result in altered PLFA diversity patterns [40, 41]. Similarly alteration of the grass metabolic status and productivity due to grazing or fertilization was also demonstrable with PLFA analysis [41, 42].

The portion of the soil system most impacted directly by plant growth dynamics is the rhizosphere. Considerable effort has been expended in determining the utility of general indicators of changes in microbial diversity in distinguishing varying degrees of alteration of the dynamics of the plant rhizosphere. Both metabolic diversity and PLFA methods have been shown to be useful in evaluating rhizosphere populations and to provide fingerprints that vary between plant species [43-45]. A study of the effects of amending soil with synthetic root exudates [46] provides evidence suggesting variation in the nature of the carbon compounds is determining, at least in part, the specific fingerprints in rhizosphere soils. An interesting nuance of studies of the rhizosphere involves evaluation of the effect of using transgenic alfalfa strains [47] or root inoculants of laboratory grown bacteria [48], both of which are anticipated to affect the composition of root exudates. Metabolic diversity analyses were useful in elucidating changes in the rhizosphere communities with the genetically modified alfalfa crop. In comparison metabolic and PLFA analysis revealed that the rhizosphere microbial community structure of the rhizosphere Dahurian wild rye (*Elymus dauricus*) was affected more by root inoculation with laboratory grown cultures than was the community structure of meadow brome (*Bromus biebersteinii*) rhizosphere.

4. WHAT IS THE RELATIONSHIP OF THE DATA PROVIDED BY THE METHOD TO REAL SOIL PROPERTIES AND PROCESSES?

Two primary methods have been considered in this study, PLFA and metabolic diversity analysis. A quick study of the sample of publications cited above leaves the reader with the conclusion that a commonly used descriptor of the data is the word "fingerprint". It is usually stated or concluded that either metabolic diversity analysis or PLFA diversity analysis or both have been shown in the particular study to provide a distinctive pattern or fingerprint representative of the particular ecosystem type or management system under study. Perhaps more interesting or even useful questions relate to the next stage of development of this "fingerprint concept". That is, how comparable are the fingerprints between similar systems in varying locations around the world and how do the fingerprints vary with time? Responses to these specific questions of the boundaries of applicability of the fingerprint data must await accumulation of more data from situations that can be directly compared.

A more basic question related to the validation of the methods involves a consideration of the relationship of the actual values measured with either of these procedures to the actual condition in the soil. Does the occurrence of a specific fatty acid in the PLFA reflect the presence of that substance in the soil sample extracted or does the metabolic activity (its simple occurrence as well as its rate) correlate directly with an activity in the field? The responses to these questions are less problematic with the PLFA assay since the procedure is based on a chemical extraction technique. Thus, the amount of change in composition of the fatty acid mixture extracted from the soil sample relies on appropriate storage and extraction of the soil sample so as to minimize any change that might occur in the fatty acid distribution in the soil sample between sample collection and analysis. As indicated above, greater difficulties associated with data interpretation of the fatty acid data are involved with assignment of changes in quantities of individual fatty acids to changes in specific groups of soil microbes.

Many more difficulties are associated with validation of the metabolic diversity data beyond simply using the results as a reasonably distinctive fingerprint associated with a specific soil sample. Perhaps the most troublesome aspect of these studies involves attempts to gain as much information as possible from the large amount of data that are collected to develop the soil fingerprint. For example, a single micro-titer plate associated with the BIOLOG procedure could include an indication of the rate of catabolism or transformation of each of the individual 95 substrates. Clearly, the desire of all experimenters is to use all meaningful data provided by the procedure. Thus, the question becomes how do the nature of the substances transformed on the BIOLOG plate and their specific metabolic rates relate to the processes catalyzed in the soil in the field? As will be noted below, responses to this question are neither simple, nor without controversy.

Abundant quantities of "numbers" can be collected for each soil sample analyzed by the BIOLOG procedure. For example, if the inoculated plates are read at appropriate time intervals, the kinetics of color development, presumably the time course of transformation of each individual carbon source, is provided. The temptation would be to relate not only the capability to catabolize the individual substrate (the pluses and minuses) but also the actual rate of metabolism to the field situation. Additionally, when the metabolic diversity data from several different field sites are analyzed using a principal component analysis, the correlation of the individual substrates contributing most to the separation of the data for the individual sites is provided (see Table 1 for an example of this type of data). Again, it is tempting to relate the identity of the substances most responsible for separation of the data to actual differences in metabolic capabilities in the field. For example, one could be tempted when noting that separations were most correlated to carbohydrate metabolism to conclude that the carbohydrate metabolic capabilities differ meaningfully between the two soil systems sampled. As will be indicated below, many problems are associated with these interpretations, not the least of which relates to the basic carbon and energy source of the soil heterotrophic soil microbial community. The primary energy source for this community is photosynthetically fixed carbon—plant biomass. Plant biomass is mostly carbohydrate [49]. Thus, it is logical to assume that carbohydrate metabolism must be a major energy producing process in all soil systems. Although, as will be documented in the following paragraphs, these limitations to data interpretation have been documented in the literature, a strong temptation still exists to relate the specific metabolic capabilities, rates, and correlation directly to the field situation. It must be concluded that such extension of the data are not supported by theoretical underpinnings of the procedure and are premature at best. As stated succinctly in studies of rhizosphere populations with synthetic root exudates and of wheat, ryegrass, bentgrass and clover [50] as well as of hybrid larch (*Larix eurolepis* A. Henry) and Sitka spruce (*Picea sitchensis* Bong. Carr.) [51] using metabolic diversity analysis, the procedure is useful in distinguishing community types but interpretation of the data regarding metabolism of specific carbon can not be accomplished without an understanding of the substrates available in the specific ecosystem studied.

Before techniques can be considered for use in analyzing data relating to metabolic capabilities as assessed by the BIOLOG procedure to *in situ* field activities, it must be determined that the specific substrate oxidized in the test procedure can reasonably be anticipated to occur and be metabolized in the field. Conditions for the assay needed to determine that there even can be a meaningful relationship between metabolic capability data collected using the BIOLOG or related procedures include the following: a) the microbial populations inoculated into each of the wells in the microtiter plate must represent the native

Table 1
Substrates with high correlation coefficients for PC1 and PC2 in PCA of microbial diversity patterns for a native New Jersey Pineland soil, Lakewood A, and two cultivated sites sampled in fall, 1997

PC1	r	PC2	r
Carbohydrates		*Carbohydrates*	
N-Acetyl-D-galactosamine	0.7232	N-Acetyl-D-glucosamine	0.6197
N-Acetyl-D-glucosamine	-0.6602	L-Rhamnose	0.6232
L-Arabinose	0.6807		
D-Arabitol	-0.6716	*Polymers*	
Cellobiose	-0.9433	Tween 80	0.6633
L-Fucose	-0.6688		
Gentiobiose	-0.8488	*Carboxylic Acids*	
m-Inositol	0.7293	Citric Acid	0.7147
D-Mannose	-0.7178	D-Galactonic acid lactone	0.6786
D-Melibiose	-0.8279	γ-Hydroxybutyric acid	0.8664
β-Methylglucoside	-0.8023	α-Ketovaleric acid	0.7327
D-Raffinose	-0.8157	D,L-Lactic acid	0.6875
L-Rhamnose	-0.6556		
Sucrose	-0.8004	*Amides*	
		Succinamic acid	0.8874
Esters			
Methylpyruvate	0.7572	*Amino Acids*	
		L-Histidine	0.6803
Polymers		L-Leucine	0.8127
Glycogen	-0.7774	Bromosuccinic acid	0.7952
Dextrin	-0.7438		
Tween 40	0.7977	*Amines*	
		2-Aminoethanol	0.6576
Carboxylic Acids		Putrescine	0.7946
Acetic Acid	0.6021		
Formic acid	0.7967		
D-Gluconic acid	-0.8106		
D-Glucosaminic acid	0.7197		
β-Hydroxybutyric acid	0.7211		
p-Hydroxyphenylacetic acid	0.7567		
α-ketobutyric acid	0.7193		
Propionic acid	0.6141		
Sebacic acid	0.8099		
Phosphorylated Chemicals			
D,L-α-Glycerol phosphate	-0.6846		
Glucose-1-phosphate	-0.7585		
Glucose-6-phosphate	-0.7454		

Table 1. Continued

Amino Acids	
L-Ornithine	0.6186
L-Phenylalanine	0.8465
L-Proline	0.8731
L-Pyroglutamic acid	0.7327
D-Serine	-0.6312
Aromatic Chemicals	
Uridine	-0.8322
Brominated Chemicals	
Omosuccinic acid	0.6566

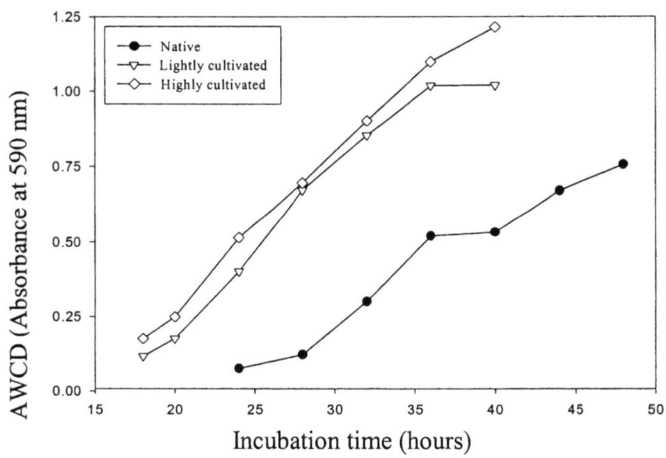

Figure 1. Rate of color development on BIOLOG microplates for native and cultivated New jersey Pine land soils. AWCD (average well color development) is the mean of the net optical density for all 95 substrate wells on microplate. Values represent the mean AWCD of three replicate BIOLOG microplates from soil extractant.

soil populations and have little variation in the active microbial populations in the discrete samples added to each test well, b) the microbial populations extracted from different soil samples must produce representative metabolic patterns at equivalent, or nearly equivalent, rates in the microtiter plates (an implication of this requirement is that standardization of the bacterial population density in the inoculum to the test wells is possible), and lastly, c) the organic carbon compounds in the microplate are substances that would reflect meaningful metabolic activities expressed in soil samples. The last point was addressed above when it was noted that there is little relationship between the substances contained in the microtiter wells and predominant organic compounds found in the soil humus fraction. The substances were selected originally for their utility in identifying axenically cultured bacteria, not for quantifying general metabolic capabilities of a mixture of soil bacteria, fungi, and enzymes. This point has been addressed [52, 53] and succinctly stated "…this approach suffers because the tested substrates do not accurately represent the types of substrates present in ecosystems" [53]. The initial two points regarding the need for representative and replicate sample inoculation into each of the test wells have also been addressed [45, 53-55]. The data support the conclusion that neither requirement is met with the procedure. The microbial populations in the individual test wells of the same microtiter plate can differ significantly. These observations as well as the underlying properties of the procedure led Garland to conclude that the validity of the procedure to be used even as a simple fingerprint representative of diversity variation in soil systems required their comparison at the same average well color development [45]. Due to variations both in the total number and the specific bacterial strains and species inoculated into each test well of the microtiter plates, the rate of color development varies meaningfully between soil samples. (See Fig. 1 for an example of this variation in soils collected from a New Jersey Pinelands forest site.) Although average well color development has been criticized [56, 57] and rebutted [58, 59], it has proven to be useful in providing distinctive fingerprints of soils, as indicated by the many studies cited above. Although alternative procedures have been proposed for the analysis of metabolic diversity data [60, 61], it must be stated that data analysis and the resulting conclusions relating to the relationship of the metabolic capabilities to "real world" microbial activities must be limited and carefully considered until more is known about the underlying principles of the metabolic diversity procedure.

5. CONCLUSIONS

Both the assessment of the capacity to metabolize a large number of organic substances (metabolic diversity) and the comparison of the variety of phospholipid fatty acids (PLFA diversity) in soil have been shown to provide distinctive indicators of general change in an ecosystem or variation in microbial properties between a variety of ecosystem types. The fingerprint indicators provided by these methods are affected by factors causing a general alteration in the soil microbial community—e.g., pollution with organic or inorganic components, alteration of soil physical or chemical properties, or major changes in the nature of fixed carbon inputs through changes in aboveground plant community composition. These procedures serve as an indicator of community change with a variety of anthropogenic interventions. Two areas of data interpretation that must await further elucidation of the underlying properties of the methods for their validation are relating data a) to specific fingerprint patterns to a specific ecosystem and b) to actual field activity and microbial diversity.

REFERENCES

1. Sundh, I., Nilsson, M., Borga, P., 1997. Variation in microbial community structure in two boreal peatlands as determined by analysis of phospholipid fatty acid profiles. Appl. Environ. Microbiol. 63, 1476-1482.
2. Steinberger, Y., Zelles, L., Bai, Q.Y., von Lutzow, M., Munch, J.C., 1999. Phospholipid fatty aids profiles as indicators for the microibal community structure in soils along a climatic transect in the Judean desert. Biol. Fertil. Soils 28, 292-300.
3. Borga, P., Nilsson, M., Tunlid, A., 1994. Bacterial communities in peat in relation to botanical composition as revealed by phospholipid fatty acid analysis. Soil Biol. Biochem. 26, 841-848.
4. Garland, J.L., Mills, A.L., 1991. Classification and characterization of heterotrophic microbial communities on the basis of patterns of community-level sole-carbon-source utilization. Appl. Environ. Microbiol. 57, 2351-2359.
5. Zak, J.C., Willig, M.R., Moorhead, D.L., Wildman, H.G., 1994. Functional diversity of microbial communities: a quantitative approach. Soil Biol. Biochem. 26, 1101-1108.
6. Torsvik, V., Goksøyr, J., Daae, F.L., 1990. High diversity in DNA of soil bacterial Appl. Environ. Microbiol. 56, 782-787.
7. Insam, H., Rangger, A., (Eds.) 1997. Microbial Communities. Functional versus Structural Approaches. Springer-Verlag. Berlin.
8. Garland, J.L., Mills, A.L. 1994. A community-level physiological approach for studying microbial communities. *In*: Ritz, K., Dighton, J., Giller, K. E. (Eds.) Beyond the Biomass. Wiley-Sayce, London. pp. 77-83.
9. Tate, R.L. III, 2000. Soil Microbiology. Second edition. John Wiley & Sons, NY.
10. Windig, A., 1994. Fingerprinting bacterial soil communities using Biolog microtitre plates. *In*: Ritz, K., Dighton, J., Giller, K.E. (Eds.) Beyond the Biomass. Wiley-Sayce, London. pp. 85-94.
11. Kelly, J.J., Tate, R.L. III, 1998. Effects of heavy metal contamination and remediation on soil microbial communities in the vicinity of a zinc smelter. J. Environ. Qual. 27, 609-617.
12. Palmborg, C., Nordgren, A., Bååth, E., 1998. Multivariate modelling of soil microbial variables in forest soil contaminated by heavy metals using wet chemical analysis and pyrolysis GC/MS. Soil Biol. Biochem. 30, 345-357.
13. Pennanen, T., Frostegard, A., Fritze, H., Bååth, E., 1996. Phospholipid fatty acid composition and heavy metal tolerance of soil microbial communities along two heavy metal-polluted gradients in coniferous forests. Appl. Environ. Microbiol. 62, 420-428.
14. Pennanen, T., Fritze, H., Vanhala, P., Kiikkila, O., Neuvonen, S., Bååth, E., 1998. Structure of a microbial community in soil after prolonged addition of low levels of simulated acid rain. Appl. Environ. Microbiol. 64, 2173-2180.
15. Rutgers, M., van't Verlaat, I.M., Wind, B., Posthuma, L., Breure, A.M., 1998. Rapid method for assessing pollution-induced community tolerance in contaminated soil. Environ. Toxicol. Chem. 17, 2210-2213.
16. Frostegard, A., Tunlid, A., Bååth, E., 1993. Phospholipid fatty acid composition, biomass, and activity of microbial communities from two soil types experimentally exposed to different heavy metals. Appl. Environ. Microbiol. 59, 3605-3617.
17. Frostegard, A., Tunlid, A., Bååth, E., 1996. Changes in microbial community structure during long-term incubation in two soils experimentally contaminated with metals. Soil Biol. Biochem. 28, 55-63.

18. Kelly, J.J., Häggblom, M., Tate, R.L. III, 1999. Changes in soil microbial communities over time resulting from one time application of zinc: a laboratory microcosm study. Soil Biol. Biochem. 31, 1455-1465.
19. Knight, B.P., McGrath, S.P., Chaudri, A.M., 1997. Biomass carbon measurements and substrate utilization patterns of microbial populations from soils amended with cadmium, copper, or zinc. Appl. Environ. Microbiol. 63, 38-43.
20. Kelly, J.J., Häggblom, M., Tate, R.L. III, 1999. Effects of land application of sewage sludge on soil heavy metal concentrations and soil microbial communities. Soil Biol. Biochem. 31, 1467-1470.
21. Bååth, E., Diaz-Ravina, M., Frostegard, A., Campbell, C.D., 1998. Effect of metal-rich sludge amendments on the soil microbial community. Appl. Environ. Microbiol. 64, 238-245.
22. Dahlin, S., Witter, E., Martensson, A., Turner, A., Bååth, E., 1997. Where's the limit? Changes in the microbiological properties of agricultural soils at low levels of metal contamination. Soil Biol. Biochem. 29, 1405-1415.
23. Rogers, B.F., 1999. Environmental factors affecting microbial community size, activity, and metabolic diversity in New Jersey Pineland soils. Ph.D. Thesis, Rutgers University, New Brunswick, NJ.
24. Zogg, G.P., Zak, D.R., Ringelberg, D.B., MacDonald, N.W., Pregitzer, K.S., White, D.C., 1997. Compositional and functional shifts in microbial communities due to soil warming. Soil Sci. Soc. Am. J. 61, 475-481.
25. Hodge, A., Paterson, E., Grayston, S.J., Campbell, C.D., Ord, B.G., Killham, K., 1998. Characterisation and microbial utilisation of exudate material from the rhizosphere of *Lolium perenne* grown under CO_2 enrichment. Soil Biol. Biochem. 30, 1033-1043.
26. Ringelberg, D.B., Stair, J.A., Almeida, J., Norby, R.J., O'Neill, E.G., White, D C., 1997. Consequences of rising atmospheric carbon dioxide levels for the belowground microbiota associated with white oak. J. Environ. Qual. 26, 495-503.
27. Rillig, M.C., Scow, K.M., Klironomos, J.N., Allen, M.F., 1997. Microbial carbon-substrate utilization in the rhizosphere of *Gutierrezia sarothrae* grown in elevated atmospheric carbon dioxide. Soil Biol. Biochem. 29, 1387-1394.
28. Pennanen, T., Perkiomaki, J., Kiikkila, O., Vanhala, P., Neuvonen, S., Fritze, H., 1998. Prolonged, simulated acid rain and heavy metal deposition: separated and combined effects on forest soil microbial community structure. FEMS Microbiol. Ecol. 27, 291-300.
29. Macalady, J.L., Fuller, M.E., Scow, K.M. 1998. Effects of metam sodium fumigation on soil microbial activity and community structure. J. Environ. Qual. 27, 54-63.
30. Fuller, M.E., Scow, K.M., Lau, S., Ferris, H. 1997. Trichloroethylene (TCE) and toluene effects on the structure and function of the soil community. Soil Biol. Biochem. 29, 75-89.
31. El Fantroussi, S., Verschuere, L., Verstraete, W., Top, E. M., 1999. Effect of phenylurea herbicides on soil microbial communities estimated by analysis of 16S rRNA gene fingerprints and community-level physiological profiles. Appl. Environ. Microbiol. 65, 982-988.
32. Fuller, M.E., Manning, J.F. Jr., 1998. Evidence for differential effects of 2,4,6-trinitrotoluene and other munitions compounds on specific subpopulations of soil microbial communities. Environ. Toxicol. Chem. 17, 2185-2195.
33. Bossio, D.A., Scow, K.M., Gunapala, N., Graham, K.J., 1998. Determinants of soil microbial communities: Effects of agricultural management, season, and soil type on phospholipid fatty acid profiles. Microbial Ecol. 36, 1-12.

34. Ibekwe, A.M., Kennedy, A.C., 1998. Phospholipid fatty acid profiles and carbon utilization patterns for analysis of microbial community structure under field and greenhouse conditions. FEMS Microbiol. Ecol. 26, 151-163.
35. Lupwayi, N.Z., Rice, W.A., Clayton, G.W., 1998. Soil microbial diversity and community structure under wheat as influenced by tillage and crop rotation. Soil Biol. Biochem. 30, 1733-1741.
36. Bossio, D.A., Scow, K.M., 1995. Impact of carbon and flooding on metabolic diversity of microbial communities in soils. Appl. Environ. Microbiol. 61, 4043-4050.
37. Bossio, D.A., Scow, K.M. 1998. Impacts of carbon and flooding on soil microbial communities: Phospholipid fatty acid profiles and substrate utilization patterns. Microbial Ecol. 35, 265-278.
38. Goodfriend, W.L., 1998. Microbial community structure of potential substrate utilization: A comparison of salt marsh, sand dune, and seawater-irrigated agronomic systems. Soil Biol. Biochem. 30, 1169-1176.
39. Staddon, W. J., Duchesne, L.C., Trevors, J.T., 1997. Microbial diversity and community structure of postdisturbance forest soils as determined by sole-carbon-source utilization patterns. Microbial Ecol. 34, 125-130.
40. Frostegard, A., Bååth, E., Tunlid, A., 1993. Shifts in the structure of soil microbial communities in limed forests as revealed by phospholipid fatty acid analysis. Soil Biol. Biochem. 25, 723-730.
41. Bardgett, R.D., Leemans, D.K., Cook, R., Hobbs, P.J., 1997. Seasonality of the soil biota of grazed and ungrazed hill grasslands. Soil Biol. Biochem. 29, 1285-1294.
42. Maire, N., Borcard, D., Laczko, E., Matthey, W., 1999. Organic matter cycling in grassland soils of the Swiss Jura mountains: Biodiversity and strategies of the living communities. Soil Biol. Biochem. 31, 1281-1293.
43. Bardgett, R.D., Lovell, R.D., Hobbs, P.J., Jarvis, S.C., 1999. Seasonal changes in soil microbial communities along a fertility gradient of temperate grasslands. Soil Biol. Biochem. 31, 1021-1030.
44. Denton, C. S., Bardgett, R. D., Cook, R., Hobbs, P. J., 1999. Low amounts of root herbivory positively influence the rhizosphere microbial community in a temperate grassland soil. Soil Biol. Biochem. 31, 155-165.
45. Garland, J.L., 1996. Patterns of potential C source utilization by rhizosphere communities. Soil Biol. Biochem. 28, 223-230.
46. Garland, J.L., 1996. Analytical approaches to the characterization of samples of microbial communities using patterns of potential C source utilization. Soil Biol. Biochem. 28, 213-221.
47. Griffiths, B.S., Ritz, K., Ebblewhite, N., Dobson, G. 1999. Soil microbial community structure: Effects of substrate loading rates. Soil Biol. Biochem. 31, 145-153.
48. Di Giovanni, G.D., Watrud, L.S., Seidler, R.J., Widmer, F., 1999. Comparison of parental and transgenic alfalfa rhizosphere bacterial communities using Biolog GN metabolic fingerprinting and enterobacterial repetitive intergenic consensus sequence-PCR (ERIC-PCR). Microbial Ecol. 37, 129-139.
49. Stevenson, F.J., 1994. Humus Chemistry. Genesis, Composition, Reactions, second edition. John Wiley & sons, NY.
50. Siciliano, S.D., Germida, J. J., 1998. Biolog analysis and fatty acid methyl ester profiles indicate that pseudomonad inoculants that promote phytoremediation alter the root-

associated microbial community of *Bromus bierbersteinii*. Soil Biol. Biochem. 30, 1717-1723.

51. Grayston, S.J. Wang, S.Q., Campbell, C.D., Edwards, A.C., 1998. Selective influence of plant species on microbial diversity in the rhizosphere. Soil Biol. Biochem. 30, 369-378.
52. Grayston, S.J., Campbell, C D., 1996. Functional biodiversity of microbial communities in the rhizospheres of hybrid larch (*Larix eurolepis*) and sitka spruce (*Picea sitchensis*). Tree Physiol. 16, 1031-1038.
53. Konopka, A., Oliver, L., Turco, R. F., 1998. The use of carbon substrate utilization patterns in environmental and ecological microbiology. Microbial Ecol. 35, 103-115.
54. Smalla, K., Wachtendorf, U., Heuer, H., Liu, W.-T., Forney, L., 1998. Analysis of BIOLOG GN substrate utilization patterns by microbial communities. Appl. Environ. Microbiol. 64, 1220-1225.
55. Haack, S.K., Garchow, H., Klug, M.J., Forney, L.J., 1995. Analysis of factors affecting the accuracy, reproducibility, and interpretation of microbial community carbon source utilization patterns. Appl. Environ. Microbiol. 61, 1458-1463.
56. Howard, P.J.A., 1997. Analysis of data from BIOLOG plates: Comments on the method of Garland and Mills. Soil Biol. Biochem. 29, 1755-1757.
57. Howard, P.J.A., 1999. Analysis of data from BIOLOG plates: Examination of relationships between samples. Soil Biol. Biochem. 31, 1197.
58. Garland, J.L., Mills, A.L., 1999. Importance of pattern analysis in community-level physiological profiles (CLPP): A reply to the letter from P.J.A. Howard. Soil Biol. Biochem. 31, 1201-1202.
59. Garland, J.L., Mills, A. L. 1999. Further comments on the interpretation of community-level physiological profiles. Soil Biol. Biochem. 31, 1203.
60. Insam, H., Hitzl, W., 1999. Data evaluation of community-level physiological profiles: A reply to the letter of P. J. A. Howard. Soil Biol. Biochem. 31, 1198-1200.
61. Lindstrom, J.E., Barry, R.P., Braddock, J.F., 1998. Microbial community analysis – a kinetic approach to constructing potential C source utilization patterns. Soil Biol. Biochem. 31, 231-239.

Developments in Soil Science, Volume 28B
Editors: A. Violante, P.M. Huang, J.-M. Bollag and L. Gianfreda

EFFECTS OF SOIL ABIOTIC PROCESSES ON THE BIOAVAILABILITY OF ANTHROPOGENIC ORGANIC RESIDUES

P. Ruggiero, M.D.R. Pizzigallo and C. Crecchio

Dipartimento di Biologia e Chimica Agroforestale ed Ambientale,
Università degli Studi di Bari, via Amendola 165/a, 70126 Bari, Italy

Soils are the ultimate sink for a large number of organic pollutants. There is a debate over the risk assessment that generally fails to effectively evaluate all the aspects of the impacts of pollutants on soils where a complex mixture of biotic and abiotic components promotes a number of processes relevant to the fate and behaviour of pollutants. This concern about pollution of organic xenobiotics gives rise to various strategies of management of contaminated sites, preferentially based on bioremediation. However, there are different confounding aspects about the bioavailability of the contaminats to microbial cells or other organisms.

Many scientific data now support some general concepts. The first relates to the heterogeneous nature of soils in which organic chemicals are not completely and immediately released to the environment to be transported to groundwater and/or taken up by bioreceptors. Some different attenuation processes can affect the concentration of an organic contaminant in the liquid phase of the soil including adsorption, sequestration, and bound residues formation.

Adsorption is the predominant process that involves polar organic contaminants and the surfaces of inorganic and organic soil components. Adsorption may be affected by the properties of the adsorbent and of the chemical, as well as by the soil physical and chemical properties. Different adsorptive chemical mechanisms may operate. Pollutants adsorbed on surfaces may be available for biodegradation, also considering that microbes may be sorbed to solids and may extract directly surface-sorbed chemicals. Nonionic organic chemicals diffuse to the relatively hydrophobic sites and surfaces rather than being attracted to them. The relative importance of organic matter surfaces to microbial events is not as well understood as that of clay minerals.

Other substances in soil, besides natural organic matter, that could act as a sink, e.g., a third phase, consisting of nonaqueous phase liquids (NAPL), may also affect the contaminant bioavailability. NAPL contaminants usually consist of hundreds of aromatic organic compounds and are often considered to be unavailable for direct microbial uptake.

The xenobiotics can be sorbed and then sequestered with time in microsites within soil matrix. Less is known about the mechanisms of sequestration than adsorption; however, researches indicate that sequestration is influenced by time, soil organic matter content, clay content, cycles of wetting and drying, and contamination type. A consensus is emerging that with increased contact time between a chemical and soil, the contaminat becomes more strongly associated with soil components. This ageing process may involve changes in the physical relationship between the chemical and soil particle including the movement of chemicals into soil micropores or within hydrophobic remote area of soil organic matter. In addition, there are chemical reactions between unaltered organic xenobiotics or their metabolites and inorganic components or, preferentially, macromolecular natural organic matter leading to the formation of stable chemical linkages. Due to the complexity of soil microhabitat, there is still much to learn about the mechanisms that contribute to sequestration. However, the risk associated with soil-aged or soil-bound chemicals may be reduced because the chemicals are unvailable for release, transport, and uptake into receptors.

Other abiotic soil processes have the potential to affect soil-xenobiotic interactions. Surface-catalysed hydrolytic and oxidation reacions degrade organic chemicals into low molecular weight products, which may partially or completely desorb or may be transformed into higher molecular weight products. Generally, abiotic reactions lead to "daughter" molecules that, maintaining their organic core, are structurally simpler and more oxigenated than the parent compounds. However, minor alterations in the structure frequently cause a drastic change in the susceptibility of biotransformation. If "daughter" molecules were still toxic, their biological degradation would complete the detoxificition process.

In conclusion, in this review we focus on the impacts of abiotic processes on bioavailability with the intent of: (1) demonstrating that the sequestration and binding of organic xenobiotics occur or have occured during the ageing of the soil- contaminant matrix; (2) identifying physical and chemical processes that are responsible for xenobiotic sequestration and the soil factors that affect them; (3) developing correlations between the occurence of sequestration and of chemical trasformation of organic xenobiotics and their bioavailability.

1. INTRODUCTION

The rapid rise in the use of agrochemicals for the control of weeds, insects, and plant diseases as well as the improper disposal and accidental release of organic compounds into the environment have brought about a widespread pollution of soil and water environments. Unnatural or man-made compounds are known as "xenobiotics". During the last decades an extensive body of information on the key processes controlling their fate and behaviour in soil environments has accumulated and understanding of potential toxicity, biogeochemical processes and their consequences, has improved substantially.

An assessment of the hazard of toxic organic contaminants deposited in the soil has to take into account several processes to which these compounds may be subjected. The main processes include biodegradation, volatilisation, leaching, sorption-desorption and sequestration within the soil matrix, abiotic transformation and plant uptake. The overall loss or disappearance of appreciable amounts of organic xenobiotics are directly controlled by these processes and result from more than one type of mechanism. The amounts of pollutant remaining in the soil constitute in the field of soil remediation a point of major concern and debate in relation to risk assessment.

Knowledge of the rates and extent of abiotic and biotic processes is a prerequisite for rational uses of the soil and for the elimination of toxic compounds that have accumulated in soil as a result of intended or unintended human activities. Field and laboratory studies demonstrate a diminishing availability of xenobiotics to microorganisms due to a sequestration of the molecules that makes them no longer available.

Clays, (hydr)oxides, and humic substances are the most important and abundant of the colloidal constituents of soils, and they influence most the interactions, whether it be sorption, binding, or degradation, between small organic pollutants and soils

Certain sorbate properties may assist in distinguishing among interaction processes. However many observations are confounded by the aggregated nature of soil colloids that exposes xenobiotics to various domains including external and internal surfaces, microvoids, and matrix heterogeneity. To untangle the various interactions, it would be useful to have direct information on the environmental fate of hazardous pollutants. Two types of abiotic sequestration can be identified: (i) a physical entrapment into sites within the soil matrix that are not readily accessed by microorganisms; (ii) a sequestration through chemical reactions, mineral surface- or enzyme-mediated, that can produce covalent binding of the xenobiotics to soil components. The bioavailabilty and toxicity of the entrapped or bound residues depend upon their release, determined by the presence of microbial, physical or chemical mechanisms able to facilitate desorption or to overcome the binding of sequestered molecules.

Sequestration involves adsorption on inorganic and organic soil surfaces that abiotically promote a number of reactions relevant to environmental impact of organic xenobiotics. The transformation products seldom reach complete mineralization. They might be themselves sorbed and sequestered through a variety of physical and chemical processes and follow the same fate as the parent compounds. However, they are generally more soluble than the parent molecules. This increases their mobility and transport through runoff to surface waters, and their access as substrates for microbial growth and metabolism, thus their bioavailabilty.

In this general context, this review provides an understanding of the effects and implications of sorption, binding, and abiotic reactions of xenobiotics with respect to bioavailability. These processes conceivably coexist in natural environments and their individual contribution to bioavailability of organics remains uncertain. The uses of soil models in batch or column experiments and of single components isolated from soils or chemically synthesized seek to understand at the molecular level the interactions that take place. In the same way, models that combine sorption and biodegradation, have been used to interpret results of laboratory studies. The reversibility of the sorption process and the extent of protection of sorbed pollutant from degradation have important practical consequences and have become in recent years the object of considerable attention among scientists. Often a very important fraction of pollutants remains non-available to remediation processes, either physico-chemical or biological. All these needed information are crucial for proper risk assessment, and for the development of strategies in remediation projects and of legislation to protect soils.

2. BASIC BIOCHEMICAL PROCESSES IN ORGANIC POLLUTANTS TRANSFORMATION

The widespread presence of microorganisms in the soil (at least 10,000 genetically different microorganisms per gram [1]), with their remarkable ability to adapt and proliferate in virtually any imaginable environment, results in a dominant role of the microbial transformation in the degradation of most man-made compounds. In soils, monitoring of organic contaminants has been chiefly restricted to pesticides that are microbially transformed either directly through metabolic processes, such as biodegradation, cometabolism, polymerization or conjugation, and accumulation, or indirectly through secondary effects of microbial activity changing environmental soil parameters such as pH, redox potential, reactive products, etc. [2]. The rate of microbial transformation depends on the physico-chemical properties of organic compound (i.e. structure, solubility, concentration, toxicity) and on the soil environmental conditions (i.e. soil characteristics, temperature, humidity, drainage etc.). A number of xenobiotic molecules or their intermediates can be transformed by microorganisms or their enzymes. The major and common processes are oxidation, reduction, hydrolysis, and synthesis reactions [2]. The principal enzymes (extracellular or intracellular) responsible for the bioconversion of xenobiotics include a variety of hydrolases and oxidoreductases and, among them, various enzymes capable of dehalogenation. The reaction catalysed are: hydroxylation, *N*-dealkylation, β-oxidation, decarboxylation, ether cleavage, oxidative coupling, ring cleavage, thioethers oxidation, sulfoxide reduction, nitrogroup reduction, reductive dehalogenation, hydrolytic dehalogenation [2].

Soil enzymatic activity is prevalently due to extracellular enzymes. They are produced principally by microorganisms, and may be defined as all enzymes that catalyse reactions outside of the organism that synthesised them [3]. These enzymes may survive in soil through stabilization as immobilised enzymes and they are responsible for transformation processes leading to the detoxification of xenobiotics.

Most organic pollutants are degraded in soil. Various products can be obtained even from the same initial compound depending on environmental conditions. Moreover some compounds may be transformed into more recalcitrant and toxic intermediates compared to their parent compounds. Bacteria are the primary group of microorganisms responsible for the detoxification of xenobiotics, whereas fungi and yeasts are less important. There are typically 10^8-10^{10} bacterial cells per gram of soil, and the diversity within this group is enormous. *Pseudomonas* and *Bacillus* species are considered to be the most efficient groups for the pollutants transformation.

Microbial degradation of small organic compounds occurs intracellularly. The substrates serve as sources of carbon, nitrogen, and energy for microbial growth and reproduction. The parent compound has first to reach the cell surface, then it has to be transported into the cell, either by an active or a passive transport. The passive diffusion is non-specific and concerns predominantly non-polar compounds, whereas more polar compounds need a transport system for uptake. These transport proteins are highly specific for single compounds and their activity, regulating the xenobiotic internal concentration, may have an important influence on the toxicity towards the degrading organisms. In the cell, the organic compound is converted to intermediates channelled directly into the central cell metabolism for energy generation and supply of carbon for anabolic processes. The presence or absence of molecular oxygen can be decisive for the choice of the biochemical pathway. Finally, inorganic end products, like carbon dioxide, diffuse out of the cell. When mineralization is not complete, the end product is actively transported out of the cell [4].

3. THE CONCEPT OF BIOAVAILABILITY

The response of a contaminant-degrading microorganism when exposed to the contaminant is an actual decrease in the contaminant mass over time as a consequence of the mass transfer to biologically accesible regions. This process may be considered as "mass bioavailability". The bioavailable fraction can also be regarded as the carrier of toxicity. Thus, a "toxicological bioavailability" can also be considered when the organism exposure to the contaminant leads to toxic response [5, 6].

Many organic compounds that are quickly biodegraded in aqueous solution have been found to persist in the presence of soils where organic xenobiotics may be entrapped and their availability to microorganisms lowered. This means that the fraction of contaminant available to any kind of organism is time dependant and has to be expressed as a rate. Thus, the bioavailability of a chemical is determined by the rate of mass transfer to the cell relative to the intrinsic biochemical activity of microorganisms [7]. A global definition of bioavailability has to conciliate different approaches linked to the numerous factors intrinsic to the biological system, including the toxic response-inducing organisms, environmental conditions (pH, water, temperature, redox potential, presence of oxygen and nutrients) and the extent of organism exposure to the contaminants. The concept of bioavailability has been critically discussed recently by Baveye and Bladon [6], who introduced the "sink theory" of the bioavailabilty of organic xenobiotics, according to which the key factor is the ability of microbial cells to act as sinks for these compounds. This means that, depending on their physiological status, the organisms adopt a number of strategies (release of extracellular enzymes, regulation of uptake systems, production of biosurfactants, systematic depletion of the xenobiotic concentration in the volume surrounding the organism) to force their environment to release the xenobiotics at a faster rate [6]. Therefore, not only the slow mass transfer to the degrading microorganisms may cause a reduced bioavailability of pollutants, but also the lack of appropriate intracellular genetic or biochemical factors.

Biodegradation processes are even more complicated to understand if one realises that a single strain may be unable to completely degrade a certain xenobiotic, which may be transformed only by the synergistic action of microbial communities that function cooperatively [2]. On the other hand, due to their remarkable genetic flexibility and physiological diversity, microorganisms have adapted to thrive on many synthetic environmental pollutants. Mutations in structural genes, rearrangement of DNA segments, and exchange of genetic information may yield genotypes with novel catabolic activities. Such activities proliferate leading to mutant strains that degrade even more recalcitrant compounds [8, 9]. Examples of the adaptation to chloroalkanes and chloroaromatics have been given [9].

4. SORPTION-DESORPTION PROCESSES OF ORGANIC XENOBIOTICS BY SOIL COLLOIDS

In general, only dissolved molecules can be transported across the cell membranes of microorganisms [10]. The substrate must be freely available for passive diffusion or for binding to the active site of the protein required for its transport. Outside of the bacterial cell, bioavailability can be affected by the molecular speciation of the organic contaminants in the aqueous phase, such as protonation, deprotonation, complex formation with metal ions and dissolved organic matter (DOM). DOM, in its turn, can be considered a continuum of substances of biotic origin, partially degraded and transformed, showing a wide range of pK_a and PZC (point of zero charge) values, molecular sizes, and functional groups [11]. Hence, the chemical characteristics of the liquid phase may render organic compounds more or less favourable for microbial degradation.

In soil, the bioavailability of an organic xenobiotic is strongly influenced by the presence of organic and inorganic fractions. The characteristics and properties of clay minerals, (hydr)oxides, and organic matter, are considered to be well known and will not be described in this chapter.

Adsorption is probably a major process that controls the concentration of an organic contaminant in the liquid phase, regulating consequently its accessibility to target organisms. Numerous reviews describe mechanisms of pesticides and other anthropogenic organic adsorption by soils, clay minerals, and humic substances [12-16] and it is not our intention to describe these processes in detail. Briefly, the extent of adsorption depends on the properties of adsorbent, which include quantitative and qualitative composition, exchange properties, moisture content, temperature, pH, and of the compound, that is molecular size and shape, nature, number and placement of functional groups influencing the polarity, polarizability, solubility, the acid, base, redox properties.

As a consequence of these properties, different adsorptive mechanisms, including ionic binding, hydrogen bonding, ligand exchange, charge-transfer complexes, covalent bonding, hydrophobic partitioning, Van der Waals forces, may operate. Small, sparingly soluble compounds tend to be attracted to relatively hydrophobic sites and surfaces. This accumulation process provides the principal mechanism of sorption of neutral, less polar, xenobiotics by humic substances [12, 14]. Sorption has often been correlated with the mass fraction of soil organic carbon and sorbate hydrophobicity. Even most studies on DOM-water partitioning of organic contaminants have focused on neutral compounds. Nevertheless, recently the binding of hydrophobic organic cations to DOM has been investigated [17]. It was found that sorption was governed by complexation of the organic cation by negatively charged ligands (i.e., carboxylate and phenolate groups) of the humic acids.

It is just useful to remember that inorganic and organic soil colloids generally are not isolated in soils but are organised to form aggregates. Clay minerals coat larger sand and silt particles, which cluster together into microaggregates and, then, aggregates that are stabilised by organic matter and inorganic components. Particle sizes and aggregate stability define pore sizes. The pore size distribution and the type and amount of clay and organic matter within the aggregates influence, in their turn, water retention against gravitational loss.

Many sorption studies on xenobiotics in soil show that sorption reactions occur in a period ranging from minutes to a few hours (instantaneously) upon contact of xenobiotic molecules with the active surface of soil matrix. During equilibration with soil, the organic compound partitions between the sorption sites and the soil solution. By constructing sorption isotherms, apparent equilibrium partition coefficients (K_p) can be calculated. In consideration of the role of organic matter in sorption processes, partition coefficients are also expressed as a function of organic carbon content (K_{oc}). K_p and K_{oc} values are generally used to estimate the sorption of a chemical to a range of soils as well as to clays and humic substances [18].

Worthy of note is that recent studies reconsider the earlier concepts of the macromolecular structure of humic substances. Conte and Piccolo [19] have suggested that, rather than being polymeric, humic substances are composed of relatively small molecules that self-associate mainly via hydrophobic forces. The accumulation of hydrophobic compounds promotes molecular association and microaggregates formation. Different opinions about "random coil" and "self-associating aggregate" models, molecular size range, hydrophobic and hydrophilic regions, have been critically reviewed in a special issue (November 1999) of Soil Science.

The micro- and macromolecular properties of humic substances have remarkable implications on the sequestration of organic xenobiotics. Evidence has been presented recently suggesting that the partition model of sorption to soil organic matter is simplistic. According to dual-mode model [20], total sorption to soil organic matter would be the sum of sorption in the dissolution (partition) domain and sorption in the hole-filling domain. The holes are postulated to be nanometer-size voids abundant within the organic matrix that provide complexation sites. These voids are accessible to organic molecules by diffusion through the solid phase but inaccessible to microbial cells and their enzymes. The dual-mode sorption model was tested for chlorinated benzenes and a possible mechanism of sequestration reducing bioavailability with time was suggested [20].

Depending on the nature of binding, a portion of the sorbed compound may be released back to the soil solution as a result of a decrease in its solution concentration. This desorption may be realised replacing xenobiotic solution in contact with the soil at the equilibrium with a more diluted or xenobiotic-free solution. Desorption is a combination of chemical kinetic and mass-transfer processes, and thus it is difficult to distinguish in soil which step controls the overall desorption rate. The reversibility of sorption reactions should give ideally superimposable sorption and desorption curves. However, hysteresis, which represents differences between sorption-desorption isotherms, has frequently been observed in isotherms for a number of pesticides. Some of the most significant explanations proposed for such apparent hysteresis include (i) experimental artefacts, (ii) chemical or biological transformation, (iii) incomplete equilibration [21].

The physicochemical character of the soil organic matter has marked influence on sorption-desorption hysteresis, desorption rates, and bioavailabilty of organic contaminant. Recent studies carried out on sorption and desorption equilibria of phenanthrene on three different geosorbents have shown that the more reduced and condensed the organic matter, the greater the extent of sorption-desorption hysteresis, the slower the desorption rate, and the less readily bioavailable the sorbed contaminant [22].

For modeling purposes, most sorption studies are carried out using batch methods involving vigorous shaking or stirring of soil suspensions. However, sorption and especially desorption in natural particles can be exceedingly slow. Pignatello and Xing [23] analysed the potential causes of slow sorption-desorption likely due to the existence of high-energy adsorption sites both within the internal matrix of organic matter and in intraparticle nanopores and to mass-transfer limitation through them. The authors outlined the implications on bioavailability and remediation of sorption processes occurring on time scales of months or longer, without a true equilibrium being reached, suggesting some approaches to promoting desorption from the slow state.

4.1. Nonaqueous phase liquid (NAPL)

The existence of solvent phases other than water further complicates the issue since hydrophobic organic chemicals can exist as nonaqueous phase liquids (NAPL). NAPL contaminants usually consist of mixtures of chemicals that differ in water solubility, density, viscosity, and sorption partition coefficient. They are often considered to be unavailable for direct microbial uptake until they break up into small blobs or begin to dissolve in the aqueous phase. Coal tar is a dense NAPL often associated with contaminated soils [24]. Coal tar is a multicomponent NAPL composed of hundreds of aromatic organic compounds, including PAH compounds such as naphtalene, phenanthrene, etc.

The bioavailability of organic substrates has been quantitatively assessed in a two phase, soil-water system. Very little is known when a NAPL is introduced into the system. The first demonstration of substantial microbial degradation and depletion of naphthalene from NAPL comes by Ghosal et al. [25]. The rate and extent of microbial degradation have been evaluated in mixed, NAPL-water batch systems. In such systems the rate of mass transfer may control the overall rate of biotransformation. The formation of semi-rigid films around NAPL with a high resistance towards NAPL-water mass transfer is considered to be the possible cause of reduced bioavailability. Recently, abiotic mass transfer tests and biomineralization studies carried out in solid-slurry and liquid-liquid dispersion systems have shown that independent equilibrium and kinetic dissolution tests aided in quantifying PAH bioavailability and potential initial biotransformation [26].

4.2. Ageing and bound residue formation

In the organic xenobiotics sorption studies, a portion of the sorbed chemical is desorbable, whereas a portion, called "bound residue", cannot be extracted from soil. The "extractability" of a compound will depend on the nature of the extractant. There is a large consensus that the term "bound residues" refers to chemical species that are unextracted by methods that do not substantially change the compounds themselves or the structure of the matrix [16].

The extent of the bound residue fraction in soils is variable, depending on many factors (contact time, soil type, method used to assess the percentage that is bound). Up to 90% of an applied pesticide may associate with the bound fraction of soil [27, 28].

Chemical reactions between unaltered organic xenobiotics or their metabolites and organic and inorganic soil components often lead to the formation of stable chemical linkages. The formed complexes may be highly resistant to microbial degradation. A list of the binding forces and of the relative interaction mechanisms has been reported in many reviews [14, 16, 29]. Bound residues are preferentially formed through ionic bonds between clay minerals and cationic organics and covalent bonds between xenobiotics and various naturally occurring compounds, including humic substances, in reactions catalysed by enzymes or abiotic soil components. Oxidative coupling reactions are of particular significance also because humic substances contain many free radicals that may become binding sites for certain organic xenobiotics. Hydrophobic bonding is the major mechanism responsible for binding water-insoluble compounds to hydrophobic subunits of humic compounds.

However, more recent studies show that the extractability of a contaminant with organic solvents decreases with increasing sorption time. This means that with increased contact time between a chemical and soil, the chemical becomes less available for solvent removal. This phenomenon is termed "ageing" and one explanation that has been provided to account for it is that the contaminant becomes more strongly associated with soil components over time [16]. For example, high resistance to extraction has been observed when phenanthrene or 4-nitrophenol was aged in an organic matter-rich soil and a loam under aseptic conditions for long periods up to 315 days [30]. The time-dependent decline in bioavailability of phenanthrene and atrazine in laboratory-aged (up to 200 days) dissimilar soils has been confirmed [31, 32]. Recent studies by Tang et al. [33] have demonstrated that PAHs aged for 140-203 days in soil become less available for bacterial degradation. Even sulfonylurea herbicides, whose high rate of degradation by hydrolysis is well known [34], have been shown to persist in soil 12-18 months after application [35].

Apparently, during prolonged residence times in soil, organic contaminants can form stronger bonds with soil or can be incorporated in structural micropores of mineral lattices or in hydrophobic remote areas of soil organic matrix [16, 23, 29, 36, 37].

Scow and Johnson [18] state that experimentally it is difficult to distinguish between chemicals that are irreversibly sorbed to soil and those that have extremely slow release kinetics. Therefore, these authors consider a chemical bound or aged if it is not removed from soil by a specified solvent extraction method. In contrast, Alexander [36] outlines that molecules behaving as aged compounds should not be confused with bound residues. Aged compounds can be extracted by some organic solvents expanding the time and/or the intensity of extraction; bound residues are not extracted by procedures that do not appreciably alter their nature.

A complete explanation of all the concurrent mechanisms governing the overall binding of a given chemical is not available. The term "sequestration" is also used in this context [16, 36,38, 39] and the xenobiotics can be sorbed and then sequestered with time in accessible microsites within the soil matrix. According to kinetics of sorption and desorption, the initial phase of sorption is rapid. This phase is followed by a considerably slower uptake allowing the molecules to diffuse to internal and more remote sites [23, 36]. Desorption likewise is initially rapid, releasing the molecules from sites very near to the particle-water interface; these molecules remain readily bioavailable. Then it becomes slow thus limiting the flux of contaminant to the aqueous phase; the contaminant is not readily bioavailable.

The uptake and release of nonionic contaminants by soil particles can be considered bimodal in that they occur in fast and slow stages due to their intraparticle or intraorganic matter diffusion [23, 36]. Consequently, the attainment of apparent equilibrium in sorption, is not predictable and may require weeks to many months.

The identification of bound xenobiotics and characterization of the binding sites are admittedly complex to investigate. The covalent nature of bonds has been confirmed by model studies in which xenobiotic chemicals, such as phenols or anilines, have been allowed to interact with monomeric constituents of humus (e.g., syringic acid or guaiacol) [29]. Methods to release bound residues (e.g., high-temperature distillation, supercritical fluid extraction or microwave extraction) make soil-bound xenobiotics available for subsequent identification by conventional analytical, spectroscopic and radioactive techniques [38].

However, much remain to be understood in this respect. Obtaining a complete explanation of all the processes that govern the soil-xenobiotics interactions is hampered also by the wide variety of xenobiotics belonging to all chemical classes and by the inherent complexity of natural porous media. The fundamental knowledge of where toxic contaminants are located, how they are associated, and what fundamental mechanisms govern their sorption, is uncertain. The future of research on these topics appears to be linked to spectroscopic techniques, particularly to the Electron Paramagnetic Resonance (EPR) and to the new technique of microprobe two-step laser desorption/laser ionization mass spectromertry (μL^2MS). EPR spectroscopy provides the opportunity to monitor and elucidate directly the different molecular-scale behaviours of paramagnetic spin probes in soils. Recent experiments, conducted also in our laboratory, have demonstrated the application of EPR to study the sorption and desorption overtime (ageing) of a hydrophobic spin probe with a benzene side group on Ca-hectorite suspensions [40] and on hectorite, humic acids, and their complexes [41]. Changes in molecular scale interaction of the spin probe with model soil systems (e.g. adsorbed, in bulk solution or in large intersticial pores) and desorption kinetics have been monitored. The technique of μL^2MS has been very recently used to identify and characterize trace hydrophobic contaminants on geosorbents [42, 43] A polycyclic aromatic hydrocarbons (PAHs) distribution across individual particles as well as from particle to particle was reported [42]. At the same time, desorption kinetic tests were used to evaluate PAH affinity for different sediment fractions [43]. Spin probe technique and complementary microscale analytic techniques appear as new and potentially useful approaches to evaluate the ageing effect on the partitioning and availability of xenobiotics in soils.

5. MICROSCOPIC DISTRIBUTION OF MICROBES IN SOIL

Abiotic processes of adsorption, binding, and ageing make xenobiotics less readily available to microorganisms and thus resistant to biodegradation. A discussion regarding whether sorbed xenobiotics can be released and become of toxicological and ecological significance, is of particular concern. The persistence of an organic contaminant is not only due to its ineherent molecular property (in this case the compound should be properly termed recalcitrant), but it is much more often a result of environmental microhabitats that affect both the mass transfer of the xenobiotic to the microorganism and the activity of microorganism itself.

The characteristics and properties of the microhabitats within which soil microbial communities exist and metabolize are dominated by the characteristics and properties of clay minerals and organic matter and of aggregates that are responsible for (the) temporary or permanent sequestration of the chemical and for water retention. Most microbial activity probably occurs less frequently in soil than in habitats wherein water is continuous. The availability of water allows microorganisms to take up nutrients and maintains cell surface integrity. Aggregates and clusters of aggregates with the retained water control the spatial distribution of microbial population and its activity.

In soil, microhabitats are variable in dimension. Their solid and soluble phases are also qualitatively and quantitatively heterogeneous in composition. Microhabitats contain single cells, small colonies, or mixed communities occurring on particulate surfaces and within soil aggregates. Independently on water potential values, many microorganisms associate themselves with solid surfaces. However, van Loosdrecht et al. [44], reviewing many studies conducted particularly in artificial systems, argued that only a very small part of the bacterial surface was in direct contact with adsorbed materials on the soil surface. It follows that for geometric reasons the direct transfer of substrate between bacterial cell and soil surface would be of limited capacity compared to transfer through the aqueous phase. According to these authors, this means that observed differences between adhered and free cells could all be attributed to an indirect feature, because the solid influences the medium rather than the bacterium [44]. Burns [45] considered some experimental facts to support the hypothesis that adhesion and growth of bacteria on surfaces resulted in development of biofilms at solid-liquid interfaces. Biofilms would be the principal sites of microbial activity in soil.

Stotzky [46, 47] discussed the great importance of surfaces to microorganisms and their activities in soil, especially in relation to the presence of clay minerals. How clay minerals affect microbial substrate utilization rates and growth yields is not always clear. Both positive (substrate and/or nutrients concentration, pH buffering, protection from predators) and negative (immobilization of cells, inactivation of enzymes, reduced O_2 availability, surface reactivity towards substrates) effects have been demonstrated [45].

Microorganisms can also occur as aqueous suspensions inside pores of soil aggregates. Pore sizes control the spatial distribution of microbial population. The intraparticle pore volume resides mainly in pores that are less than 100 nm in diameter [48]. Considering that most soil bacteria range from 0.5 to 1.0 μm [49], bacteria will be physically excluded from most of the intraparticle pores. It has been reported that soil bacteria reside primarily in pores with sizes of 0.8-3 μm [7] and that only 6% of inner volume of soil aggregates was occupied by microorganisms [50]. Small pores are saturated by water most of the time, but are subjected to some disadvantages for activity, such as lack of available O_2 and limitation in substrate accessibility.

Various factors affect the structure of microhabitats and the spatial distribution of microbes in soil. The permanent effects of some cultural practices (e.g., cultivation, tillage, irrigation) may be among these factors. In addition, the microorganisms themselves contribute to the building of their own microhabitats, modifying the organization and aggregation levels of soil constituents [51]. Most of the available results have been obtained with model systems; noninvasive measurement techniques at the scale of soil pores are needed to discriminate between microorganisms within soil aggregates and to investigate whether they are active or dormant in such microhabitats.

6. BIOAVAILABILITY OF ADSORBED AND AGED RESIDUES

The preponderence of the experimental evidence suggests that the free bulk water concentration of organic contaminant determines its rate of uptake and consequently its bioavailability. In presence of a contaminated porous system, however, biovailability is modified. Pollutants and microorganisms have a different microscopic distribution. Firstly, pollutants as well as microbes can be adsorbed on soil surfaces. Pollutants adsorbed on surfaces may be subjected to microbial degradation. Indeed the adsorbed pollutant is in equilibrium in trace amounts with the soil solution and thus available for degradation. Moreover, it has not been excluded so far that microorganisms may take up directly surface-sorbed compounds. As a consequence of microbial activity, a steeper concentration gradient will exist near the surface and the actual desorption rates will be greater than the rate of desorption in sterile systems. This "driven" desorption can even be more pronounced if microbial cells are attached to the surface [52]. Direct attachment of microorganisms to soil surfaces leads to biofilms formation at soil-liquid interfacial region that may also influence the extent and rate of contaminant sorption as a result of either physical blockage of sorption sites or creation of diffusional barriers altering the sorption equilibria [53].

Secondly, pollutants diffuse into smaller pores accessible to bacteria too, where, however, slow mass transfer of nutrients and/or organic substrate might limit the extent of microbial growth. Yet, pollutants are also slowly sequestered into extremely small pores inaccessible to even the smallest of bacteria, where the direct contact between microbes and contaminants, which is necessary for biodegradation to occur, is hindered. Increasingly "remote" sites appear to become supplied with pollutants with time. For example, the soil fumigant 1,2-dibromoethane (EDB) was found in agricultural topsoils up to 19 years after its last known application [54]. It was suggested that the residue of EDB was a result of a slow diffusion in remote micropore sites as a consequence of long exposure times and high initial field-applied concentration [54]. Aged residues of simazine, applied for 20 consecutive years, were also shown to be biologically unavailable to sugarbeet and to microbial degraders [55]. Ageing soil-naphthalene slurries (up to 3 yr equilibration) progressively diminished both the rate and extent of naphthalene degradation by *Pseudomonas putida.* Exposure of soil interior sorption sites by sonication of 3 yr old slurries increased naphthalene availability to this organism [56].

Finally, part of the organic contaminant may be present as NAPL. NAPL compounds are by definition immiscible with water. These separate-phase pollutants may be used as a substrate in instances where they break up into fine droplets surrounded by water or begin to dissolve in the aqueous phase [57]. Thus, mass transfer rates have been described in terms of water velocity, the contact area between NAPL and water, and the physical properties of porous medium [58]. Ageing of NAPL in aqueous systems has been also observed. A semigelatinous interfacial film by hydrogen bonding between water and aromatic π-electron in PAHs was thought to have formed [59], and to have undergone a transition to a structured or condensed material, likely changing interfacial mass transfer rates [58].

Bioavailability depends on the time frame in which the above mechanisms are reversible. Before degradation of a target substrate can occur, the substrate must be freely soluble in water, two effects of sorption on biodegradation should be considered: dilution effects and desorption rate effects.

The removal of an organic contaminant from the bulk water phase reduces its concentration in water and at higher K_p values the dissolved fraction at sorption equilibrium decreases. At low substrate concentration insufficient energy and carbon may be available for microbial growth. That is, a threshold concentration may exist below which biomass growth cannot be sustained. At high concentrations, on the other hand, many organic compounds may be inhibitory or toxic to microbial growth. In this case sorption may lower the aqueous phase level of these harmful compounds, thus reducing their toxicity.

Abiotic desorption includes the mass transfer of contaminant from regions inaccessible to organisms and even to extracellular enzymes of microorganisms to biologically accessible regions. These two steps occur sequentially and end with the biological uptake and transformation of contaminant. If the diffusion out of these remote regions is extremely slow, the biodegradation will be governed by the very slow rate of desorption. Some experimental results on the biodegradation of toluene [60] and α-hexachlorocyclohexane [61] in soil slurries and aromatic hydrocarbon in sediments [62] demonstrated that the overall degradation was controlled by the slow rate of release, not by the activity of degrading microorganisms. Nevertheless, desorption of sorbed (but not necessarily aged) substrate may also prolong the availability of the substrate, thus sustaining the microorganisms for a longer period of time [44, 63].

Interestingly enough is the role of humic materials as a possible cause of sequestration. The entrapment of hydrophobic chemicals in pores (nanometer size) in the organic matrix (similar to those in glassy polymers), other than within its solid phase, has been proposed to account for desorptional retardations. Tests carried out with model sorbents on the desorption kinetics of phenanthrene, PCBs, and chlorobenzenes showed that sequestration and reduced bioavailability occured when hydrophobic compounds entered into nanopores in the organic matter or pores coated with organic material. Conversely, the chemicals were more quickly desorbed and metabolized by microorganisms if the pore surfaces were not hydrophobic [64, 65].

Though many investigations suggest that sorption reduces bioavailability, some results of opposite sign appear to show that sorbed compounds may be bioavailable without prior desorption. Some chlorophenols were available for biotransformation in sediments, likely in sorbed state [63], whereas Guerrin and Boyd [66] reported that the utilization of sorbed naphtalene by microorganisms in a soil-water system depended on the type of microorganism present. The relationship of microorganisms with solid surfaces is admittedly complex. Surface attachment may influence cell size, shape, spore formation [67] without necessarily alter their metabolic activities. On the other hand, even the extensive research on the microbial activity in the presence of pure clay systems leads to contradictory results [47] and witnesses the difficulty in studying this activity in even more complex systems such as soil, where pollutants are also physically trapped or fixed by strong chemical bonds to soil particles.

Moreover, desorption itself may require biological intervention (biotic desorption) as microorganisms produce surfactants or other emulsifiers that facilitate the desorption of organic compounds from soil. More recently, attention has been also turned towards the effects of dissolved organic matter (DOM) on the desorption of hydrophobic contaminants from soil in different soil-aqueous solution systems and after prolonged periods of ageing. The decisive factor controlling desorption PAHs was the concentration and sorptive properties of the DOM. Interestingly, the desorption of PAHs from soil was controlled by the time elapsed since contamination. An effect of ageing was found for relatively short periods in which a decrease of the PAH concentration that could be desorbed in the solution phase was observed. No further effect of ageing on PAH desorption was shown, suggesting a fraction of PAH that was not or very slowly desorbing from soil [68].

Direct chemical attachment and incorporation of pollutants to reactive sites on or into colloidal soil components (bound residues) and the consequences on their bioavailabilty will be discussed in the next sections.

The declining bioavailability of aged compounds has practical consequences on the estimation of risk from toxic chemical in contaminated sites. Inasmuch as "the magnitude of reduction in bioavailability resulting from ageing is different for a single compound in different soils, for different compounds in the same soil, and different periods of time that a compound has remained in soil", M. Alexander very recently asked himself how the degree of exposure and risk from an aged compound might be predicted [69]. Current analytical methods measure total and not bioavailable concentrations, thus leading to an overestimation of toxicological risk and to an uncorrected evaluation of polluted sites. The author suggested new assay methods by using chemical or physical assay whose results must correlate well with the results of bioassays with the particular organism of concern [69]. These methods are urgently needed for the further development of legislation to protect the soil.

Whatever the process of adsorption and ageing, much remains to be learned in the coupled biodegradation and sorption processes. Many factors control these processes (Table 1); to untangle them, it would produce reliable estimates of the sorption processes and their consequences for the environment. Rate models that couple sorption/desorption-related mass transfer processes and biodegradation have been used to interpret laboratory results of microbial degradation of sorbed organic solutes in soil or clay suspensions [7, 18, 23, 57, 70, 71]. Shelton and Doherty [70] proposed a model for describing pesticide-substrate bioavailability and rates of biodegradation in soil. The model, verified for 2,4-D degradation, accounted for sorption to soil surfaces, diffusion into the internal matrix of soil organic matter or aggregates, and microbial growth. A study conducted on the effects of adsorption on kinetics of aldicarb degradation [71] indicated that, although degradation of aldicarb occurred primarily in the soil solution, it was non-negligible for sorbed chemicals. By a mathematical approach, it was shown that the relative contributions of the two phases to the total degradation depended on both the adsorption coefficient and the relative degradation rate constants for the dissolved and sorbed chemicals. Specifically Bosma et al. [7] introduced a dimensionless bioavailability number, *Bn*, to quantify the importance of mass transfer relative to intrinsic activity of the cell. By combining, under well-defined conditions, the Michaelis-Menten equation, which describes quantitatively the intrinsic microbial conversion, and mass transfer equations, via pathways including various mechanisms, bioavailability number was obtained as:

$$Bn = \frac{k}{(q_{max}/K_m)}$$

where k represents the mass transfer rate constant and q_{max}/K_m the microbial specific affinity. If a Bn value is less than unity, the microbial degradation is controlled by the mass transfer. Conversely, if a Bn value is greater than unity, microbial degradation is the rate-limiting step.

Table 1
Factors controlling coupled biodegradation and sorption of xenobiotics (summarized from [18])

Microbial population density
Differences among microorganisms
Availability of nutrients and electron acceptors
Distribution (surfaces, pores, aggregates) of xenobiotics in soil
Distribution (surfaces, pores, aggregates) of microorganisms in soil
Sorption partition coefficient
Diffusion rate coefficient
Soil moisture
Ageing and bound residue formation
Predation of biodegrading populations
Competition between biodegrading and other organisms
Production of surfactants or other emulsifies
Binding to soil collodial components

This number has been applied to explain the observed kinetics of the biotransformation of α–hexachloocyclohexane, 3-chlorodibenzofuran [7] and phenanthrene [22] in soils. The Bn values were always significantly smaller than unity, indicative of limitation due to mass transfer, and not to intrinsic microbial activity. This result is expected to be common for highly hydrophobic organic contaminants, owing to their propensity towards sorption and correspondingly retarded rates of diffusion.

Different modeling approaches to predict experimental observations are beginning to offer some insight into the mechanisms and relationships among sorption, desorption, and biodegradation at laboratory scale. However, many questions remain and predictive modeling appears to be elusive, especially in the context of complicated soil-water systems. Most existing models describing sorption, desorption, and mass transfer kinetics are enough sophisticated. Nevertheless, they do not include those chemicals fractions that are irreversibly sorbed to soil (bound residues) and those that have release kinetics so slow (aged residues) that, on a time scale of the process, can be considered to reside in biologically inaccessible sites. On the other hand, the models describing biodegradation and cell growth appear simplistic. Microorganisms are assumed to occur only outside the aggregates and biodegradation is described by the Michaelis-Menten or Monod equations and the expression of first-order kinetics that can be derived from both equations [7, 18, 23]. Other biological elements are in general omitted; for instance, bioaugumentation, biosurfactants production, cell growth, adaptation, and efficiency of substrate utilization. Thus, a more powerful conceptual framework, within which results obtained with more complex systems find a natural place, is needed.

7. ABIOTIC TRANSFORMATIONS OF ORGANIC XENOBIOTICS IN SOIL

Laboratory and field studies, focused on adsorption and ageing processes and conducted with the aim to improve the assessment and subsequent remediation of contaminated sites, often ignore that xenobiotics may become irreversibly bound to soil components with time and may partially transform in soil by combined chemical and biological processes. Conceivably, the bioavailability of transformation products could be different from that of the parent compounds. Partial degradation might lead to chemisorption-reactive daughter products, and xenobiotics and their metabolites may oligomerize or may be sequestered by humic fractions via enzymatic or mineral-surface catalysed oxidative reactions, which ultimately yield more stable and higher molecular weight polymers (nondesorbable and nonbioavailable).

The identification and the mechanisms of the transformation processes in soil are complex and involve a broad range of issues. The soil can be considered as a catalytic system containing a complex mixture of biotic (enzymes, viruses, bacteria, fungi, algae, the fauna) and abiotic components (clay minerals, oxides, hydroxides, noncrystalline components, organic matter) [3]. Two or more of them are often involved in the same reaction and the relative importance of individual processes in natural habitat is very difficult to determine. The additional interaction of these various components creates a heterogeneous and complex system that is difficult to analyze.

Microbial metabolism is believed to be the primary force in xenobiotic transformation and a great deal of laboratory work emphasizes the metabolic activity of single organisms and communities in batch cultures supplied with high concentrations of pure substrates. However, there is a profound difference between laboratory conditions and natural environments like soil, where barriers to movement and diffusion and stressed conditions can bring about unfavorable conditions to the activity of microbial communities. Whenever this occurs, abiotic transformations can dominate the fate of xenobiotics. Even when intense biological activity takes place, however, nonbiological processes, such as photolysis, hydrolysis, oxidation reactions, may still be important.

Abiotic transformation rates are strongly influenced by a wide variety of environmental factors. The chemical structure of the compound determines to some extent its rate of transformation. A set of features contributes to the ease or difficulty of transformation: molecular size, shape, charge, functional groups, solubility in water, lipid/water partition coefficient. Even if a given structure is degradable, further or different substitutions, branching and polymerization of the molecule can lead to greater recalcitrance. In a number of instances, transformation leads to the production of a compound more toxic to the natural community than the original substrate. The effect of a particular toxicant may be temporary.

The persistence of the toxicant and the ability of the microorganisms to adapt to its presence determine the duration of this effect.

Environmental factors control the types and rates of the chemical reactions. These results come out from many studies, generally conducted with pure soil components in controlled and reproducible conditions. This influence is mainly due to the involvement of surfaces in the transformation reactions for which catalytically active adsorption sites must be available. Therefore, soil particle sizes, mineralogy, organic matter content, reflecting the variability of the surfaces, play a significant role in adsorption processes. Most xenobiotics display a strong affinity towards soil organic matter. Nevertheless, the prevalence of clay minerals in the solid phase and their large surface area and high charge make the contribution of the mineral clay fraction to the surface-enhanced abiotic transformations in soils as important as that of the organic matter.

Humus-mineral complexes are also components of soil colloids. The role of mineral-bound humic substances in pollutants sorption is almost unknown. Bound humic materials may alter the surface properties of the sorbent affecting the type and charge of surface functional groups [72] or the relative distribution of hydrophobic sorption sites [73]. For instance, the results of quinoline sorption on kaolinite-humic acid complexes indicated that the mineral-sorbed organic fraction diminished retention of cationic quinoline despite an increase in overall cation-exchange capacity [74].

Some others important factors responsible for the enhancement of transformation processes of xenobiotics are composition of the exchange complex, nature of the exchangeable cations, polarization and dissociation of water molecules in the hydration shell of exchangeable cations, ability of heavy metal cations to act as complexants and carriers of electrons. Soil pH is fundamental in controlling adsorption, ion exchange, redox, complexation, precipitation reactions, all directly or indirectly linked to the rate of many abiotic transformations. Many transformations are pH-catalyzed reactions and the water content determines their rate and kinetics. Soils subjected to fluctuations in water content come under the influence of another variable: the redox potential. Under most conditions, the soil redox potential is controlled by the concentration of O_2 in the gas and liquid phases. Redox reactions in soils cause appreciable redistribution of soluble and insoluble forms of some minerals, the surfaces of many of which are directly involved in reactions with pollutants.

In soils, chemical transformations are accelerated; some occur rapidly, others more slowly and even a simple reaction can involve a series of steps, each of them characterized by a specific rate. The chemical change occurs at the interface between phases, usually a solid-liquid interface. The presence of a diffuse liquid phase, in which many organic and inorganic solutes are solubilized, and of a colloidal fraction, which offers a large surface area to substrate interactions, confers intense catalytic effects to soil reactions.

The concurrent processes of soil catalysis can be summarized as: (a) adsorption of the reactants on the surface of the catalyst; (b) chemical reactions of the activated forms on the surface; (c) desorption of the products of the reaction and their diffusion into homogeneous phase. Any one of the three steps above could be rate controlling. Processes (a) and (c), are diffusion phenomena and they are fundamentally governed by diffusion laws. The result of all this will be an uncertain and less definitive knowledge of the catalysis mechanisms compared to those in homogeneous phase. Reaction kinetics will be controlled by the slowest step in the sequence of chemical and physical events. All these limitations make the kinetics of surface catalysed reactions more difficult to describe by rate equations than the kinetics of homogeneous reactions. The kinetics of reactions at solid-liquid interfaces has been discussed by Mingelgrin and Prost [75] and by Wolfe et al. [76].

7.1. Reactions of xenobiotics involving soil inorganic components

Catalytic effects of clay mineral and metal (hydr)oxide surfaces promote a number of abiotic reactions relevant to the environmental impact of organic xenobiotics. One of the present authors has recently reviewed how catalyst may influence reactions and several examples have been presented to illustrate the most important mechanisms [77]. Adsorption is the primary stage determining the transformation of organic chemicals, followed by chemical reactions of the activated forms on the surface. Hydrolytic, oxidation and polymerization reactions catalysed by clay minerals and metal (hydr)oxides dominate the abiotic transformation of xenobiotics (Table 2). The efficiency of the catalytic process depends on mineral structure, mineral chemistry, and surface reactivity and on structure and properties of the xenobiotic as well as on the reaction conditions.

In studies with clay minerals, to distinguish the effect of the constituent metal ions (Al, Si, Fe) from that of the exchangeable cations, homoionic minerals are often employed. This allows one to develop an understanding of the reaction mechanisms of many environmentally important reactions and to assess the availability of a suitable mineral surface towards a xenobiotic. Both Brönsted acidity and Lewis acidity are involved in the adsorption and surface hydrolysis of xenobiotic compounds [77]. The chemical degradation appears sharply dependent on the polarizing power of exchangeable cations and a number of more soluble products are obtained.

Clays demonstrate the ability to catalyse electron transfer reactions (Table 2, and references therein). The systems that have been investigated most are Fe(III) and Cu(II) smectites. Smectite clays can shuttle electrons between structural Fe(II)/Fe(III) centres or exchangeable Fe(III) and Cu(II) and organic molecules adsorbed on the external and internal surfaces. The most investigated substrates are phenols and aromatic amines; the resulting products include quinones and oligomers of the parent compounds. Chlorophenols, including PCP, dibenzo-*p*-dioxin, and chloroethenes are ubiquitous environmental pollutants, which are dechlorinated and polymerized on Cu(II)-smectite via radical-cation intermediates, yielding products significantly less toxic than the parent compounds [113-116]. Abiotic reduction of nitrobenzenes (TNT) by Fe(III) adsorbed on montomorillonite has also been demonstrated [145].

Catalytic effects of metal (hydr)oxides of high surface area have been known for a long time. Adsorption of xenobiotics by mineral surfaces exposes them to a chemical microenvironment that can exert a profound effect on pathway and rates of transformation. Some important studies describe the susceptibility of carboxylate, thionate, and oxonate esters to surface-catalysed hydrolysis by Al_2O_3, FeOOH and TiO_2 surfaces [104-106]. Various mechanisms cause the accumulation of these chemicals at the oxide/water interface (electrostatic attraction, outer-sphere complexes, inner-sphere chelates), promoting hydrolysis. The magnitude of surface catalysis is significantly different among the three (hydr)oxides examined and different products are observed depending on the pH and dissolved oxygen amount. The formation of 1,2-bis(ethylthio)ethane as a hazardous, persistent hydrolysis product of the insecticide disulfuton was discovered [105].

The role of metal (hydr)oxides, such as Fe(III) and especially Mn(III/IV) oxides, common components of many soils, in catalysing the oxidation of a number of adsorbed xenobiotic organic compounds might be used to achieve the remediation of contaminated environments.

Over the last decade, the mode of the very rapid oxidation of substituted phenols and aromatic amines has been elucidated [77]. The mechanism described involved the quick adsorption of the aromatic substrate onto oxide surface sites by forming an inner or outer sphere complex. Subsequent electron transfer formed the oxidized molecule, in general a phenoxyl or anilino radical, which was then released into solution and further oxidized to dimers and oligomeric oxidation products by oxidative coupling reactions. Finally, the reduced metal was released into solution and reoxidized by O_2 to Mn(III/IV) or Fe(III).

The oxidation of phenols and phenolic acids and the formation of nitrogeneous polymers in phenol-amino acid systems by Mn oxides were considered very active reactions in promoting the formation of humic acids [77]. In parallel, however, detailed research has been devoted to abiotic transformations of aromatic pollutants. Oxidative coupling reactions involving chlorophenols and chloroanilines released chloride ions [135]. Self-coupling was possible with the formation of oligomers insoluble in water; but cross-coupling with other molecules, and especially with humic acids, was responsible for the binding of chlorinated phenols and anilines to humic acids [135, 137]. The dechlorination and oligomerization may be important for detoxification and removal of these compounds from the environment. Likewise, soils themselves induce irreversible reactions of aromatic amines. The results obtained on sorption and abiotic transformation of aromatic amines and α-naphtylamine by various soils gave support to the catalytic role of Mn-minerals in whole soils to cause the start of the oxidation reaction [143, 144].

It is worth mentioning the unique reference by Cheney et al. [127] to birnessite-catalysed degradation of phenoxyacetic herbicides, such as 2,4-D. Experiments conducted showed: i) the solid-state degradation occurred at significant rates in the presence of birnessite; ii) CO_2 was a major product of degradation; iii) the reaction proceeded by 2,4-D-assisted reductive dissolution of birnessite to produce Mn(II).

Recently, the approach of Cheney et al. has been also applied to investigate the short time, surface catalysed degradation of atrazine on birnessite [146, 147]. The products were mainly mono and didealkylatrazine. Mn oxide reduction was a minor reaction, and the alkyl chains were converted mainly to the alkenes, in a non-redox process. *N*-dealkylation is a significant detoxification reaction for *s*-triazines and, until then, has been classified strictly as a biological transformation, with time scales on the order of weeks to months in field soils [146].

Increasing interest in abiotic remediation technologies has led to consider Fenton reaction systems for on-site and in situ treatment. Fenton's reagent, a well-known strong oxidant, catalyses the decomposition of hydrogen peroxide by Fe(II), which results in the generation of hydroxyl radicals and other species such as superoxide and hydroperoxyl radicals [150]. Hydroxyl radicals react with most environmental contaminants via either hydroxylation or hydrogen atom abstraction. Many biorefractory compounds have been effectively degraded by hydroxyl radicals within minutes (Table 2, and references therein). Some environmental applications of Fenton chemistry have some modifications, including Fe(III) chelates [148, 149], iron minerals goethite, hematite, and magnetite [150, 156], pyrophosphate [154].

Table 2
Abiotic reactions of xenobiotics by soil inorganic components

Inorganic component(s)	Chemical(s)	Type of reaction	References
Homoionic kaolinites and bentonites	Parathion, pirimiphos-ethyl	Hydrolysis	78-80
Montmorillonite, kaolinite, attapulgite	Organophosphorous pesticides	Hydrolysis	81, 82
Homoionic bentonites	Chlorbufame, phenylcarbamates, quinalphos	Hydrolysis	83-85
Homoionic hectorite	Azinphosmethyl	Hydrolysis	86, 87
Homoionic montmorillonites	Alachlor, asulam, azinphosmethyl, diclofop methyl, fenarimol, fenoxaprop ethyl, isoxaben, isoxazole, nisulam, phosmet, triazines	Hydrolysis	86-102
Clay minerals	Polydimethylsiloxanes	Hydrolysis	103
Al, Fe, and Ti oxides	Carboxylate esters, phosphate esters, phosphorothionate esters	Hydrolysis	104-106
Hectorite	Benzidines	Oxidation	107, 108
Allophane, homoionic montmorillonites	*o*-, *m*-,*p*-chlorophenols	Oxidation	109
Pillared bentonites	Substituted anisoles	Oxidation	110
Homoionic smectites	Benzene, chloroanisoles, chloroethenes, chlorophenols, dibenzo-*p*-dioxins, phenol, substituted phenols	Oxidation	111-117

Table 2. Continued

Al oxides	Catechol, pyrogallol	Oxidation	118-120
Mn oxides	Aromatic and nonaromatic compounds, 2,4-D, naphthalenediol, phenols, substituted anilines	Oxidation	121-129
Mn and Fe oxides	Chlorophenols, phenolic acids, phenols	Oxidation	130-142
Soils	Aromatic amines, α-naphthylamine	Oxidation	143, 144
Fe(II)-montmorillonite	TNT	Reduction	145
Mn oxide	Atrazine	No net redox reaction	146, 147
H_2O_2/Fe(III) chelates	Atrazine, baygon, carbaryl, 2,4-D picloram, 2,4,5-T, trifluralin	Oxidation	148, 149
H_2O_2/Fe(II) (Fenton's reagent)	Chlorinated aromatic compounds, dieldrin, diesel fuel, hexadecane, methylnaphthalene, PCP, tetrachloethene, trifluralin	Oxidation	150-155
H_2O_2/Fe oxides	PCP	Oxidation	156

Results obtained suggest that naturally occurring iron oxyhydroxides can serve as effective Fenton catalysts [150, 155]. Fenton-like reaction has been also applied to degradation of hydrophobic contaminants. The oxidation and mineralization of sorbed and NAPL hexadecane was investigated using high concentration of hydrogen peroxide and an iron(II) catalyst. More significantly, the results showed that even a highly hydrophobic compound that is sorbed, or in a NAPL-phase, could be oxidized in < 72h [150, 155].

Photocatalysis is presented as a separate class of abiotic catalysis because reactions are only catalysed when the system is illuminated with light of suitable wavelength. We speculate that photocatalysis may play an important role in regulating the concentrations of redox-sensitive compounds in natural water bodies. In this environment the degradation of pesticides is attributed mainly to photolysis by direct sunlight or by sensitized photolysis. Direct photodegradation in solution of various halobenzonitriles [157], chlorophenols [158] and other pesticide groups have been recently reported. Few examples of indirect photoenhanced reactions were the photosensitized reaction of 2,4-D and 2,4,5-T involving H_2O_2 or Fe [159], of atrazine involving low-molecular weight organic compounds [160], and of polydimethylsiloxane, present in a wide range of consumer products, involving nitrate and nitrite ions [103]. Photocatalytic processes using minerals with semiconducting properties, such as titanium oxide (TiO_2), are receiving increasing attention. Extremely efficient degradation processes of bentazon and triazines were shown [161-163].

In the soil compartment, only surfaces exposed to solar radiation can contribute to photodegradation. The effects are limited to very shallow depth and are not easy predictable. Due to the limited number of studies involving photocatalysis in soils there is a paucity of good examples to illustrate the importance of this type of catalysis. Both humic substances and mineral surfaces of soil can potentially provoke phototransformation of organic xenobiotics. Examples were the enhanced phototransformation of the organophosphorous fungicide tolcofos methyl [164], the pyrethroid insecticide esfenvalerate [165, 166], the herbicide metolachlor [167], chlorinated dioxins [168-169]. The role of sunlight in soil xenobiotic degradation remains uncertain. The products obtained are often similar to those derived from other biotic and abiotic processes. For a more detailed discussion the reader can refer to a recent review of one of the present authors [77].

7.2. Reactions of xenobiotics involving soil organic components

Soil organic matter consists of hydrophilic and hydrophobic regions, which results in a sorption of polar and non-polar xenobiotics, thus affecting their toxicity and degradation and their bioavailability. An emerging view for some researchers is that sorption is prevalently due to intraorganic matter diffusion through two distinct domains (rubber-like or glass-like regions) of soil organic matter having different sorption properties (partitioning and hole-filling) [20, 23].

Regardless of the need to assess the heterogeneous distribution of these domains in soil organic matter, where chemicals can be physically sequestered, some xenobiotics can be chemically bound to humic polymers. The presence of reactive functional groups as well as the presence of stable free radicals would indicate that humic polymers could affect transformation of xenobiotics. However, only a limited group of such chemicals appears susceptible to covalent linking processes. Among them, xenobiotic phenols and anilines have been assumed to form covalent linkages with soil humic substances. However, the intricate and heterogeneous structure of humic substances makes it difficult to achieve direct insight into the mechanisms of covalent binding. With the aid of simple models in which the xenobiotics have been allowed to interact with monomeric constituents of humic acids (caffeic, ferulic, gallic, protocatechuic, syringic, vanillic acids, guaiacol, catechol, syringaldheide, resorcinol), it was demonstrated that substituted phenols and aromatic amines were transformed to reactive intermediates. These intermediates could then bind to humic monomers by oxidative coupling (free radical coupling or *o*-quinone coupling) in reactions catalysed by enzymes or birnessite [170-173]. Mass spectra of the resulting products allowed the identification of oligomers obtained.

Recently, some authors have shown that the peroxidase-catalysed oxidation of PCP in the presence of a potential humic precursor such as *p*-coumaric acid can remove up to 90% of PCP. The major part of PCP was incorporated into the polymer from *p*-coumaric acid by the release of a chlorine atom from the PCP molecule to form a diphenylether linkage with the *p*-hydroxy group of the *p*-coumaric acid polymer [174].

Further studies have involved natural humic acids. This new approach has been made possible by the application of ^{13}C- or ^{15}N-labeled chemicals in combination with ^{13}C- or ^{15}N-NMR spectroscopy [38, 173, 175, 176]. Some significant cross-coupling reactions of xenobiotics with humic substances were identified. A free radical reaction between 2,4-dichlorophenol (2,4-DCP) and humic acid, catalysed by horseradish peroxidase, resulted in the formation of C-C and C-O linkages [175]. If the position with the impaired electron on 2,4-DCP molecule was substituted with the chlorine atom, the coupling of such a radical resulted in a release of chloride ions [135, 173]. The dehalogenation is mediated also by an abiotic catalyst (birnessite) and is a crucial step for detoxification and removal of these compounds from the environment. Thorn et al. [176] gave direct spectroscopic evidence for the covalent binding of aniline to the functional groups in humic substances. Aniline underwent nucleophilic addition reactions with the carbonyl functionality of the humic substances and became incorporated in the form of anilinohydroquinone, anilinoquinone, anilide, imine, and heterocyclic nitrogen. In contrast to chlorophenols, aniline did not require the presence of a biotic or abiotic catalyst to mediate the binding [176]. Very recently, Park et al. [137], investigating the transformation of chlorinated phenols and anilines with oxidoreductive catalysts in the presence of humic acid, have shown that 4-chlorophenol (4-CP) was mainly bound to humic acid and formed by self-coupling few oligomers, whereas 4-chloroaniline (4-CA) was largely subjected to oligomerization with less binding to humic acid. The effect of humic acid depended also on its concentration and source.

^{13}C-NMR spectra of soil-bound residues of a mixture of ^{13}C- and ^{14}C-labeled fungicide anilazine and cyprodinil, extracted and purified from soil, indicated that anilazine was abiotically bound to humic acid through the formation of strong ether and ester linkages between the dehalogenated triazine ring to anilazine and the humic matrix [177]. The experiments with cyprodinil showed a cleavage of the cyprodinil molecule between the aromatic rings followed by covalent binding of the separated phenyl and pyrimidyl moieties to humic acid [178]. The resolution of NMR spectra has been greatly improved by silylation of the soil samples and application of ^{13}C-depleted humic material [38, 179, 180]. Silylation involves the reaction between trimethylchlorosilane, or other donors of the trimethylsilyl group, with active hydrogens of functional groups (-OH, =NH, $-NH_2$, -SH, -COOH) present in organic compounds. Upon silylation, significant amounts of soil humus, together with bound residues, become soluble in organic solvents. NMR spectroscopy in combination with silylation is of value not only to the investigation of binding mechanisms, but also to the evaluation of physically sequestered chemicals [38, 180].

Beside the mentioned reactions, considered in the last years the most important for their potential detoxification function in relation to anthropogenic chemicals, other abiotic chemical processes involving xenobiotics and soil organic matter should be considered. The most known reactions were the hydroxylation of the chloro-*s*-triazines as a result of the formation of H-bonding between a ring atom of triazine and a protonated carboxyl group of the humic acid [181], the binding of diquat and paraquat to humic acids by ion exchange mechanisms [182], the incorporation of 2,4-D, chorpropham, and methoxychlor into humic-like polymers [183, 184]. Of concern are the recent results on the effects of humic acids on the oxidative transformation of 1-naphthol, a primary metabolite of naphthalene and carbaryl [185]. The formation of secondary products more reactive than those obtained in the absence of humic acids suggested that weak complexes were formed and that functional groups in humic acids might participate directly in electron-transfer reactions with 1-naphthol. Also in PCB-contaminated soil treated using a chemical decontamination process, PCBs underwent partial dechlorination and degradation, during which organic matter in the soil performed as hydrogen donors [186].

8. ABIOTIC REACTIONS AND BIOAVAILABILITY

There is no doubt whatsoever that microorganisms remain the most environmentally acceptable way of transforming or removing, or even completely mineralizing, xenobiotics, provided the soil habitat supports their growth and activity. Nevertheless, the ability of the soil to retain xenobiotics and to cause chemical reactions, occurring on the active surfaces on mineral particles and humic substances, sheds light on more complex strategies for the bioavailabilty and detoxification of xenobiotics in soil environments. It has become clear that the abiotic reactions responsible for the evolution of the bioavailable fraction of a contaminant in soil are of two main kinds: hydrolytic and oxidation reactions. They will conceivably produce degradation products more soluble than the parent molecules, and bound residues formation including parent and degradation products.

Abiotically catalysed hydrolytic and oxidative reactions can lead to the complete mineralization of xenobiotics. For instance, Cheney et al. [127] demonstrated that a Mn oxide, similar to the oxides that are widespread in soils, degraded the herbicide 2,4-D to CO_2. But more frequently, abiotic reactions lead to structurally simple and/or more oxygenated molecules, which nevertheless maintain their organic nature. Even if complete mineralization of the organic compound is not attained, this does not reduce the importance of the role of abiotic components in detoxification reactions. Minor alterations in the structure of chemical compounds frequently cause a drastic change in the susceptibility of biotransformation. This means that abiotic degradation makes available small molecules that can be utilized as nutrient and energy sources by microorganisms. If these molecules were still toxic, their biological degradation would complete the detoxification process.

So far, however, no studies have systematically examined the fate of transformed products. This gap is in part due to the variety of the chemical structures of the products obtained and to the inability to predict general dissipation patterns. "Daughter" molecules, however, can be sorbed to soil as soon as they come in contact with the surfaces of minerals and organic matter through a variety of chemical processes depending also on their characteristics (*e.g.*, structure, functional groups, size, configuration, solubility). Sorption to soil will affect their availability for microbial degradation. It is worth noting, however, that in the surface-catalysed reactions, bonds between participating groups are formed and broken on clay- or metal oxide- or organic matter-surface and the appearance of reaction products in solution may be delayed or suppressed by slow desorption. The bioavailability of adsorbed molecules is dependent upon the extent of the adsorption and binding of xenobiotics to soil minerals and organic matter.

Some significant results can aid in understanding the importance of the combined abiotic and biotic reactions in detoxification processes. Soil microorganisms are known to partially degrade atrazine and other *s*-triazines by dehalogenation and *N*-dealkylation reactions followed by a complete mineralization of the triazine ring [187-189]. Ring cleavage apparently occurs only after hydroxylation, but the microbial formation of hydroxyatrazine and dealkylated products was considered a relatively slow process [147]. Conversely, photolytic and proton-catalysed degradation of atrazine to hydroxyatrazine on mineral and humic surfaces and its Mn oxide-catalysed *N*-dealkylation were very rapid [146, 147, 160, 181]. A different distribution of products among *N*-dealkylated forms was also observed [146]. Thus, abiotic degradation could shorten atrazine residence times in soil, and could affect biodegradation pathways by altering the nature of the substrates available for microbial metabolism.

Polychlorinated aromatic compounds are dehalogenated before microbial ring cleavage. Biological dehalogenation is considered a very slow reaction, particularly for highly chlorinated phenols, such as PCP. It has been shown that the initial rate-limiting step for the microbes can be bypassed by the addition of chemical oxidant. PCP was degraded in the mineral-silica sand systems with H_2O_2 [150, 151] without iron addition. Naturally occurring iron in the form of amorphous or crystalline iron oxides may supply sufficient iron for Fenton-like reactions. The dechlorination and partial mineralization of PCP was also verified. Thus the biodegradation of PCP could be enhanced by abiotic preliminary oxidation and, conceivably, the resulting partially oxidized products would be more available to microbial consortia than the parent toxic xenobiotic. The Fenton reaction, with or without UV radiation is a viable process to convert many pesticides and other xenobiotics to less toxic or more biodegradable intermediates.

Abiotic surface-catalysed hydrolysis of organophosphorus pesticides significantly affects their fate in the soil environment. Different products, more or less bioavailable and hazardous, can be formed [104, 105]

Sometimes soil abiotic components have opposite effects stabilizing xenobiotics against degradation. For instance, sulfonylureas degrade rapidly in soil, in particular in acidic soils, in which chemical hydrolysis is the primary degradation pathway. The hydrolysis products are then easily degraded by microbial-mediated metabolism [190]. As the main goal of the biological activity and low toxicity of sulfonylureas is a relatively rapid decomposition, results on the formation with clay minerals of stable chelate complexes, which remain adsorbed, appear of great concern, particularly in dry soil conditions [191, 192].

Finally, the following is a significant example of the degradation of a persistent xenobiotic by a combination of abiotic and biotic mechanisms. Polydimethylsiloxane (PDMS) fluids are polymers present in a wide range of household and personal care products. Due to their very low water solubility, they are tightly adsorbed to organic matter during the sewage treatment process. The sewage sludge generated is generally recycled as soil amendment. As a consequence of the high molecular weight, PDMS may be perceived as not biodegradable and such products could present a potential threat to the environment. The combined abiotic and biotic degradation of PDMS has been recently shown. According to the results presented by Stevens [103], the following scheme can be drawn:

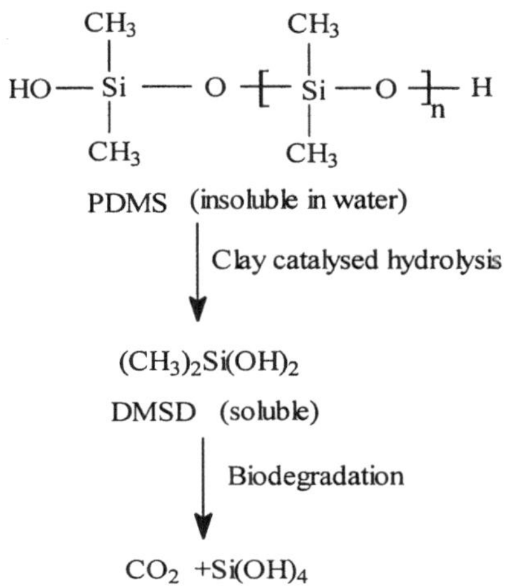

The relationship with water was the crucial step to PDMS degradation in soil. Clay mineral surfaces catalysed the hydrolysis of PDMS in contact with soil. Catalytic activity depended on the surface acidity of the clays and, thus, on the nature of exchangeable cation and the moisture content. A breakdown to a water soluble degradation product, dimethylsilanediol (DMSD) occurred. This transformed a nonbiodegradable compound in a form, which was now biologically available to microorganisms. Two microorganisms, a fungus, *Fusarium oxysporum,* and a bacterium, *Arthrobacter* species, isolated from soils, were shown to metabolize DMSD to silicic acid and carbon dioxide. There was sufficient evidence to indicate that PDMS degradation occurred in soils and that the organisms capable of metabolizing DMSD were fairly widespread [103].

Another question is the decrease of bioavailability and the possible loss of toxicity with ageing of chemicals as a consequence of chemical reactions. Currently, this is a central issue in the study and management of contaminated soils. Investigations on decomposition of xenobiotics show that a large fraction of the residues can be tightly bound to soil. These residues form irreversibly bound compounds over time by covalent bonding or chemical incorporation in humic substances. The discussion regards whether bound residues are retained permanently or can be released and become a long-term threat to the environment. Aged compounds were considered to belong to the category of poorly extractable. The term "bound residues" was proposed to relate to all residues retained in a physical or chemical way in the soil, which were non-extractable without altering their chemical nature [36]. There is no sharp borderline concerning which compounds are available to biology, but it is considered that the probability of bioavailability decreases from the aged to the bound compounds.

To draw a general inference from these assessments is hard to do. Lerch et al. [193], for instance, refuted the current view of covalent binding as the primary mechanism for bound xenobiotic residues in soil. The authors showed that hydroxylated and dealkylated atrazine degradation products were sorbed to soil by a mixed-mode binding mechanism (cation exchange and hydrophobic) occurring simultaneously. Only about 40% of residues were recovered from aged soils. It was therefore conceivable that the soil-bound hydroxylated atrazine degradation products were the primary sink for atrazine in the environment prior to complete mineralization.

Many xenobiotics, either polymerized or co-polymerized with humic substances through biotic or abiotic reactions, are considered to be resistant against microbial remobilization or biodegradation and their bioavailabilty is considered to be small. On the other hand, we have emphasized above the ability of clay minerals and Mn oxides to alter the toxicity of some pollutants, such as chlorophenols and substituted aromatic amines, whose concentration was decreased by polymerization and/or cross-coupling with humic constituents. Nevertheless, it is not clearly known if polymerization is safe and an irreversible pathway for removal of contaminants. Whether the xenobiotic residues, once bound, are stable with respect to microbial action and remain associated with humic matter is uncertain. The bound residues might be released at a later time or might follow the fate of humic matter whose phenolic core is highly recalcitrant, with half-lives measured in centuries. Few studies have examined to date the possibility of the release of bound residues involving microorganisms using organic matter as a nutrient source. Hsu and Bartha [194] and Saxena and Bartha [195] gave evidence relative to the release of radioactivity and evolution of $^{14}CO_2$, at a rate of 1% per week, from ^{14}C-3,4-dichloroaniline-humic acid complex exposed to the activity of microorganisms. Dec and Bollag [196] and Dec et al. [197] incorporated ^{14}C-labeled catechol, 4-chlorophenol, 2,4-dichlorophenol, 2,4,5-trichlorophenol, and pentachlorophenol with synthetic or natural humic acids. After 13 weeks of incubation with microbial soil populations only small amounts of radioactivity were released from the polymers (up to 10% of the initially bound radioactivity as $^{14}CO_2$ and up to 12% as unchanged compound). Most radioactivity (up to 90%) remained bound to humic acids, suggesting that once chlorophenols were incorporated into humus, their release was very slow. Following Dec and Bollag [173], these results suggested that microbial release involved only some *surface* fraction, while the remainder that was incorporated into a *core* humic polymer was essentially inaccessible to microorganisms. At present, the binding of xenobiotics is considered beneficial, since all available data indicate only minimal release of bound residues. However, the formation of bound residues may prevent xenobiotics form being mineralized or transformed into less hazardous byproducts.

The importance of abiotic soil reactions in governing the bioavailability of organic pollutants in soil merits great attention. Coexisting abiotic-biotic reactions, which heretofore have been scarcely considered, appear to be of great concern, also because they may have application in *in situ* and *ex situ* soil remediation.

9. CONCLUSIONS

Many anthropogenic organic compounds that are quickly biodegraded in aqueous solution have been found to persist in the presence of soils. Their fate in soil is governed by an array of non-biological, as well as biological, processes. These include sorption-desorption and chemical and biological transformation, that are perhaps the two most important processes as the bulk of the applied xenobiotics is either sorbed by organic and inorganic soil constituents, or chemically and microbially transformed and degraded. The risk from organic pollutants has been questioned by recent research. The exposure of an organism to a toxic compound in soil is not related to the total concentration of that substance in soil but rather to the amount that is actually available. As a result, methods currently used for determining total concentrations of organic pollutants overestimate exposure, and thus risk, from toxic chemicals in contaminated sites [69].

The bioavailabilty of xenobiotics in soils declines with time and a lot of abiotic processes concur to make them less available for uptake by organisms and for biodegradation and bioremediation by microorganisms.

The chemical properties of the xenobiotic (i.e., structure, solubility, concentration, toxicity) and of the soil (i.e. texture, organic C and clay content, clay mineralogy, CEC, surface area, nanoporosity) and environmental conditions, affecting the fate of the pollutant, as well as of the microbial activity itself, all influence the bioavailability and the ability of the soil to degrade it. Sorption of organic contaminants tends to prevent the direct contact between microorganisms and contaminants, which is necessary for biodegradation to occur. However, microbes and enzymes appear to interact directly with soil particulate determining at mineral and organic surfaces a concentration of substrates and their "predators" that affects the rate and the extent of bioavailabilty.

Ageing may result from the formation of chemical bonds with soil components or physical entrapment of the compound in the soil organic matter or mineral lattice. Depending on the nature of the interaction, sequestered compounds may be released back to soil solution. The formation of stronger bonds with increased residence times in soil can lead to bound residues that are considered nonbioavailable or available at time scales of years or longer.

Many hydrophobic organic contaminants can exist as NAPL or become rapidly and irreversibly associated with humic substances and partially decomposed biomass in soil. Organic matter remains the major accumulation phase of nonpolar organic contaminants. However, not only does the partitioning of contaminant between water and soil vary with the organic carbon content, but the molecular structure, surface area and composition of organic matter may all influence the sorption as well. Consequently, compounds of different chemical classes apparently bind to different portions of the humic matter and this may also affect their bioavailability [198]. The incorporation of toxic residues into soil humus might reduce the movement and the bioavailability of the chemicals. If subsequent release of the bound residue is limited and poses no hazard, the use of binding can be considered as a potentially valuable strategy for the decontamination of polluted soil. The formation of non-extractable bound residues is indeed a significant sink of organic pollutants in soils.

Prediction of biodegradation rates of sorbed xenobiotics (parent and degradation products) is more complex than is the case of readily available molecules. The characteristics of the sorption/desorption and mass transfer processes to biologically accessible regions have major implications for assessments of the availability and toxicity of pollutants in soils. The overall rate of their mineralization or transformation can be limited or even controlled by mass transfer kinetics, not by the activity of the degrading microorganisms. Limits to the biodegradation can be also set by some soil microenvironmental factor (e.g., temperature, pH, redox potential), that limits physiological activity of the appropriate microbial consortia, by some intracellular biochemical or genetic factor (e.g., enzymes or expression of genes), that limits utilization of the compound, and by possible environmental stresses independent of the target compound.

Such a conclusion in regard to abiotic processes also involves the abiotically generated degradation products. "Daughter" molecules will be more soluble and thus more bioavailable, but even more easily sorbed and transformed to the surfaces of soil minerals and organic matter than the parent molecules. Sequestering within clay minerals, humic substances, nonaqueous phase liquids, and inaccessible micropores in the soil aggregates or transforming to irreversibly bound forms are considered to reduce mobility and bioavailabilty of anthropogenic chemicals and their daughter products in soils. The abiotic processes are interrelated and their effects on bioavailability should be studied and understood in concert. Nevertheless, some inferences can be made. Not all organic xenobiotics in soil are available to groundwater and living organisms and those that are, consist of a fraction that is rapidly released and a fraction that is slowly released. This means that new soil pollution is different from old soil pollution and that the mere presence of organic xenobiotics in soils does not translate to unacceptable risk. Managing only the fraction of contaminant rapidly released in soil is an acceptable strategy for contaminated sites.

REFERENCES

1. Torsvik, V.L., Goksoyr, J., Daae, F.L., 1990. High diversity in DNA of soil bacteria. Appl. Environ. Microbiol. 56, 782-787 .
2. Bollag, J.-M., Liu, S.-Y., 1990. Biological transformation processes of pesticides. *In*: Cheng, H.H. (Ed.), Pesticides in the soil environment: processes, impacts, and modelling. Soil Science Society of America, Madison, WI, pp. 169-211.
3. Ruggiero, P., Dec, J., Bollag, J.-M., 1996. Soil as catalytic system. *In*: Stotzky, G., Bollag, J.-M. (Eds.), Soil Biochemistry, Vol. 9, Marcel Dekker Inc., New York, pp.79-122.
4. Zehnder, A.J.B., 1999. Bioremediation of environments contaminated with organic xenobiotics: putting microbial metabolism to work. *In*: Baveye, P., Block, J.-C., Gonchurak, V.V. (Eds.), Bioavailability of organic xenobiotics in the environment, NATO ASI Series 2. Environment , Vol. 64, Kluwer Academic Publishers, Dordrecht, pp.79-92.
5. Alexander, R.R., Alexander, M., 2000. Bioavailability of genotoxic compounds in soils. Environ. Sci. Technol. 34, 1589-1593.
6. Baveye, P., Bladon, R., 1999. Bioavailability of organic xenobiotics in the environment: a critical perspective. *In*: Baveye, P., Block, J.-C., Gonchurak, V.V. (Eds.), Bioavailability of organic xenobiotics in the environment, NATO ASI Series 2. Environment, Vol. 64, Kluwer Academic Publishers, Dordrecht, pp 227-248.
7. Bosma, T.N.P., Middeldorp, P.J.M., Schraa, G., Zehnder, A.J.B., 1997. Mass transfer limitation of biotransformation: quantifying bioavailability. Environ. Sci. Technol. 31, 248-252.
8. Van der Meer, J.R., de Wos, W.M., Harayama, S., Zehnder, A.J.B., 1992. Molecular mechanisms of genetic adaptation to xenobiotic compounds. Microbiol. Rev., 56, 677-694.
9. Baldi, F,. Janssen, D.B., Reineke, W., 1997. Adaptation: dynamics of genes, enzyme activities and populations. *In*: Wild, J.R., Varfolomeyev, S.D., Scozzafava, A. (Eds.), Perspectives in bioremediation, NATO ASI Series 3. High Technology, Vol. 19, Kluwer Academic Publishers, Dordrecht, pp. 93-117.
10. Ogram, A.V., Jessup, R.E., Ou, L.T., Rao, P.S.C., 1985. Effects of sorption on biochemical degradation rates of (2,4-dichlorophenoxy)acetic acid in soils. Appl. Environ. Microbiol. 49, 582-587.

11. Weigand, H., Totsche, K.U., 1998. Flow and reactivity effects on dissolved organic matter transport in soil columns. Soil Sci. Soc. Am. J. 62, 1268-1274.
12. Hayes, M.H.B., Mingelgrin, U., 1991. Interactions between small organic chemicals and soil colloidal constituents. *In*: Bolt, G.H., De Boodt, M.F., Hayes, M.H.B., McBride, M.B. (Eds.), Interactions at the soil colloid-soil solution interface, NATO ASI Series E: applied science,Vol. 190, Kluwer Academic Publishers, Dordrecht, pp. 323-407.
13. Weber, J.B., Best, J.A., Gonese, J.U., 1993. Bioavailability and bioactivity of sorbed organic chemicals. *In*: Sorption and degradation of pesticides and organic chemicals in soil, SSSA Special publication n.32, Soil Science Society of America, Madison, WI, pp.153-196.
14. Senesi, N., Miano, T.M., 1995. The role of abiotic interactions with humic substances on the environmental impact of organic pollutants. *In*: Huang, P.M., Berthelin, J., Bollag, J.-M, McGill, W.B., Page, A.L. (Eds.), Environmental impact of soil component interactions, Vol 1, CRC Lewis Publ., Boca Raton, FL, pp. 311-330.
15. Gennari, M., Gessa, C., 1999. Evaluation of soil adsorption-desorption capacity for the assessment of pesticide bioavailability. *In*: Baveye, P., Block, J.-C., Gonchurak, V.V. (Eds.), Bioavailability of organic xenobiotics in the environment, NATO ASI Series 2. Environment, Vol. 64, Kluwer Academic Publishers, Dordrecht, pp. 207-225.
16. Gevao, B., Semple, K.T., Jones, K.C., 2000. Bound pesticide residues in soils: a review. Environ. Poll. 108, 3-14.
17. Arnold, C.G., Ciani, A., Müller, S.R., Amirbahman, A., Schwarzenbach, R.P., 1998. Association of triorganotin compounds with dissolved humic acids. Environ. Sci. Technol. 32, 2976-2983.
18. Scow, K.M., Johnson, C.R., 1997. Effect of sorption on biodegradation of soil pollutants. Advances in Agronomy 58, 1-56.
19. Conte, P., Piccolo, A., 1999. Conformational arrangement of dissolved humic substances. Influence of solution composition on association of humic molecules. Environ. Sci. Technol. 33, 1682-1690.
20. Xing, B., Pignatello, J.J., 1997. Dual-mode sorption of low-polarity compounds in glassy poly(vinyl chloride) and soil organic matter. Environ. Sci. Technol. 31, 792-799.
21. Pignatello, J.J., 1989. Sorption dynamics of organic compounds in soils and sediments. *In*: Sawhney, B.L., Brown, K. (Eds.), Reaction and movement of organic chemicals in soils, SSSA Special publication n.22, Soil Science Society of America, Madison, WI, pp.45-80.
22. Luecking, A.D., Huang, W., Soderstrom-Schwarz, S., Kim, M., Weber Jr., W.J., 2000. Relationship of soil organic matter characteristics to organic contaminant sequestration and bioavailability. J. Environ. Qual. 29, 317-323.
23. Pignatello, J.J., Xing, B., 1996. Mechanisms of slow sorption of organic chemicals to natural particles. Environ. Sci. Technol. 30, 1-11.
24. Luthy, R.G., Dzombak, D.A., Peters, C.A., Roy, S.B., Ramaswami, A., Nakles, D.V., Nott, B.R., 1994. Remediating tar-contaminated soils at manufactured gas plant sites.Environ. Sci. Technol. 28, 266-277.
25. Ghoshal, S., Ramaswami, A., Luthy, R.G., 1996. Biodegradation of naphthalene from coal tar and heptamethylnonane in mixed batch systems. Environ. Sci. Technol. 30, 1282-1291.

26. Ramaswami, A., Ghoshal, S., Luthy, R.G., 1997. Mass transfer and bioavailability of PAH compounds in coal tar NAPL-slurry systems. 2. Experimental evaluations. Environ. Sci. Technol. 31, 2268-2276.
27. Khan, S.U., 1982. Bound pesticide residues in soils and plant, Residue Rev. 84, 1-25.
28. Calderbank, A., 1989. The occurrence and significance of bound pesticide residues in soil. Rev. Environ. Contam. Toxicol. 108, 71-103.
29. Bollag, J.-M., Dec, J., Huang, P.M., 1998. Formation mechanisms of complex organic structures in soil habitats. Adv. Agron., 63, 237-266.
30. Hatzinger, P.B., Alexander, M., 1995. Effect of aging of chemicals in soil on their biodegradability and extractability. Environ. Sci. Technol. 29, 537-545.
31. Chung, N., Alexander, M., 1998. Differences in sequestration and bioavailability of organic compounds aged in dissimilar soils. Environ. Sci. Technol. 32, 855-860.
32. Radosevich, M., Traina, S.J. Tuovinen, O.H., 1997. Atrazine mineralization in laboratory-aged soil microcosms inoculated with *s*-triazine-degrading bacteria. J. Environ. Qual. 26, 206-214.
33. Tang, J., Carroquino, M.J., Robertson, B.K., Alexander, M., 1998. Combined effect of sequestration and bioremediation in reducing the bioavailability of polycyclic aromatic hydrocarbons in soil. Environ. Sci. Technol. 32, 3586-3590.
34. Dinelli, G., Vicari, A., Accinelli, C., 1998. Degradation and side effects of three sulfonylurea herbicides in soil. J. Environ. Qual. 27, 1459-1464.
35. Nicholls, P.H., Evans, A.A., 1985. Adsorption and movement in soils of chlorsulfuron and other weak acids, Proceedings 1985 British Crop Protection Conference-Weeds, pp. 333-339.
36. Alexander, M., 1995. How toxic are toxic chemicals in soil? Environ. Sci. Technol. 29, 2713-2717.
37. Burgos, W.D., Novak, J.T., Berry, D.F., 1996. Reversible sorption and irreversible binding of naphthalene and α-naphthol to soil: elucidation of processes. Environ. Sci. Technol. 30, 1205-1211.
38. Dec, J., Bollag, J.-M., 1997. Determination of covalent and noncovalent binding interactions between xenobiotic chemicals and soil. Soil Sci. 162, 858-874.
39. Kelsey, J.W., Alexander, M., 1997. Declining bioavailability and inappropriate estimation of risk of persistent compounds. Environ. Toxicol. Chem. 16, 582-585.
40. Dumestre, A., McBride, M. B., Baveye, P. 2000. Use of EPR to monitor the distribution and availability of organic xenobiotics in model soil systems. Environ. Sci. Technol. 34, 1259-1264.
41. Spagnuolo, M., Baveye, P., McBride, M. B., Ruggiero, P., 2000. EPR study of partitioning and availability of a hydrophobic spin probe in model soil systems: long-term sorption experiments. *In*: Violante A., Gianfreda, L. (Eds.), Soil Mineral-Organic Matter-Microorganism Interactions and Ecosystem Health. Abstracts, 3^{d} Int. Symp. of the IUSS Working Group MO, Naples-Capri, Italy, p. 62.
42. Gillette, J. S., Luthy, R.G., Clemett, S.J., Zare, R.N., 1999. Direct observation of polycyclic aromatic hydrocarbons on the geosorbents at the subparticle scale. Environ. Sci. Technol. 33, 1185-1192.
43. Ghosh, U., Gillette, J.S., Luthy, R.G., Clemett, S.J., Zare, R.N., 2000. Microscale location, characterization, and association of polycyclic aromatic hydrocarbons on harbor sediment particles. Environ. Sci. Technol. 34, 1729-1736.

44. van Loosdrecht, M.C.M., Lyklema, J., Norde, W. Zehnder, A.J.B., 1990. Influence of interfaces on microbial activity. Microb. Rev. 54, 75-87.
45. Burns, R.G., 1989. Microbial and enzymatic activities in soil biofilms. *In*: Characklis, W.G., Widerer, P.A. (Eds.), Structure and function of biofilms, John Wiley and Sons, Chichester, pp 333-349.
46. Stotzky, G., 1997. Soil as an environment for microbial life. *In*: van Elsas, J.D., Trevors, J.T., Wellington, E.M.H. (Eds.), Modern soil microbiology, Marcel Dekker Inc., New York, pp. 1-20.
47. Stotzky, G., 1986. Influence of soil mineral colloids on metabolic processes, growth, adhesion, and ecology of microbes and viruses. *In*: Huang, P.M., Schnitzer, M. (Eds.), Interaction of soil minerals with natural organics and microbes, SSSA Special publication n. 17, Soil Science Society of America, Madison, WI, pp. 305-428.
48. Chung, N., Alexander, M., 1999. Relationship between nanoporosity and other properties of soil. Soil Sci. 164, 726-730.
49. Alexander, M., 1977. Introduction to soil microbiology. 2nd ed., John Wiley and Sons New York.
50. Kilbertus, G., 1980. Etudes des microhabitats contenus dans les agrégats du sol. Leur relation avec la biomasse bactérienne présente et la taille des procaryotes. Rev. Ecol. Biol. Sol. 17, 543-557.
51. Guckert, A., Breisch, H., Reisinger, O.,1975. Interface sol-racine. I. Etude au microscope électronique des relations mucigel-argiles-microorganismes. Scil Biol. Biochem. 7, 241-250.
52. Verstraete, W., Devliegher, W., 1996. Formation of non bioavailable organic residues in soil: perspectives for site remediation. Biodegradation 7, 471-485.
53. Rao, P.S.C., Bellin, C.A., Lee, L.S., 1995. Sorption and biodegradation of organic contaminants in soils: conceptual representations of process coupling. *In*: Huang, P.M., Berthelin, J., Bollag, J.-M., McGill, W.B., Page, A.L. (Eds.), Environmental impact of soil component interactions, Vol 1, CRC Lewis Publ.,Boca Raton, FL, pp. 263-274.
54. Steinberg, S.M., Pignatello, J.J., Sawhney, B.L., 1987. Persistence 1,2-dibromoethane in soils: entrapment in intraparticle micropores. Environ. Sci. Technol. 21, 1201-1208.
55. Scribner, S.L., Benzing, T.R., Sun, S., Boyd, S.A., 1992. Desorption and bioavailability of aged simazine residues in soil from a continuous corn field. J. Environ. Qual. 21, 115-120.
56. Guerin, W.F., Boyd, S.A., 1997. Bioavailability of naphthalene associated with natural and synthetic sorbents. Wat. Res. 31, 1504-1512.
57. Mihelcic, J.R., Lueking, D.R., Mitzell, R.J., Stapleton, J. M., 1993. Bioavailability of sorbed- and separate-phase chemicals. Biodegradation 4, 141-153.
58. Luthy, R.G., Aiken, G.R., Brusseau, M.L., Cunningham, S.D., Gschwend, P.M., Pignatello, J.J., Reinhard, M., Traina, S.J., Weber Jr., W.J., Westall, J.C., 1997. Sequestration of hydrophobic organic contaminants by geosorbents. Environ. Sci. Technol. 31, 3341-3347.
59. Nelson, E.C., Ghoshal, S., Edwards, J.C., Marsh, G.X., Luthy, R.G., 1996. Chemical characterization of coal tar-water interfacial films. Environ. Sci. Technol. 30, 1014-1022.
60. Robinson, K.G., Farmer, W.S., Novak, J.T., 1990. Availability of sorbed toluene in soils for biodegradation by acclimated bacteria. Wat. Res. 24, 345-350.
61. Rijnaarts, H.H.M., Bachmann, A., Jumelet, J.C., Zehnder, A.J.B., 1990. Effect of desorption and intraparticle mass transfer on the aerobic biomineralization of α-

hexachlorocyclohexane in a contaminated calcareous soil. Environ. Sci. Technol. 24, 1349-1354.

62. Zhang, W., Bouwer, E., Wilson, L., Durant, N., 1995. Biotransformation of aromatic hydrocarbons in subsurface biofilms. Wat. Sci. Technol. 31, 1-14.
63. Remberger, M., Allard, A.S., Neilson, A.H., 1986. Biotransformations of chloroguaicols, chlorocatechols, and chloroveratroles in sediments. Appl. Environ. Microbiol. 51, 552-558.
64. Nam, K., Alexander, M., 1998. Role of nanoporosity and hydrophobicity in sequestration and bioavailabilty: tests with model solids. Environ. Sci. Technol. 32, 71-74.
65. Cornelissen, G., van Noort, P.C.M., Govers, H.A.J., 1998. Mechanism of slow desorption of organic compounds from sediments: a study using model sorbents. Environ. Sci. Technol. 32, 3124-3131.
66. Guerin W.F., Boyd, S.A., 1992. Differential bioavailability of soil-sorbed naphthalene to two bacterial species. Appl. Environ. Microbiol. 58, 1142-1152.
67. Fletcher, M., 1985. Effect of solid surfaces on the activity of attached bacteria. *In*: Savage, D.C., Fletcher, M. (Eds.), Bacterial Adhesion, Plenum Press, New York, pp. 339-362.
68. Kögel-Knabner, I., Totsche, K.U., Raber, B., 2000. Desorption of polycyclic aromatic hydrocarbons from soil in the presence of dissolved organic matter: effect of solution composition and aging. J. Environ. Qual. 29, 906-916.
69. Alexander, M., 2000. Aging, bioavailability, and overstimation of risk from environmental pollutants. Environ. Sci. Technol. 34, 4259-4265.
70. Shelton, D.R., Doherty, M.A., 1997. A model describing pesticide bioavailability and biodegradation in soil. Soil Sci. Soc. Am. J. 61, 1078-1084.
71. Guo, L., Wagenet, R.J., Jury, W.A., 1999. Adsorption effects on kinetics of aldicarb degradation: equilibrium model and application to incubation and transport experiments. Soil Sci. Soc. Am. J. 63, 1637-1644.
72. Kretzschmar, R., Hesterberg, D., Sticher, H., 1997. Effects of adsorbed humic acid on surface charge and flocculation of kaolinite. Soil Sci. Soc. Am. J. 61, 101-108.
73. Schlautman, M.A., Morgan, J.J., 1993. Binding of a fluorescent hydrophobic organic probe by dissolved humic substances and organically-coated aluminium oxide surfaces. Environ. Sci. Technol. 27, 2523-2532.
74. Chorover, J., Amistadi, M.K., Burgos, W.D., Hatcher, P.G., 1999. Quinoline sorption on kaolinite-humic acid complexes. Soil Sci. Soc. Am. J. 63, 850-857.
75. Mingelgrin, U., Prost, R., 1989. Surface interactions of toxic organic chemicals with minerals. *In*: Gerstl, Z., Chen, Y., Mingelgrin, U., Yaron, B., (Eds.). Toxic Organic Chemicals in Porous Media, Spinger-Verlag, New York, pp. 91-135.
76. Wolfe, N.L., Mingelgrin, U., Miller, G.C., 1990. Abiotic transformations in water, sediments, and soil. *In*: Cheng, H.H. (Ed.), Pesticides in the Soil Environment: Processes, Impacts, and Modelling, SSSA Book Series: 2, Soil Science Society of America, Madison, WI, pp. 103-168.
77. Ruggiero, P., 1999. Abiotic transformation of organic xenobiotics in soils: a compounding factor in the assessment of bioavailability. *In*: Baveye, P., Block, J.-C., Gonchurak, V.V., (Eds.), Bioavailability of organic xenobiotics in the environment, NATO ASI Series 2. Environment, Vol. 64, Kluwer Academic Publishers, Dordrecht, pp. 159-205.

78. Mingelgrin, U., Saltzman, S., 1979. Surface reactions of parathion on clays. Clays Clay Min. 27, 72-78.
79. Saltzman, S., Yaron, B., Mingelgrin, U., 1974. The surface catalyzed hydrolysis of parathion on kaolinite. Soil Sci. Soc. Am. Proc. 38, 231-234.
80. Mingelgrin, U., Gerstl, Z., Yaron, B., 1975. Pirimiphos ethyl-clay surface interactions. Soil Sci. Soc. Am. Proc. 39, 834-837.
81. Mingelgrin, U., Saltzman, S., Yaron, B., 1977. A possible model for the surface-induced hydrolysis of organophosphorus pesticides on kaolinite clays. Soil Sci. Soc. Am. J. 41, 519-523.
82. Yaron, B., 1978. Some aspects of surface interactions of clays with organophosphorus pesticides. Soil Sci. 125, 210-216.
83. Sabadie, J., Bastide, J., Coste, C.M., 1985. Dégradation catalitique de carbamates herbicides déposés sur bentonites homoioniques. II. Cas du chlorbufame. Weed Res. 25, 121-127.
84. Sabadie, J., Coste, C.M., 1988. Dégradation catalytique de carbamates herbicides déposés sur bentonites homoioniques. VI. Cas du N-phenylcarbamate de propyne-2 yle. Weed Res. 28, 13-18.
85. Pusino, A., Gessa, C., Kozlowski, H., 1988. Catalytic hydrolysis of quinalphos on homoionic clays, Pest. Sci. 24, 1-8.
86. Sánchez Camazano, M., Sánchez Martín, M.J., 1991. Hydrolysis of azinphosmethyl induced by the surface of smectites, Clays Clay Min. 39, 609-613.
87. Sánchez Camazano, M., Sánchez Martín, M.J., 1984. Aspects of the adsorption of azinphos-methyl by smectites, J. Agr. Food Chem. 32, 720-725.
88. Sánchez Camazano, M., Sánchez Martín, M.J., 1983. Montmorillonite catalyzed hydrolysis of phosmet, Soil Sci. 136, 89-93.
89. Skipper, H.D., Volk, V.V., Mortland, M.M., Raman, K.V., 1978. Hydrolysis of atrazine on soil colloids, Weed Sci. 26, 46-51.
90. Nguyen, T.T., 1986. Infrared spectroscopic study of the formamide-Na-montmorillonite complex. Conversion of *s*-triazine to formamide, Clays Clay Min. 34, 521-528.
91. Brown, C.B., White, J.L., 1969. Reactions of 1,2-*s*-triazines with soil clays, Soil Sci. Soc. Am. Proc. 33, 863-867.
92. White, J.L., 1975/76. Determination of susceptibility of *s*-triazine herbicides to protonation and hydrolysis by mineral surfaces, Arch. Environ. Contam. Toxicol. 3, 461-469.
93. Fusi, P., Ristori, G.G., Malquori, A., 1980. Montmorillonite-asulam interactions: I. Catalytic decomposition of asulam adsorbed on H- and Al-clay. Clay Min. 15, 147-155.
94. Ristori, G.G., Fusi, P., Franci, M., 1981. Montmorillonite-asulam interactions: II. Catalytic decomposition of asulam adsorbed on Mg-, Ba-, Ca-, Li-, Na-, K- and Cs-clay. Clay Min. 16, 125-137.
95. Fusi, P., Ristori, G.G., Franci, M., 1982. Adsorption and catalytic decomposition of 4-nitrobenzenesulphonylmethylcarbamate by smectite. Clays Clay Min. 30, 306-310.
96. Fusi, P., Ristori, G.G., Cecconi, S., Franci, M., 1983. Adsorption and degradation of fenarimol on montmorillonite. Clays Clay Min. 31, 312-314.
97. Fusi, P., Franci, M., Bosetto, M., Calamai, L., 1990. Isoxazole complexes with homoionic montmorillonite. Agrochimica 34, 103-110.
98. Fusi, P., Franci, M., Ristori, G.G., 1989. Interactions of isoxaben with montmorillonite. Appl. Clay Sci. 4, 235-245.

99. Bosetto, M., Arfaioli, P., Fusi, P., 1993. Interactions of alachlor with homoionic montmorillonites. Soil Sci. 155, 105-113.
100. Pusino, A., Gessa, C., 1990. Catalytic hydrolysis of diclofop-methyl on Ca-, Na- and K-montmorillonites. Pest. Sci. 30, 211-216.
101. Pusino, A., Micera, G., Gessa, C., Petretto, S., 1989. Interaction of diclofop and diclofop-methyl with Al^{3+}-, Fe^{3+}-, and Cu^{2+}-saturated montmorillonite. Clays Clay Min. 37, 558-562.
102. Pusino, A., Petretto, S., Gessa, C., 1996. Montmorillonite surface-catalyzed hydrolysis of fenoxaprop-ethyl. J. Agr. Food Chem. 44, 1150-1154.
103. Stevens, C., 1998. Environmental degradatiopn pathways for the breakdown of polydimethylsiloxanes. J. Inorg. Biochem. 69, 203-207.
104. Torrents, A., Stone, A.T., 1994. Oxide surface-catalyzed hydrolysis of carboxylate esters and phosphorothioate esters. Soil Sci. Soc. Am. J. 58, 738-745.
105. Smolen, J.M., Stone, A.T., 1998. Metal (hydr)oxide surface-catalyzed hydrolysis of chlorpyrifos-methyl, chlorpyrifos-methyl oxon, and paraoxon. Soil Sci. Soc. Am. J. 62, 636-643.
106. Dannenberg, A., Pehkonen, S.O., 1998. Investigation of the heterogeneously catalyzed hydrolysis of organophosphorus pesticides. J. Agr. Food Chem. 46, 325-334.
107. McBride, M.B., 1979. Reactivity of adsorbed and structural iron in hectorite as indicated by oxidation of benzidine. Clays Clay Min. 27, 224-230.
108. McBride, M.B., 1985. Surface reactions of 3,3',5,5'-tetramethyl benzidine on hectorite. Clays Clay Min.33, 510-516.
109. Soma, Y., Soma, M., 1989. Formation of hydroxydibenzofurans from chlorophenols adsorbed on Fe-ion exchanged montmorillonite. Chemosphere 18, 1895-1902.
110. Carrado, K.A., Hayatsu, R., Botto, R.E., Winans, R.E., 1990. Reactivity of anisoles on clay and pillared clay surfaces. Clays Clay Min. 38, 250-256.
111. Mortland, M.M., Halloran, L.J., 1976. Polymerization of aromatic molecules on smectite. Soil Sci. Soc. Am. J. 40, 367-370.
112. Sawhney, B.L., 1985. Vapor-phase sorption and polymerization of phenols by smectite in air and nitrogen. Clays Clay Min. 33, 123-127.
113. Sawhney, B.L., Kozloski, R.K., Isaacson, P.J., Gent, P.N., 1984. Polymerization of 2,6-dimethylphenol on smectite surfaces. Clays Clay Min. 32, 108-114.
114. Boyd, S.A., Mortland, M.M., 1986. Radical formation and polymerization of chlorophenols and chloroanisole on copper(II)-smectite. Environ. Sci. Technol. 20, 1056-1058.
115. Govindaraj, N., Mortland, M.M., Boyd, S.A., 1987. Single electron transfer mechanism of oxidative dechlorination of 4-chloroanisole on copper(II)-smectite. Environ. Sci. Technol. 21, 1119-1123.
116. Boyd, S.A., Mortland, M.M., 1985. Dioxin radical formation and polymerization on Cu(II)-smectite. Nature 316, 532-535.
117. Mortland, M.M., Boyd, S.A., 1989. Polymerization and dechlorination of chloroethenes on Cu(II)-smectite via radical-cation intermediates. Environ. Sci. Technol. 23, 223-227.
118. McBride, M.B., Wesselink, L.G., 1988. Chemisorption of catechol on gibbsite, boehmite, and noncrystalline alumina surfaces. Environ. Sci. Technol. 22, 703-708.
119. McBride, M.B., Sikora, F.J., Wesselink, L.G., 1988. Complexation and catalyzed oxidative polymerization of catechol by aluminum in acidic solution. Soil Sc. Soc. Am. J. 52, 985-993.

120. Wang, T.S.C., Wang, M.-C., Huang, P.M., 1983. Catalytic synthesis of humic substances by using aluminas as catalysts. Soil Sci. 136, 226-230.
121. Laha, S., Luthy, R.G., 1990. Oxidation of primary aromatic amines by manganese dioxide. Environ. Sci. Technol. 24, 363-373.
122. Adrian, P., Lahaniatis, E.S., Andreux, F., Mansour, M., Scheunert, I., Korte, F., 1989. Reaction of the soil pollutant 4-chloroaniline with the humic acid monomer catechol. Chemosphere 18, 1599-1609.
123. Whelan, G., Sims, R.C., 1995. Mn-catalyzed oxidation of naphtalenediol. Hazard. Waste Hazard. Mat. 12, 381-394.
124. Ukrainczyk, L., McBride, M.B., 1992. Oxidation of phenol in acidic aqueous suspension of manganese oxides. Clays Clay Min. 40, 157-166.
125. Stone, A.T., Morgan, J.J., 1984. Reduction and dissolution of manganese(III) and manganese(IV) oxides by organics. 1. Reaction with hydroquinone. Environ. Sci. Technol. 18, 450-456.
126. Stone, A.T., Morgan, J.J., 1984. Reduction and dissolution of manganese(III) and manganese(IV) oxides by organics: 2. Survey of the reactivity of organics. Environ. Sci. Technol. 18, 617-624.
127. Cheney, M.A., Sposito, G., McGrath, A.E., Criddle, R.S., 1996. Abiotic degradation of 2,4-D (dichlorophenoxyacetic acid) on synthetic birnessite: a calorespirometric method. Coll. Surf. 107, 131-140.
128. Klausen, J., Haderlein, S.B., Schwarzenbach, R.P., 1997. Oxidation of substituted anilines by aqueous MnO_2: effect of co-solutes on initial and quasi-steady-state kinetics. Environ. Sci. Technol. 31, 2642-2649.
129. Majcher, E.H., Chorover, J., Bollag, J.-M., Huang, P.M., 2000. Evolution of CO_2 during birnessite-induced oxidation of ^{14}C-labeled catechol. Soil Sc. Soc. Am. J. 64, 157-163.
130. Shindo, H., Huang, P.M., 1982. Role of Mn(IV) oxide in abiotic formation of humic substances in the environment. Nature 298, 363-365.
131. Shindo, H., Huang, P.M., 1984. Catalytic effects of manganese(IV), iron(III), aluminum, and silicon oxides on the formation of phenolic polymers. Soil Sci. Soc. Am. J. 48, 927-934.
132. Wang, T.S.C., Huang, P.M., Chou, C.-H., Chen, J.-H., 1986. The role of soil minerals in the abiotic polymerization of phenolic compounds and formation of humic substances. *In*: Huang, P.M., Schnitzer, M. (Eds.), Interactions of Soil Minerals with Natural Organic and Microbes, SSSA Special Publication No.17, Soil Science Society of America, Madison, WI, pp. 251-281.
133. Lehmann, R.G., Cheng, H.H., 1988. Reactivity of phenolic acids in soil and formation of oxidation products. Soil Sci. Soc. Am. J. 52, 1304-1309.
134. Lehmann, R.G., Cheng, H.H., Harsh, J.B., 1987. Oxidation of phenolic acids by soil iron and manganese oxides. Soil Sci. Soc. Am. J. 51, 352-356.
135. Dec, J., Bollag, J.-M., 1994. Dehalogenation of chlorinated phenols during oxidative coupling. Environ. Sci. Technol. 28, 484-490.
136. Pizzigallo, M.D.R., Ruggiero, P., Spagnuolo, M., 1998. Oxidation of pentachlorophenol and dinitro-orthocresol by manganese and iron oxides. Fresenius Environ. Bull. 7, 552-557.
137. Park, J.-W, Dec, J., Kim, J.-E., Bollag, J.-M., 2000. Transformation of chlorinated phenols and anilines in the presence of humic acid. J. Environ. Qual. 29, 214-220.

138. Pizzigallo, M.D.R., Ruggiero, P., 1992. Catalytic transformation of chlorophenols by manganese oxides. Fresenius Environ. Bull. 1, 428-433.
139. Pizzigallo, M.D.R., Ruggiero, P., Crecchio, C., Mininni, R., 1995. Manganese and iron oxides as reactants for oxidation of chlorophenols. Soil Sci. Soc. Am. J. 59, 444-452.
140. Ukrainczyk, L., McBride, M.B., 1993 Oxidation and dechlorination of chlorophenols in dilute aqueous suspension of manganese oxides: reaction products. Environ. Toxicol. Chem. 12, 2015-2022.
141. Ukrainczyk, L., McBride, M.B., 1993. The oxidative dechlorination reaction of 2,4,6-trichlorophenol in dilute aqueous suspension of manganese oxides. Environ. Toxicol. Chem. 12, 2005-2014.
142. Ulrich, H.-J., Stone, A.T., 1989. Oxidation of chlorophenols adsorbed to manganese oxide surfaces. Environ. Sci. Technol. 23, 421-428.
143. Li, H. Lee, L. S., Jafvert, C.T., Graveel, J.G., 2000. Effect of substitution on irreversible binding and transformation of aromatic amines with soils in aqueous systems. Environ. Sci. Technol. 34, 3674-3680.
144. Li, H., Lee, L. S., 1999. Sorption and abiotic transformation of aniline and α–naphthylamine by surface soils. Environ. Sci. Technol. 33, 1864-1870.
145. Brannon, J.M., Price, C.B., Hayes, C., 1998. Abiotic transformation of TNT in montmorillonite and soil suspensions under reducing conditions. Chemosphere 36, 1453-1462.
146. Cheney, M.A., Shin, J.Y., Crowley, D.E., Alvey, S., Malengreau, N., Sposito, G., 1998. Atrazine dealkylation on a manganese oxide surface. Coll. Surf. 137, 267-273.
147. Wang, D., Shin, J.Y., Cheney, M.A., Sposito, G., Spiro, T.G., 1999. Manganese dioxide as a catalyst for oxygen-independent atrazine dealkylation. Environ. Sci. Technol. 33, 3160-3165.
148. Sun, Y., Pignatello, J.J., 1992. Chemical treatment of pesticides wastes. Evaluation of Fe(III) chelates for catalytic hydrogen peroxide oxidation of 2,4-D at circumneutral pH. J. Agr. Food Chem. 40, 322-327.
149. Sun, Y., Pignatello, J.J., 1993. Activation of hydrogen peroxide by Fe(III) chelates for abiotic degradation of herbicides and insecticides in water. J. Agr. Food Chem. 41, 308-312.
150. Tyre, B.W., Watts, R.J., Miller, G.C., 1991. Treatment of four biorefractory contaminants in soils using catalyzed hydrogen peroxide. J. Environ. Qual. 20, 832-838.
151. Lee, S.H., Carberry, J.B., 1992. Biodegradation of PCP enhanced by chemical oxidation pre-treatment. Water Environ. Res. 64, 682-690.
152. Koyama, O., Kamagata, Y., Nakamura, K., 1994. Degradation of chlorinated aromatics by Fenton oxidation and methanogenic digester sludge. Water Res. 28, 895-899.
153. Chen, C.T., Tafuri, A.N., Rahman, M., Foerst, M.B., 1998 Chemical oxidation treatment of petroleum contaminated soil using Fenton's reagent. J. Environ. Sci. Health A 33, 987-1008.
154. Wang, X., Brusseau, M.L., 1998. Effect of pyrophosphate on the dechlorination of tetrachloroethene by the Fenton reaction. Environ. Toxicol. Chem. 17, 1689-1694.
155. Watts, R.J., Stanton, P.C., 1999. Mineralization of sorbed and NAPL-phase hexadecane by catalyzed hydrogen peroxide. Water Res. 33, 1405-1414.
156. Watts, R.J., Udell, M.D., Kong, S., Leungh, S.W., 1999. Fenton-like soil remediation catalyzed by naturally occurring iron minerals. Environ. Engin. Sci. 16, 93-103.

157. Millet, M., Palm, W.-U., Zetzsch, C. 1998. Abiotic degradation of halobenzonitriles: investigation of the photolysis in solution. Ecotox. Environ. Saf. 41, 44-50.
158. Svenson, A., Hynning, P.-A., 1997. Increased aquatic toxicity following photolytic conversion of an organochlorine pollutant. Chemosphere 34, 1685-1692.
159. Pignatello, J.J., 1992. Dark and photoassisted Fe^{3+}catalyzed degradation of chlorophenoxy herbicides by hydrogen peroxide. Environ. Sci. Technol. 26, 944-951.
160. Hapeman, C.J., Bilboulian, S., Anderson, B.G., Torrents, A., 1998. Structural influences of low-molecular-weight dissolved organic carbon mimics on the photolytic fate of atrazine. Environ. Toxicol. Chem. 17, 975-981.
161. Chiarenzelli, J.R., Scrudato, R.J., Rafferty, D.E., Wunderlich, M.L., Roberts, R.N., Pagano, J.J., Yates, M., 1995. Photocatalytic degradation of simulated pesticide rinsates in water and water plus matrices. Chemosphere 30, 173-185.
162. Pelizzetti, E., Maurino, V., Minero, C., Carlin, C., Pramauro, V., Zerbinati, O., Tosato, M.L, 1990. Photocatalytic degradation of atrazine and other s-triazine herbicides. Environ. Sci. Technol. 24, 1559-1565.
163. Pelizzetti, E., Maurino, V., Minero, C., Zerbinati, O., Borgarello, E., 1989. Photocatalytic degradation of bentazon by TiO_2 particles. Chemosphere 18, 1437-1445.
164. Katagi, T., 1990. Photoinduced oxidation of the organophosphorus fungicide tolclofos-methyl on clay minerals. J. Agr. Food Chem. 38, 1595-1600.
165. Katagi, T., 1991. Photodegradation of the pyrethroid insecticide esfenvalerate on soil, clay minerals and humic acid surfaces, J. Agr. Food Chem. 39, 1351-1356.
166. Katagi, T., 1993. Photodegradation of esfenvalerate in clay suspensions. J. Agr. Food Chem. 41, 2178-2183.
167. Mathew, R., Khan, S.U., 1996. Photodegradation of metolachlor in water in the presence of soil mineral and organic constituents. J. Agr. Food Chem. 44, 3996-4000.
168. Kleatiwong, S., Nguyen, L.V., Hebert, V.R., Hackett, M., Miller, G.C., 1990. Photolysis of chlorinated dioxins in organic solvents and on soils. Environ. Sci. Technol. 24, 1575-1580.
169. Miller, R.G., Hebert, V.R., Millie, M.J., Mitzel, R., Zepp, R.G., 1989. Photolysis of octachlorodibenzyl-*p*-dioxin on soils: Production of 2,3,7,8-TCDD. Chemosphere 18, 1265-1274.
170. Berry, D.F., Boyd, S.A., 1985. Reaction rates of phenolic humus constituents and anilines during cross-coupling. Soil Biol. Biochem. 17, 631-636.
171. Dec, J., Bollag, J.-M., 1996. Effect of various factors on dehalogenation of chlorinated phenols and anilines during oxidative coupling. Environ. Sci. Technol. 29, 657-663.
172. Park, J.-W., Dec, J. Kim, J.-E., Bollag, J.-M., 1999. Effect of humic constituents on the transformation of chlorinated phenols and anilines in the presence of oxidoreductive enzymes or birnessite. Environ. Sci. Technol. 33, 2028-2034.
173. Dec, J., Bollag, J.-M., 2000. Phenoloxidase-mediated interactions of phenols and anilines with humic materials. J. Environ. Qual. 29, 665-676.
174. Morimoto, K., Tatsumi, K., Kuroda, K.-I., 2000. Peroxidase catalyzed co-polymerization of pentachlorophenol and a potential humic precursor. Soil Biol. Biochem. 32, 1071-1077.
175. Hatcher, P.G., Bortiatynsky, J.M., Minard, R.D., Dec, J., Bollag, J.-M., 1993. Use of high resolution ^{13}C NMR to examine the enzymatic covalent binding of ^{13}C-labeled 2,4-dichlorophenol with humic substances. Environ. Sci. Technol. 27, 2098-2103.

176. Thorn, K.A., Pettigrew, P.J., Goldenberg, W.S., 1996. Covalent binding aniline to humic substances. 2. ^{15}N NMR studies of nucleophilic addition reactions. Environ. Sci. Technol. 30, 2764-2775.
177. Haider, K., Spiteller, M., Wais, A., Fild, M., 1993. Evaluation of the binding mechanism of anilazine and its metabolites in soil organic matter. Int. J. Environ. Anal. Chem. 53, 125-137.
178. Dec, J., Haider, K., Benesi, A., Rangaswamy, V., Schäffer, A., Plücken, U., Bollag, J.-M., 1997. Analysis of soil-bound residues of ^{13}C-labeled fungicide cyprodinil by NMR spectroscopy. Environ. Sci. Technol. 31, 1128-1135.
179. Dec, J., Haider, K., Schäffer, A., Fernandes, E., Bollag, J.-M., 1997. Use of a silylation procedure and ^{13}C-NMR spectroscopy to characterize bound and sequestered residues of cyprodinil in soil. Environ. Sci. Technol. 31, 2991-2997.
180. Haider, K., Spiteller, M., Dec, J., Schäffer, A., 2000. Silylation of soil organic matter extraction of humic compounds and soil-bound residues. *In*: Bollag, J.-M., Stotzky, G. (Eds.), Soil Biochemistry, Vol. 10, Marcel Dekker Inc., New York, pp. 139-170.
181. Amstrong, D.E., Konrad, J.G., 1974. Nonbiological degradation of pesticides. *In*: Guenzi W.D. (Ed.), Pesticides in Soil and Water, Soil Science Society of America, Madison, WI, pp. 123-130.
182. Khan, S.U., 1974. Adsorption of bipyridylium herbicides by humic acid. J. Environ. Qual. 3, 202-206.
183. Mathur, S.P., Morley, H.V., 1978. Incorporation of methoxichlor-^{14}C in model humic acids prepared from hydroquinone. Bull. Environ. Contam. Toxicol. 20, 268-274.
184. Wolf, D.C., Martin, J.P., 1976. Decomposition of fungal mycelia and humic-type polymers containing carbon-14 from ring and side-chain labelled 2,4-D and chlorpropham. Soil Sci. Soc. Am. Proc. 40, 700-704.
185. Karthikeyan, K.G., Chorover, J., 2000. Effects of solution chemistry on the oxidative transformation of 1-naphthol and its complexation with humic acid. Environ. Sci. Technol. 34, 2939-2946.
186. Taniguchi, S., Murakami, A., Hosomi, M., Miyamura, A., Uchida, R., 1997. Chemical decontamination of PCB-contaminated soil. Chemosphere 34, 1631-1637.
187. Mandelbaum, R.T., Wackett, L.P., Allan, D.L., 1993. Rapid hydrolysis of atrazine to hydroxyatrazine by soil bacteria. Environ. Sci. Technol. 27, 1943-1946.
188. Mandelbaum, R.T., Allan, D.L., Wackett, L.P., 1995. Isolation and characterization of a *Pseudomonas* sp. that mineralizes the *s*-triazine herbicide atrazine. Appl. Environ. Microbiol. 61, 1451-1457.
189. Radosevich, M., Traina, S.J., Hao, Y.L., Tuovinen, O. H., 1995. Degradation and mineralization of atrazine by a soil bacterial isolate. Appl. Environ. Microbiol. 61, 297-302.
190. Bichat, F., Sims, G.K., Mulvaney, R.L., 1999. Microbial utilization of heterocyclic nitrogen from atrazine. Soil Sci. Soc. Am. J. 63, 100-110.
191. Pantani, O., Pusino, A., Calamai, L., Gessa, C., Fusi, P., 1996. Adsorption and degradation of rimsulfuron on Al hectorite. J. Agric. Food Chem. 44, 617-621.
192. Calamai, L., Pantani, O., Pusino, A., Gessa, C., Fusi, P., 1997. Interaction of rimsulfuron with smectites. Clays Clay Min. 45, 23-27.
193. Lerch, R.N., Thurman, E.M., Kruger, E.L., 1997. Mixed-mode sorption of hydroxylated atrazine degradation products to soil: a mechanism for bound residue. Environ. Sci. Technol. 31, 1539-1546.

194. Hsu, T.-S., Bartha, R., 1974. Biodegradation of chloroaniline-humus complexes in soil and culture solution. Soil Sci. 118, 213-220.
195. Saxena, A., Bartha, R., 1983. Microbial mineralization of humic acid-3,4-dichloroaniline complexes. Soil Biol.Biochem. 15, 59-62.
196. Dec. J., Bollag, J.-M., 1988. Microbial release and degradation of catechol and chlorophenols bound to synthetic humic acid. Soil Sci. Soc. Am. J. 52, 1366-1371.
197. Dec, J., Shuttleworth, K.L., Bollag, J.-M., 1990. Microbial release of 2,4-dichlorophenol bound to humic acid or incorporated during humification. J. Environ. Qual. 19, 546-551.
198. Kukkonen, J., Landrum, P.F., 1996. Distribution of organic carbon and organic xenobiotics among different particle-size fractions in sediments. Chemosphere 32, 1063-1076.

Developments in Soil Science, Volume 28B
Editors: A. Violante, P.M. Huang, J.-M. Bollag and L. Gianfreda

INTERACTIONS OF MINERALS-ORGANIC MATTER-LIVING ORGANISMS ON THE FATE OF ALLELOCHEMICALS AND XENOBIOTICS IN SOIL: A METHODOLOGICAL EVALUATION

H. H. Cheng[a] and W. C. Koskinen[b]

[a]Department of Soil, Water, & Climate, University of Minnesota, St. Paul, Minnesota 55108, U.S.A.

[b]USDA-Agriculture Research Service, St. Paul, Minnesota 55108, U.S.A.

Allelochemicals are secondary metabolites exuded by plants to protect themselves from adverse environmental conditions and from invasion of surrounding plants and microbes. These chemicals are, by nature, biologically reactive. They are often exuded from roots into the surrounding soil (i.e., the rhizosphere), and they interact actively with both the mineral and organic components of the soil. Because of their reactivity, they could be ideal compounds for characterizing the dynamics of the active phase of soil organic matter transformation and the processes of carbon immobilization (or sequestration) and humification. However, the high reactivity of these chemicals also means that they are unstable and could be readily degraded or transformed into other types of chemicals. Thus, even though the biological phenomenon of allelopathy has been reported extensively in the literature, e.g. [1], the identity of most of the chemicals involved in allelopathy (i.e., allelochemicals) still has not been determined with certainty. Thus, before allelochemicals can serve as model chemicals for studying the interaction of organic chemicals with minerals, organic matter, and living organisms, there is need to understand the behavior and fate of these chemicals in the soil.

1. GENERAL CONSIDERATIONS

To ascertain the identity of allelochemicals in the soil environment, it is necessary to first establish the cause-and-effect relationship between the observed allelopathy and the chemicals involved in causing the observed plant responses. As Cheng [2] pointed out, although many chemicals have been suspected to be allelopathic, the evidence for establishing them as allelochemicals has been circumstantial and may not meet the criteria for establishing the cause-and-effect relationship. Currently, there is strong evidence that when allelopathy is observed, the host plant has been found to be capable of exuding specific chemicals from their roots and when the roots of affected plants in the surroundings are exposed to these chemicals, they display stunted growth or other adverse symptoms characteristic of allelopathy. However, this evidence needs to be further verified, so that the chemical identified as being exuded from the roots of the host plant is actually the chemical that caused allelopathic symptom of affected plants [3].

Based on these considerations, Cheng [2] specified criteria that are needed to establish the cause-and-effect relationship of allelopathy that defines the identity of allelochemicals in the soil environment. Only when the chemical, which is exuded from the roots, is transported across the soil medium without being degraded and reaches the roots of affected plants in sufficient amount over sufficient time to cause plant damage, would it be termed "allelochemical". Because of the elusive nature of these chemicals, they have posed a great methodological challenge to researchers wishing to study the processes affecting their fate and transport in the soil environment. However, there are similarities between the processes affecting the allelochemicals and those affecting such xenobiotics as pesticides, of which a great deal are known, as a result much can be learned from understanding the methodology used for characterizing the behavior of pesticides in the soil environment [4].

2. COMPARISON OF ALLELOCHEMICALS AND XENOBIOTICS

How the process pathways on the fate and transport of pesticides and allelochemicals in the soil environment parallel each other is graphically depicted in Figure 1. Furthermore, chemicals that have been suspected to possess allelopathic properties have similar characteristics to many synthetic pesticides and xenobiotics. A comparison of these characteristics is summarized in Table 1. Since many xenobiotics, especially pesticides, have been studied extensively, these chemicals can serve as useful model compounds for characterization of the fate of allelochemicals in the soil environment.

Table 1
A comparison of the characteristics of allelochemicals and xenobiotics.

Allelochemicals	Xenobiotics
Organic chemicals of biological origin	Organic chemicals of mostly chemical origin
Bioreactive	Bioreactive
Chemical structure mostly unknown	Chemical structure mostly known
Mostly water-soluble	Solubilities vary
In the rhizosphere	Where-ever applied

Before the knowledge base on pesticides can be applied to characterize the fate and behavior of allelochemicals in soil, a better understanding of the influence of soil components on the behavior of allelochemicals and xenobiotics will be important. On the other hand, research and analytical methods used to study the processes involved in the transport and transformation of allelochemicals in the soil must be carefully scrutinized. Many of the methods developed for characterizing the fate of xenobiotics have been applied to allelochemical studies. While these cross-overs in methodology adaptation have been useful, there is also need to appreciate the a)

b)

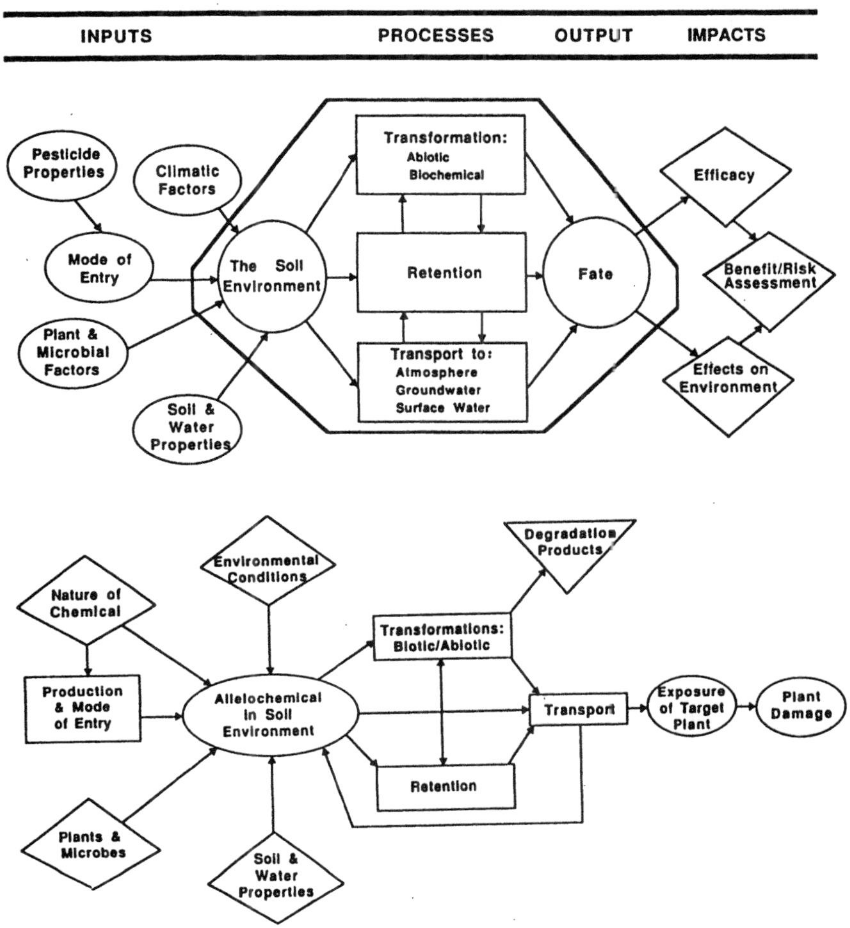

Figure 1. Processes governing the fate and transport of (a) pesticides (after Cheng [4]. Reproduced with permission. Soil Science Society of America); and (b) allelochemicals (after Cheng [5]. Reproduced with permission. Academia Sinica, Taipei, ROC) in the soil environment.

differences between the allelochemicals and the xenobiotics in relation to their interaction with the mineral and organic components of soil, especially in the presence of living organisms. As experience with studying the behavior of xenobiotics in the soil environment indicates, unawareness of many interactions between allelochemicals (or xenobiotics and other similar chemicals) and soil minerals, organic matter, and living organisms can lead to spurious experimental results which, can easily be misinterpreted leading to erroneous conclusions.

3. METHODOLOGICAL CONSIDERATIONS

Allelochemicals and xenobiotics are both subject to the same factors affecting the processes of their transport and fate. Figure 1 shows a conceptual model with the inputs, processes, and ultimate fate of allelochemicals and pesticides. While all chemicals are subjected to the same general processes, the specific pathways each chemical follows can be significantly different. These differences would in turn require different methodological considerations for studying their overall fate. For instance, consider the mode of entry of the chemical into the soil-water environment: Pesticides are usually applied to soil one or a few times per year, which result in less than parts-per-million levels in the surface 15 cm of soil.

In contrast, allelochemicals enter the soil more or less continually as root exudates or plant and debris leachates in conjunction with other organic molecules and inorganic ions, including plant and microorganism nutrients throughout the growing season. Also the sphere of influence of allelochemicals is limited to the rhizosphere and the effective scale for studying their fate is in terms of millimeters.

3.1. Sources

Root exudates may account for as much soil organic carbon as the root mass itself [6-8]. Root exudates consist of a variety of organic chemicals including terpenoids, alkaloids, and phenolic acids. Phenolic acids are also the building blocks of lignin and are released when lignin is degraded [9, 10]. The most common phenolic acids found in soil are ferulic, *p*-coumaric, vanillic, *p*-hydroxybenozic, caffeic, and salicylic acids [11]. Several studies have shown that phenolic acids can reduce plant growth, water utilization, hydraulic conductivity, and nutrient uptake [12-15]. The extent to which these occur is directly related to the amount of root system in contact with the phenolic acids[11].

Since root contact is the primary way receiving plants are potentially affected by phenolic acids and pesticides, understanding the mobility and persistence of these chemicals in the soil is a key in managing their environmental fate. Mobility and persistence are directly or indirectly controlled by retention (sorption-desorption) and transformation (degradation or decomposition) processes in soil. In order to characterize and quantify these processes, special attention must be paid to the methods used in the experiments, even to the point of how soils were collected and stored prior to use in their characterization and how the chemicals are extracted and analyzed.

3.2. Sample Handling

Sample handling has been shown to affect the chemical and biological properties of soils. For instance, changes in predominant microbial species and losses of metabolically active soil microbial biomass of up to 20% occur within 3 to 4 months in soil stored at 2 to 4 °C, while storage at room temperature for 30 days results in changes in the dominance of microbial species in soil and losses of up to 70% of microbial biomass [16]. Carbon mineralization and pesticide degradation studies using frozen, and then thawed soils, give comparable data to those using freshly collected soil, however nitrogen transformation data tend to be erratic. Also, in a study on the influence of air-drying of soils on sorption and abiotic oxidation of phenolic acids, after addition to soil, more ferulic and vanillic acid was found in solution of air-dried soil than field moist soil [17]. The increases in phenolic acid levels in air-dried soils were attributed to a decrease in abiotic soil oxidative activity.

3.3. Extraction

Prior to characterizing processes affecting allelochemicals and pesticides in soil we must have methods to extract them from soil for analysis. Polar solvents based on aqueous methanol and aqueous acetonitrile have been used routinely to extract pesticides from soil that had chemicals applied up to 18 months previously [18]. Small amounts of acid or base added to polar solvents can increase extraction efficiency for acidic and basic pesticides. For instance, acetonitrile:water:ammonium hydroxide has been used successfully to extract clopyralid, picloram, and phenoxypropanoic acids, while acetonitrile:water:acetic acid has been used to extract aliphatic and benzoic acids, benzonitriles, diclofop acid, dinitroanilines, sulfonylureas, and thiocarbamates [18].

While there does not appear to be a universal extraction solvent, if the mechanism of pesticide binding to the soil is known, a solvent system can be designed to optimize the extraction of the pesticide residue from soil [19]. For instance, in contrast to the above basic mixed solvent system previously mentioned for picloram extraction from soil, 2M KCl also has been used to extract picloram with little dissolution of soil organic matter [20]. This approach can be applied to extraction of allelochemicals.

Aqueous soil extracts have been used to determine the soluble fraction of phenolic acids in soils while a variety of extractants, most commonly EDTA, have been used to extract the sorbed phenolic fraction [21, 22]. More recently, it was reported that citrate was superior to EDTA [23]. However, the Folin and Ciocalteu's reactive-phenolic carbon concentrations in citrate extracts of a phenolic acid mixture-amended soils were not significantly different from water control soil extracts, indicating that the commonly used Folin and Ciocalteu's method is not suitable with soil citrate extractions [24]. The large quantity of organic matter extracted by citrate would probably mask absorbance due to any extracted phenolic compounds resulting in overestimation of the phenolic acids. Alternative methods for phenolic acid extraction have also met with mixed success. For instance, supercritical fluid extraction has been used with mixed results to extract phenolic acids from soil minerals [25, 26].

To extract phenolics from soil, complex solvents are necessary because complex binding mechanisms are involved. An aqueous mixed solvent system consisting of ascorbic acid, citric acid, and acetone has proved to be an effective extractant for phenolic compounds [27]. It was assumed that ascorbic acid helped maintain the soil under reduced conditions to minimize oxidation reactions; citric acid, a stronger metal complexing agent than the phenolic compounds, helped break the metal-phenolic complexes; and acetone lowered the hydrogen bonding tendencies. These reactions in concert with one another improved the phenolic compound extraction.

3.4. Retention

Retention (sorption-desorption) of allelochemicals, pesticides, and related organic chemicals by soil is arguably the most important process affecting the fate of these chemicals in the environment. It controls the amount of chemical in solution, which would be available for transport to the site of action and for transformation. Although there have been few methods available to characterize specific bonds or mechanisms involved in retention, there have been a variety of methods to characterize the overall magnitude and strength of binding to soil.

Customarily, studies on the characterization of organic chemical sorption-desorption in soil are carried out using the batch equilibration method, where the amount of chemical sorbed to soil is assumed to be that removed from the solution phase. The solution concentration in batch slurry sorption-desorption studies, including aqueous $CaCl_2$ extractable, has been considered as 'readily available' [28] for uptake by soil microorganisms and plants and for transport. However, losses of both pesticides and allelochemicals from solution can also occur by volatilization, chemical decomposition, microbial degradation, or sorption on the glassware during the sorption-desorption equilibrations. For instance, Koskinen et al. [29] demonstrated that 2,4,5-T was degraded during desorption equilibrations. Cecchi [30] found that *p*-coumaric acid sorbed to glass centrifuge tubes during sorption-desorption equilibrations.

Errors resulting from the loss of chemical by transformation during equilibration can result in under- or over-estimation of sorption depending on the method of measurement of loss of chemical from solution. If chemicals were analyzed by usual chromatographic methods, calculating sorption-desorption using the traditional batch method could result in an overestimation of the amount of chemicals sorbed to soil. If ^{14}C-labeled chemicals and the loss of ^{14}C from solution were measured in order to calculate sorbed chemical concentration, the error would depend on the location of the ^{14}C-label, the mechanism of degradation, and the sorbability of the breakdown product.

A phenolic degradation product of 2,4,5-T was found to be sorbed to a greater extent than 2,4,5-T [31]. Sorption of hydroxyatrazine to soil was greater than atrazine sorption [32]. In contrast, sorption of the hydroxylated metabolite of imazethapyr was less than imazethapyr in a range of soils [33]. If ring-labeled-^{14}C chemicals were used to characterize sorption-desorption and transformation occurred, in the case of 2,4,5-T and atrazine, sorption would be overestimated because the ^{14}C-labeled degradate would be sorbed, while in the case of imazethapyr, sorption would be underestimated because the ^{14}C-labeled degradate would remain in solution.

Assuming the transformation of the allelochemical or pesticide during equilibration could be quantified, the question would arise as to how to correct the sorption-desorption of the chemical for the amount of chemical transformed. The correction can be made to either the sorbed phase of the solution phase, which would result in different sorption-desorption isotherms. The validity of correcting either phase assumes that the rate of transformation is either much faster or much slower than the rate of desorption, which may or may not be the case. It would be better to prevent transformation of the chemical, if at all possible. For instance, those chemicals that are aerobically degraded, sorption-desorption could be characterized under anaerobic conditions [34].

The characterization of sorption-desorption using the batch equilibration method typically results in a sorption coefficient to characterize both processes. This assumes that the allelochemical or pesticide sorption-desorption is a reversible and equilibrium process, independent of time. However, removal of chemical from the solution phase can result in a desorption isotherm that is different from the sorption isotherm, usually in the direction of higher retention of the chemical by the soil than would be predicted by the sorption isotherm. This effect is known as hysteresis. The availability of the chemicals for transport or biological uptake by soil microorganisms and plants would be overestimated if hysteresis is not accounted for in sorption-desorption characterization. The overestimation would depend on the degree of hysteresis exhibited by the chemicals. A variety of classes of pesticides have exhibited hysteresis during desorption from soil. The observation of weak acid pesticides such as derivatives of benzoic, phenoxy (2,4,5-T) [29], and picolinic (picloram) [35] acids exhibiting hysteresis is surprising for these chemicals are often only slightly sorbed by soil.

In contrast, allelochemicals, such as phenolic acids, have been shown to be extensively sorbed to soil; >70% of five different cinnamic and benzoic acid derivatives were sorbed to seven soils with a range in physical and chemical soil properties [30]. Reversibility of sorption of these chemicals may depend on the soil type. For instance, in one study ferulic and *p*-coumaric acids did not desorb from seven different soils [30]. But, in another study Blum [36] showed that >80% of applied did desorb from soil using aqueous solution. However, the amount of desorbed phenolics may have been overestimated by the method used for extraction and analysis. The lack of aqueous desorbability of the phenolic acids indicate that the chemicals may have the potential of residing in soil while remaining unavailable to soil microorganisms and plants. This is in contrast to observations from sorption studies, which would indicate that these chemicals are available by assuming sorption to be a reversible and equilibrium reaction.

3.5. Transformation

Pesticides are susceptible to a variety of transformations in the environment, including both microbial metabolism and chemical decomposition. Studies indicate that for many pesticides, microbial metabolism is the dominant transformation process, although the influence of plants to chemical transformation should not be ignored, e.g., see [37]. A wide variety of pesticide-degrading microorganisms have been identified, indicating the extraordinary metabolic diversity of microorganisms, as well as the wide variety in pesticide structural chemistry. The transformations can vary from minor alterations to complete mineralization to CO_2 and H_2O. The majority of pesticides, although susceptible to enzymatic transformations, are cometabolized by microorganisms, i.e. they are not utilized as sole sources of C, N, and/or energy for growth. The majority of pesticides are susceptible to some form of attack by oxidative enzymes, which result in ether cleavage, aryl-hydroxylation, aryl dealkylation, and sulfoxidation [38]. A number of pesticides are susceptible to reductive reactions, particularly reduction of nitro groups and reductive dechlorination. Many pesticides possess bonds that are susceptible to hydrolytic attack, such as hydrolysis of carboxylic acid esters, carbamates, and organophosphates.

Spatial and temporal variability in the rates of biodegradation of specific pesticides is frequently observed because of differences in numbers of degrading organisms, pesticide availability (which is controlled by sorption-desorption processes), and soil parameters such as temperature, moisture, and pH (which affect the activity of the degrading organisms). Pesticides vary greatly in persistence; those with half-lives <10 days are considered relatively nonpersistent, while those with half-lives >100 days are considered relatively persistent.

In contrast to research on pesticide degradation, very little work has been done on degradation of allelochemicals in soil. As opposed to the case with pesticides, which have known structures and are applied to soil at known rates, neither the structures of many allelochemicals in field soil are known, nor is the concentration of the chemicals in soil known. For instance, dehydromatricaria ester (DME) is an allelochemical produced by *Solidago altissima* L., which inhibits growth of rice seedlings when present at 100 $\mu g\ g^{-1}$ in soil [39]. However, laboratory studies have shown that it degrades rapidly and DME has only been found at levels up to 4.5 $\mu g\ g^{-1}$ in field soil. As a result DME released by *Solidago altissima* L. in actuality would not be allelopathic. Also, most mechanistic research on pesticide degradation used ^{14}C-labeled pesticides, which facilitated analyses and calculation of mass balances of applied chemical. The lack of available ^{14}C-labeled allelochemicals has limited mechanistic research on these chemicals.

Numerous factors must be considered if degradation of allelochemicals is to be characterized. Many of these factors have already been studied to characterize pesticide degradation. As mentioned previously, pesticides are usually applied to soil one or a few times per year, which result in ppm levels in the surface 15 cm of soil, whereas allelochemicals enter the soil throughout the growing season. Repeated applications of pesticides have been shown to stimulate microbial degraders, i.e. to induce production of degrading enzymes, resulting in accelerated degradation of the pesticide. Repeated application/exudation of low concentrations of allelochemicals could also stimulate microbial degraders resulting in accelerated degradation of these chemicals. Although, this effect has not been reported, it must be evaluated if kinetics of degradation can be determined.

Effects of concentration of allelochemicals on soil processes must be determined. Since these chemicals are generally strongly sorbed, high concentrations in soil may be necessary to maintain enough chemical in solution to observe degradation taking place. If the soil concentration in solution is too low, the chemical may not be available to the degrading organisms. If the solution concentration is too low, there may not be enough in solution to induce degrading enzymes, whereas if the solution concentration is too high, the chemicals may be toxic to the degrading organism.

Most pesticides are applied to bare soil. Therefore in laboratory studies, adding pesticides to soil can reflect to a degree what happens in the field. In contrast, allelochemicals are exuded from plant roots or released from decomposing plant materials along with other organic chemicals and nutrients, which can be used as growth substrates for degrading organisms. This in turn would result in faster degradation than would be the case if the allelochemical were applied to soil, as in the pesticide degradation experiments.

In soil, phenolic acids may be degraded by microorganisms by conversion to smaller molecules and CO_2, similar to pesticides. However, they can also be converted into biomass [40] and transformed by extracellular enzymes to small dimers and ogliomers of the phenolic acids [41]. Lehmann et al. [42] showed that phenolic acids reacted with soil Fe and Mn oxides in an oxidation-reduction reaction and that the reaction was primarily an abiotic process. Soluble Fe(II) and Mn (II) were formed by a reduction reaction while the phenolic acids were oxidized. Oxidized products of reactions of phenolic acids with MnO_2 were rapidly sorbed to the MnO_2 surfaces and became less extractable by acidified organic solvents except after acid or alkaline hydrolysis treatment [43]. Characterization of these abiotic processes requires the use of ^{14}C-labeled pesticides.

While use of ^{14}C-labeled allelochemicals facilitates elucidation of mechanisms, as has been the case with pesticides, they must be used with caution. Measurement of the rate of degradation depends on the location of the label in relation to the mechanism of degradation. For instance, $^{14}CO_2$ evolution from side-chain labeled 2,4,5-T was much faster than from ring-labeled chemical [29]. The mechanism of degradation was for organisms to mineralize the side chain first leaving a metabolite with the ring intact, which was strongly sorbed to sorbed to soil.

4. ADDITIONAL CONSIDERATIONS

The purpose of the above discussion on methodological considerations for studying the fate of allelochemicals in the soil environment is not intended to be an exhaustive treatise. Only a few of the methodological considerations were discussed here. However, by pointing out some of the common pitfalls, it is hoped that this paper will help researchers new to the field to become aware of many such pitfalls, which could lead to erroneous results in their research. This is especially true for those researchers who are well versed in physico-chemical approaches and prefer to use them to tackle problems of biological or biochemically-involved systems. While attempting to define the chemical identity of such systems, it is essential that experimental approaches must respect the biological integrity of these systems.

Furthermore, because of the complexity of such systems, characterization of the components of such systems is often carried out at different scales. When data from these component studies are pulled together in synthesizing the systems as a whole, data generated at different scales can also lead to erroneous results. This is especially true for allelochemical studies and for study of interactions at the inorganic-organic interfaces in the soil.

REFERENCES

1. Rice, E.L., 1984. Allelopathy. Second Edition. Academic Press, Inc. Orlando, FL.
2. Cheng, H.H., 1992. A conceptual framework for assessing allelochemicals in the soil environment. *In*: Rizvi, S.J.H., Rizvi, V. (Eds.), Allelopathy: Basic and Applied Aspects. Chapman & Hall, London. pp. 21-29.
3. Fisher, R.F., 1979. Allelopathy. *In*: Horsfall, J.G., Cowling, E.B. (Eds.), Plant disease: An Advanced treatise. Vol. IV. Academic Press, New York. pp. 313-330.
4. Cheng, H.H., 1990. Pesticides in the soil environment: An overview. *In*: Cheng, H.H. (Ed.) Pesticides in the soil environment: Processes, impacts, and modeling. SSSA Book Ser. No. 2. Soil Science Society of America, Madison, WI. pp. 1-5.
5. Cheng, H.H., 1989. Assessment of the fate and transport of allelochemicals in the soil. *In*: Chou, C.H., Waller, G.R. (Eds.) Phytochemical ecology: Allelochemicals, mycotoxins, and insect pheromones and allomones. Institute of Botany, Academia Sinica Monogr. Ser. No. 9, Taipei. pp. 209-216.
6. Helal, H.M., Sauerbeck, D.R., 1986. Effect of plant roots on carbon metabolism of soil microbial biomass. Z. Pflanzenernaehr. Bodenk. 149:181-188.
7. Merchx, R.A., Dijkstra, D.M., den Hartog, A., van Veen. J.A., 1987. Production of root derived material and associated microbial growth in soil at different nutrient levels. Biol. Fert. Soils 5, 126-132.
8. Bottner, P., Pansu, M., Sallih, Z., 1999. Modeling the effect of active roots on soil organic matter turnover. Plant Soil 216, 15-25.
9. Flaig, W. 1966. The chemistry of humic substance. *In*: The use of isotopes in soil organic matter studies. Suppl. J. Appl. Radioisotop., Pergamon Press, Oxford. pp. 103-127.
10. Siqueira, J.O., Nair, M.G., Hammerschmidt, R., Safir, G.R., 1991. Significance of phenolic compounds in plant-soil-microbial systems. Crit. Rev. Plant Sci. 10, 63-121.
11. Blum, U., Shaffer, S.R., Lehman, M.E., 1999. Evidence for inhibitory allelopathic interactions involving phenolic acids in field soils: Concepts vs. an experimental model. Crit. Rev. Plant Sci. 18, 673-693.
12. Lyu, S.W., Blum, U., 1990. Effects of ferulic acid an allelopathic compound, on net P, K, and water uptake by cucumber seedlings in a split-root system. J. Chem. Ecol. 16, 2429-2439.
13. Lyu, S.W., Blum, U., Gerig, T.M., O'brien, T.E.,1990. Effects of mixtures phenolic acids on phosphorus uptake by cucumber seedlings. J. Chem. Ecol. 16, 2559-2567
14. Bergmark, C.L., Jackson, W.A., Volk, R.J., Blum, U., 1992. Differential inhibition by ferulic acid of nitrate and ammonium uptake in *Zea mays* L. Plant Physiol. 98, 639-645.
15. Booker, F.L., Blum, U., Fiscus, E.L., 1992. Short term effects of ferulic acid on ion uptake and water relations in cucumber seedlings. J. Exp. Bot. 43, 649-655.
16. Anderson, J.P.E., 2000. Handling and storage of soils for experiments with crop protection

chemicals. Abstr. 3rd Intern. Symp. Environ. Aspects Pest. Microbiol., p. 17
17. Makino, T., Tkahashi, Y., Sakurai, Y., Kamimura, Y., 1998. Influence of air-drying on adsorption and oxidation phenolic acids in andosols and non-andosols. Soil Sci. Plant Nutr. 44, 281-288.
18. Smith, A.E., 1992. A review of the extraction of herbicide residues from aged Saskatchewan field soils. Intern. J. Environ. Anal. Chem. 46, 111-116.
19. Cheng, H.H., 1990. Organic residues in soils: Mechanisms of retention and extractability. Intern. J. Environ. Anal. Chem. 9, 165-171.
20. Cheng, H.H., 1969. Extraction and colorimetric determination of picloram in soil. J. Agric. Food Chem. 17, 1174-1177.
21. Blum, U., Gerig, T.M., Worsham, A.D., Holappa, L.D., King, L.D., 1992. Allelopathic activity in wheat-conventional and wheat no-till soils: Development of soil extract bioassay. J. Chem. Ecol. 18, 2191-2221.
22. Blum, U., Worsham, A.D., King, L.D., Gerig T.M., 1994. Use of water and EDTA extractions to estimate available (free and reversibly bound) phenolic acids in Cecil soils. J. Chem. Ecol. 20, 341-359.
23. Blum, U., 1997. Benefits of citrate over EDTA for extracting phenolic acids from soils and plant debris, J. Chem. Ecol. 23, 347-362.
24. Ohno, T., First P.R., 1998. Assessment of the Folin and Ciocalteu's method for determining soil phenolic carbon. J. Environ. Qual. 27, 776-782.
25. Cheng, H.H., Gan, J., Koskinen, W.C., Jarvis, L.J., 1995. Potential of the supercritical fluid extraction technique for characterizing organic-inorganic interactions in soil. *In*: Huang, P. M., Berthelin, J., Bollag, J.-M., McGill, W. B., Page, A. L. (Eds.), Environmental Impact of Soil Component Interactions: Natural and Anthropogenic Organics. Lewis Publishers, Boca Raton, FL. pp. 191-197.
26. Koskinen, W.C., Cheng, H.H., Jarvis, L.J., Sorenson, B.A., 1995. Characterization of mechanisms of pesticide retention in soils using the supercritical fluid extraction technique. Intern. J. Environ. Anal. Chem. 58, 379-385.
27. Cheng, H.H., Haider, K., Harper, S. S., 1983. Catechol and chlorocatechol in soil: Degradation and extractability. Soil Biol. Biochem. 3, 311-317.
28. Stalder, L., Pestemer, W., 1980. Availability to plants of herbicide residues in soil. Part I. A rapid method for estimating potentially available residues of herbicides. Weed Res. 20, 341-347.
29. Koskinen, W.C., O'Connor, G.A., Cheng, H.H., 1979. Characterization of hysteresis in the desorption of 2,4,5-T from soils. Soil Sci. Soc. Am. J. 43, 871-874.
30. Cecchi, A.M., 2001. Sorption-desorption of phenolic acids in soil. M. S. Thesis. University of Minnesota.
31. Koskinen, W.C., 1980. Evaluation of the batch equilibration method for characterization of adsorption-desorption of 2,4,5-T. Ph.D. Thesis. Washington State Univ. Microfilms. Ann Arbor, Mich. (Diss. Abstr. 41:20B).
32. Clay, S.A., Koskinen, W., 1990. Adsorption and desorption of atrazine, hydroxyatrazine, and S-glutathione atrazine on two soils. Weed Sci. 38, 262-266.
33. Gan, J., Weimer, M.R., Koskinen, W.C., Buhler, D.D., Wyse, D.L., Becker, R.L., 1994. Sorption desorption of imazethapyr and 5-hydroxyimazethapyr in Minnesota soils. Weed Sci. 42, 92-97.
34. Koskinen, W.C., Cheng, H.H., 1982. Elimination of aerobic degradation during characterization of pesticide adsorption-desorption. Soil Sci. Soc. Am. J. 46, 256-259.

35. Farmer, W.J., Aochi Y., 1974. Picloram sorption by soils. Soil Sci. Soc. Am. Proc. 38, 418-423.
36. Blum, U., 1998. Effects of microbial utilization of phenolic acids and their phenolic acid breakdown products on allelopathic interactions. J. Chem. Ecol. 24, 685-708.
37. Cheng, H.H., Fuehr, F., Mittelstaedt W., 1975. Fate of methabenzthiazuron in the plant-soil system. *In*: Coulston, F., Korte, F.(Eds). Pesticides. Environ. Qual. Safety, Suppl. Vol. III, pp. 271- 276.
38. Kearney, P.C., Shelton, D.R., Koskinen, W.C., 1998. Soil chemistry of pesticides. Encycl. Chem. Technol. 22, 419-451.
39. Ito, I., Kobayashi, K., Yoneyama T., 1998. Fate of dehydromatricaria ester added to soil and its implication for allelopathic effect of *Solidago altissima*. L. Ann. Bot. 82, 625-630.
40. Martin, J.P., Haider K., 1976. Decomposition of specifically carbon-14 labeled ferulic acid: Free and linked into model humic acid-type polymers. Soil Sci. Soc. Am. J. 40, 377-380.
41. Bollag, J.-M., Liu, S.-Y., Minard R.D., 1982. Enzymatic oligomerization of vanillic acid. Soil Biol. Biochem. 14, 157-163.
42. Lehmann, R.G., Cheng, H.H., Harsh, J.B, 1987. Oxidation of phenolic acids by soil iron and manganese oxides. Soil Sci. Soc. Am. J. 51, 352-356.
43. Lehmann, R.G., Cheng, H.H., 1988. Reactivity of phenolic acids in soil and formation of oxidation products. Soil Sci. Soc. Am. J. 52, 1304-1309.

Developments in Soil Science, Volume 28B
Editors: A. Violante, P.M. Huang, J.-M. Bollag and L. Gianfreda

ENVIRONMENTAL EFFECT OF METSULFURON-METHYL ON THE SIZE OF MICROBIAL BIOMASS AND N-MINERALIZATION IN A LOAMY SAND SOIL

A. M. El-Ghamry [a], C. Y. Huang [b] and J. M. Xu [b]

[a] Mansoura University, Faculty of Agriculture, Soil Science Dept., El-Mansoura, Egypt

[b] College of Environmental and Resource Sciences, Zhejiang University, Hangzhou 310029, China

Effects of metsulfuron-methyl, a sulfonylurea herbicide, on microbial biomass and N-mineralization were investigated in a loamy sandy soil in the laboratory. The herbicide was added at four levels: control (0.0), 0.01, 0.10, and 1.00 μg g^{-1} soil, and the soil was incubated at 25°C. Soil microbial biomass-C (C_{mic}), soil microbial biomass-N (N_{mic}), and N-mineralization (N_{min}) contents were estimated 1, 3, 5, 7, 10, 15, 25, and 45 days after the application of the herbicide. Fumigation-extraction (FE) methods were used for determination of C_{mic} and N_{mic}. In comparison to untreated soil, the C_{mic} and N_{mic} decreased significantly ($P < 0.05$) in soils treated with herbicide at concentrations of 0.1 and 1.0 μg g^{-1} soil, particularly within the initial 7 days of incubation. A greater enhancement in the soil microbial biomass C:N ratio was found in the treated soil than the non-treated control soil. The N_{min} declined significantly in the soil treated with herbicide at 1.0 μg g^{-1} soil within the first 3 days of incubation. Among the different treatments (levels) tested, the metsulfuron-methyl showed their relative toxicity in the order of 1.0 > 0.10 > 0.01 μg g^{-1} soil. The effects were found to be transitory and were significant only at the higher rates of herbicide application.

1. INTRODUCTION

Application of pesticides to improve crop productivity is an essential component of present day agricultural practices. This has resulted in either stimulatory or inhibitory effects on the soil microflora, including nitrogen-fixing organisms. Most of the pesticides are degraded by very few microorganisms and are toxic for others [1].

Pesticides, by design, are biocides; their utility lies in their ability to kill noxious or unwanted organisms, but they are rarely selective. Most pesticides act by interfering with fundamental biochemical and physiological processes that are common to a wide range of pests and non-target organisms [2]. Assessing the hazards of pesticide use relies on a balance of evidence accumulated through various lines of inquiry. Thresholds are set and codes of safe practice established on the basis of biochemical studies, toxicity tests on a limited number of organisms, and field monitoring [3].

The maintenance of soil fertility depends on the size and the activity of soil microbial biomass [4], which is of fundamental importance in the biological cycles of almost all major plant nutrients [5]. It is generally recognized that the microbial biomass is the eye of the needle through which all organic matters that enter the soil must pass [6]. Although the soil microbial biomass represents only a small fraction of the total amount of soil C, N, P and S, it has a relatively rapid turnover [7]. Thus the soil microbial biomass is considered an active nutrient pool to plants.

Pesticide use in less developed countries has tended to increase faster than it has in the industrialized countries, and that trend is causing concern [8]. Herbicides affect various soil microbial processes [9], inhibit decomposition [10], and depending on the type and rate of application, can alter the biomass quantitatively and qualitatively in both the short- and long–term [11]. Much work concerning the toxicity of sulfonylurea herbicides on weeds and crops has been reported [12, 13], but there are fewer reports about their effects on microbial biomass and its activities.

The sulfonylurea herbicides are a group of compounds that control broad-leaved weeds and some grasses in cereal crops [14], at very low application rates (2–75 g ha^{-1}) [15]. Metsulfuron-methyl is a sulfonylurea herbicide. It displays high herbicidal activity and very low mammalian toxicity and is widely used in agriculture [16]. The half-life for metsulfuron-methyl at different soil water contents and temperatures ranges from 8 to 36 days [17].

There are many contrasting reports on the impact of herbicides on soil microbial biomass. For example, the use of pesticides is thought to be harmful to microorganisms and their activities contributing to soil fertility and therefore avoided [18]. Ismail et al. [19] observed that microbial biomass in a clay loam soil increased with the herbicide (metsulfuron-methyl) treatment during the first 9 days of incubation but declined from day 19 onward. However, in a sandy loam soil, the biomass decreased with an increase of herbicide concentration on day 1 but increased thereafter. Perucci and Scarponi [20] showed that rimsulfuron herbicide reduced the microbial biomass carbon at 10 FR (Field rate) and 100 FR within the first 10 days.

Marsh and Davies [21] found that the mineralization of nitrogen in triangle soil was reduced from week 3 and that the herbicide inhibited oxidation of NH_4^+-N to NO_3^--N throughout the test period, indicating a considerable inhibition of nitrification. In contrast, Hart and Brookes [22] showed that the mineralization of soil organic N to ammonium and then nitrate was mostly unaffected by the pesticide treatmentand therefore had no measurable long-term harmful effects on the soil microbial activity, as assessed by N-mineralization.

Junnila et al. [23] reported that metsulfuron-methyl had little effect on microbial dehydrogenase and nitrification activities. The half-life of metsulfuron-methyl varied from 5 days in the acidic soil to 69 days in the alkaline soil. Under sterile conditions, the half-life increased in alkaline soil to 139 days but was not changed in the acidic soil. Metsulfuron-methyl degradation mainly resulted in the formation of the amino-triazine. In the acidic soil, degradation was characterized by rapid hydrolysis giving two specific unidentified metabolites, which were not detected during incubations in the alkaline soil [24].

The present investigation aimed at assessing the influence of metsulfuron-methyl on microbial biomass carbon (C_{mic}), microbial biomass nitrogen (N_{mic}), the ratio of C_{mic}/N_{mic}, and N-mineralization in a loamy sand soil.

2. MATERIALS AND METHODS

A laboratory incubation experiment was conducted using a loamy sand soil having 1.76% total organic carbon, 0.158% total nitrogen, 22.4% soil moisture at -33kPa, and pH 6.27. The loamy sand soil was collected from the surface layer (0-20 cm) from Hangzhou, Zhejiang province, China.

After sampling and preparation of soil, the soil sample was divided into four sub-samples. One sub-sample was used as a control, and the others were treated with metsulfuron-methyl herbicide.

Methanol solutions of metsulfuron-methyl were prepared at three different concentrations: 0.5, 5.0, and 50 $\mu g\ ml^{-1}$.

The herbicide was incorporated into the soil sub-sample as follows: 48 ml of the methanolic solution of metsulfuron-methyl were added to 120 g of air-dried soil of each sub-sample and the control soil.

After removal of the methanol by evaporation at room temperature, each of the 120-g soil was divided into 24 portions (8 stages × 3 replicates; 5g each), transferred into beakers that each contained 95 g fresh soil (oven-dry basis) and homogenized.

Three application rates corresponding to 0.01 (L1), 0.10 (L2), and 1.00 (L3) $\mu g\ g^{-1}$ soil were used.

Soil moisture was adjusted to 60% water content at -33 kPa and incubated in the dark at 25±1°C. The beakers were removed from the incubator every day and brought to the original weight by adding the required amount of distilled water. Three beakers each for control and treated soils were removed and submitted to analysis for C_{mic} and N_{mic} at 1, 3, 5, 7, 10, 15, 25 and 45 days after metsulfuron-methyl treatment.

Soil samples for the determination of microbial biomass-C were extracted by a fumigation-extraction (FE) method [25], and the organic carbon in the soil extracts was measured using an automated total organic carbon analyzer [26]. Soil samples for the determination of microbial biomass-N were extracted by an FE method [27], and the total N in the soil extracts was measured using the Kjeldahl digestion method [28].

Anaerobic N-mineralization was determined in 10 g soil; the samples were flooded with 25 ml deionized water in a 30 ml glass bottles. The bottles were gently tapped for 30 s to remove air bubbles, sealed with rubber stoppers, then incubated at 25±1°C. After the incubation period, the samples were transferred to 125 ml bottles and extracted with 25 ml 4 M KCl with shaking for 1 h at 150 rpm and subsequent gravity filtering using prewashed Whatman No. 5 paper [29]. Ammonium was determined colorimetrically [30].

Water contents at an applied pressure of –33 kPa were determined using a pressure membrane system similar to that described by Heining [31]. The pH (in water, 1:2.5) of the soil was measured with a pH meter. Total soil N was determined by the Kjeldahl method and total organic carbon by the Walkley-Black procedure [32].

Data were analyzed following C.R. Design and Duncan's multiple range tests using statistix software [33].

3. RESULTS

3.1. The effect of metsulfuron-methyl on microbial biomass carbon (C_{mic})

The responses of the microbial biomass-C content to the herbicide treatments are shown in Table 1. The data showed that the microbial biomass-C contents were not significantly affected at the 0.01 $\mu g\ g^{-1}$ application rate. But when metsulfuron-methyl was applied at the rate of 0.1 and 1.0 $\mu g\ g^{-1}$, the reduction in microbial biomass-C became significant, especially within the first 7 days, relative to the control.

Table 1
Effect of metsulfuron-methyl on microbial biomass-C (C_{mic})

Incubation period (day)	Herbicide Treatment ($\mu g\ g^{-1}$soil)[a]				Reduction (%)			LSD 0.05
	Control	0.01	0.10	1.00	0.01	0.10	1.00	
	$\mu g\ g^{-1}$	$\mu g\ g^{-1}$	$\mu g\ g^{-1}$	$\mu g\ g^{-1}$				
1	255.52 a	248.25 a	216.69 b	205.16 b	2.9	15.2	19.7	21.87
3	251.51 a	240.10 ab	209.29 bc	193.47 c	4.5	16.8	23.1	26.78
5	232.55 a	217.22 ab	201.54 bc	185.89 c	6.6	13.3	20.1	17.87
7	228.67 a	216.69 ab	197.67 bc	182.01 c	5.2	13.6	20.4	15.47
10	216.69 a	220.92 a	201.38 ab	182.31 b	-1.9*	7.1	15.9	18.97
15	209.46 a	213.17 a	209.63 a	194.26 a	-1.8	-0.1	7.3	14.15
25	215.41 a	219.47 a	211.95 a	200.06 a	-1.9	1.6	7.1	15.38
45	216.16 a	216.69 a	212.47 a	213.52 a	-0.3	1.7	1.2	12.62

[a]Means with different letters, within rows, differ significantly according to LSD ($P < 0.05$).
*Increase.

The results revealed that metsulfuron-methyl treatment, applied at the rate of 0.01 $\mu g\ g^{-1}$ soil, reduced the C_{mic} contents by 2.9, 4.5, 6.6, and 5.2 at 1, 3, 5, and 7 days of incubation, respectively, against the control. A nominal increase of 1.9, 1.8, 1.9, and 0.3% was noticed at 10, 15, 25, and 45 days of incubation, respectively, compared with the control. However, when the metsulfuron-methyl treatment was applied at 0.1 $\mu g\ g^{-1}$, the decrease in C_{mic} was 15.2, 16.8, 13.3, 13.6, 7.1, 0.1(increase), 1.6, and 1.7%, respectively, against the control at the same incubation periods. The soil treated with 1.0 $\mu g\ g^{-1}$ of herbicide resulted in 19.7, 23.1, 20.1, 20.4, 15.9, 7.3, 7.1, and 1.2% decline in C_{mic}, respectively, with reference to the control at the same incubation periods.

3.2. The effect of metsulfuron-methyl on microbial biomass nitrogen (N_{mic})

The effects of metsulfuron-methyl treatments on soil microbial biomass-N are shown in Table 2. When metsulfuron-methyl was applied at the rate of 0.01 $\mu g\ g^{-1}$ soil, there was no significant change in soil microbial biomass-N contents. It resulted in 6.1, 10.8, 8.8, 6.4, 2.7, 2.9, 2.7 and 2.6% reduction in N_{mic} at 1, 3, 5, 7, 10, 15, 25, and 45 days of incubation, respectively, against the control. However, the 0.1 $\mu g\ g^{-1}$ dosage caused significant reduction in N_{mic} contents, where it was reduced by 20.9, 18.8, 14.7, 21.9, 17.1, 11.4, 5.4, and 2.8% at the same incubation periods, respectively, relative to the control. The N_{mic} reduction with 1.0 $\mu g\ g^{-1}$ soil application rate became highly significant against the control within the initial 7 days after incubation. Here, the N_{mic} decreased by 31.3, 32.5, 29.5, 28.2, 22.7, 17.0, 11.1, and 10.9, at the same incubation periods, respectively, as compared to the control.

Table 2
Effect of metsulfuron-methyl on microbial biomass-N (N_{mic})

Incubation period (day)	Herbicide Treatment ($\mu g\ g^{-1}$soil)[a]				Reduction (%)			LSD 0.05
	Control	0.01	0.10	1.00	0.01	0.10	1.00	
	$\mu g\ g^{-1}$	$\mu g\ g^{-1}$	$\mu g\ g^{-1}$	$\mu g\ g^{-1}$				
1	43.38 a	40.73 ab	34.31 bc	29.81 c	6.1	20.9	31.3	4.66
3	33.41 a	29.82 ab	27.13 bc	22.57 c	10.8	18.8	32.5	3.29
5	30.75 a	28.06 ab	26.23 b	21.69 c	8.8	14.7	29.5	2.55
7	28.94 a	27.09 ab	22.61 bc	20.79 c	6.4	21.9	28.2	3.60
10	31.60 a	30.75 a	26.21 ab	24.44 b	2.7	17.1	22.7	4.42
15	31.68 a	30.75 a	28.08 ab	26.29 b	2.9	11.4	17.0	2.96
25	32.31 a	31.45 a	30.57 a	28.73 a	2.7	5.4	11.1	3.59
45	31.53 a	30.70 a	30.65 a	28.08 a	2.6	2.8	10.9	3.89

[a] Means with different letters, within rows, differ significantly according to LSD ($P < 0.05$).

3.3. The effect of metsulfuron-methyl on (C_{mic}/N_{mic}) ratio

The addition of metsulfuron-methyl to the soil produced an increase in the C_{mic}/N_{mic} ratio (Figure 1). Results indicated that the increase in C_{mic}/N_{mic} ratio was related to the increased levels of herbicide.

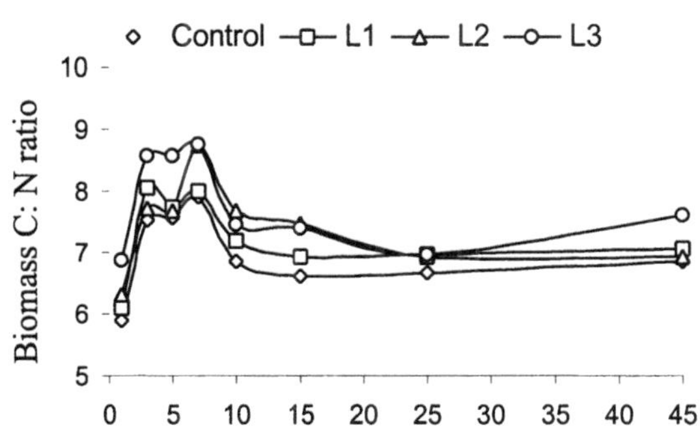

Figure 1. Effect of metsulfuron-methyl on C_{mic}/N_{mic} ratio.L1 ($0.01\mu g\ g^{-1}$), L2 ($0.1\mu g\ g^{-1}$) and L3 ($1.0\ \mu g\ g^{-1}$).

The data indicated that herbicide application at the rate of 0.01 $\mu g\ g^{-1}$ soil promoted C_{mic}/N_{mic} ratio by 3.5, 6.9, 2.4, 1.2, 4.8, 4.8, 4.7, and 2.9% at 1, 3, 5, 7, 10, 15, 25, and 45 days of incubation, respectively, relative to the control. At 0.10 $\mu g\ g^{-1}$ soil herbicide dosage, the ratio was improved by 7.2, 2.5, 1.6, 10.6, 12.1, 12.9, 3.9, and 1.1% at the same incubation periods, respectively, compared with the control. For the 1.00 µg g^{-1} soil application rate, it was increased by 16.8, 13.9, 13.3, 10.8, 8.8, 11.8, 4.4 and 10.9% at the same incubation periods, respectively, against the control.

3.4. The effect of metsulfuron-methyl herbicide on N-mineralization (N_{min})

The dynamic responses of N-mineralization to the herbicide treatments are shown in Table 3. It was found that N-mineralization was not significantly modified in the 0.01 and 0.10 µg g^{-1}soil. When metsulfuron-methyl was applied at 1.00 µg g^{-1} soil, the decrease in the N-mineralization was significant, especially within the first 3 days, as compared to the control.

Table 3
Effect of metsulfuron-methyl on N-mineralization (N_{min})

Incubation period (day)	Herbicide Treatment (µg g^{-1}soil)[a]				Reduction (%)			LSD 0.05
	Control	0.01	0.10	1.00	0.01	0.10	1.00	
	µg g^{-1}	µg g^{-1}	µg g^{-1}	µg g^{-1}				
1	17.73 a*	16.33 ab	14.47 ab	14.00 b	7.9	18.4	21.0	2.64
3	19.13 a	19.23 ab	17.73 ab	15.87 b	-0.5*	7.3	17.0	2.36
5	21.00 a	21.28 a	20.53 a	19.23 a	-1.3	2.2	8.4	3.14
7	23.80 a	24.27 a	23.33 a	22.40 a	-2.0	2.0	5.9	3.72
10	28.00 a	28.47 a	26.60 a	25.67 a	-1.7	5.00	8.3	3.65
15	32.01 a	32.20 a	31.73 a	30.33 a	-0.6	0.9	5.3	3.15
25	32.67 a	32.85 a	31.92 a	31.27 a	-0.6	2.3	4.3	2.78
45	35.00 a	34.72 a	34.53 a	33.60 a	0.8	1.3	4.0	2.45

[a] Means with different letters, within rows, differ significantly according to LSD ($P<0.05$).
* Increase.

The results further indicated that the maximum increase (1.97%) by metsulfuron-methyl application at 0.01 µg g^{-1} was observed at the 7th day of incubation. The reduction was highest (7.90%) at the first day of incubation and lowest (0.80%) at the 45th day of incubation compared to control. However, when applied at 0.10 µg g^{-1}, the highest reduction (18.39%) in N-mineralization was noted at the first day of incubation and the least reduction (0.87%) at the 15th day of incubation against the control. The soil treated with 1.00 µg g^{-1} resulted in 21.04% as the highest reduction at the first day of incubation, while the lowest reduction (4.00%) was recorded at the 45th day of incubation with reference to the control. So the general observation noted is that the highest reduction of all herbicide treatments occurred on the first day of incubation compared to the control and this reduction decreased with the passage of time, suggesting the decline of the herbicidal effect with time.

4. DISCUSSION

The soil microbial biomass-C and -N decreased with increasing levels of herbicide application; this decrease was significant during the first 7 days and 10 days at 0.10 and 1.00 $\mu g\ g^{-1}$ soil, respectively. However, at the lower rate (0.01 $\mu g\ g^{-1}$), the decrease was not significant.

The reason for the decrease of microbial biomass as affected by metsulfuron-methyl may be related to the toxicity effect of the herbicide and its low adsorption coefficient. The fate of sulfonylurea herbicides in soils is directly related to their chemical structure and mainly to the ionization of the sulfonylurea bridge. These compounds are weak acids with pK_a ranging from 3 to 5 (3.3 for metsulfuron-methyl) and in soils are mainly in the ionized form. This explains their low sorption coefficients, which are pH-dependent [15, 34]. It is possible that because the herbicide was not fixed in the soil matrix by sorption, its toxicity remained high to microbes. Another possible explanation is that the microorganisms did not easily adapt to metsulfuron-methyl, which has a tendency to persist for a fairly long time. Junnila et al. [23] reported that metsulfuron persisted longer in sandy soils than in other soils. The phytotoxicity of the 4 g a.i.(active integrated) ha^{-1} dose persisted in the 0-5 cm layer of all soils for 1 month on average. The phytotoxicity of the 12 g a.i. ha^{-1} dose persisted for at least 1 month and generally for 1 year [23]. Perucci and Scarponi [20] have observed reduced microbial biomass in clay loam soil following application of rimsulfuron herbicide.

The half-life of metsulfuron–methyl is between 8 and 36 days [17]. This gives a possible explanation as to why the highest concentration caused a significant decrease in microbial biomass in the first 7 days. During that time, the herbicide was almost unchanged qualitatively and quantitatively and maintained its inherent toxicity.

The noted increase of the C_{mic}/N_{mic} ratio was related to the decrease of microbial biomass-C and -N and the increase in the levels of metsulfuron-methyl. The increase in the level of metsulfuron-methyl caused subsequent increase in toxicity of the herbicide, resulting in greater reduction in N_{mic} in treated soils than in the control. The ratio of biomass C_{mic}/N_{mic} was, therefore, more in favor of C_{mic} than N_{mic}. The increase in biomass C/N ratio can highlight changes in microbial populations, which could be a good indicator of the changes in the microbial community structure [35].

The current results further revealed that the application at 1.00 $\mu g\ g^{-1}$ soil significantly ($P < 0.05$) reduced the soil N-mineralization within the first 3 days after treatment as compared to the control. However, there was no significant effect for other doses at any time. Similar observations have been reported by Junnila et al. [23] that metsulfuron-methyl had only a very small effect on microbial activity. In some soils, slight inhibition, particularly of nitrification, was found initially, but later there was either no effect or even some stimulation.

The decrease in N-mineralization by metsulfuron-methyl at high rates may be related to the toxic effect of the herbicide, since its effect on soil fauna is either a direct toxic effect or through the influence on predator-prey interactions. This could partly be explained by the death of microorganisms in soil. Hassink et al. [36] suggested that microbial activity, rather than the size of the microbial biomass, is a better indicator of N-mineralization, particularly in sandy-textured soil. Marumoto et al. [37] reported that the mineralization of microbial N is rapid and contributed approximately 30-75% of the N in the ammonium pool at the end of experiments, using air-dried soils.

There are numerous difficulties in the interpretation of the data available from various research programs carried out on the effect of herbicides on N-mineralization in soils. In field trials, Kasper and Fischbeck [38] indicated that the N-mineralization rate and potentially mineralizable N are not affected by formulations such as 2,4-D amine or ester at rates of 1.12 kg ha^{-1}, but urease was temporarily depressed by both formulations. Nitrification is also reduced temporarily by the ester formulation. Harper et al. [39] showed that nitrogen in the clover crop increased until anthesis and then declined slightly prior to desiccation with herbicides.

5. CONCLUSIONS

It has been observed that application of metsulfuron-methyl was followed by temporary reduction of microbial biomass and N-mineralization. These results have shown that this herbicide had no detrimental effect on soil microflora, hence on soil fertility, when used at the recommended rate. It can be concluded that changes in microbial biomass occurred only at herbicide concentrations much higher than that which occurs following field application at the recommended rates. The side effects of these chemicals on soil microorganisms are probably of little ecological significance.

ACKNOWLEDGMENTS

Financial support of the National Natural Science Foundation of China (No.49871044), the Zhejiang Provincial Natural Science Foundation of China (No.RC99032) and NKBRSF (No.G1999011809) was gratefully appreciated.

REFERENCES

1. Chalam, A. V., Sasikaia, C., Ramana, C.V., Uma, N.R., Rao, P.R., 1997. Effect of pesticides on the diazotrophic growth and nitrogenase activity of purple nonsulfur bacteria. Bull. Environ. Contam. Toxicol. 58, 463-468.
2. Conway, G. R., Pretty, J.N., 1991. Unwelcome Harvest: Agriculture and Pollution. Earthscan Publications, London.
3. Pingall, P. L., Roger, P.A., 1995. Impact of Pesticides on Farmer Health and the Rice Environment. P 664 IRRI (International Rice Research Institute), Kluwer Academic Publishers.
4. Alexander, M., 1977. Introduction to Soil Microbiology. John Wiley, New York.
5. Robert, M., Chenu, C., 1992. Interactions between soil minerals and microorganisms. *In*: Stotzky, G., Bollag, J.-M. (Eds.), Soil Biochemistry. Vol. 7. Marcel Dekker, New York, pp. 307-379.
6. Jenkinson, D. S., 1988. Determination of microbial biomass C and N in soil. *In*: Wilson, J.R. (Ed.), Advances in Nitrogen Cycling in Agricultural Ecosystems. CAB International, Wallingford, England, pp. 368-386.
7. Amato, M., Ladd, J.N., 1980. Studies of nitrogen immobilization and mineralization in calcareous soils. V. Formation and distribution of isotope labelled biomass during decomposition of ^{14}C and ^{15}N labelled plant material. Soil Biol. Biochem. 12, 405-411.
8. Moody, K., 1990. Pesticide use in rice culture: general aspects. Workshop on the Environmental and Health Impacts of Pesticide Use, 28-30 May 1990, International Rice Research Institute (IRRI), Los Banos, Philippines.

9. Tyler, G. 1980. Metals in sporophores of basidiomycetes. Transact. Brit. Mycolog. Soc. 74, 41-49.
10. Grossbard, E., Wingfield, G.I., 1978. Effects of paraquat, aminotriazole and glyphosate on cellulose decomposition. Weed Res. 18, 347-353.
11. Anderson, J. P. E., Armstrong, R.A., Smith, S.N., 1981. Methods to evaluate pesticide damage to the biomass of the soil microflora. Soil Biol. Biochem. 13, 149-153.
12. Bayer, E. M., Duffy, M.J.V., Schlueter, D.D., 1987. Herbicides, degradation and mode of action. *In*: Kearney, P.C. and Kaufman, D.D. (Eds.), Marcel Dekker Inc., Yew York, pp. 117-180.
13. John S. F., Thomas, G.P., Hilman, C.R., 1993. Potential environmental risks associated with the new sulfonylurea herbicides. Environ. Sci. Technol. 27, 2250-2256.
14. Blair A. M., Martin, T.D.A., 1988. Review of the activity, fate and mode of action of sulfonylurea herbicides. Pestic. Sci. 22, 195-219.
15. Brown, H. M., 1990. Mode of action, crop selectivity and soil relations of the sulfonylurea herbicides. Pestic. Sci. 29, 263-281.
16. 16 James V. H. 1990. Chemistry of sulfunylurea herbicides. Pestic. Sci., 29, 247.
17. James, T. K., Klaffenbach, P., Holland, P.T., Rahman, A., 1995. Degradation of primisulfuron-methyl and metsulfuron-methyl in soil. Weed Res. 35, 113-120.
18. Wolf, R. (Ed.), 1977. Organic Farming: Yesterday's and Tomorrow's Agriculture. Rodale Press, Emmaus, PA.
19. Ismail, B. S., Goh, K.M., Kader, J., 1996. Effects of metsulfuron-methyl on microbial biomass and populations in soils. J. Environ. Sci. Health 31, 987-999.
20. Perucci, P., Scarponi, L., 1996. Side effects of rimsulfuron on the microbial biomass of a clay-loam soil. J. Environ. Qual. 25, 610-613.
21. Marsh, J. A. P., Davies, H.A., 1981. Effects of dichloropop and mecoprop on respiration and transformation of nitrogen in two soils. Bull. Environm. Contam. Toxicol. 26, 108-115.
22. Hart, M. R., Brookes, P.C., 1996. Soil microbial biomass and mineralization of soil organic matter after 19 years of cumulative field applications of pesticides. Soil Biol. Biochem. 28, 1641-1649.
23. Junnila, S., Heinonen, A.H., Ervio, L.R., Laitinen, P., 1994. Phytotoxic persistence and microbiological effects of chlorsulfuron and metsulfuron in finnish soils. Weed Res. 34, 413-423.
24. Pons, N., Barriuso, E., 1998. Fate of metsulfuron-methyl in soils in relation to pedopedo-climatic conditions. Pestic. Sci. 53, 311-323.
25. Vance, E. D., Brookes, P.C., Jenkinson, D.S., 1987. An extraction method for measuring soil microbial biomass C. Soil Biol. Biochem. 19, 703-707.
26. Wu, J., Joergensen, R.G., Pommerening, B., Chaussod, R., Brookes, P.C., 1990. Measurement of soil microbial biomass by fumigation-extraction: an automated procedure. Soil Biol. Biochem. 20, 1167-1169.
27. Brookes, P. C., Landman, A., Pruden, G., Jenkinson, D.S., 1985. Chloroform fumigation and the release of soil N: a rapid extraction method to measure microbial biomass N in soil. Soil Biol. Biochem. 17, 837-842.
28. Brookes, P. C., Kragt, J.F., Powlson, D.S., Jenkinson, D.S., 1985. Chloroform fumigation and the release of soil nitrogen: the effects of fumigation time and temperature. Soil Biol. Biochem. 17, 831-835.

29. Barrios, E., Buresh, R.J. Sprent, J.I., 1996. Nitrogen mineralization in density fractions of soil organic matter from maize and legume cropping systems. Soil Biol. Biochem. 28, 1459-1465.
30. Anderson, J. M., Ingram, J.S.I., 1993. Tropical Soil Biology and Fertility: A Handbook of Methods. CAB International. Wallingford, England.
31. Heining, B., 1963. A pressure membrane apparatus. J. Agric. Eng. Res. 8, 48-49.
32. Jackson, M. L., 1958. Nitrogen determinations for soils and plant tissue. *In*: Jackson, M.L. (Ed.), Soil Chemical Analysis. Constable, London.
33. CoStat Statistical Software, 1990. CoStat. Manual Revision 4.2, p.271.
34. Walker, A., Welch, S.J., 1989. The relative movement and persistence in soil of chlorsulfuron, metsulfuron-methyl and triasulfuron. Weed Res. 29, 375-383.
35. Khan, K. S., Huang, C.Y., 1998. Effect of lead-zinc interaction on size of microbial biomass in red soil. Pedosphere 8, 143-148.
36. Hassink, J., Bouwman, L.A., Zwart, K.B., Brussaard, L., 1993. Relationships between habitable pore space, soil biota and mineralisation rates in grassland soils. Soil Biol. Biochem. 25, 47-55.
37. Marumoto, T., Anderson, J.P.E., Domsch, K.H., 1982. Mineralisation of nutrients from the soil microbial biomass. Soil Biol. Biochem. 14, 469-475.
38. Kasper, J., Fischbeck, G., 1989. Competition between winter wheat and weeds with different methods of sowing. Mitteil. Gesel. Pflanzen. 2, 108-111.
39. Harper, L. A, Hendrix, P.F., Langdale, G.W., Coleman, D. C., 1995. Clover management to provide optimum nitrogen and soil water conservation. Crop Sci. 35, 176-182.

Developments in Soil Science, Volume 28B
Editors: A. Violante, P.M. Huang, J.-M. Bollag and L. Gianfreda

TRANSFORMATION AND BINDING OF 2,4,6-TRINITROTOLUENE IN SOIL AMENDED WITH ACTIVATED CARBON

G.K. Vasilyeva[a], V.D. Kreslavski[b], P.J. Shea[c] and J.-M. Bollag[d]

[a]Institute of Physicochemical and Biological Problems in Soil Science, RAS, Pushchino, Moscow Region, 142290 Russia*

[b]Institute of Basic Biological Problems, RAS, Pushchino, Moscow Region, 142290 Russia

[c]Institute of Agriculture and Natural Resources, University of Nebraska, Lincoln, NE, 68583, USA

[d]Laboratory of Soil Biochemistry, The Pennsylvania State University, University Park, PA 16802, USA

The explosive 2,4,6-trinitrotoluene (TNT) is a toxic and persistent contaminant at sites of munitions manufacturing and handling. Laboratory experiments were conducted to determine mechanisms of TNT transformation and detoxification in soil. Under aerobic conditions, TNT was simultaneously reduced and oxidized to products that polymerize and become strongly bound to the soil matrix. The persistence of TNT in highly contaminated soils can be attributed to toxicity to microorganisms and plants, which decreased the binding of residues. The readily available fraction (extractable with 3 mM $CaCl_2$) was most critical to TNT toxicity in soil. The remaining TNT (extractable with acetonitrile) did not cause much toxicity but was available for chemical and microbial transformation. Activated carbon introduced into the soil maintained a low concentration of readily available TNT, resulting in decreased TNT toxicity and sharply accelerated microbially mediated reductive transformation and subsequent binding in soil. In addition, activated carbon promoted rapid binding of TNT through oxidation. These observations indicate the potential use of activated carbon to promote bioremediation of highly contaminated soils.

1. INTRODUCTION

The explosive 2,4,6-trinitrotoluene (TNT) is a widespread contaminant at sites of munitions manufacturing and handling [1]. TNT is toxic and persistent in the environment [1-6].

* This research was supported in part by RFFI no. 01-05-97006, NSF/EPSCoR cooperative agreement no. 925522, the University of Nebraska Water Center/Environmental Programs and is a contribution of project nos. NEB 12-239 and 251.

Bioremediation of TNT-contaminated soils is usually based on creating favorable conditions for reduction and binding of TNT and its products to soil; however, the binding mechanism(s) is not clear and some toxic products may accumulate [7-13]. The objective of our research is to elucidate mechanisms of TNT transformation and detoxification in soil and to suggest an approach to accelerate this process through the addition of activated carbon.

2. MATERIALS AND METHODS

Sharpsburg soil (NE, USA, clay loam, pH 7.5, 3.4% C_{org}) and powdered activated carbon (DARCO, particle size 35 μm) were used in the experiments. To study the accumulation of bound residues, uniformly labeled [ring-^{14}C]TNT was used. TNT (dissolved in acetone) was added to the soil to obtain 500, 1000 and 2000 mg TNT/kg soil. The acetone was removed by exposure to air for 2 h. Activated carbon was added at 0.1, 0.25, 0.5, 0.75 or 1.0% (w/w). The treated soil (150 g dry weight) was incubated in 10 × 12 cm pots, which were covered with polyethylene film and placed in a controlled environment chamber (Conviron, Winnipeg, Manitoba, Canada) at 22°C and 60% relative humidity. After 18 d, 7-day-old tall fescue (*Festuca arundinaceae*) or corn (*Zea mays*) seedlings were planted (6 plants per pot) and grown for 52 d. After the plants were harvested, the soil was incubated for an additional 50 d. Controls included native soil, TNT-contaminated soil without activated carbon, and uncontaminated soil treated with acetone and 0.25, 0.75 or 1.0% (w/w) activated carbon only.

TNT and transformation products were extracted from soil or activated carbon with 3 mM $CaCl_2$ and then with acetonitrile. Concentrations of TNT and its transformation products in soil were determined 6 h after adding activated carbon and periodically during the incubation. Duplicate, 3-g soil samples were extracted with 15 ml of 3 mM $CaCl_2$ by vortexing for 0.5 min, then extracted twice with 20 ml of acetonitrile by sonicating for 7 h. The soil was separated from each extract by centrifuging for 30 min at 7000 *g*. The soil was then combusted to $^{14}CO_2$ using a biological oxidizer (Harvey Biological Oxidizer OX 600, Hillsdale, NJ) to determine bound ^{14}C. High-performance liquid chromatography (HPLC), gas chromatography-mass spectrometry (GC/MS), and liquid scintillation counting (LSC) were used for analysis and identification of TNT and transformation products, as described previously [14].

After an initial 7-d equilibration, the influence of activated carbon addition on TNT loss in the soils during the remaining 113 d of incubation was determined by comparing first-order rate constants (k): $k = (\ln C_o - \ln C)/(t-t_o)$, where C = TNT concentration in the 3 mM $CaCl_2$ or acetonitrile extracts (mg/l) at time t (d) and C_o = TNT concentration (mg/l) at t_o (7 d).

Heterotrophic soil microorganisms were counted by plating on Tryptic Soy Agar (TSA) [15]. Soil microorganisms were enumerated immediately before planting and 10 d after harvesting the corn plants. Samples included duplicates from six replications of the treatments.

3. RESULTS

3.1. Transformation of TNT in soil

Transformation of TNT in the more highly contaminated soil (2000 and 1000 mg TNT/kg) was slow compared to soil containing 500 mg TNT/kg (Figure 1A). After a 120-d incubation,

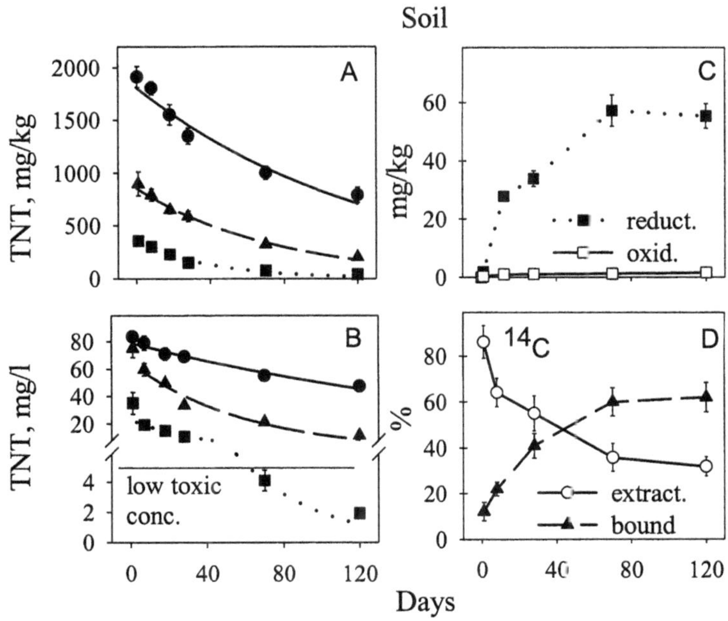

Figure 1. Transformation of [^{14}C] TNT in soil containing 500 (-■-), 1000 (-▲-) and 2000 (-●-) mg TNT/kg (A - total extractable TNT in soil, B - TNT concentration in $CaCl_2$ extract) and at 2000 mg TNT/kg (C - concentration of reduction and oxidation products, D - content of total extractable and bound ^{14}C).

up to 40, 20 and 8% of the added TNT was extractable from the respective soils. During that time, up to 7% was detected by HPLC as extractable transformation products. The concentration of TNT in 3 mM $CaCl_2$ extracts was relatively high (usually 20-80 mg/l), except in comparatively less contaminated soil after 30 d incubation, where it was reduced to 5 mg/l and less (Figure 1B). TNT transformation products primarily consisted of reduction derivatives (Figure 6: hydroxylamino (II), monoamino (III), diamino, and azoxy (IV) derivatives, up to 60 mg/kg total); however small amounts (up to 5 mg/kg) of oxidation products (2,4,6-trinitrobenzaldehyde (TNBAld, VI), 1,3,5-trinitrobenzene (TNB, IX) and 3,5-dinitroaniline (3,5DNA) were also detected (Figure 1C). Most of the remaining [^{14}C]TNT was transformed to bound residues (Figure 1D). After the 120-d incubation, mineralization to $^{14}CO_2$ did not exceed 1.5%.

3.2. Influence of activated carbon on TNT transformation in soil

Soil amendment with activated carbon sharply accelerated TNT transformation and binding in soil containing 2000 and 1000 mg TNT/kg. Within 6 h after mixing with activated carbon, extractable TNT was reduced by 20 or 40% (depending on the initial TNT concentration), compared to less than 10 or 15% decrease in unamended soil (Figure 2A).

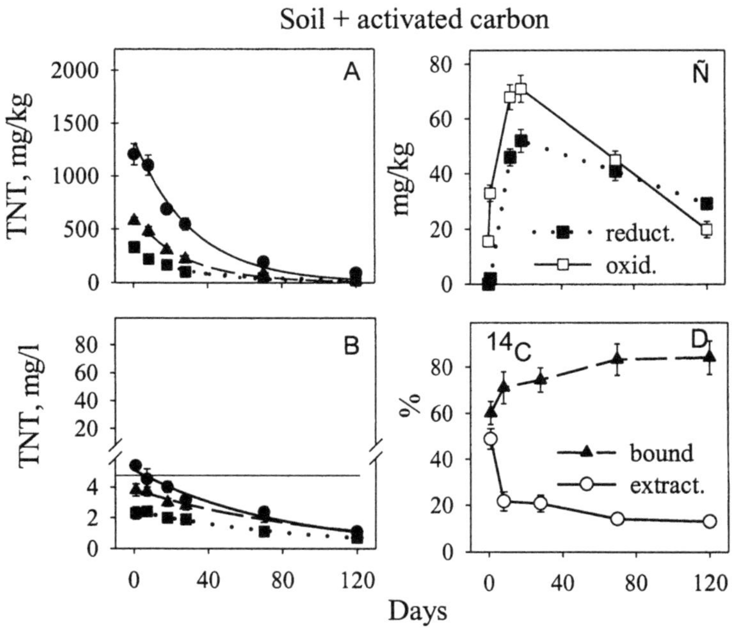

Figure 2. Transformation of [^{14}C] TNT in soil amended with activated carbon. Legend is similar to Figure 1.

The difference between TNT loss in amended and unamended soil containing 500 mg TNT/kg during the first few hours did not exceed 10%. After about 7 d, the process slowed but continued until only 2 to 4% of the initially added TNT was detected in the activated carbon-amended soil by 120 d. Final concentrations of total extractable TNT in soils containing 500, 1000 and 2000 mg/kg TNT and amended with activated carbon were 24, 42 and 134 mg/kg, respectively. Corresponding concentrations in soils without activated carbon were 39, 201 and 793 mg/kg, respectively.

The concentration of TNT in 3 mM $CaCl_2$ extracts from soil amended with activated carbon was less than 5 mg/l (Figure 2B), resulting in low toxicity to soil microorganisms and plants [14]. The maximum accumulation of TNT reduction products in unamended and amended soils was similar; however, these products accumulated and disappeared more rapidly in the presence of the activated carbon (Figure 2C). In contrast, transient accumulation of oxidation products (up to 50 mg/kg) was significantly greater compared to unamended soil.

Similar to unamended soil, almost all of the [^{14}C]TNT was present as bound ^{14}C. Two phases of ^{14}C-binding can be differentiated: initial rapid binding followed by slower binding during the incubation (Figure 2D).

First-order equations provided a good fit to TNT concentration in the $CaCl_2$ extracts and to total extractable TNT in soil between 7 and 120 d of incubation in both unamended and activated carbon-amended soils. First-order rate constants indicated that TNT transformation was accelerated in the presence of activated carbon (Table 1). As a result, the accumulation of bound ^{14}C residues was greater in soil amended with activated carbon than in the unamended soil, especially at 2000 mg TNT/kg (92% compared to 55%). There was also a smaller content of residual TNT and transformation products in the amended than unamended soil (11% compared to 45%).

Table 1
First-order rate constants (k) and (goodness of fit) R^2 for total extractable TNT in unamended contaminated soils (no AC) and in soils amended with activated carbon (AC)

Amendment	500 mg TNT/kg		1000 mg TNT/kg		2000 mg TNT/kg	
	$k \times 10^{-2}\ d^{-1}$	R^2	$k \times 10^{-2}\ d^{-1}$	R^2	$k \times 10^{-2}\ d^{-1}$	R^2
No AC	2.48 (0.43)[a]	0.97	1.32 (0.10)	0.99	0.77 (0.12)	0.95
AC	3.07 (0.49)	0.98	3.55 (0.47)	0.99	3.20 (0.49)	0.98

[a]Parenthetic values indicate standard deviations.

3.3. Toxicity of TNT to plants and microorganisms

Toxicity of TNT to tall fescue in soil amended with activated carbon depended on the amount added, and plant growth mirrored the initial TNT concentration in $CaCl_2$ extracts from soils amended with various doses of activated carbon (Figure 3A and B). Tall fescue growth was similar to the control only in the amended soils, where the TNT concentration in the $CaCl_2$ extract was less than 5 mg/l, resulting in low toxicity to the plants. Activated carbon alone did not influence plant growth. Similar results were obtained with corn (data not shown).

Microbial toxicity of TNT was indicated by a greater than 50% decrease in platable microorganisms in soils containing 1000 or 2000 mg TNT/kg (Figure 4). This contrasted with the greater number of microorganisms in all TNT-contaminated soils amended with activated carbon, which were not influenced by the initial TNT concentration in soil.

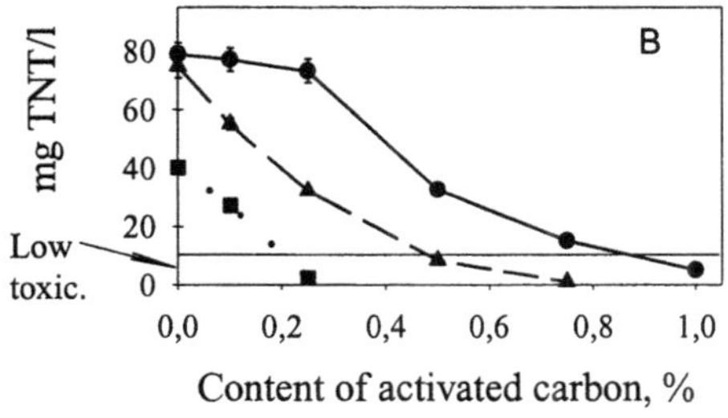

Figure 3. Influence of activated carbon on growth of tall fescue (A) and concentration of TNT in $CaCl_2$ extract (B) at 500 (-■-), 1000 (-▲-) and 2000 (-•-) mg TNT/kg in comparison to control (-◊-) soil.

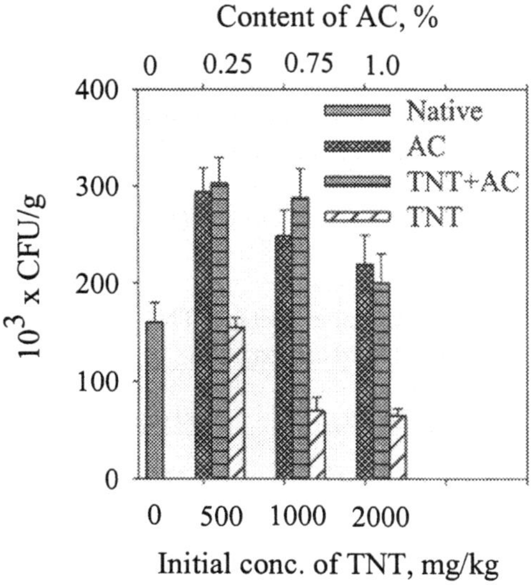

Figure 4. Microbial count in soil contaminated with TNT and amended with activated carbon (separately and simultaneously) compared to native soil.

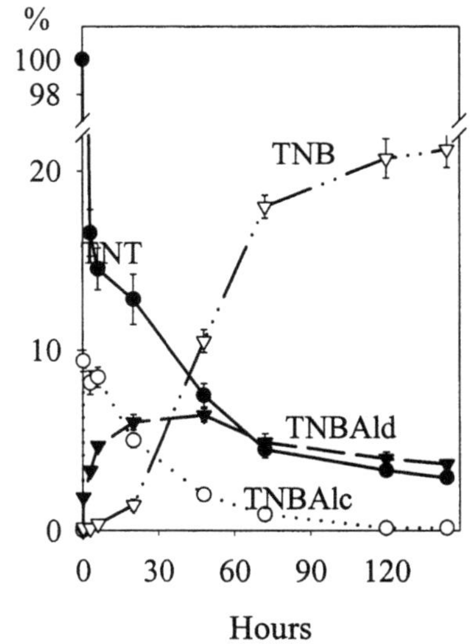

Figure 5. Transformation of TNT in suspension of activated carbon.

3.4. TNT transformation in an activated carbon suspension

In activated carbon suspension, TNT was rapidly and irreversibly bound (up to 79%) to the sorbent. Most of the remaining TNT (extractable with acetonitrile) was transformed to TNB through transient accumulation of alcohol and aldehyde derivatives (Figure 5). The presence of TNB and TNBAld in the suspension was indicated by the similarity of retention times and mass spectra to those of standards analyzed by HPLC and GC/MS. Based on the mass spectrum, a third HPLC peak was postulated as TNBAlc. Although the parent ion was not observed, fragment ions at m/z 210 and 211 can be attributed to splitting of [O + OH] and [CH_2OH + H] from the parent molecule. The identity of the product was confirmed by its appearance in suspension immediately after adding activated carbon and disappearance accompanied by accumulation of TNBAld and then TNB. Results indicated stoichiometric conversion of (presumably) TNBAlc to TNB under oxic conditions. No accumulation of TNT reduction products (2HADNT, 4HADNT, 2ADNT, 4ADNT, 2,2'AZT, 2,4'AZT or 4,4'AZT) was indicated by HPLC or GC/MS in the activated carbon suspension. In contrast, most of the TNB was reversibly adsorbed to the activated carbon, and no TNB products accumulated in the activated carbon suspension.

4. DISCUSSION

TNT degradation decreased with increasing initial concentration. After 4 months, up to 40 and 7% of the initial TNT (2000 mg/kg) were present as TNT and its HPLC-detectable products, respectively. The persistence of TNT in highly contaminated soils can be attributed in part to toxicity to microorganisms and plants. The TNT fraction that is extractable by 3 mM $CaCl_2$ was most critical to toxicity. The soils, which yielded 20-80 mg/l in the $CaCl_2$ extracts, were toxic to tall fescue and corn as well as to soil bacteria. In less contaminated soil (500 mg TNT/kg), TNT was reduced below 5 mg/l after about a month, permitting weak plant growth; the plants died at higher TNT concentrations. Our observations are consistent with the responses of other terrestrial plants to TNT contamination [16,17]. We also confirmed the reported toxicity of TNT to microorganisms [2-6].

Minimal doses of activated carbon amendments (0.25, 0.75 and 1.0%, w/w for 500, 1000 and 2000 mg TNT/kg, respectively) maintained the TNT concentration in soil solution below 5 mg/l, resulting in low toxicity to plants and microorganisms. Moreover, addition of activated carbon alone increased the number of soil microorganisms, probably due to the resulting higher water-holding capacity and/or greater surface area available for microbial growth.

Thus, TNT extractable with 3 mM $CaCl_2$, which simulates a soil solution [14], can be considered "readily available," as suggested by Comfort et al. [1]. Toxicity from the remaining extractable TNT was low, but the TNT was potentially available for chemical and microbial transformation. The comparatively rapid and extensive transformation of TNT indicates controlled availability of reversibly adsorbed (extractable) TNT, consistent with the microbial availability of chlorophenols and chloroanilines adsorbed to activated carbon [18, 19].

The rapid decrease in TNT concentration immediately after adding activated carbon was due to strong binding, with a gradual increase in bound (unextractable) residues of TNT and its transformation products during the subsequent incubation. After 120 d, the activated carbon amendment decreased total TNT and transformation products by 89% compared to 55% in unamended soil. Experiments with [^{14}C]TNT confirmed that most of the transformed TNT became strongly bound residues, with small amounts of HPLC-detectable products. First-order rate constants indicated that TNT transformation was also accelerated in the presence of activated carbon, apparently due to greater microbial activity in soil amended with activated carbon.

Postulated mechanism(s) of TNT transformation and binding in soil are derived from our experimental results and those of other researchers (Figure 6). Chemical oxidation and microbial/chemical processes occur simultaneously. Chemical oxidation of the methyl group results in polymerization and/or covalent binding of the 2,4,6-trinitrobenzyl-radical and TNBAld to soil. A small amount of the TNT is oxidized to TNB. Microbial and chemical reduction of TNT nitro groups results in binding of the hydroxylamino and amino derivatives to the soil matrix. In soil amended with activated carbon, similar processes contribute to TNT transformation and detoxification; however rates are substantially higher compared to native soil. Rapid accumulation of [^{14}C]TNBAld in soil treated with [^{14}C]TNT and simultaneous accumulation of unextractable ^{14}C immediately after mixing with activated carbon may indicate abiotic TNT oxidation and binding to the soil-activated carbon matrix. Catalytic oxidation of the TNT methyl group and strong binding appears associated with the activated carbon amendment because similar oxidation products (TNBAlc, TNBAld and TNB) were detected in the carbon suspension. This contrasted with the reversible adsorption of TNB and its stability to further oxidation in the presence of activated carbon.

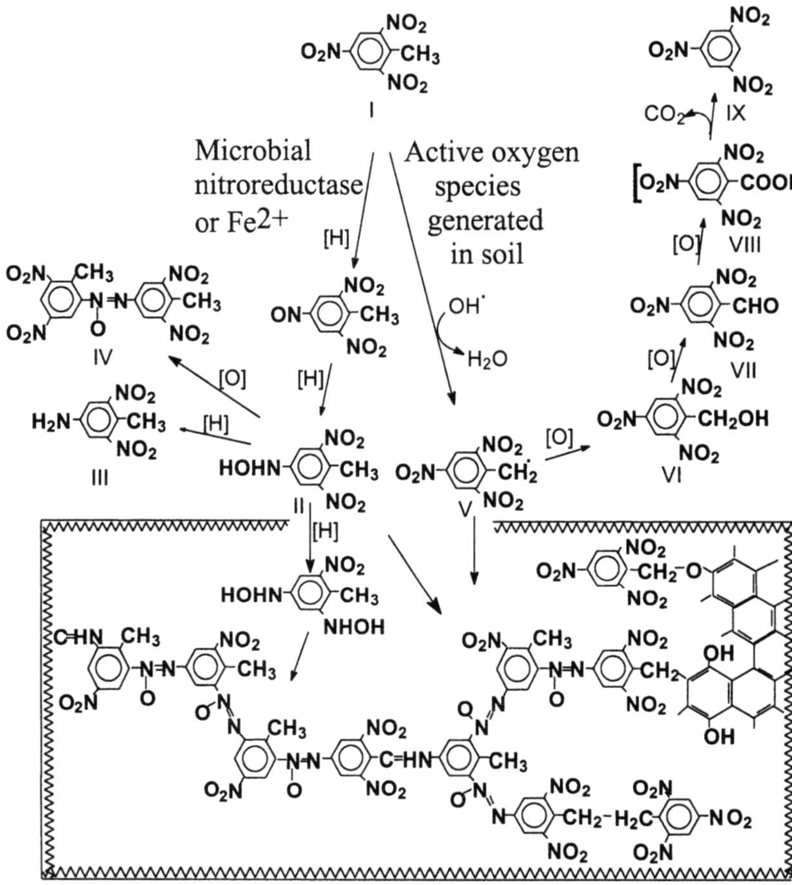

Figure 6. Proposed mechanism of TNT transformation and binding in soil.

Catalytic oxidation of TNT likely proceeds through a free radical mechanism(s) and transient accumulation oftrinitrobenzyl radical ($PiCH_2^{\cdot}$, where Pi is 2,4,6-trinitrophenyl). The process can be attributed to surface oxic groups of activated carbon (such as quinone) and metallic ions known to catalyze oxidation, following a scheme similar to that suggested for butyl mercaptan oxidation [20]. Although the TNB aromatic ring is very persistent and resists oxidation due to the three symmetric electron-withdrawing nitro groups, the TNT methyl group can be attacked by reactive oxygen species, forming the trinitrobenzyl radical in which an unpaired electron is delocalized on the aromatic ring [21].

Trinitrobenzyl radicals accumulating on the carbon surface may form covalent bonds with active surface groups (phenyl and hydroquinone), as demonstrated for 3,5-dinitrofluorobenzene and 4-nitroaniline [20]. In addition, the trinitrobenzyl radicals can react with TNT, other oxidized products, and with other trinitrobenzyl radicals to produce various oligomers and polymers. Dimers may include 2,2',4,4',6,6'-hexanitrobibenzyl ($PiCH_2$-CH_2Pi) and 2,2',4,4',6,6'-hexanitrostilbene (PiCH=CHPi), as well as 2,2'-dicarboxy-3,3',5,5'-tetranitroazoxybenzene resulting from intramolecular redox reactions catalyzed by the carbon [20]. These and other polymerized products of TNT may become unextractable from soil micropores.

All of the dimer products, along with TNBAlc, TNBAld, TNBAcid and TNB, were detected during TNT heating [22] and photodecomposition in aqueous solution in the presence of H_2O_2 [23], O_3 [24], or TiO_2 [25], as well as in reaction with aryl halides [26]. Transient formation of the monomeric oxidation products of TNT was also demonstrated during transformation through the Fenton reaction [27]; Ho [28] postulated a similar pathway for 2,4-dinitrotoluene photolysis in the presence of H_2O_2.

Catalyzed oxidation on the surface of activated carbon has also been demonstrated for several substituted phenols, anilines, polychlorinated biphenyls and humic acids. Oxygen significantly increased the adsorptive capacity of the activated carbon for these compounds [29-32].

Detection of small amounts of TNBAld (0.4 mg/kg) and TNB (1.6 mg/kg) and rapid binding of 5 to 20% of the TNT in unamended soil may be due in part to similar mechanisms of reaction with hydroxyl radicals and other active oxygen species that have been detected at low concentrations in soil [33]. Whereas Esteve-Núñez and Ramos [34] reported little accumulation of oxidized products (TNBAld, TNB, and 3,5DNA) during TNT transformation by *Pseudomonas* sp. under anaerobic conditions, Vanderberg et al. [35] detected other oxidized products (2-amino-4,6-dinitrobenzoic acid and 2,4-diamino-6-nitrobenzyl methyl ester) during reduction by *Mycobacterium vaccae* in the presence of propane.

In contrast, TNT is easily reduced in soil (chemically and by various microorganisms), under both anaerobic and aerobic conditions. The detected TNT reduction products (2ADNT, 4ADNT, small amounts of hydroxylaminodinitrotoluenes (HADNTs) and tetranitroazoxytoluenes) are commonly found in soil at all munitions-contaminated sites and usually accumulate in TNT-contaminated soils during bioremediation [7-14]. More rapid accumulation and disappearance of the reduced TNT products in soil amended with activated carbon compared to unamended soil indicates the primary role of microbial reduction of TNT in the second step of TNT transformation and detoxification through binding. Recent studies [36, 37] indicate rapid and strong binding of TNT reduction products in soil during sequential anaerobic/aerobic treatment. The hydroxylamino derivatives evidently were key products that became bound to the soil matrix. Formation of -NO=N- and -CH=N- bonds in soil has been demonstrated through NMR spectroscopy in experiments with [$^{15}N_3$]TNT [36]. In our experiments, the hydroxylamino derivatives were probably adsorbed to the activated carbon-soil matrix and became bound by dimerization to azoxy compounds or formed higher oligomers with other TNT molecules.

Although a substantial amount of TNT (up to 80%) can be transformed to tetranitroazoxytoluenes (AZTs) in some culture media [6], little AZTs accumulated in soil. This fact may indicate that HADNTs preferentially interact with soil organic matter or that higher oligomers are most commonly formed. TNT binding via formation of polymerized TNT products with poly azoxy bonds is particularly favored during sequential anaerobic/aerobic treatment because dihydroxylamino derivatives are formed simultaneously. In addition, Shiff base bonds may form between nitrogen of TNT and carboxy groups of soil humus or other covalent bonds. Similar bonds were described in laccase-catalyzed reactions between 2,4-diamino-6-nitrotoluene and the model compound guaiacol [38]. Shiff base bonds are also possible between TNT transformation products with reduced NHOH or NH_2 groups and the oxidized CHO group.

5. CONCLUSIONS

Adding activated carbon to the soil decreased TNT concentrations in soil solution and created favorable conditions for growth of microorganisms and plants. It accelerated binding because the TNT reversibly adsorbed to the activated carbon was gradually available for microbial transformation. Moreover, the activated carbon sharply increased TNT binding through an oxidative mechanism(s). These observations indicate the potential use of activated carbon to minimize toxicity of TNT and its products, thus promoting further TNT transformation and bound residue formation in highly contaminated soils.

ACKNOWLEDGMENTS

We thank Drs. Rhae Drijber and Steve Comfort for valuable suggestions and use of their laboratory facilities.

REFERENCES

1. Comfort, S.D., Shea, P.J., Hundal, L.S., Li, Z., Woodbury, B.L., Martin, J.L., Powers, W.L., 1995. TNT transport and fate in contaminated soil. J. Environ. Qual. 24, 1174-1182.
2. Drzyzga, O., Gorontzy, T., Schmidt, A., Blotevogel, K-H., 1995. Toxicity of explosives and related compounds to the luminescent bacterium *Vibrio fischerii* NRRL-B-11177. Arch. Environ. Contam. Toxicol. 28, 229-235.
3. Rieger, P.G., Knackmuss, H.J., 1995. Basic knowledge and perspectives on biodegradation of 2,4,6-trinitrotoluene and related nitroaromatic compounds in contaminated soil. *In*: Spain, J.C. (Ed.), Biodegradation of Nitroaromatic Compounds. Environ. Res. Ser. 49, Plenum Press, NY. pp. 1-18.
4. Fuller, M.E., Manning, J.F., 1998. Evidence of differential effects of 2,4,6-trinitrotoluene and other munitions compounds on specific subpopulations of soil microbial communities. Environ. Toxicol. Chem. 17, 2185-2188.
5. Gong, P., Siciliano, S.D., Greer, C.W., Paquet, L., Hawari, J., Sunahara, G.I., 1999. Effects and bioavailability of 2,4,6-trinitrotoluene in spiked and field-contaminated soils to indigenous microorganisms. Environ. Toxicol. Chem. 18, 2681-2688.
6. Vasilyeva, G.K., Oh, B.-T., Shea, P.J., Drijber, R.A., Kreslavski, V.D., Minard, R., Bollag, J.-M., 2000. Aerobic TNT reduction via 2-hydroxylamino-4,6-dinitrotoluene by *Pseudomonas aeruginosa* strain MX isolated from munitions-contaminated soil. Biorem. J. 4, 111-124.
7. Colli, S.L. Donnelly, K.C., 1995. Degradation of 2,4,6-trinitrotoluene (TNT) in an aerobic reactor. Chemosphere 31, 3025-3032.
8. Funk, S.B., Crawford, D.L., Roberts, D.J., Crawford, R.L., Mead, G., Davis-Hoover, W.,

1995. Full-scale anaerobic bioremediation of trinitrotoluene (TNT) contaminated soil. Appl. Biochem. Biotech. 51/52, 625-633.
9. Pasti-Grigsby, M.B., Lewis, T.A., Crawford, D.L., Crawford, R.T., 1996. Transformation of 2,4,6-trinitrotoluene (TNT) by actinomycetes isolated from TNT-contaminated and uncontaminated environments. Appl. Environ. Microbiol. 62, 1120-1123.
10. Thorne, P.G., Leggett, D.C., 1997. Hydrolytic release of bound residues from composted soil contaminated with 2,4,6-trinitrotoluene. Environ. Toxicol. Chem. 16, 1132-1134.
11. Widrig, D.L., Boopathy, R., Manning, J.F., 1997. Bioremediation of TNT contaminated soil: a laboratory study. Environ. Toxicol. Chem. 16, 1141-1148.
12. Daun, G., Lenke, H., Reus, M., Knackmuss, H.-J., 1998. Biological treatment of TNT-contaminated soil. 1. Anaerobic cometabolic reduction and interaction of TNT and metabolites with soil components. Environ. Sci. Technol. 32, 1956-1963.
13. Lenke, H., Warrelmann, J., Daun, G., Hund, K., Sieglen, U., Walter, U., Knackmuss, H.-J., 1998. Biological treatment of TNT-contaminated soil. 2. Biologically induced immobilization of the contaminants in full-scale application. Environ. Sci. Technol. 32, 1964-1971.
14. Vasilyeva, G.K., Kreslavski, V.D., Oh, B.-T., Shea, P.J., 2001. Potential of activated carbon to decrease 2,4,6-trinitrotoluene toxicity and accelerate soil decontamination. Environ. Toxicol. Chem. 20, 965-971.
15. Lorch, H.J., Benckieser, G., Ottow, J.C.G., 1995. *In*: Alef, K., Nanipieri, P. (Eds.), Methods in Applied Soil Microbiology and Biochemistry, Academic Press, San Diego, CA. pp. 146-191.
16. Peterson, M.M., Horst, G.L., Shea, P.J., Comfort, S.D., Peterson, R.K.D., 1996. TNT and 4-amino-2,6-dinitrotoluene influence on germination and early seedling development of tall fescue. Environ. Pollut. 93, 57-62.
17. Thompson, P.L., Ramer, L.A., Guffey, A.P., Schnoor, J.L., 1998. Decreased transpiration in poplar trees exposed to 2,4,6-trinitrotoluene. Environ. Toxicol. Chem. 17, 902-906.
18. Speitel, G.E., Lu, C.J., Zhu, X.J., 1989. Biodegradation of trace concentrations of substituted phenols in granular activated carbon columns. Environ. Sci. Technol. 23, 68-74.
19. Vasilyeva, G.K., Bakhaeva, L.P., Surovtseva, E.G., 1996. The use of in situ soil adsorptive bioremediation following an accidental spill of propanil in the Krasnodar region of Russia. Land. Contam. Reclam. 4, 263-268.
20. Bansal, R.C., Donnet, J.-B., Stoeckli, F. 1988. Active Carbon. Marcel Dekker, Inc., NY. 482 P.
21. Nouhebel. D.C.,Walton. J.C., 1974. Free Radical Chemistry. University Press, Cambridge. 260 P.
22. Dacons, J.C., Adolph, H.G., Kamlet, M., 1970. Some novel observations concerning the thermal decomposition of 2,4,6-trinitrotoluene. J. Phys. Chem. 74, 3035-3040.
23. Burlison, N.E., Sitzman, M.E., Kaplan, L.A., Kayser, E., 1979. Photochemical generation of the 2,4,6-trinitrobenzyl anion. J. Org. Chem. 44, 3695-3698.
24. Wang, A., Yin, J., 1988. Removal of aromatic compounds from water by ozonation. Ozone. Sci. Eng. 10, 1-23.
25. Dillert, R., Brandt, M., Fornefett, I., Siebers, U., Bahnemann, D., 1995. Photocatalytic degradation of trinitrotoluene and other nitroaromatic compounds. Chemosphere 30, 2333-2341.

26. Shipp, K.G., Kaplan, L.A., Sitzman, M.E., 1972. Reaction of o-substituted polynitrotoluenes. III. 2,4,6-trinitrobenzyl anion as a nucleophile at aromatic carbon J. Org. Chem. 37, 1966-1970.
27. Li, Z.M., Comfort, S.D., Shea, P.J., 1997. Destruction of 2,4,6-trinitrotoluene by Fenton oxidation. J. Environ. Qual. 26, 480-487.
28. Ho, P.C.. Photooxidation of 2,4-dinitrotoluene in the presence of hydrogen peroxide. Environ. Sci. Technol. 20, 260-267.
29. Yonge, D.R., Keinath, T.M., Poznanska, K., Jiang, Z.P., 1985. Single-solute irreversible adsorption on granular activated carbon. Environ. Sci. Technol. 19, 690-694.
30. Vidic, R.D., Suidan, M.T., 1991. Role of dissolved oxygen on the adsorptive capacity of activated carbon for synthetic and natural organic matter. Environ Sci Technol. 25, 1612-1618.
31. Vidic, R.D., Suidan, M.T., Brenner, R.C., 1993. Oxidative coupling of phenols on activated carbon: Impact on adsorption equilibrium. Environ. Sci. Technol. 27, 2079-2085.
32. Tessmer, C.H., Vidic, R.D., Uranowski, L.J., 1997. Impact of oxygen-containing surface functional groups on activated carbon adsorption of phenols. Environ. Sci. Technol. 31, 1872-1878.
33. Ruggiero, P., Dec, J., Bollag, J.-M., 1996. Soil as a catalytic system. *In*: Stotzky, G., Bollag, J-M. (Eds). Soil Biochemistry. Vol. 9, Marcel Dekker, NY. pp. 79-120.
34. Esteve-Núñez, A., Ramos, J.L., 1998. Metabolism of 2,4,6-trinitrotoluene by *Pseudomonas* sp. JLR11. Environ. Sci. Technol. 32, 3802-3808.
35. Vanderberg, L.A., Perry, J.J., Unkefer, P.J., 1995. Catabolism of 2,4,6-trinitrotoluene by *Mycobacterium vaccae.* Appl. Microbiol. Biotechnol. 43, 937-945.
36. Achtnich, C., Fernandes, E., Bollag, J.-M., Knackmuss, H.-J., Lenke, H., 1999. Covalent binding of [^{15}N]TNT to soil organic matter during a bioremediation process analyzed by 15N NMR spectroscopy. Environ. Sci. Technol. 33, 4448-4456.
37. Achtnich, C., Sieglen, U., Knackmuss, H.-J., Lenke, H., 1999. Irreversible binding of biologically reduced 2,4,6-trinitrotoluene to soil. Environ. Toxicol. Chem. 18, 2416-2423.
38. Dawel, G., Kastner, M., Michels, J., Popitz, W., Gunther, W., Fritsche, W., 1997. Structure of laccase-mediated product of coupling of 2,4-diamino-6-nitrotoluene to guaiacol, a model for coupling of 2,4,6-trinitrotoluene metabolites to a humic organic soil matrix. Appl. Environ. Microbiol. 63, 2560-2565.

Developments in Soil Science, Volume 28B
Editors: A. Violante, P.M. Huang, J.-M. Bollag and L. Gianfreda

PHYTOREMEDIATION OF OIL-CONTAMINATED SOILS

J.J. Germida, C.M. Frick and R.E. Farrell

Department of Soil Science, University of Saskatchewan, 51 Campus Drive, Saskatoon, Saskatchewan S7N 5A8, Canada

Phytoremediation (i.e., plant–assisted bioremediation) of hydrocarbon-contaminated soils is a steadily evolving technology that holds promise as an effective and low-cost alternative to most engineering techniques and traditional bioremediation methods. It is essentially *ecological engineering,* that capitalizes on the naturally occurring, synergistic relationships among plants, microorganisms, and the environment that have evolved over millions of years. Phytoremediation takes advantage of the fact that plants have extensive root systems that explore large volumes of soil, support larger bacterial populations in the rhizosphere (the region immediately surrounding the root) than in the surrounding bulk soil, and produce exudates which can directly affect the activity of the rhizobacterial populations. Given that petroleum hydrocarbons are amenable to microbial degradation and that phytoremediation often involves enhanced microbial degradation of organic contaminants in the rhizosphere, oil contaminated soils are considered to be good candidates for phytoremediation. Nevertheless, despite the flexibility and adaptability that various plant-associated remediation pathways provide, interactions between these pathways as well as the biochemical and ecological interactions between the plant/microbe/environment continuum give rise to a high level of complexity surrounding phytoremediation. Understanding this complexity is crucial to the success of any phytoremediation effort. Here, we discuss the key processes involved in phytoremediation, its promises and limitations, and its applicability to the harsh environments (semi-arid, short growing season, and long, cold winters) characteristic of the oil and gas producing regions of western Canada.

1. INTRODUCTION

Petroleum hydrocarbons are naturally occurring chemicals exploited by humans for a wide variety of activities. Representative types of petroleum hydrocarbons include alkanes (e.g., methane, ethane, propane), aromatics (e.g. benzene, toluene, ethylbenzene, and xylene, collectively known as BTEX), polycyclic aromatic hydrocarbons (PAHs; e.g., naphthalene, phenanthrene, anthracene, benzo[a]pyrene), and creosote [1, 2]. As our reliance on petrochemicals has increased, so has the number of sites contaminated with petroleum hydrocarbons [3]. In Saskatchewan alone, for example, it is estimated that there currently are more than 3000 sites contaminated with petroleum hydrocarbons [4]. Today's environmental managers can choose from a variety of approaches to remediate petroleum-contaminated soil and groundwater. These range from intensive engineering techniques to natural attenuation–a "hands-off" approach that relies entirely on natural processes to remediate sites with no human intervention. Phytoremediation is the name given to a set of emerging environmental cleanup technologies that use vegetation to enhance the dissipation or stabilization of environmental contaminants [5]. As phytoremediation employs human initiative to enhance natural, plant-assisted degradation processes, it represents a technology that is intermediate between engineering and natural attenuation.

Though generally considered a long-term remediation option, limited to soils where the contamination is shallow and occurs at low to medium concentrations, phytoremediation has been used effectively to remediate a variety of organic contaminants (including petroleum hydrocarbons) in soil and groundwater. For example, hybrid poplar trees (*Populus deltoides* x *nigra*) can take up, and degrade, both trichloroethylene (TCE) [6] and the herbicide atrazine [7] from contaminated soil. Siciliano and Germida [8] demonstrated that forage grasses inoculated with bacteria can degrade a variety of chlorinated benzoic acids–both alone and in mixtures. Various grasses and leguminous plants, together with their associated microorganisms, also have been shown to increase the removal of petroleum hydrocarbons from contaminated soil [9–13].

The objective of this paper is to showcase the potential of phytoremediation as an effective tool for cleaning up soils contaminated with petroleum hydrocarbons. We have approached this objective by reviewing the literature dealing with both the fundamental and applied aspects of phytoremediation as it pertains to petroleum hydrocarbons.

2. PHYTOREMEDIATION MECHANISMS

Plant-assisted remediation of soils contaminated with organics (including petroleum hydrocarbons) is generally believed to occur through one or more of the following mechanisms: phytostabilization, phytodegradation, phytovolatilization, and rhizodegradation (Figure 1) [14–16]. A brief description of these key degradation processes follows.

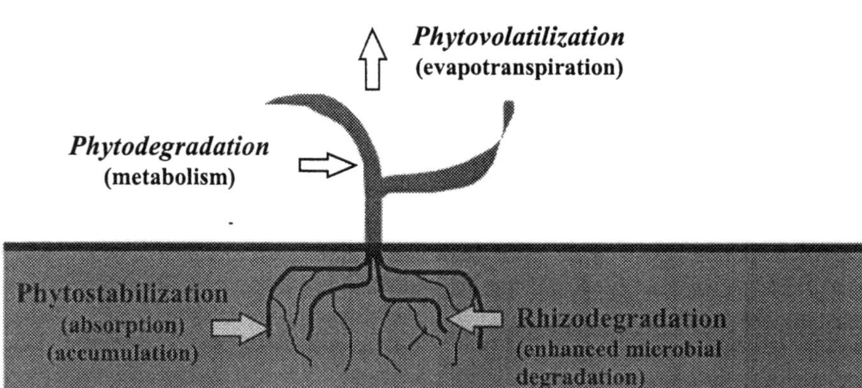

Figure 1. Mechanisms involved in the phytoremediation of petroleum hydrocarbons.

2.1. Phytostabilization

Phytostabilization involves the use of plants to contain, or immobilize, contaminants in the soil or groundwater. The mechanisms involved may include absorption and accumulation by roots, adsorption onto root surfaces, or precipitation within the root zone. Plant uptake and accumulation of petroleum hydrocarbons from contaminated soil, however, is generally quite small (e.g., 2 to 8% accumulation of soil-applied [^{14}C]benzene in alfalfa shoots and less than 2% in roots) [17, 18]. Thus, in the case of petroleum hydrocarbons, phytostabilization may simply involve the establishment of a vegetative cover to minimize any potential migration of the contaminant through erosion, leaching, or soil dispersion. Plants (especially trees) also can act as organic pumps, transpiring water, which in turn keeps the contaminant in the root zone and helps prevent it from migrating off-site [9, 19].

The phytostabilization of petroleum hydrocarbons (and other organic contaminants) also can involve incorporation of the contaminant into humic materials in the rhizosphere. That is, the plants may supply enzymes that bind a contaminant into soil organic matter (humus) or may increase the organic matter content of the soil, thus allowing for increased humification [14]. Indeed, Aprill and Sims [9] attributed the disappearance of four PAHs in soils planted with deep rooted prairie grasses to humification of the PAHs.

Finally, lipophilic organic contaminants (including some petroleum hydrocarbons) can bind with soil organic matter. Thus, the addition of plant-derived organic matter (in the form of exudates, mucigels, lysates, or residues) to the soil may reduce the bioavailability of a contaminant and, hence, its biodegradability [14].

2.2. Phytodegradation

Phytodegradation (sometimes referred to as phytotransformation) involves the breakdown of contaminants either internally, through metabolic processes, or externally, through the release of plant-produced enzymes into the soil. To date, however, there have been only a few studies in which plants have been shown to degrade petroleum hydrocarbons directly [20–22]. Plants can play an indirect role in the degradation of petroleum hydrocarbons by releasing enzymes capable of transforming these contaminants by catalyzing chemical reactions in soil. Schnoor et al. [19] identified plant enzymes as the causative agents in the transformation of contaminants mixed with sediment and soil. Isolated enzyme systems included dehalogenase, nitroreductase, peroxidase, laccase, and nitrilase. These findings suggest that plant enzymes might have significant catalytic effects well beyond the plant and also support the idea that the influence of phytoremediation may be felt well past the death of a single plant [14].

2.3. Phytovolatilization

Phytovolatilization refers to the uptake and transpiration of a contaminant by a plant. In this way, the contaminant itself, or a metabolite of the contaminant (if phytodegradation is also involved), is released into the atmosphere. For example, Watkins et al. [23] found that the volatilization of [^{14}C]naphthalene was enhanced in a sandy loam soil planted to bell rhodesgrass (*Cloris gayana*) compared to unvegetated soil. They suggested that the naphthalene was taken up by the roots of the grass, translocated within the plant and volatilized through the stems and leaves. The authors noted that this mechanism of removal would reduce the amount of naphthalene available for biodegradation, but may have implications for air quality monitoring and regulatory compliance. Likewise, Wiltse et al. [24] observed leaf burn in alfalfa plants growing in crude oil-contaminated soil, and suggested that an unidentified compound from the contaminated soil was being translocated and volatilized through the plant. The leaf burn gradually disappeared as the experiment progressed, indicating dissipation of the contaminants responsible for the effect.

2.4. Rhizodegradation

Rhizodegradation (also referred to as rhizosphere biodegradation, enhanced rhizosphere biodegradation, and plant-assisted biodegradation) involves the breakdown of contaminants in the soil as a result of microbial activity that is enhanced in the presence of the rhizosphere. Plants and microorganisms are involved–both directly and indirectly–in the degradation or transformation of petroleum hydrocarbons into products that are generally less toxic and less persistent in the environment than the parent compounds [25]. However, though plants and microorganisms can degrade petroleum hydrocarbons independently of each other, it is the interaction between plants and microorganisms in the rhizosphere that is considered the primary mechanism by which petroleum hydrocarbons are degraded in soils [10]. Enhanced rhizosphere biodegradation is discussed in greater detail in the following section.

3. ENHANCED RHIZOSPHERE BIODEGRADATION

3.1. The rhizosphere effect

The rhizosphere is the zone of soil under the direct influence of roots and is generally considered to extend 1–5 mm from the root surface. Plant roots excrete a variety of organic compounds (i.e., root exudates) into the rhizosphere. These compounds–which include sugars, amino acids, organic acids, vitamins, tannins, alkaloids, sterols, enzymes, growth factors, and other unidentified substances [19, 26]–provide a rich source of nutrients for the microbial populations in the rhizosphere. In fact, root exudates can amount to 20% or more of plant photosynthesis annually [19] and can provide sufficient carbon and energy to support large numbers of microbes (e.g., approximately $10^8 - 10^9$ vegetative microbes per gram of soil in the rhizosphere) [27]. Due to the presence of these exudates, microbial populations and activities are 5 to 100 times greater in the rhizosphere than in the bulk soil [11, 28–30]. This plant-induced enhancement of microbial populations and activity in the root zone is referred to as the *rhizosphere effect* [29].

Living plants have extensive root systems that help to bring microbes, nutrients, and contaminants into contact with each other [14]. Thus, the presence of plants in contaminated soils greatly increases the volume of soil in which active microbial degradation can be stimulated. Several studies serve as examples of the rhizosphere effect in the phytoremediation of petroleum hydrocarbons. Gunther et al. [11] found higher microbial numbers and activity coupled with increased degradation in hydrocarbon-contaminated soil planted to ryegrass (*Lolium perenne* L.) compared to unvegetated soil. The authors suggested that plant roots stimulated microbes, which enhanced the degradation of the hydrocarbons. Jordahl et al. [31] reported that populations of microorganisms capable of degrading benzene, toluene, and xylene were five times more abundant in the rhizosphere of poplar trees (*Populus deltoides* x *nigra* DN-34, Imperial Carolina) than in the bulk soil. Nichols et al. [32] found higher numbers of organic chemical degraders in rhizosphere and contaminated soils compared to bulk and uncontaminated soils, respectively. Plants creating the rhizosphere in this experiment included alfalfa (*Medicago sativa* L.) and alpine bluegrass (*Poa alpina* L.), while the contaminants included hexadecane,

(2,2-dimethylpropyl) benzene, *cis*-decahydronaphthalene, benzoic acid, phenanthrene, and pyrene. More recently, Miya and Firestone [33] reported that increased degradation of phenanthrene in the rhizosphere of slender oat (*Avena barbata*) was associated with increased numbers of phenanthrene-degraders in the rhizosphere soil compared to bulk soil.

It is important to note, however, that not all investigators have found a strong link between the rhizosphere effect and the mineralization of petroleum hydrocarbons from soil. For example, Ferro et al. [17] found that the rate and extent of mineralization of [^{14}C]benzene in artificially contaminated soils planted with alfalfa (*Medicago sativa* Mesa, var. Cimarron VR) was not significantly different from that in unplanted soils. Similarly, crested wheatgrass [*Agropyron desertorum* (Fisher ex Link) Schultes] did not increase the rate or extent of mineralization of [^{14}C]phenanthrene when planted and unplanted systems were compared [34].

3.2. Role of the plant in rhizodegradation

Plants play a key role in the rhizodegradation of petroleum hydrocarbons–namely, without the plant there is no rhizosphere and, hence, no rhizosphere effect. In addition to supplying root exudates and nutrients that increase the size and activity of the microbial community, plants also play a role in the rhizodegradation of petroleum hydrocarbons by releasing organic compounds into the soil that affect the bioavailability of the contaminant, producing contaminant analogs that stimulate the growth of hydrocarbon-degrading species, and releasing exudates that serve as primary substrates for the cometabolic degradation of petroleum hydrocarbons.

Perhaps the most important role played by plants in rhizosphere biodegradation is production and release of root exudates that induce the rhizosphere effect. Root exudate patterns are known to be dependent upon plant species and the stage of plant development. Hedge and Fletcher [35] found that the release of phenolics by the roots of red mulberry (*Morus rubra* L.) increased continuously from an early vegetative stage to leaf senescence– with a massive release at the end of the season accompanying leaf senescence. Thus, the production of root exudates and the types of microbial communities they support should be considered to be a site–time–plant specific phenomenon [36]. Site and time factors include variables such as soil type, nutrient levels, pH, water availability, temperature, oxygen status of soil, light intensity, and atmospheric carbon dioxide concentration; all of which significantly affect the type and quantity of root exudates [36].

The biodegradation of any hydrophobic contaminant (e.g., petroleum hydrocarbons) is frequently limited by the amount of contaminant that is accessible to the degrading microorganisms; i.e., contaminant bioavailability. In addition to soil humic substances, root exudates may potentially interact with petroleum hydrocarbons (PAHs in particular) to alter their bioavailability [37, 38]. However, whether root exudates are produced in sufficient quantity to significantly affect the bioavailability of petroleum hydrocarbons in soils remains to be seen.

It is often recommended that growth promoting 'contaminant analogs' be used to produce enrichment cultures for studying the biodegradation of recalcitrant organic compounds [39]. Moreover, there is evidence to suggest that plant-derived hydrocarbons can be released into the rhizosphere, and that these compounds may act to stimulate the biodegradation of contaminant hydrocarbons. For example, Atlas and Cerniglia [40] attributed the rapid biodegradation of pristane and phytane in soils contaminated as a result of the *Exxon Valdez* oil spill to microbes that had evolved to consume naturally occurring terpenes produced by the surrounding pine trees.

Cometabolism is the process by which a compound that cannot support the growth of microorganisms can be modified or degraded when another growth-supporting substrate is present [41]. Organic molecules, including root exudates and petroleum hydrocarbons, can provide energy to support populations of microbes that co-metabolize specific petroleum hydrocarbons. Sims and Overcash [42] reported a significant increase in the biodegradation of benzo[*a*]pyrene in contaminated soils following the addition of phenanthrene as a primary carbon source. Such a rhizosphere effect also was observed in experiments with [^{14}C]pyrene [17] in which the authors hypothesized that plant exudates, serving as co-metabolites, may have accelerated the biodegradation of the pyrene. Likewise, Kanaly et al. [43] found that benzo[*a*]pyrene was almost completely degraded (95% degradation) by microbes in soil containing suitable co-substrates present in a crude oil mixture; however, the specific compounds acting as co-substrates in the crude oil mixture were not identified.

3.3. Role of the microorganism in rhizodegradation

Microorganisms are used to destroy or immobilize organic contaminants in the absence of plants in a process referred to as *bioremediation*. In this section, we focus on issues concerning the role of the microorganism in the biodegradation of petroleum hydrocarbons in the presence of plants–a mechanism of phytoremediation.

Microbial degradation of an organic contaminant normally occurs because the degrading microorganisms are able to use the contaminant for their own growth and reproduction [44]. Organic contaminants not only provide the microorganisms with a source of carbon, one of the basic building blocks of new cell components, they also provide electrons that the organisms use to obtain energy. Basic microbial metabolism of contaminants involves aerobic respiration (i.e., respiration in the presence of oxygen). Variations in metabolism include anaerobic respiration, secondary utilization and cometabolism, using inorganic compounds as electron donors, fermentation, and reductive dehalogenation [26].

In general, microorganisms act on a wider range of substrates, carry out more difficult degradative steps, and generally take the contaminant to a molecularly simpler end point than plants [41]. However, not all microorganisms degrade organic contaminants in the same manner. Aerobic degradation of PAHs by prokaryotic microorganisms, such as bacteria, involves a dioxygenase enzyme, the incorporation of two atoms of molecular oxygen into the contaminant, and the production of less toxic compounds such as acids, alcohols, carbon dioxide and water [25, 45]. In contrast, degradation by eukaryotic fungi initially involves the incorporation of only one atom of oxygen into the PAH, which is similar to the degradation mechanism found in mammals [46–48]. Whereas most fungal transformations result in compounds that are less toxic than the parent molecule [48], some of the minor metabolites produced during the fungal degradation of PAHs are more toxic than the parent compound.

There are several points other of interest with regards to the microbial communities involved in the phytoremediation of organic contaminants. First, soil microorganisms may experience selective enrichment of contaminant-tolerant species when exposed to a contaminant for prolonged periods of time. Such selective enrichment may, in turn, result in enhanced degradation of the contaminant [28]. Second, some species of bacteria appear to be able to degrade a wide variety of rarely-occurring compounds without having to first adapt to contaminated conditions [36]. For example, catabolic pathways in pseudomonads allow these bacteria to degrade a variety of aromatic contaminants, such as toluene, *m*-xylene, and naphthalene, without having to synthesize a large number of different enzymes [49]. Third, evidence suggests that the degradation of certain contaminants occurs only when a specific consortium of microbes is present together at a contaminated site [28]. Lappin et al. [50] isolated five species of bacteria from the rhizosphere of wheat that could grow on, and degrade, the herbicide mecoprop [2-(2-methyl 4-chlorophenoxy)propionic acid] as long as two or more species occurred together. Individually, however, none of the species could degrade mecoprop. Fourth, White and Alexander [51] found a consortium of microbes that could utilize phenanthrene sorbed to soil without first desorbing it. This has important implications for bioavailability, since it appears that some bacteria may not require certain contaminants to be in the aqueous phase before degradation can occur. Finally, the composition and size of the microbial community in the rhizosphere is dependent on the species and age of the plant, as well as on soil type [29, 52, 53]. It also may depend on the exposure history of the plant roots to contaminants [28].

Another role played by microbes involves their ability to reduce the phytotoxicity of contaminants to the point where plants can grow in adverse soil conditions and thereby stimulate the degradation of other non-phytotoxic contaminants [36]. Radwan et al. [54] found that the plant *Senecio glaucus* grew along the polluted border of an oil lake in the Kuwaiti desert. Interestingly, the plant roots and adhering sand particles were white and clean, while the surface of the transitional zone between the root and shoot was black and polluted. The authors suggested that microbes detoxified contaminants in the rhizosphere, which allowed for the survival of the plants in the oil-contaminated soils. Rasolomanana & Balandreau [55] found improved growth of rice in soil to which oil residues were applied. The authors hypothesized that the increased growth resulted from the removal of the oil residues by various bacterial species of the genus *Bacillus*, which used plant exudates to cometabolize the oil residues in the rhizosphere.

4. SPECIAL CONSIDERATIONS WITH PHYTOREMEDIATION

Several considerations must be met when implementing a phytoremediation system to reclaim/restore contaminated soils. The three primary considerations are (i) type and concentration of contaminant, (ii) establishment of appropriate plants and microorganisms, and (iii) influence of environmental factors on phytoremediation potential.

4.1 Type and concentration of petroleum hydrocarbons

The inherent degradability of a hydrocarbon depends on its chemical nature; thus, certain petroleum hydrocarbons are easier to phytoremediate than are others. In general, BTEX compounds are relatively easy to remediate because they are rapidly degraded in the presence of oxygen; are relatively soluble, thus making them bioavailable; and can serve as the primary electron donor for a wide variety of soil bacteria [44]. In general, weathering processes involving volatilization, photomodification, hydrolysis, leaching, and biotransformation, selectively reduce the concentration of easily-degradable components of a contaminant (e.g., BETX compounds); thus making older, highly weathered sites more difficult to phytoremediate than younger, less weathered sites [14, 52]. Likewise, large lipophilic molecules (such as the four and five-ring PAHs) are more difficult to remediate than smaller molecules. The difficulty in degrading these compounds is a reflection of their limited bioavailability (a consequence of their strong adsorption to soil organic matter and clay) as well as their limited ability to cross cellular membranes, which prevents their entry into microbes and plants [56]. Nevertheless, cometabolism by microorganisms has been shown to result in the degradation of some fairly large PAHs, such as benzo[*a*]pyrene [43].

In general, the hydrocarbon accumulation in plants is related to the lipid content of the plant tissue–the greater the lipid content the greater the accumulation [18, 57-59]. Attempts to model the uptake and accumulation of petroleum hydrocarbons in plants are generally based on chemical characteristics of the hydrocarbon–in particular, the affinity of the compound for lipids. Models using the octanol-water partition coefficient (K_{ow}; a measure of a chemical's affinity for water versus lipids or fats) [2] have been developed and tested to varying degrees for the uptake of organic chemicals by plants [60–63]. One general model indicates chemicals that are highly water soluble (i.e., hydrophilic compounds with a log K_{ow} <0.5) are not sufficiently sorbed to roots or actively transported through plant membranes [19]. Conversely, hydrophobic chemicals (log K_{ow} >3.0) are not easily transported within the plant because they are strongly bound to, and may not pass beyond, the surface of the root where the proportion of lipids is high [36]. Moderately hydrophobic organic chemicals (log K_{ow} = 0.5–3.0), on the other hand, are effectively taken up by plants; these chemicals include BTEX compounds, some chlorinated solvents, and short-chain aliphatic chemicals [19, 36].

Whether the effect of a contaminant is adverse or beneficial depends to a certain degree on its concentration. For example, it is well known that, within certain concentration ranges, plants and microbes can tolerate petroleum hydrocarbons. Indeed, it is this capability that provides the foundation for plant-assisted bioremediation of oil-impacted soils. Indeed, nearly 80 years ago Carr [64] reported that the addition of small amounts of crude oil (0.75%, w/w) to soil enhanced the growth of soybean and that soybean plants could 'tolerate' fairly large amounts of crude oil (4%, w/w). More recently, Radwan et al. [54] reported finding plants (primarily belonging to the family Compositae) growing in black, oil-contaminated sands in Kuwait. Growth of native, wild plants was observed at oil concentrations up to 10% (w/w).

The phytoremediation of petroleum hydrocarbons may be ineffective if concentrations of the contaminants are either too high or too low. Concentrations of contaminants that are too high will result in a toxic response, which may include the death of the plant or soil microorganisms. Contaminants generally exhibit a treatable concentration range, above which the contaminant prevents or slows metabolic activity. This, in turn, prevents the growth of new microbial biomass needed to stimulate rapid removal of the contaminant [44, 65]. For example, Rogers et al. [66] found that the growth of white clover, tilesy sage, Bering hairgrass, and alpine bluegrass was enhanced by exposure to 0.1% (w/w) of a mixture of organic chemicals, but was severely limited by exposure to higher concentrations of the same mixture.

Conversely, low concentrations of contaminants may limit the extent to which phytoremediation can further reduce the contaminant level. That is, degrader-microorganisms may be physiologically incapable of reducing contaminant concentrations to very low levels because the uptake and metabolism of the contaminant stops at low concentrations–even when environmental conditions are optimal. Low contaminant concentrations also may cause microbes capable of degrading the contaminant to switch to alternative substrates or may result in the death of the microbes due to lack of sustenance [44, 65].

4.2 Establishment of appropriate plant and microorganisms

The successful implementation of any phytoremediation system will require the establishment of appropriate plants (and/or microorganisms) at the contaminated site. Factors to consider include: (i) selection of an appropriate plant species, (ii) the influence of contaminants on seed germination, (iii) the use of native versus non-native plants, and (iv) the effectiveness of inoculating contaminated soils with microorganisms.

Plant selection is considered to be site-specific. Nevertheless, many plants have a wide habitat and/or geographic distribution. Thus, it may be possible to select suitable 'candidate' plants from the growing list of phytoremediator plants and–based on environmental and phytological characteristics–'apply' that plant (or a close relative) to a different location. To assist in the selection of appropriate phytoremediator plants, Farrell et al. [67] constructed a database of plants that play a role in the phytoremediation of petroleum hydrocarbons. The *PhytoPet©* database (ver. 2.1) is essentially an inventory of plants with a demonstrated ability to phytoremediate or, at the very least, tolerate soils contaminated with petroleum hydrocarbons. Used in conjunction with other tools (e.g., the *Phytoremediation Decision Tree* [68]), the *PhytoPet©* database can be used with site-specific information to evaluate whether phytoremediation is appropriate for a particular site. A list of plants with demonstrated phytoremediation potential is presented in Table 1.

Table 1
Plants that tolerate or phytoremediate petroleum hydrocarbons

Hydrocarbon Tolerant Plants	Ref.	Phytoremediator Plants	Ref.
Lotus corniculatus (birdsfoot trefoil)	88, 89	*Medicago sativa* L. (alfalfa)	13, 17, 24, 32, 41, 66, 90
Medicago lupulina (black medick)	88	*Andropogon gerardi* (big bluestem)	9
Trifolium repen (white clover)	88	*Bouteloua gracilis* (blue grama)	9
Melilotus altissima	88	*Elymus canadensis* (canada wild-rye)	9
Psoralea bituminosa	88	*Buchloe dactyloides* (common buffalograss)	12
		Buchloe dactyloides var Prairie (prairie buffalograss)	12
Robinia pseudacacia	88	*Festuca ruba* (red fescue)	89
Vicia tetrasperma	88	*Sorghastrum nutans* (indiangrass)	9
		Schizachyrium scoparius (little bluestem)	9, 90
		Zoysia japonica var. Meyer (meyer zoysiagrass)	12
		Lolium perenne L. (Ryegrass)	11
		Bouteloua curtipendula (side oats grama)	9
		Sorghum vulgare L. (sudangrass)	13
		Panicum virgatum (switchgrass)	9, 13
		Agropyron smithii (western wheatgrass)	9

[1] Adapted from Farrell et al. [67].

In general, establishing plants at a phytoremediation site involves getting seeds to germinate in the contaminated soil. Thus, Cunningham et al. [14] suggest that prior to planting at a given site, tests be conducted to determine whether the plant(s) being considered for phytoremediation will germinate successfully in the contaminated soil. To date, germination tests have been conducted for a variety of plants and petroleum-contaminated soils. For example, Chaineau et al. [69] assessed the germination of lettuce (*Lactuca sativa* L.), barley (*Hordeum vulgare* L.), clover (*Trifolium repens* L.), maize (*Zea mays* L.), bean (*Phaseolus vulgaris* L.), wheat (*Triticum aestivum*), and sunflower (*Helianthus annuus*) in soils contaminated with fuel oil. The authors determined that, following 8 days of exposure, the lethal concentration or LC_{50} (i.e., the concentration of fuel oil that prohibited 50% of seeds from germinating) varied from 0.3 to 4% (w/w) for lettuce, barley, clover, and maize and from 4 to 9% (w/w) for bean, wheat, and sunflower. Wiltse et al. [24] reported that alfalfa seed germinated in soil contaminated with up to 5% (w/w) crude oil. Germination of oat (*Avena sativa*) decreased from 90% to 25% as the crude oil content of an artificially contaminated soil was increased from 0 to 6% (v/w) [70]. Epuri and Sorensen [71] evaluated the germination of four species of grass (tall fescue, perennial rye grass, reliant hard fescue, and nassau Kentucky bluegrass) in soils contaminated with PAHs (ca. 0.022% w/w total PAH) and PCBs (ca. 0.0026% w/w total PCB in soil) and reported that tall fescue yielded the highest germination rate (individual rates were not reported). Though more costly and labor-intensive than planting seeds, the transplanting of seedlings or mature plants may help speed the establishment of a healthy plant community on-site.

Whatever establishment method is chosen for phytoremediation, the plant species must be well adapted to the soil and climate of the region–making soil characteristics, length of growing season, average temperature, and annual rainfall important considerations in phytoremediation planning [14, 72]. Native plants have an inherent advantage in that they already are adapted to the climatic and soil conditions of a given region. Nevertheless, situations may arise where the most effective plants (or microorganisms) may not be native to the contaminated site. A decision must then be made as to whether non-native species should be introduced. The introduction of a non-native (exotic) species into any ecosystem should not be taken lightly. Indeed, research indicates that 4 to 19% of exotic organisms introduced into natural and agricultural ecosystems in the United States have had severe adverse effects on both the environment and economy [73]. Not surprisingly, therefore, there is much concern among ecologists and naturalists about the use of exotic species as vehicles for the phytoremediation of contaminated sites. Nevertheless, the use of exotic species should not be ruled out without due consideration of the threat they actually pose. (This should include discussions with appropriate regulatory agencies and concerned public organizations about species suitability.) Genetically modified (engineered) organisms (GMOs) are considered a type of non-native organism; consequently, similar issues arise when considering the release of GMOs into the environment [73].

The utility and effectiveness of inoculating oil-impacted sites with contaminant-degrading microorganisms is the subject of much debate. For example, Cunningham et al. [14] state that it is a common experience for soil and plant inoculants to be out-competed by native microflora. However, this is true even for some symbiotic relationships (e.g., soybean/*Bradyrhizobium*). Heitkamp and Cerniglia [74] found that competition with indigenous microorganisms did not adversely affect the degradation of pyrene by a *Mycobacterium* species inoculated into sediments; that is, unless organic nutrients (glucose and peptone) were added to the sediments, in which case there was an overgrowth of indigenous bacterial species. Conversely, Radwan et al. [75] found that a population of oil-degrading *Arthrobacter* species (strain KCC201) indigenous to a sand treated with weathered crude oil (20% w/w) decreased dramatically following inoculation with non-indigenous, oil-degrading *Arthrobacter* and *Candida* species. Decreases in the number of indigenous *Arthrobacter* were thought to be due to competition with the inoculated species and the authors suggested that seeding with microbial 'cocktails' (irrespective of their origin) may be useless, or even harmful if the introduced species out-compete any indigenous, oil-degrading species.

Nevertheless, several studies have identified plant inoculation with microorganisms as a means of enhancing the degradation of organic contaminants. Schwendinger [70] identified drastically reduced adaptation times and greater total CO_2 production over a 7-week period after seeding *Cellulomonas* (a cellulose-decomposing bacterium) into sandy loam soil with relatively high levels of oil (100 mL kg^{-1} soil). Grosser et al. [76] determined that the isolation, propagation, and reintroduction of *Mycobacterium* species (at 10^7 cfu g^{-1} soil) resulted in enhanced mineralization of phenanthrene, anthracene, and pyrene; i.e., relative to mineralization by the indigenous microbial population. Likewise, Madsen and Kristensen [77] determined that soil inoculation with phenanthrene-degrading bacteria enhanced the degradation of [^{14}C]phenanthrene–particularly in soils that had no prior history of hydrocarbon contamination. However, a fairly large quantity of inoculum ($\geq 10^6$ cells g^{-1} soil) was required to establish the introduced bacteria.

In general, the bacteria *Pseudomonas*, *Arthrobacter*, *Alcaligenes*, *Corynebacterium*, *Flavobacterium*, *Achromobacter*, *Micrococcus*, *Nocardia*, and *Mycobacterium*, are reported as the most active bacterial species in the degradation of hydrocarbons in soil [52]. However, a number of fungi also degrade hydrocarbons in soil, including *Aspergillus ochraceus*, *Cunninghamella elegans*, *Phanerochaete chrysosporium*, *Saccharomyces cerevisiae*, and *Syncephalastrum racemosum* [48]. A list of hydrocarbon-degrading soil microorganisms is presented in Table 2.

4.3 Influence of environment on phytoremediation

A variety of environmental conditions affect or alter the mechanisms of phytoremediation. Soil type, structure and organic matter content can limit the bioavailability of petroleum contaminants [14, 78–80]. Plants that accumulate PAHs can experience toxic effects, such as reduced growth, resulting from modifications and photosensitizations of PAHs exposed to solar ultraviolet radiation, which penetrates into plant tissues [81–83]. Water content in soil and wetlands affects plant/microbial growth and the availability of oxygen required for aerobic respiration [25] Fertilizers increase degradation by increasing availability of and reducing competition for limited nutrients in oil-contaminated soil [70, 84]. Finally, temperature changes affect the rates at which various processes take place, with the rate of microbial degradation generally doubling for every 10 °C increase in temperature [25, 85, 86]. These various environmental factors, together with biodegradation, cause weathering, or the loss of certain fractions of the contaminant mixture, with the end result being that the more resistant compounds remaining in the soil [52].

Table 2
Soil microorganisms that degrade hydrocarbons[1]

-------- Bacteria --------		-------- Fungi --------	
Achromobacter	*Micrococcus*	*Acremonium*	*Monilia*
Acinetobacter	*Mycobacterium*	*Aspergillus*	*Mortierella*
Alcaligenes	*Norcardia*	*Aureobasidium*	*Paecilomyces*
Arthrobacter	*Proteus*	*Beauveria*	*Penicillium*
Bacillus	*Pseudomonas*	*Botrytis*	*Phoma*
Brevibacterium	*Rhodococcus*	*Candida*	*Phanerochaete*
Chromobacterium	*Sarcina*	*Chrysosporium*	*Rhodotorula*
Corynebacterium	*Serratia*	*Cladosporium*	*Saccharamyces*
Cytophaga	*Spirillum*	*Cochliobolus*	*Scolecobasidium*
Erwinia	*Streptomyces*	*Cunninghamella*	*Sporobolomyces*
Flavobacterium	*Vibrio*	*Cylindrocarpon*	*Sprotrichum*
	Xanthomonas	*Debaryomyces*	*Spicaria*
		Fusarium	*Syncephalastrum*
		Geotrichum	*Tolypocladium*
		Gliocladium	*Torulopsis*
		Graphium	*Trichoderma*
		Humicola	*Verticillum*

[1] Adapted from Frick et al. [87].

5. CONCLUSIONS

Phytoremediation, the use of plants and their associated microorganisms for the *in situ* treatment of contaminated soils, continues to emerge as a 'green' technology with potential for the effective and inexpensive cleanup of a broad range of organic and inorganic wastes. Initial indications are that phytoremediation is effective at degrading and containing petroleum hydrocarbons in soil as well as transferring these compounds from soil to the atmosphere. Phytoremediation is an *in situ*, solar driven technique that limits environmental disturbance, reduces costs, and is well-suited to the treatment of large areas of surface contamination, when other methods may not be cost-effective. In general, phytoremediation is looked upon favorably because it exploits the natural ability of the environment to restore itself [14].

Indirect benefits of phytoremediation include improvement of soil quality by improving soil structure (aggregates and peds), increasing porosity/aggregation and, therefore, water infiltration, providing nutrients (nitrogen-fixing legumes), accelerating nutrient cycling, and increasing soil organic carbon. The use of plants to remediate contaminated soils also stabilizes the soil, thus preventing erosion and direct human exposure (i.e., by preventing the consumption of contaminated soil by children and the inhalation of soil particles carried in the wind). Another indirect benefit is that the growth of certain hardy plants in a contaminated soil can allow for the growth of other, less hardy plants. An experiment outlined by Cunningham et al. [14] indicated that a tolerant grass species (*Vetiveria zizanioides*) thrived in a clay soil contaminated with up to 3% total petroleum hydrocarbons. The same soil was initially extremely phytotoxic to a variety of crop plants tested. However, after a 1-year period when the soil was cropped to only *V.*

zizanioides, several crop species could be grown together with *V. zizanioides*–even though there was no detectable change in the quantity of contaminants.

Because phytoremediation is a relatively new technology, not all of its mechanisms are completely understood. Moreover, phytoremediation has its limitations and is not appropriate for all situations. For example, phytoremediation is inappropriate when the depth of contamination extends beyond the root zone. Though some plants, such as trees, may have root systems that can extend to a depth of 60 m, most plants do not produce roots to anywhere near this depth and root density generally decreases with depth. Consequently, as depth increases beyond one or two meters, relatively immobile contaminants–those that cannot migrate to the plant roots during water uptake–are increasingly unlikely to be affected by phytoremediation.

Contaminants are tightly bound to soil particles or organic matter, may not be available to plants or microbes for degradation. Environmental conditions, such as soil texture, pH, salinity, temperature and presence of non-hydrocarbon contaminants (e.g., metals) must all be within the limits tolerated by plants. In addition, plants will not grow if concentrations of the target contaminant are too high. The effectiveness of phytoremediation also will depend on the chemical nature of the contaminants themselves. For example, there is the potential for water-soluble contaminants to leach away before phytoremediation can reclaim the area. Similarly, evaporation of volatile petroleum hydrocarbons into the air from the soil directly or through the plant is simply a transfer of the contaminant from one environmental medium to another. Therefore, air quality issues also must be considered when evaluating phytoremediation.

Finally, phytoremediation is slower than *ex situ* methods, typically requiring several seasons for site clean-up. This is particularly true for hydrophobic pollutants (including petroleum hydrocarbons) that are tightly bound to soil particles. Thus, phytoremediation is not an appropriate solution where the target contaminant presents an immediate danger to human health or the environment.

REFERENCES

1. Lyons, W. C., 1996. Standard Handbook of Petroleum and Natural Gas Engineering. Gulf Publishing Co., Houston,TX.
2. Mackay, D., 1991. Multimedia Environmental Models: The Fugacity Approach. Lewis Publishers, Chelsea, MI.
3. Bauman, B., 1991. Research needs: motor fuel contaminated soils. *In*: Calebrese, E.J., Kostecki, P.T. (Eds.), Hydrocarbon Contaminated Soils. Lewis Publishers, Chelsea, MI, pp. 41-56.
4. Han, T., 2000. Personal communication.
5. Kulakow, P., 2000. Annual Report of the RTDF Phytoremediation Action Team – TPH Subgroup Cooperative Field Trials. Available on-line (in pdf format–1368K) at: http://www.engg.ksu.edu/HSRC/RTDFtphrp1.html.
6. Newman, L.A., Strand, S.E., Choe, N., Duffy, J., Ekuan, G., Ruszaj, M., Shurtleff, B.B., Wilmoth, J., Heilman, P., Gordon, M.P., 1997. Uptake and biotransformation of trichloroethylene by hybrid poplars. Environ. Sci. Technol. 31, 1062-1067.
7. Burken, J.G., Schnoor, J.L., 1997. Uptake and metabolism of atrazine by poplar trees. Environ. Sci. Technol. 31, 1399-1406.
8. Siciliano, S.D., Germida, J.J., 1998. Degradation of chlorinated benzoic acid mixtures by plant-bacteria associations. Environ. Toxicol. Chem. 17, 728-733.

9. Aprill, W., Sims, R.C., 1990. Evaluation of the use of prairie grasses for stimulating polycyclic aromatic hydrocarbon treatment in soil. Chemosphere 20, 253-265.
10. Banks, M.K., Govindaraju, R.S., Schwab, A.P., Kulakow, P., 2000. Field demonstration. *In*: Fiorenza, S., Oubre, C.L., Ward, C.H. (Eds.), Phytoremediation of Hydrocarbon-Contaminated Soil. Lewis Publishers, Boca Raton, FL, pp. 3-88.
11. Gunther, T., Dornberger, U., Fritsche, W., 1996. Effects of ryegrass on biodegradation of hydrocarbons in soil. Chemosphere 33, 203-215.
12. Qiu, X., Leland, T.W., Shah, S.I., Sorensen, D.L., Kendall, E.W., 1997. Field study: grass remediation for clay soil contaminated with polycyclic aromatic hydrocarbons. *In*: Kruger, E.L., Anderson, T.A., Coats, J.R. (Eds.), Phytoremediation of Soil and Water Contaminants. ACS Symposium Series 664. American Chemical Society, Washington, DC, pp. 186-199.
13. Reilley, K.A., Banks, M.K., Schwab, A.P., 1996. Organic chemicals in the environment: dissipation of polycyclic aromatic hydrocarbons in the rhizosphere. J. Environ. Qual. 25, 212-219.
14. Cunningham, S.D., Anderson, T.A., Schwab, A.P., Hsu, F.C., 1996. Phytoremediation of soils contaminated with organic pollutants. Adv. Agron. 56, 55-114.
15. Davis, L.C., Banks, M.K., Schwab, A.P., Narayanan, M., Erickson L.E., 1998. Plant-based bioremediation. *In*: Sikdar, S.K., Irvine R.L. (Eds.), Bioremediation: Principles and Practice, Vol. 2. Biodegradation Technology Developments. Technomic Publ. Co., Lancaster, PA, pp. 183-217.
16. United States Environmental Protection Agency, 1999. Phytoremediation Resource Guide. Office of Solid Waste and Emergency Response, Washington, DC. EPA 52-B-99-003.
17. Ferro, A., Kennedy, J., Doucette, W., Nelson, S., Jauregui, G., McFarland, B., Bugbee, B., 1997. Fate of benzene in soils planted with alfalfa: uptake, volatilization, and degradation. *In*: Kruger, E.L., Anderson, T.A., Coats, J.R. (Eds.), Phytoremediation of Soil and Water Contaminants. ACS Symposium Series 664. American Chemical Society, Washington, DC, pp. 223-237.
18. Edwards, N.T., 1983. Polycyclic aromatic hydrocarbons (PAH's) in the terrestrial environment - a review. J. Environ. Qual. 12, 427-441.
19. Schnoor, J.L., Licht, L.A., McCutcheon, S.C., Wolfe, N.L., Carreira, L.H., 1995. Phytoremediation of organic and nutrient contaminants. Environ. Sci. Technol. 29, 318-323.
20. Durmishidze, S.V., 1977 (Transl. 1978). Metabolism of certain air-polluting organic compounds in plants (review). Appl. Biochem. Microbiol. 13, 646-653.
21. Edwards, N.T., Ross-Todd, B.M., Garver, E.G., 1982. Uptake and metabolism of ^{14}C anthracene by soybean (*Glycine max*). Environ. Exp. Bot. 22, 349-357.
22. Edwards, N.T., 1988. Assimilation and metabolism of polycyclic aromatic hydrocarbons by vegetation - an approach to this controversial issue and suggestions for future research. In: Polycyclic Aromatic Hydrocarbons: A Decade of Progress – 10th International Symposium, Columbus, OH, Battelle Press.
23. Watkins, J.W., Sorensen, D.L., Sims, R.C., 1994. Volatilization and mineralization of naphthalene in soil-grass microcosms. *In*: Anderson, T.A., Coats, J.R. (Eds.), Bioremediation Through Rhizosphere Technology. ACS Symposium Series 563. American Chemical Society, Washington, DC, pp. 123-131.
24. Wiltse, C.C., Rooney, W.L., Chen, Z., Schwab, A.P., Banks, M.K., 1998. Greenhouse evaluation of agronomic and crude oil-phytoremediation potential among alfalfa genotypes. J. Environ. Qual. 27, 169-173.

25. Eweis, J.B., Ergas, S.J., Chang, D.P.Y., Schroeder, E.D., 1998. Bioremediation Principles. McGraw-Hill, Inc., Toronto.
26. Sylvia, D.M., Fuhrmann, J.J., Hartel, P.G., Zuberer, D.A., 1998. Principles and Applications of Soil Microbiology. Prentice Hall, Upper Saddle River, NJ.
27. Erickson, L.E., Davis, L.C., Muralidharan, N., 1995. Bioenergetics and bioremediation of contaminated soil. Thermochim. Acta 250, 353-358.
28. Anderson, T.A., Guthrie, E.A., Walton, B.T., 1993. Bioremediation in the rhizosphere. Environ. Sci. Technol. 27, 2630-2636.
29. Atlas, R.M., Bartha, R., 1998. Microbial Ecology: Fundamentals and Applications. Benjamin/Cummings Publishing Company Inc., Don Mills, ON.
30. Paul, E.A., Clark, F.E., 1989. Occurrences and distribution of soil organics. In: Paul, E.A., Clark, F.E. (Eds.). Soil Microbiology and Biochemistry. Academic Press, San Diego, pp. 81-84.
31. Jordahl, J.L., Foster, L., Schnoor, J.L., Alvarez, P.J.J., 1997. Effect of hybrid poplar trees on microbial populations important to hazardous waste bioremediation. Environ. Toxicol. Chem. 16, 1318-1321.
32. Nichols, T.D., Wolf, D.C., Rogers, H.B., Beyrouty, C.A., Reynolds, C.M., 1997. Rhizosphere microbial populations in contaminated soils. Water Air Soil Pollut. 95, 165-178.
33. Miya, R.K., M.K. Firestone., 2000. Phenanthrene degrader community dynamics in rhizosphere soil from a common annual grass. J. Environ. Qual. 29, 584-592.
34. Ferro, A.M., Sims, R.C., Bugbe, B., 1994. Hycrest crested wheatgrass accelerates the degradation of pentachlorophenol in soil . J. Environ. Qual. 23, 272-279.
35. Hegde, R.S., J.S. Fletcher., 1997. Influence of plant growth stage and season on the release of root phenolics by mulberry as related to development of phytoremediation technology. Chemosphere 32, 2471-2479.
36. Siciliano, S.D., Germida, J.J., 1998. Mechanisms of phytoremediation: biochemical and ecological interactions between plants and bacteria. Environ. Rev. 6, 65-79.
37. Eadie, B.J., Morehead, N.R., Landrum, P.F., 1990. Three-phase partitioning of hydrophobic organic compounds in Great Lakes (USA) waters. Chemosphere 20, 161-178.
38. Kordel, W., Dassenakis, M., Lintelmann, J., Padberg, S., 1997. The importance of natural organic material for environmental processes in water and soils. Pure Appl. Chem. 69, 1571-1600.
39. Focht, D.D., 1994. Microbiological procedures for biodegradation research. *In*: Weaver, R.W., Angle, S., Bottomley, P., Bezdicek, D., Smith, S., Tabatabai, A., Wollum, A. (Eds.), Methods of Soil Analysis, Part 2. Microbiological and Biochemical Properties, Soil Science Society of America, Madison, WI, SSSA Book Series No. 5, pp. 407-426.
40. Atlas, R.M., C.E. Cerniglia., 1995. Bioremediation of petroleum pollutants: Diversity and environmental aspects of hydrocarbon biodegradation. BioScience 45, 332-339.
41. Cunningham, S.D., Berti, W.R., 1993. Remediation of contaminated soils with green plants: An overview. In vitro cell. dev. biol., Plant 29P, 207-212.
42. Sims, R.C., Overcash, M.R., 1983. Fate of polynuclear aromatic compounds (PNAs) in soil-plant systems. Residue Rev. 88, 1-68.
43. Kanaly, R., Bartha, R., Fogel, S., Findlay, M., 1997. Biodegradation of [^{14}C]benzo[a]pyrene added in crude oil to uncontaminated soil. Appl. Environ. Microbiol. 63, 4511-4515.

44. Committee on In Situ Bioremediation, Water Science and Technology Board, Commission on Engineering and Technical Systems, and National Research Council., 1993. *In-Situ* Bioremediation: When Does It Work? National Academy Press, Washington, DC.
45. Gibson, D.T., V. Subramanian., 1984. Microbial degradation of aromatic hydrocarbons. *In*: Gibson, D.T. (Ed.), Microbial Degradation of Organic Compounds. Marcel Dekker, Inc., New York, NY, pp.181-252.
46. Cerniglia, C.E., Gibson, D.T., 1979. Oxidation of benzo[a]pyrene by the filamentous fungus *Cunninghamella elegans*. J. Biol. Chem. 254, 12174-12180.
47. Cerniglia, C.E., Kelly, D.W., Freeman, J.P., Miller, D.W., 1986. Microbial metabolism of pyrene. Chem. Biol. Inter. 57, 203-216.
48. Sutherland, J.B., 1992. Detoxification of polycyclic aromatic hydrocarbons by fungi. J. Indust. Microbiol. 9, 53-62.
49. Houghton, J.E., Shanley M.S., 1994. Catabolic potential of pseudomonads: A regulatory perspective. *In*: Chaudbry, G.R. (Ed.), Biological Degradation and Bioremediation of Toxic Chemicals, Timber Press, Portland, OR, pp. 11-32.
50. Lappin, H.M., Greaves, M.P., Slater, J.H., 1985. Degradation of the herbicide mecoprop [2-(2-methyl-4-chlorophenoxy) propionic acid] by a synergistic microbial community. Appl. Environ. Microbiol. 49, 429-433.
51. White, J.C., Alexander, M., 1996. Reduced biodegradability of desorption-resistant fractions of polycyclic aromatic hydrocarbons in soil and aquifer solids. Environ. Tox. Chem. 15, 1973-1978.
52. Bossert, I., Bartha, R., 1984. The fate of petroleum in soil ecosystems. *In*: Atlas R.M. (Ed.), Petroleum Microbiology. MacMillan, New York, NY, pp. 435-473.
53. Campbell, R., 1985. Plant Microbiology. Edward Arnold, Baltimore, MD.
54. Radwan, S., Sorkhoh, N., El-Nemr, I., 1995. Oil biodegradation around roots. Nature 376, 302.
55. Rasolomanana, J.L., Balandreau, J., 1987. Rôle de la rhizosphère dans la biodégradation de composés récalcitrants: cas d'une rizire polluée par des résidus pétroliers. Rev. Ecol. Biol. Sol 24, 443-457.
56. Cookson, H.T.J., 1995. Bioremediation Engineering: Design and Application. McGraw-Hill Inc., Toronto.
57. Schwab, A.P., Al-Assi, A.A., Banks, M.K., 1998. Adsorption of naphthalene onto plant roots. J. Environ. Qual. 27, 220-224.
58. Simonich, S.L., Hites, R.A., 1994. Importance of vegetation in removing polycyclic aromatic hydrocarbons from the atmosphere. Nature 370, 49-51.
59. Simonich, S.L., Hites, R.A., 1994. Vegetation-atmosphere partitioning of polycyclic aromatic hydrocarbons. Environ. Sci. Technol. 28, 939-943.
60. Briggs, G.G., Bromilow, R.H., Evans, A.A., 1982. Relationships between lipophilicity and root uptake and translocation of non-ionised chemicals by barley. Pestic. Sci. 13, 495-504.
61. Paterson, S., Mackay, D., McFarlane, C., 1994. A model of organic chemical uptake by plants from soil and the atmosphere. Environ. Sci. Technol. 28, 2259-2266.
62. Ryan, J.A., Bell, R.M., Davidson, J.M., O'Connor, G.A., 1988. Plant uptake of non-ionic organic chemicals from soils. Chemosphere 17, 2299-2323.
63. Trapp, S., Matthies, M., Scheunert, I., Topp, E.M., 1990. Modeling the bioconcentration of organic chemicals in plants. Environ. Sci. Technol. 24, 1246-1252.
64. Carr, R. H., 1919. Vegetative growth in soils containing crude petroleum. Soil Sci. 8, 67-

68.
65. Hrudey, S.E., Pollard, S.J., 1993. The challenge of contaminates sites: remediation approaches in North America. Environ. Rev. 1, 55-72.
66. Rogers, H.B., Beyrouty, C.A., Nichols, T.D., Wolf, D.C., Reynolds, C.M., 1996. Selection of cold-tolerant plants for growth in soils contaminated with organics. J. Soil Contam. 5, 171-186.
67. Farrell, R.E., Frick, C.M., Germida, J.J., 2000. PhytoPet© – A database of plants that play a role in the phytoremediation of petroleum hydrocarbons. *In*: Proceedings of the Second Phytoremediation Technical Seminar – June 13, 2000, Vancouver, BC. Environment Canada, Ottawa, pp. 29-40.
68. Interstate Technology and Regulatory Cooperation (ITRC) Work Group – Phytoremediation Work Team, Phytoremediation Decision Tree, 1999. Available on-line (in pdf format – 139K) at: http://clu-in.org/techpubs.htm.
69. Chaineau, C.H., Morel, J.L., Oudot, J., 1997. Phytotoxicity and plant uptake of fuel oil hydrocarbons. J. Environ. Qual. 26, 1478-1483.
70. Schwendinger, R.B., 1968. Reclamation of soil contaminated with oil. J. Inst. Petrol. 54, 182-197.
71. Epuri, V., Sorensen, D.L., 1997. Benzo(a)pyrene and hexachlorobiphenyl contaminated soil: phytoremediation potential. *In*: Kruger, E.L., Anderson, T.A., Coats, J.R. (Eds.), Phytoremediation of Soil and Water Contaminants. ACS Symposium Series 664. American Chemical Society, Washington, DC, pp. 200-222.
72. Newman, L. A., Doty, S. L., Gery, K. L., Heilman, P. E., Muizieks, I., Shang, T. Q., Siemieniec, S. T., Strand, S. E., Wang, X., Wilson, A. M., Gordon, M. P., 1998. Phytoremediation of organic contaminants: a review of phytoremediation research at the University of Washington. J. Soil Contam. 7, 531-542.
73. Office of Technology Assessment (OTA), 1993. Harmful Non-indigenous Species in the United States. U.S. Government Printing Office, Washington, DC.
74. Heitkamp, M.A., Cerniglia, C.E., 1989. Polycyclic aromatic hydrocarbon degradation by a Mycobacterium sp. in microcosms containing sediment and water from a pristine ecosystem. Appl. Environ. Microbiol. 55, 1968-1973.
75. Radwan, S.S., Al-Awadhi, H., Sorkhoh, N.A., El-Nemr, I.M., 1997. Rhizospheric hydrocarbon-utilizing microorganisms as potential contributors to phytoremediation for the oily Kuwaiti desert. Microbiol. Res. 153, 247 – 251.
76. Grosser, R.J., Warshawsky, D., Vestal, J.R., 1995. Mineralization of polycyclic and n-heterocyclic aromatic compounds in hydrocarbon-contaminated soils. Environ. Toxicol. Chem. 14, 375-382.
77. Madsen, R., Kristensen, P., 1997. Effects of bacterial inoculation and nonionic surfactants on degradation of polycyclic aromatic hydrocarbons in soil. Environ. Toxicol. Chem. 16, 631-637.
78. Alexander, M., Hatzinger, P.B., Kelsey, J.W., Kottler, B.D., Nam, K., 1997. Sequestration and realistic risk from toxic chemicals remaining after bioremediation. Ann. NY Acad. Sci. 829, 1-5.
79. Boyle, J.J., Shann, J.R., 1998. The influence of planting and soil characteristics on mineralization of 2,4,5-T in rhizosphere soil. J. Environ. Qual. 27, 704-709.
80. Otten, A., Alphenaar, A., Pijls, C., Spuij, F., de Wit, H., 1997. *In Situ* Soil Remediation. Kluwer Academic Publishers, Boston.

81. Ren, L., Huang, X.-D., McConkey, B.J., Dixon, D.G., Greenberg, B. M., 1994. Photoinduced toxicity of three polycyclic aromatic hydrocarbons (fluoranthene, pyrene, and naphthalene) to the duckweed *Lemna gibba* L. G-3. Exotoxicol. Environ. Saf. 28, 160-171.
82. Huang, X.-D., Dixon, D.G., Greenberg, B.M., 1993. Impacts of UV radiation and photomodification on the toxicity of PAHs to the higher plant *Lemna gibba* (duckweed). Environ. Toxicol. Chem. 12, 1067-1077.
83. Duxbury, C.L., Dixon, D.G., Greenberg, B.M., 1997. Effects of simulated solar radiation on the bioaccumulation of polycyclic aromatic hydrocarbons by the duckweed *Lemna gibba*. Environ. Toxicol. Chem. 16, 1739-1748.
84. Steffensen, W.S., Alexander, M., 1995. Role of competition for inorganic nutrients in the biodegradation of mixtures of substrates. Appl. Environ. Microbiol. 61, 2859-2862.
85. Dibble, J.T., Bartha, R., 1979. Rehabilitation of oil-inundated agricultural land: A case history. Soil Sci. 128, 56-60.
86. Wright, A.L., Weaver, R.W., Webb, J.W., 1997. Oil bioremediation in salt marsh mesocosms as influenced by N and P fertilization, flooding, and season. Water Air Soil Pollut. 95, 179-191.
87. Frick, C.M., Germida, J.J., Farrell, R.E., 1999. Assessment of Phytoremediation as an In-Situ Technique for Cleaning Oil-Contaminated Sites. Petroleum Technology Alliance Canada, Calgary, AB, 82 pp.
88. Gudin, C., Syratt, W.J., 1975. Biological aspects of land rehabilitation following hydrocarbon contamination. Environ. Pollut. 8, 107-112.
89. Reynolds, C.M., Wolf, D.C., 1999. Microbial based strategies for assessing rhizosphere-enhanced phytoremediation. *In*: Proceedings of the Phytoremediation Technical Seminar – May 31-June 1, 1999, Calgary, AB, pp. 125-135. Environment Canada, Ottawa.
90. Pradhan, S.P., Conrad, J.R., Paterek, J.R., Srivastava, V.J., 1998. Potential of phytoremediation for treatment of PAHs in soil at MGP sites. J. Soil Contam. 7, 467-480.

Developments in Soil Science, Volume 28B
Editors: A. Violante, P.M. Huang, J.-M. Bollag and L. Gianfreda

SECONDARY MOBILIZATION OF HEAVY METALS IN POLLUTED SOILS UNDER MICROBIAL INFLUENCE (MODEL EXPERIMENT)

A.Yu. Bespalova, G.V. Motuzova and O.E. Marfenina

Moscow State M.V. Lomonosov University, Soil Science Faculty,
119189, Vorobievy Hills, Moscow, Russia

The role of the soil microbial community in heavy metals mobility change was investigated in background and strongly polluted podzols of the Kola Peninsula. It is shown that microbiological activity in polluted soils leads to copper mobilization, mostly due to organic matter decomposition and release of copper bound to it. Two common soil microfungi (*Mucor hiemalis* Wehmer and *Trichoderma viride* Pers. ex Gray), was shown to be able to modify copper availability in soil. Copper mobilization under their influence occurs in different time intervals corresponding to their growth rates.

1. INTRODUCTION

The influence of industrial pollution on the environment increases the fluxes of heavy metals in ecosystems. Most metals contaminating soil remain in the upper organic layer, which is a geochemical barrier to their further migration. Transformation of heavy metal compounds here can occur due to biotic and abiotic factors; the influences are difficult to separate. Higher plants are known to be an important biotic factor affecting metal mobility and availability in soil. Intensive copper and nickel mining at the Kola Peninsula led to the disturbance of the original vegetation over relatively large areas and development of industrial barrens and sparse forests [1].

Under specific industrial conditions, soil microorganisms can play the main role in metal compounds transformation between soluble and insoluble phases, as they are considered to be more resistant to metal contamination than other components of the biocenosis. Soil microorganisms can exhibit different detoxification mechanisms to high doses of heavy metals. Namely they are [2, 3]:

1) Maintenance of constant, relatively low heavy metal content in the cell due to:
   - membrane permeability reduction and selective ion uptake;
   - active efflux of heavy metals from the cell.
2) Accumulation of heavy metals by the cell in slightly toxic forms:
   - biosorption at the cell surface;
   - active metabolic uptake and intracellular accumulation:
     · compartmentation in subcellular structures (vacuoles, for instance);
     · storage in the form of non-toxic compounds (bound to thioles, metalloproteins, etc).
3) Decrease of heavy metal toxicity due to their binding to extracellular metabolites.

These mechanisms of interaction of microbial cells with metal compounds can play an important role in metal transformation between soluble and insoluble states.

Microscopic fungi are known to be the dominant part of the microbial community in northern soils, especially in upper organic horizons [4, 5]. They are also the most tolerant organisms of the soil microbiota to heavy metal contamination and low pH values that occur in these soils as a result of industrial pollution [6, 7]. The capacity of soil fungi to modify metal availability is shown in numerous investigations [3, 8]. This capacity has been well explored due to the possible application to metal removal from solid and liquid substrates and industrial wastes [3, 9]. It has been shown that fungi can immobilize high amounts of metals inside and on the surface of mycelium or by means of extracellular metabolites [3, 9, 10]. On the other hand, fungi can release metals from organic matter during its decomposition or leach them from inorganic substrates and increase their availability [3, 11]. However, most of investigations have been carried out with pure cultures and on synthetic media.

The problem of interaction of microorganisms with heavy metal compounds directly in soil is not sufficiently covered in the literature. The reason is that there are a lot of soil factors and environmental conditions that affect metal transformation, whose influences are difficult to distinguish. Moreover, there are difficulties in separating microorganisms from soil and in determining their particular role in metal transformation. However, there are some investigations concerning the assessment of the amount of metals associated with microorganisms in soil [8, 12] and their contribution to metal transformation in soil [8, 13].

The goals of this investigation were i) to evaluate the possible contribution of microbiota to heavy metals transformation between soluble and insoluble phases and ii) to study the effects of two common soil microfungi on copper availability in polluted podzols.

2. MATERIALS AND METHODS

2.1. Soil sampling

Sampling sites are situated in the northern taiga zone of the Kola Peninsula. Soils were sampled along the southwest gradient of heavy metal pollution from a copper-nickel smelter, "Severonickel", that has been functioning since 1938. Two sites were chosen: 1) a background pine forest – 100 km from the smelter; 2) a strongly polluted area of industrial barrens and sparse forests – 10 km from the smelter.

Samples of the upper organic layer from podzols (0-5 cm) were analyzed because they accumulate much of the metals contaminating soil and are characterized by the highest microbial biomass and activity [14].

2.2. Sample incubation

Soil samples were wetted to 60% of water holding capacity with sterile distilled water and incubated over sterile K_2SO_4-saturated solution for constant humidity maintenance. The temperature of incubation was 18-20^0 C.

Three variants of soil incubation were established.

1) Soil with natural microbial community without any pretreatment.

2) Soils with removed microbial community which were γ-sterilized before the incubation. The irradiation dose was 2,5 Mrad.

3) Sterilized soil samples with the addition of spore suspension of (a) *Mucor hiemalis* Wehmer or (b) *Trichoderma viride* Pers. ex Gray. Approximately one million spores were

added per gram of dry soil. Both microfungi are active decomposers of soil organic matter and have relatively high growth rates.

2.3. Soil analyses

Soil pH was measured in 1:25 distilled water suspension. Soil organic matter was estimated by $K_2Cr_2O_7$ oxidation method. Copper and zinc available compounds were extracted by 0.05N $CaCl_2$ and 1N CH_3COONH_4 solutions. Copper associated with organic matter was extracted by 0.1M $Na_4P_2O_7$ solution [15]. Metal contents in extracts were determined by AAS method. Chemical characteristics of the studied soils are given in Table 1. Fungal mycelium was separated with membranous filters. *Mucor* and *Trichoderma* growth was studied in membranous chambers placed in sterilized soil samples [16]. All measurements were made in three replications.

Table 1
Chemical characteristics of the studied soils

	pH	Total carbon, %	Cu bound to organic matter, ppm	Available Zn, ppm		Available Cu, ppm	
				0.05 N $CaCl_2$	1 N CH_3COONH_4	0.05 N $CaCl_2$	1 N CH_3COONH_4
Background podzol	5.21	27	62.76	14.9	20.4	4.8	11.9
Polluted podzol	4.38	21.5	1192	10.6	13.9	201	401

3. RESULTS AND DISCUSSION

The dynamics of copper and zinc available compound concentrations in background and polluted podzols were observed during microbial succession. It is shown that background soils are characterized by an initial increase in copper and zinc availability during the incubation (Figure 1). During the first 15 days of the experiment, the concentrations of available metal compounds increase by about 1.15-1.25-fold. But afterwards their availability decreases, and the concentrations of both metals in the soil solution become even lower than in the beginning of the incubation. The same processes we observed for zinc in polluted soil samples, whose concentration there is close to the background samples (Table 1). Thus, metals whose concentrations in soil are close to background (zinc in both soils and copper in background soil), after 40 days of incubation are partly immobilized. For copper, whose concentration in polluted soil is much higher than the background level (Table 1), the increase of availability takes place during all 40 days of the experiment (Figure 2).

Wetting of dry soil samples initiated soil microbial succession development, which can result in solubilization or immobilization of heavy metals, depending on the metal, its concentration in soil solution, and the stage of microbial succession development. In polluted soils, wetting leads to mobilization of copper that is present in soil solution in high concentrations. However, these changes in concentrations of available metal compounds during the microbial succession can occur due both to change of soil sample moisture (abiotic agent) or as a result of microbial activity initiated by soil wetting (biotic agent). To analyze the influence of these two agents separately, we observed the dynamics of copper availability in sterilized polluted soil samples. It is shown that in sterilized soil samples wetting also leads to increased copper availability but less in comparison with microbial community (Figure 3).

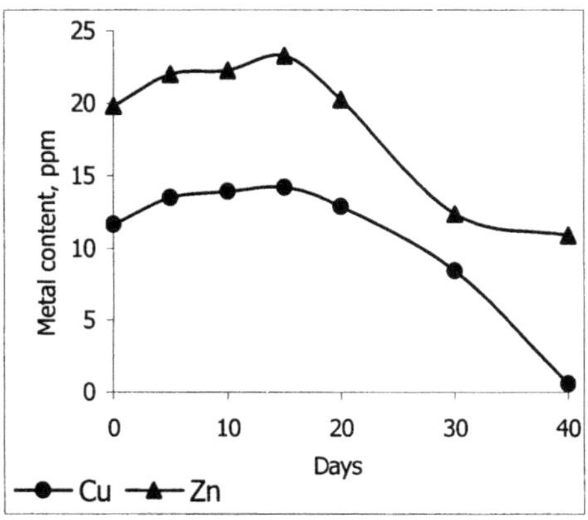

Figure 1. Copper and zinc availability dynamics in background podzol (1 N CH_3COONH_4-extracted).

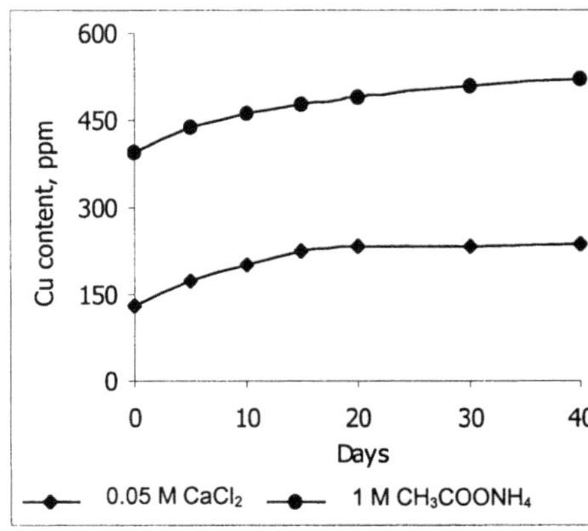

Figure 2. Copper availability dynamics in polluted podzol.

Wetting of sterilized soil samples leads to slight copper availability increase, while in the natural soil samples the concentration of available compounds practically doubles. The difference in these dynamics in natural and sterilized soil samples indicates the effect of the soil microbial community. It is shown that both wetting and microbial activity lead to copper availability increase in polluted podzols, but the contribution by the microbial community is more significant than only wetting.

Increases in active copper availability in polluted soils mainly occurs during the first 15 days of incubation. During this period, active fungal mycelium growth and organic matter decomposition take place. The correlation between the concentration of available copper compounds and the length of fungal mycelium in soil is noted. The concentration of copper associated with soil organic matter was also investigated during the experiment with polluted soil samples. During the first 15 days of the experiment, the content of copper bound to soil organic matter decreased by 1.5-fold. It is suggested that this amount of copper was distributed throughout the microbial biomass capable of binding high amounts of heavy metals [8, 10] and soil compounds with different degrees of mobility.

The composition of the soil microbial community can influence greatly the direction of metal compounds mobility change, for various microorganisms have different mechanisms of interaction with heavy metals [3]. Fungi are considered to be dominant in the microbial community of northern soils, especially under heavy metal pollution [4 - 6].

It is shown that the contamination by both fungi, *Mucor hiemalis* and *Trichoderma viride*, to sterilized soil leads to a significant increase of copper mobility (Figure 4). In the case of

Mucor the greatest difference from the control was observed at the first part of the experiment (3-24 days). After this period, the concentration values of available copper compounds did not differ significantly from the control. In the case of *Trichoderma* the greatest difference from the control was observed during the latest part of the experiment (24-66 days). Before this period, no significant differences from the control were observed. These data agree with the growth rates of these microfungi. The maxima in concentration of copper available compounds correspond to the maxima of growth intensity for these microfungi as measured by the membranous chambers method (Figure 5).

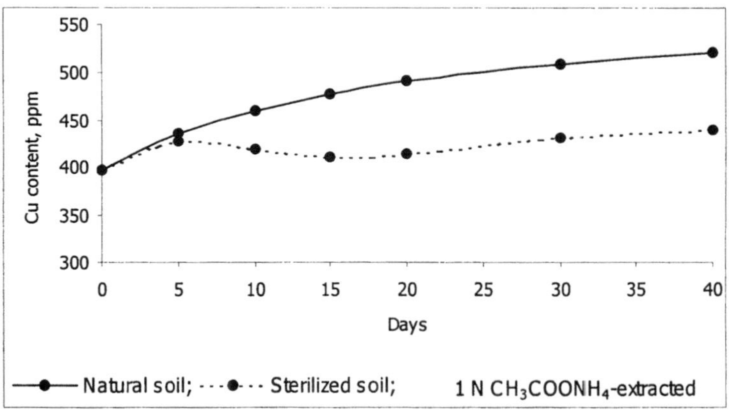

Figure 3. Copper availability dynamics in natural and sterilized polluted podzol.

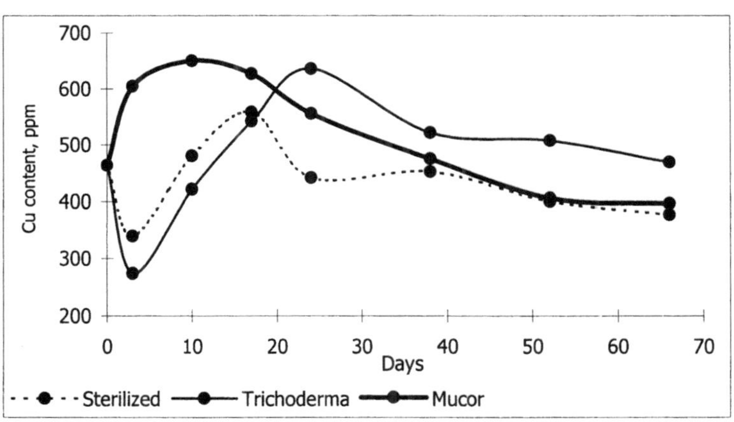

Figure 4. Copper availability dynamics in sterilized, *Mucor*-, and *Trichoderma*-inoculated samples of polluted podzol (1 N CH_3COONH_4-extracted).

Thus, it is shown that two common soil microfungi can affect copper availability in polluted podzols. Their activity leads to copper availability increases, but this process occurs in different time intervals depending on the rate of fungus development in soil.

The amount of copper associated with soil organic matter is also examined during the experiment. There are no significant changes in its values in the control. Both *Trichoderma viride* and *Mucor hiemalis* addition leads to a decrease in the amount of copper bound to organic matter (by about 1.1-1.2-fold). In the case of *Mucor,* it occurs mainly during the first period of incubation, and in the case of *Trichoderma,* during the latest period. The decrease in amount of copper bound to organic matter corresponds to the maxima of copper availability in both variants.

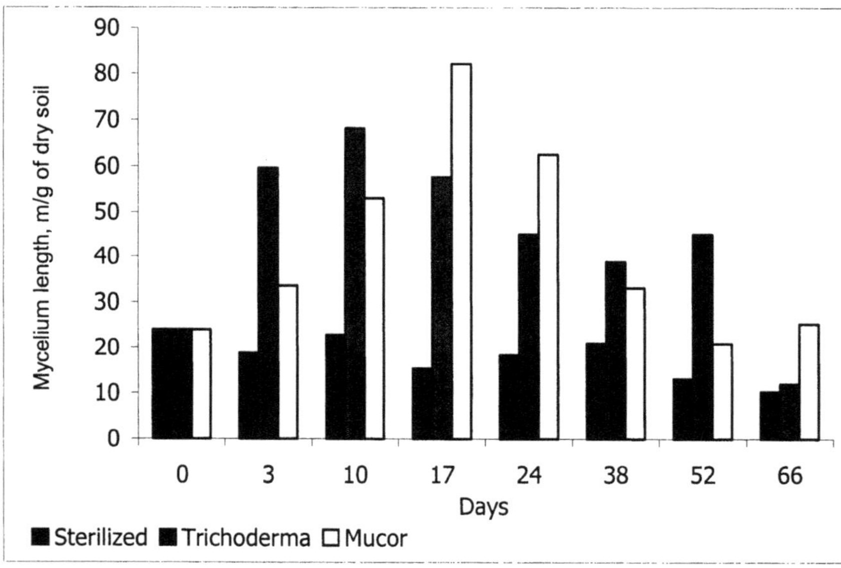

Figure 5. Mycelium growth measured by membranous chambers method.

4. CONCLUSIONS

The difference in copper availability dynamics in natural and sterilized soil samples demonstrates the possible effect of microbiota on heavy metals availability in soil. The activity of the soil microbial community may lead to increases or decreases of heavy metals mobility, depending on the metal, its concentration in soil solution and the stage of microbial succession development. In the polluted soil, microbiological activity leads to the increase of availability of metals that are present in soil solution in high concentrations.

The possibility of two common soil microfungi (*Mucor hiemalis* and *Trichoderma viride*) influencing copper availability in polluted podzols is shown. Their activity also leads to increased copper availability, but this process occurs in different time intervals depending on the rate of fungal growth in soil. The increased availability of copper occurs mostly as a result of soil organic matter decomposition and the release of copper bound to it.

This study demonstrates that natural microbiological processes in the investigated soils can result in heavy metals secondary mobilization, especially in the areas of industrial barrens and sparse forests where the influence of higher plants on metals mobility is reduced. This means that in the elutriative water regime of the Kola Peninsula soils, metals can migrate down the soil profile and penetrate to groundwater and non-contaminated areas.

REFERENCES

1. Lukina, N.V., Nikonov, V.V., 1996. Biogeochemical Cycles in the Northern Forests Subjected to Air Pollution. Apatity, Russia.
2. Cervantes, C., Guiterrezcorona, F., 1994. Copper resistance mechanisms in bacteria and fungi. FEMS Microbiol. Rev. 14, 867-874.
3. White, C., Sayer, J.A., Gadd, G.M., 1997. Microbial solubilization and immobilization of toxic metals: key biogeochemical processes for treatment of contamination. FEMS Microbiol. Rev. 20, 503-514.
4. Anderson, J.P.E., Domsh, K.H., 1975. Measurements of bacterial and fungal contribution to respiration of selected agricultural and forest soils. Can. J. Microbiol. 21, 314-319.
5. Neely, C., Beare, M.H., Hardowe, W.L., Coleman, D.C., 1991. Relationships between fungal and bacterial substrate-induced respiration, biomass and plant residue decomposition. Soil Biol. Biochem. 23, 947-953.
6. Bååth, E., 1989. Effects of heavy metals in soil on microbial processes and populations (a review). Water Air Soil Pollut. 47, 335-352.
7. Giller, K.E., Witter, E., McGrath, S.P., 1998. Toxicity of metals to microorganisms and microbial processes in agricultural soils (a review). Soil Biol. Biochem. 30, 1389-1404.
8. Ledin, M., Krantz-Rülker, C., Allard, B., 1999. Microorganisms as metal sorbents: comparison with other soil constituents in multi-compartment systems. Soil Biol. Biochem.. 31, 1639-1645.
9. Puranik, P.R., Parnical, K.M., 1997. Biosorption of lead and zinc from solutions using *Streptoverticillium cinnamoneum* Waste Biomass. J. Biotechn. 55, 113-121.
10. White, C., Wilkinson, S.C., Gadd, G.M., 1995. The Role of microorganisms in biosorption of toxic metals and radionuclides. Int. Biodeterior. Biodegrad. 35, 17-23.
11. Sayer, J.A., Gadd, G.M., 1997. Solubilization and transformation of insoluble inorganic metal compounds to insoluble metal oxalates by *Aspergillus niger*. Mycol. Res. 101, 653-659.
12. Sitaula, B.K., Almås, Å., Bakken, L.R., Singh, B.R., 1999. Assessment of heavy metals associated with bacteria in soil. Soil Biol. Biochem. 31, 315-316.
13. Bruner, H., Klauser, T., Schinner, F., 1996. Die laugung von schwermetallen aus boden mit Penicillium sp. Mitteilungen der Deutschen Bodenkundlichen Gesellschaft. 81, 311-315.
14. Soderstrom, B.E., 1975. Vertical distribution of microfungi in a spruce forest soil in a south of Sweden. Transact. Brit. Mycol. Soc. 65, 419-427.
15. Arinushkina, E.V., 1970. Manual of Chemical Analyses of Soils. Moscow State University Press, Moscow, Russia.
16. Zviagintsev, D.G. (Ed.), 1991. Methods of Soil Microbiology and Biochemistry. Moscow State University Press, Moscow, Russia.

Developments in Soil Science, Volume 28B
Editors: A. Violante, P.M. Huang, J.-M. Bollag and L. Gianfreda

MICROBIAL BIOMASS RESPONSE TO HEAVY METALS IN THE FIELD

C. Gigliotti and A. Farini

Department of Crop Production, University of Milan, Via Celoria 2, 20133 Milan, Italy

An evaluation was made of several biochemical parameters (microbial biomass, respiratory activity, and cellulase activity) and ecophysiological indexes (microbial biomass C/soil organic C; respiratory C/soil organic C; respiratory C/microbial biomass C per unit time, and cellulase activity/microbial biomass C) to verify the possibility of applying these parameters not only in the laboratory but also in the field. Sixteen soil samples of different metal contents were studied from sites in Carpiano, northern Italy.

It was confirmed in the field that stress to the biomass, resulting from the interference of metals in microflora metabolism, can be revealed by ecophysiological indexes and cellulase activity.

1. INTRODUCTION

Our previous research [1] in 1991 carried out on soils from Carpiano (northern Italy) that were contaminated by heavy metals showed how the ecophysiological indexes normally used in the laboratory [2-5] are able to reveal, also in the field, the negative influence of metals on soil microflora.

Given the possibility of using ecophysiological indexes as a measure of soil quality and the particular distribution of the metals in the area of the previous study, we decided to repeat the investigation, 4 years after the first investigation (1991), at the same sites to:

- Verify the effective suitability of ecophysiological indexes to determine the influence of metals on microflora in the field and hence to confirm the physiological state of the microbial biomass of these soils;
- Verify whether cellulase activity, because cellulose is involved in the carbon cycle and was not included in the previous study, can be a additional index to evaluate the state of the microflora of the soil.

2. MATERIALS AND METHODS

2.1. Description of the investigated area and the choice of sampling sites

The municipality of Carpiano, 16 km south of Milan (north Italy), was the study site; the soils used were from an area lying within the irrigation network of the human-made Certosa canal. This canal is about 14 km long and receives most of its waters from the Lambro River, which is heavily polluted with metals. In the Carpiano municipality, the canal divides into three branches, the waters are thus in contact with many of agricultural soils.

An investigation carried out by Genevini et al. [6] reported the high metal content of some of the soils in this irrigation network, especially in those where there were the water meadows. On the basis of the above study we took 16 soil samples, selected in such a way as to have a range of samples of different heavy metal content (low, medium, high) and, at the same time, sufficiently representative of the area in terms of metal distribution.

2.2. Sample collection and preparation

From each site we collected five soil samples (0-20 cm depth). The samples were placed in polyethylene bags and then transported to the laboratory in an ice box; the soils were then sieved (4 mm) and all visible plant debris and soil fauna were removed manually. The water holding capacity of each soil was determined, and the moisture content adjusted to 60%. The soils were then stored moist at 4°C in closed polyethylene bags, storage time being no longer than 3 weeks. Subsamples for the determination of chemical parameters were air-dried and sieved (2 and 0.5 mm) before analysis.

2.3. Analytical methods

2.3.1. Chemical parameters

Soil pH in water was measured by a potentiometer (1:2.5 w:v), and water content was determined as weight loss after oven-drying moist samples (105°C, 24 h). Soil organic carbon was determined by wet oxidation with 0.167 M $K_2Cr_20_7$ and subsequent titration with 0.1 M $(NH_4)_2Fe(S0_4)_2.6H_20$; total nitrogen was determined with Kjeldahl digestion. N digestion and distillation were performed with a Buchi 426 and a Buchi 315 apparatus, respectively (Buchi Laboratoriums-Technik AG, Swizerland). Measurement of cation exchange capacity and soil texture were carried out according to Italian regulations (Italian Ministry of Agriculture and Forestry, 1992).

Heavy metals (Zn, Cu, Pb, Cr, Ni, and Cd) were extracted with 0.5 M NH_4-acetate/0.02 M ethylenediaminetetraacetic acid (EDTA) at a dry weight soil: solution ratio of 1:5, with shaking for 30 min, and the extracts filtered on Whatman no. 42 filter paper (Italian Ministry of Agriculture and Forestry 1992). Metal determination was carried out by atomic absorption spectrophotometer SP9 (Pye Unicam Ltd., Cambridge, England).

2.3.2. Microbial population

Microbial biomass C was determined by the chloroform fumigation-extraction method [7]; 25 g (oven-dry basis) of moist soil was fumigated with ethanol-free chloroform for 24 h. After removal of the $CHCl_3$, C was extracted from fumigated and unfumigated samples with 0.5 M K_2SO_4 for 30 min on a rotating shaker, using a soil extraction ratio of 1:4 (w:v). Organic C in the filtered soil extracts was determined by wet oxidation with 0.0417 M $K_2Cr_20_7$ and subsequent titration with 0.1 M $(NH_4)_2Fe(S0_4)_2.6H_20$ on 20-ml extract. Biomass C values were calculated as 2.64 Ec, where Ec is the extractable C flush, i.e., the difference in extractable C from fumigated and non fumigated samples.

Respiratory activity was determined according to the Pochon and Tardieux method [8], slightly modified, by placing subsamples of moist soil (corresponding to 50 g oven-dry soil) in 1.0 liter jars equipped with a 50-ml beaker containing 10 ml 0.25 M KOH to trap the evolved CO_2. The CO_2 trapped in the KOH solution was measured by back-titration with 0.25 M HCl to end-point pH values of 8.00 and 4.10, with a Radiometer DTS 800 multi-titration system

(Radiometer, Copenhagen, Denmark). The CO_2 determination was carried out after 10 days of soil incubation in the dark at 25°C.

Cellulase activity was determined according to Kanazawa and Miyashita [9], but with a modified method; 2.5 g moist soil samples were put into test tubes and 0.25 ml of toluene were added. The mixture was left to stand for 10 min, and then 25 ml of a 0.2 M sodium acetate buffer solution containing 250 mg of carboxymethylcellulose were added to it; this was incubated at 50°C in a shaking incubator for 4 h. The reaction solution was separated by centrifugation and the glucose content determined by the D-Glucose test, after optimizing the method.

2.4. Calculation of ecophysiological parameters

Some of the previously determined parameters were used to calculate four ratios: microbial biomass C to soil organic C (MBC:OC); respiratory C to soil organic C (soil respiration rate) (RAC:OC); respiratory C to microbial biomass C per unit time (specific respiratory activity) (RAC:MBC); and cellulase activity to microbial biomass C (specific cellulase activity) (CLC:MBC).

2.5. Statistical analysis

All results are expressed on an air-dry soil basis as the mean of five replicate determinations. All data, previously transformed according to the equation $y = \text{arcs in} \sqrt{p \cdot n^{-1}}$, were subjected to Pearson correlation analysis with Statgraphics (Statistical Graphics Corporation, Rockville, MD, USA).

3. RESULTS AND DISCUSSION

Table 1 shows the chemical properties of the studied soils. The pH of most of the soils was between 6.0 and 6.7, the exceptions being soils 2, 8, and 13 (pH 7.0, 7.0, 7.1, respectively, neutral soils) and 11, 14, and 15 (pH 5.6, 5.6, 5.7, respectively, acidic soils). About half the soils were classified as loam, the rest as silty-loam or sandy-loam. The organic C content varied from 1.20% to 6.40%; it was extremely high in soils 4 to 9, which had the highest heavy metal content. The cation exchange capacity followed the pattern of the organic matter and clay content.

The heavy metal content is expressed as the sum of Zn, Cu, Ni, Pb, Cr, and Cd, as reported in Figure 1 and compared with the 1991 data. Although there were slight variations, it was always samples 4, 5, 6, 7, 8, and 9 that showed the highest metal contents.

The organic carbon content (Figure 2) showed little variation with respect to the 1991 study; the most metal-contaminated samples were those that had the highest organic carbon content. This positive relationship between the organic carbon and the metal content was statistically significant (Table 2).

The soils richest in metals were characterized by the highest respiratory activity (Figure 3); this relationship was statistically significant. Furthermore, the respiratory activity showed a close positive relation to organic carbon.

In terms of biomass functionality, this situation is the normal condition: high organic carbon, high activity. Nevertheless, it is known that unless the conditions are unfavorable, there is a corresponding high microbial biomass.

Table 1
Chemical and physical charateristics of the soils studied.

Soil No.	pH in H_2O	pH in KCl	Organic carbon %	Total Kjeldahl N %	CEC* cm(+)/kg	Texture
1	6.4	5.4	1.43	0.20	12.0	Silty-loam
2	7.0	5.8	1.73	0.20	18.0	Loam
3	6.9	5.8	2.08	0.20	11.2	Sandy-loam
4	6.8	5.7	3.36	0.35	22.5	Silty-loam
5	6.8	5.9	4.82	0.50	21.3	Sandy-loam
6	6.9	6.2	5.23	0.45	28.5	Silty-loam
7	6.5	5.6	6.36	0.60	27.1	Loam
8	7.0	6.0	2.96	0.30	22.1	Loam
9	6.2	5.2	2.29	0.30	10.2	Loam
10	6.9	5.8	1.59	0.15	14.8	Loam
11	5.6	4.3	1.02	0.15	9.1	Sandy-loam
12	6.6	5.3	1.80	0.20	20.2	Loam
13	7.1	5.8	1.24	0.15	11.0	Sandy-loam
14	5.9	4.7	1.55	0.20	9.5	Silty-loam
15	5.7	4.4	1.20	0.10	10.1	Sandy-loam
16	6.0	4.8	1.28	0.15	10.3	Loam

*CEC cation exchange capacity

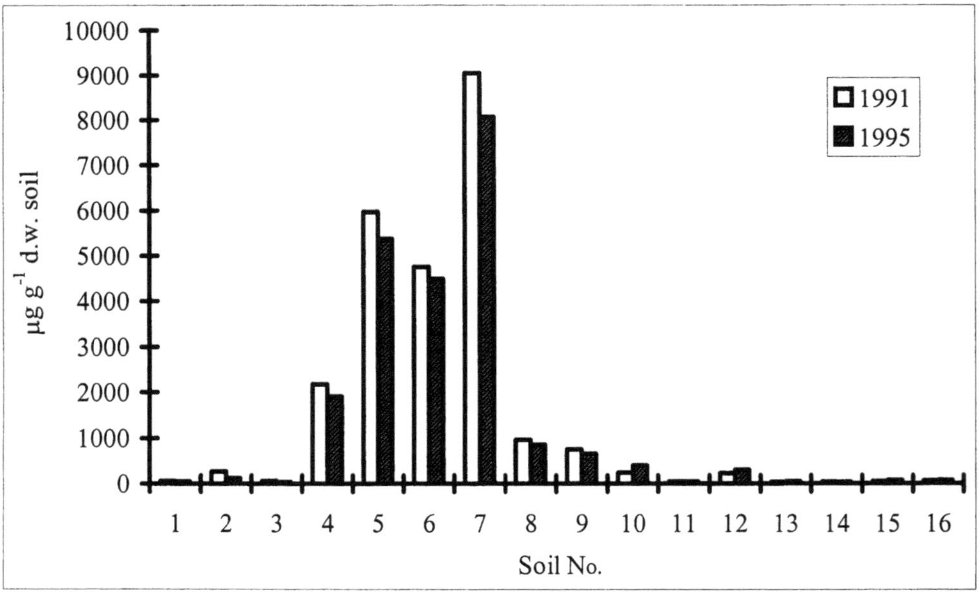

Figure 1. Sum of exchangeable Zn, Cu, Ni, Pb, Cr and Cd.

Table 2
Matrix of linear correlation coefficients and significance levels between sum of extractable heavy metals, organic C (OC), respiratory C (RAC), microbial biomass C (MBC), the ratio of respiratory C to soil organic C (RAC:OC), the ratio of microbial biomass C to soil organic C (MBC:OC) the ratio of respiratory C to microbial biomass C (RAC:MBC), and the ratio of cellulase activity to microbial biomass C (CLC:MBC).

	Metals	OC	RAC	MBC	CLC	RAC : OC	MBC : OC	RAC : MBC	CLC : MBC
Metals	-								
OC	0.9689***	-							
RAC	0.8137***	0.8643***	-						
MBC	-0.4171	-0.3047	-0.1189	-					
CLC	0.9615***	0.9555***	0.7392**	-0.3925	-				
RAC : OC	-0.8455***	-0.8903***	-0.5673*	0.3242	-0.9125***	-			
MBC : OC	-0.8345***	-0.8128***	-0.5794*	0.7790***	-0.8402***	0.8018***	-		
RAC : MBC	0.7778***	0.7276**	0.6561**	-0.8070***	0.7302**	-0.5951*	-0.9132***	-	
CLC : MBC	0.8699***	0.8191***	0.6849**	-0.5522*	0.8913***	-0.7436***	-0.8297***	0.8459***	-

*(*P≤0.05, **P≤0.01, ***P≤0.001)

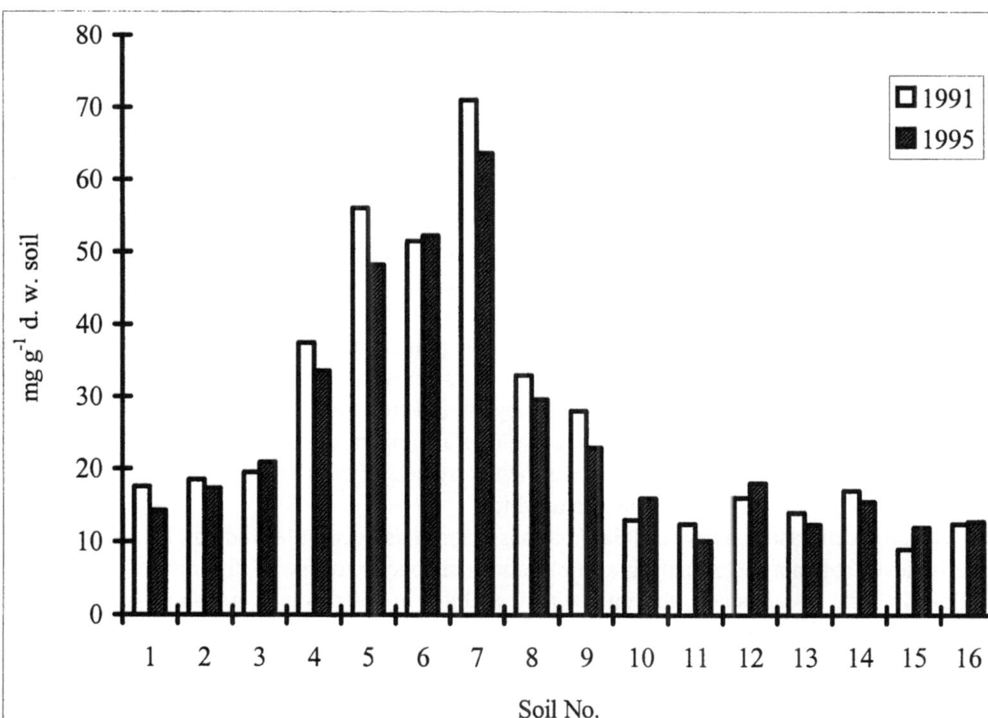

Figure 2. Organic carbon.

An analysis of the graph relative to the microbial biomass (Figure 4) clearly shows how, on the average, samples 4, 5, 6, 7, 8, and 9 had a reduced biomass compared to the other soils. Therefore, for high respiratory activity and low biomass, it can be hypothesised that heavy metals cause a state of stress in the microbial biomass. However, the relationship between the biomass and the metal was not statistically significant. This trend permits the formulation of two hypotheses: i) the metals do not influence the microflora and there are other causes for the stress, or ii) the metals affect the microbial metabolism, but such an effect, evaluated only through respiration and microbial biomass, is barely detectable, because we analyzed a field situation in which the variables that can influence the functionality of the biomass are multiple and can mask or minimize the effect of the metals.

The correlation between the ecophysiological indexes data and the content in metals confirms the second hypothesis.

With regard to the soil respiration rate (Figure 5), the soils with higher metal content had a lower soil respiration rate, hence the biomass of these soils, though having greater respiration than that of uncontaminated soils, had less respiration proportionally to the amount of organic carbon available to it. In other words, the biomass should have had greater respiration to avoid the accumulation of organic carbon [3-5].

The stressed state of the biomass is even more evident from data relating to the proportion of the biomass in total soil organic matter, the MBC:OC ratio. Such a ratio allows the evaluation of the proportion of the biomass C in the total organic carbon. Under normal conditions, microbial C comprises about 1% to 4% of the total organic carbon. In our samples, the MBC:OC ratio (Figure 6) of the contaminated soils was lower than in the uncontaminated soil and less than 1%. Hence, a high amount of metal caused a reduction in the amount of microbial biomass within the total organic matter pool [10].

Further confirmation of the adverse effects of the metals on the biomass can be found in specific respiratory activity. This ratio is a recognized index of stress in that it indicates the distribution of the metabolic energy of the microbial biomass between maintenance and growth. The lower this ratio, the greater the energy transferred to growth [4, 11]. In both 1991 and 1995, the soils characterized by a high metal contents had respiratory quotients that, on the average, were higher than in low-metal soils (Figure 7). What does this indicate? The biomass of these metal-enriched soils respires, and therefore burns, a lot more substrate for its maintenance and uses less for its growth. In fact, from the previous graphs (see Figures 3 and 4), it can be seen how the respiratory activity is raised and the biomass tends to reduce.

The cellulase activity, like the ecophysiological indexes, confirms the adverse effects of the metals on the biomass metabolism. Figure 8 allows a comparison of the data of cellulase activity and metal content. Note that the samples rich in metals always showed the greater cellulase production, a statistically significant relationship.

Specific cellulase activity (cellulase:biomass ratio) has been little studied [12]; nevertheless, we believe that this index can give information analogous to that of the respiratory quotient index. In fact, in our study, the ratio was very high in soils rich in metals (Figure 9). This confirms how, in the presence of metals, the biomass modifies its metabolism toward a greater production of cellulase, thus increasing the catabolic processes, because the cellulase takes part in the degradation of cellulose, making carbon available for later transformation to CO_2. This observation was confirmed by the close and, moreover, statistically significant, positive relationship between cellulase and respiratory activity (Figure 10); in this relationship, it can be seen that when respiratory activity is high, so is the cellulase activity. Therefore, once again, it

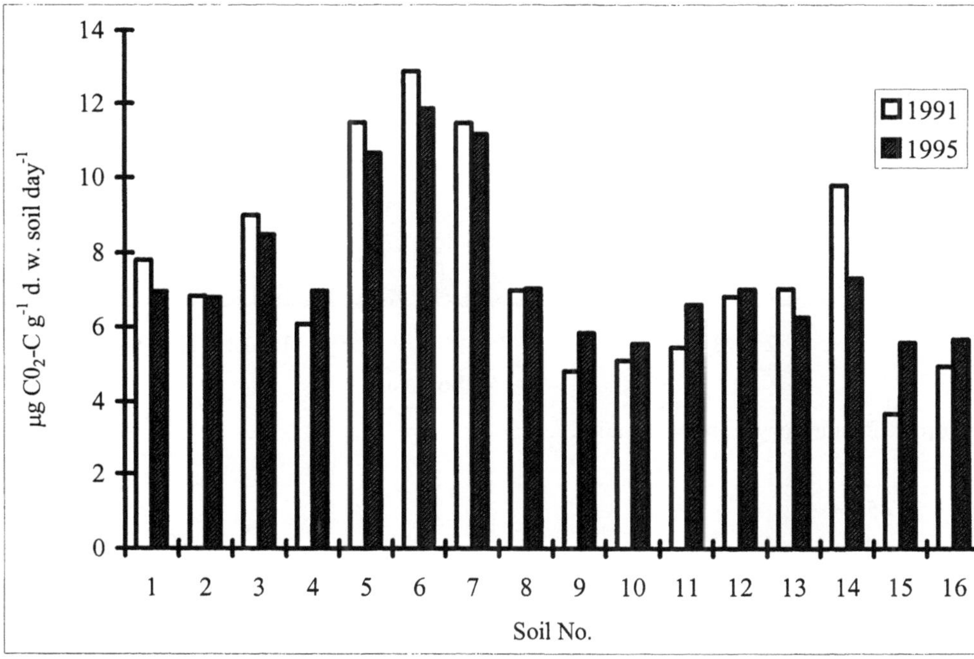

Figure 3. Respiratory activity.

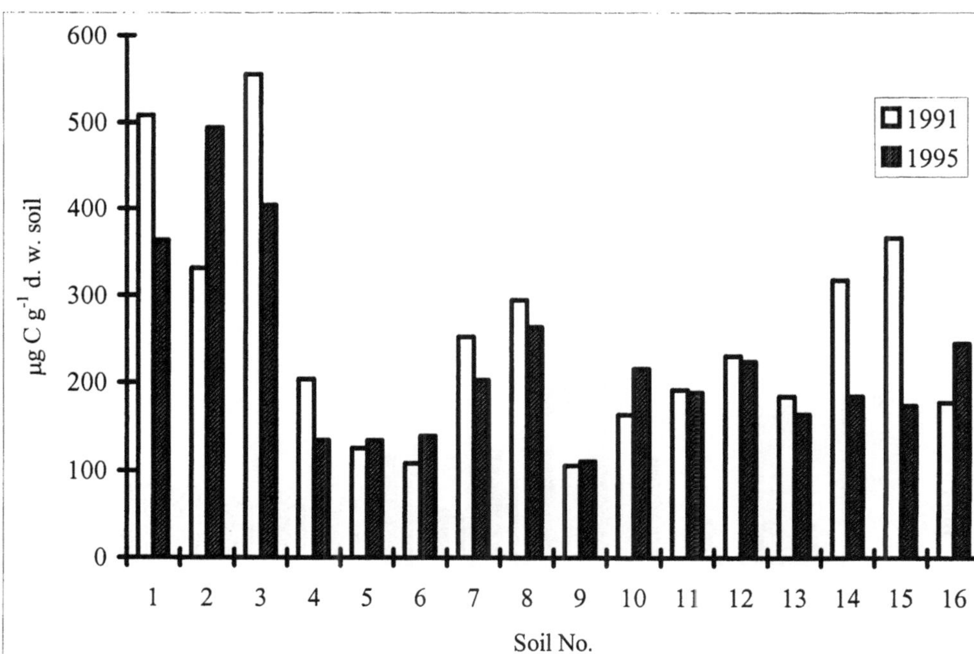

Figure 4. Microbial biomass.

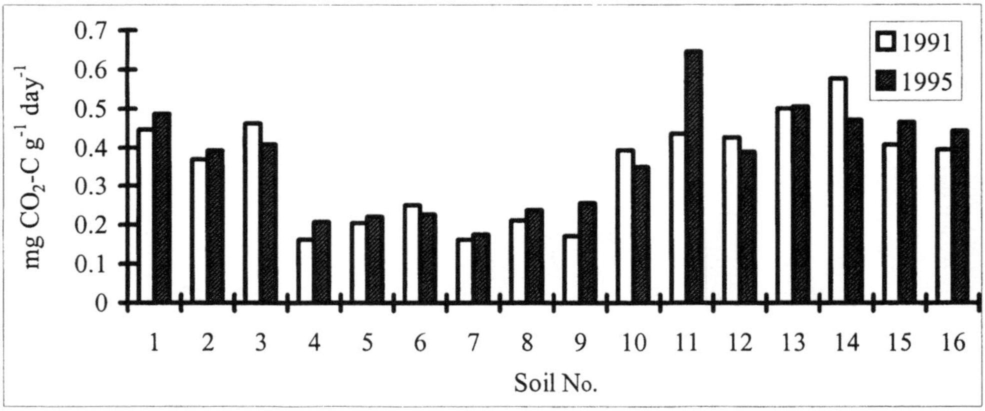

Figure 5. Respiratory activity:organic C.

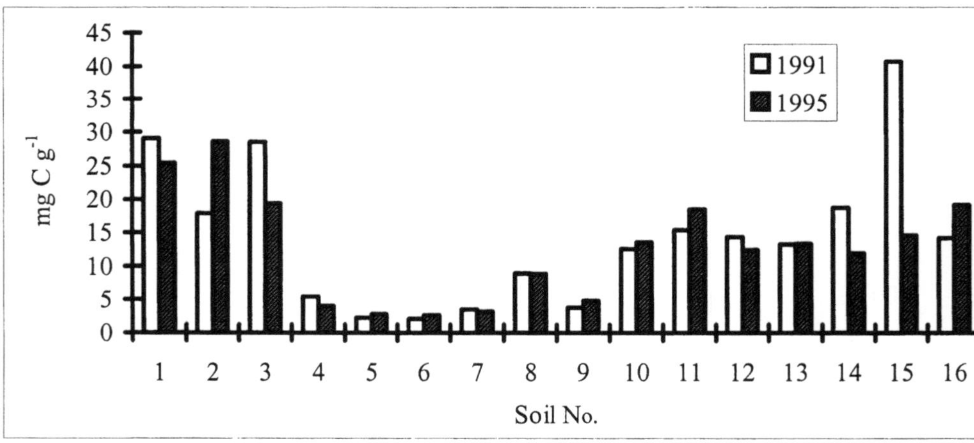

Figure 6. Microbial biomass:organic carbon.

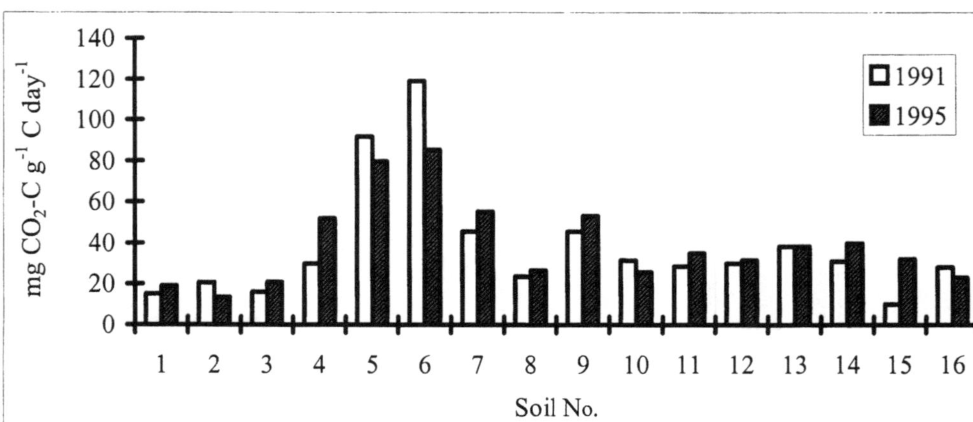

Figure 7. Respiratory activity:microbial biomass.

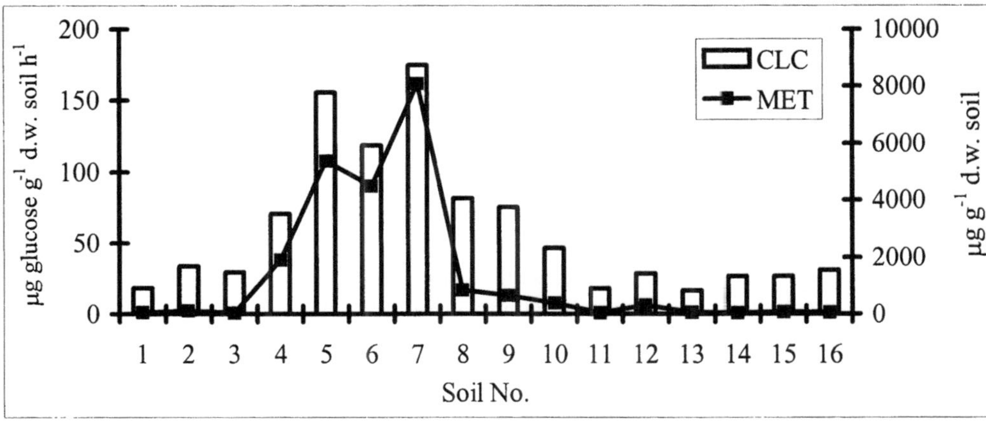

Figure 8. Cellulase activity (CLC) and heavy metals (MET).

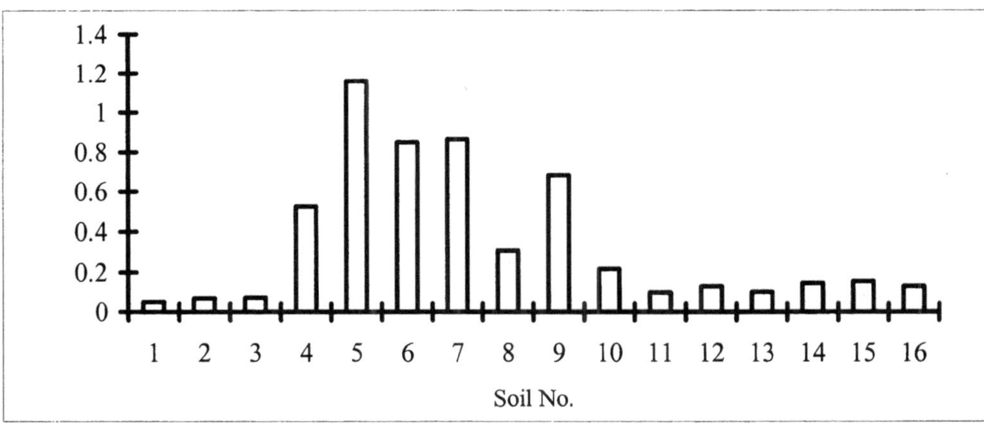

Figure 9. Cellulase activity:microbial biomass.

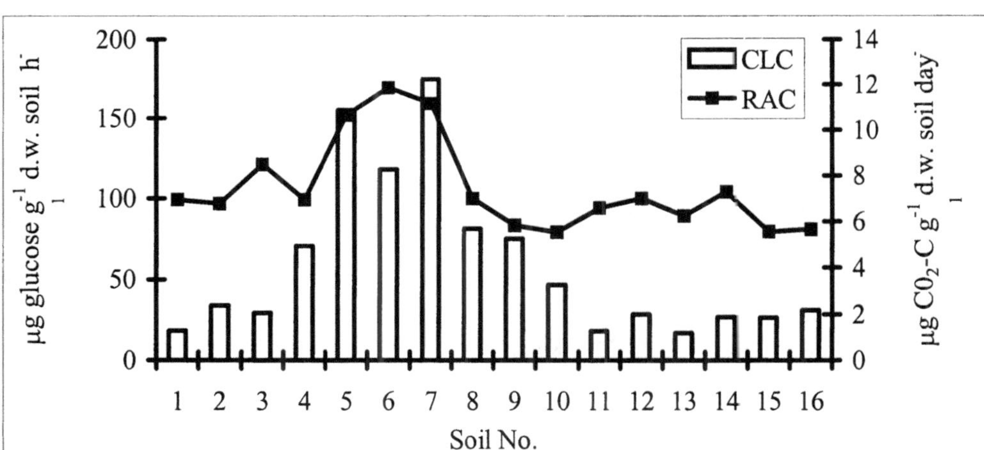

igure 10. Cellulase activity (CLC) and respiratory activity (RAC).

is shown that a biomass in unfavorable environmental conditions (the presence of metals) uses more substrate to maintain itself, producing more cellulases and CO_2 and growing less.

4. CONCLUSIONS

There has been confirmation of the applicability of ecophysiological indexes in soils from a field situation. In fact, the indexes more clearly reveal the negative interactions of metals and biomass in soils of different agronomic history.

Cellulase activity can be used, together with other parameters, as an index of environmental risk, given that such activity influences the turnover of the microbial biomass and the carbon cycle.

REFERENCES

1. Valsecchi, G., Gigliotti, C., Farini A., 1995. Microbial biomass, activity, and organic matter accumulation in soils contaminated with heavy metals. Biol. Fertil. Soils. 20, 253-259.
2. Anderson, T.H., Domsch, K.H., 1986. Carbon link between microbial biomass and soil organic matter. *In*: Megusar, F., Gantar M. (Eds.), Proceedings of the Fourth International Symposium on Microbial Ecology. Slovene Society of Microbiology. Ljubljana. pp. 467-471.
3. Chander, K., Brookes, P.C., 1991. Microbial biomass dynamics during the decomposition of glucose and maize in metal-contaminated and non-contaminated soils. Soil Biol. Biochem. 23, 917-925.
4. Chander, K., Brookes, P.C., 1991. Effects of heavy metals from past applications of sewage sludge on microbial biomass and organic matter accumulation in a sandy loam and silty loam UK soil. Soil Biol. Biochem. 23, 927-932.
5. Chander, K., Brookes, P.C., 1991. Plant input of carbon to metal-contaminated soil and effects on the soil microbial biomass. Soil Biol. Biochem. 23, 1169-1177.
6. Genevini, P.L., Sommariva, F., Del Fabbro, D., 1995. Distribuzione dei metalli in tracce nel comprensorio della Roggia Certosa (Mi). Acqua Aria. 1, 91-100.
7. Vance, E.D., Brookes, P.C., Jenkinson, D.S., 1987. An extraction method for measuring microbial biomass C. Soil Biol. Biochem. 19, 703-707.
8. Pochon, J., Tardieux, P., 1962. Techniques d'analyse en microbiologie du sol. De La Tourelle. St Maundè. Seine.
9. Kanazawa, S., Miyashita, K., 1986. A modified method for determination of cellulase activity in forest soil. Soil Sci. Plant Nutr. 32, 71-79.
10. Brookes, P.C., McGrath, S.P., 1984. Effects of metal toxicity on the size of the soil microbial biomass. J. Soil Sci. 35, 341-346.
11. Chander, K., Brookes, P.C., 1993. Residual effects of zinc, copper and nickel in sewage sludge on microbial biomass in a sandy loam soil. Soil Biol. Biochem. 25, 1231-1239.
12. Ohtonen, R., Lahdesmaki, P., Markkola, A.M., 1994. Cellulase activity in forest humus along an industrial pollution gradient in Oulu, northern Finland. Soil Biol. Biochem. 26, 97-101.

Developments in Soil Science, Volume 28B
Editors: A. Violante, P.M. Huang, J.-M. Bollag and L. Gianfreda

ROLE OF FIRE ON SOIL ORGANIC MATTER TURNOVER AND MICROBIAL ACTIVITY IN A MEDITERRANEAN BURNED AREA

F. A. Rutigliano[a], A. R. Fierro[b], R. A. De Pascale[a], A. De Marco[b] and A. Virzo De Santo[b]

[a]Dipartimento di Scienze Ambientali, Seconda Università di Napoli
Via Vivaldi 43, 81100 Caserta, ITALY

[b]Dipartimento di Biologia Vegetale, Università di Napoli Federico II
Via Foria 223, 80139 Napoli, ITALY

This study examined the effects of light and intense experimental fires and of wildfire on the soil microbial biomass and activity and on organic carbon turnover in a coastal area of southern Italy. Wildfire, which mainly affected plant cover, caused a rise in the input of organic matter to the soil but did not significantly influence total microbial biomass or fungal mycelia. In contrast, experimental fires developing on the soil surface caused a decrease in the organic carbon as well as in total microbial biomass and fungal mycelia. All types of fires resulted in an increased soil potential respiration and metabolic quotient (qCO_2: mg CO_2-C mg^{-1} C_{mic} 10 d^{-1}). Moreover, increases in the coefficient of endogenous mineralization (CEM: mg CO_2-C g^{-1} C_{org} 10 d^{-1}) after experimental fires and increases in the soil CO_2 emission after wildfire were also observed. The data suggest that burned soils may be an important source of carbon for the atmosphere for several months after fire occurrence.

1. INTRODUCTION

Soil organic matter plays an important role in the ecosystem because it represents the reserve of organic carbon and nutrients for the biotic community and, moreover, stabilizes the soil by improving its structure. As an active fraction of soil organic matter, the microbial biomass is responsible for organic matter decomposition [1] and therefore plays a fundamental role in the short-term turnover of nutrients [2].

Soil microbial biomass and organic matter turnover are influenced by anthropogenic disturbance, including burning; the extent of this effect depends on the intensity of ecosystem disturbance. In fact, Pietikainen and Fritze [3] reported a reduction in the humus layer thickness and microbial biomass as fire became more intense.

Few studies have been made to quantify the microbial biomass and activity in soils disturbed by fire [3, 4]. Studies performed so far on this topic are relative to pine wood, whilst virtually no information is presently available on the effects of fire on soil in the Mediterranean environment.

The objective of this study was to evaluate the effects of different types of fires (light and intense experimental fires and wildfire) on soil microbial biomass, activity and organic matter turnover in a Mediterranean area.

2. MATERIALS AND METHODS

2.1. Study area

The research was carried out in Castel Volturno Nature Reserve, a flat coastal area of ca. 268 ha north of Naples in southern Italy, dominated by Mediterranean maquis, sclerophyllous evergreen shrubs typical of the Mediterranean basin [5].

The climate is typically Mediterranean with dry summers and rainy autumns and winters; the mean annual temperature is 13.6°C and the mean annual rainfall is 761 mm at the Ischitella meteorological station (15 meters above sea level), close to the experimental area [6].

As a consequence of several fire events that occurred at different times in the past, the vegetation of the area is a mosaic of different aged plant patches. It is dominated by the sclerophyllous evergreen species *Quercus ilex* L., *Myrtus communis* L., *Arbutus unedo* L., *Pistacia lentiscus* L., and *Phillyrea* sp. pl. [6]. Along with native Mediterranean maquis species several *Pinus* species are also present, including *P. halepensis* Miller, *P. pinaster* Aiton, and *P. pinea* L., planted by foresters after fires. Locally, small gaps in the woody canopy cover (canopy gap) show dominance of herbs and bryophytes [6].

The soil of the study area is a *Calcaric Arenosol* according to the FAO system of soil classification [7].

The study was performed on soils from sites affected by: (i) low intensity (light) and (ii) high intensity (intense) experimental fires; and (iii) wildfire.

<u>*Experimental fires*</u> were carried out in April 1993 in an 8-year-old canopy gap in the Mediterranean maquis at two different intensities, light and intense. Light fire, produced by burning herbaceous materials, lasted about 2 hours and caused incomplete mineralization of the biomass and necromass, leaving black ash on the soil. Intense fire, produced by burning woody materials collected in nearby areas, lasted about 6 hours and caused complete mineralization of organic matter, leaving white ash on the soil. Both types of experimental fires developed on the soil surface. Soil temperatures measured at 5 cm depth 24 h after extinction were 20°C and 40°C, respectively, for light and intense fires. No temperature measurements were made during the experimental fires, however, temperature changes were monitored by thermocouples in nearby sites with similar vegetation during an intense fire in August 1988 [8] and during a light fire in July 2000 (Rutigliano, unpubl.). Temperatures above 900°C at the soil surface and about 200°C at a depth of 5 cm were measured during the 1988 fire [8]. Like the intense fire in our study, the 1988 fire left white ash on the soil. Lower temperatures (about 460°C at the soil surface and 37°C at 5 cm depth) were measured (Rutigliano, unpubl.) during the 2000 light fire, which left black ash on the soil.

For both experimental fires, soil samplings were performed 4 days before the experimental fires (used as control) and 4, 59, 180 and 353 days after the fires.

A <u>*wildfire*</u> started accidentally in July 1995 and burned an area of about 500 m^2. Before the fire, the area was covered by 15-year-old vegetation, including Mediterranean maquis species and pines introduced by foresters. The wildfire seemed mainly to affect plant cover and left black ash on the soil. The effect of wildfire was studied for 16 months. A nearby unburned area, similar to the burned area, was used as the control.

2.2. Sample collection and analyses

Soil samples, corresponding to unburned and/or burned areas, were collected with a sampling corer (diameter 8 cm), after removal of the litter, when present, from the 0-5 cm layer. Each sample consisted of 5 evenly distributed sub-samples. Samples were mixed and sieved through a 2-mm mesh. Soil analyses included: 1) N and S content, measured on dried soil samples by a Carlo Erba Analyzer NA1500; 2) K, Ca, Mg, Mn and Fe content, determined by atomic absorption spectrometry (Varian SpectrAA 20) after digestion of soil samples in a 2:1 mixture of 65% nitric acid and 50% hydrofluoric acid; 3) organic carbon (C_{org}), measured on dried (75°C) soil by ashing at 550° for 2 h; 4) microbial carbon (C_{mic}), determined within 24 h in fresh samples (stored at 4°C) using the fumigation-incubation method [9]; 5) fungal mycelia (including living and dead mycelia), estimated by the intersection method using an optical microscope [10]; and 6) potential microbial respiration, measured in the laboratory as CO_2 evolution from fresh soil samples at 55% of water-holding capacity and incubated in sealed containers for 10 days at 25°C. The CO_2 output from the soil was determined by NaOH absorption followed by two-phase titration with HCl [11]. The metabolic quotient (qCO_2) was calculated from potential respiration and C_{mic} (mg CO_2-C mg^{-1} C_{mic} 10 d^{-1}). The coefficient of endogenous mineralization (CEM) was obtained from potential respiration and C_{org} (mg CO_2-C g^{-1} C_{org} 10 d^{-1}). CO_2 emissions by the soil in the area affected by wildfire and the respective control were measured *in situ* besides in the laboratory by the closed chamber technique and gas chromatography using a thermoconductivity detector. The significance of differences between burned and unburned soils for each parameter was evaluated with the Mann-Whitney test using the package Sigma Stat 1.0.

3. RESULTS AND DISCUSSION

The three fire regimes produced different effects on soil nutrients, the organic carbon pool, carbon turnover, microbial activity and biomass. For example, during the first year following either light or intense experimental fires, the nutrient content did not change significantly (Table 1), with the exception of N content, which decreased after intense fire. In contrast, higher N, S and Fe contents ($P < 0.0001$) and generally lower Ca ($P < 0.0001$) and K ($P < 0.005$) contents compared to controls were observed in the soil affected by wildfire (Table 2).

During the first year after light and intense experimental fires, a decrease was generally observed in the organic ($P = 0.05$; Figure 1) and microbial carbon ($P < 0.05$ only for light fire; Figure 2) compared to the values measured in the respective soils before the fire. The organic carbon reduction (calculated as the mean for the 1-year period) was 57% and 21%, respectively, for soils burned by light and intense fires, whilst the microbial carbon reduction was about 35% in both cases. Compared to unburned soils, burned soils exhibited a decline in fungal mycelia (Figure 2) throughout the first six months following both light ($P < 0.05$) and intense ($P < 0.0001$) experimental fires. The decline in fungal mycelia content for the six-month period after light and intense fires was 13% and 37%, respectively. Our results are confirmed by Pietikainen and Fritze [3], who reported a decrease in microbial carbon and length of fungal hyphae after a prescribed fire in coniferous forests. Similarly, Carballas et al. [4] also reported microbial biomass reduction after wildfire.

Table 1
Soil nutrient content (mg g^{-1} dw) before (control) and at several intervals (days) after experimental fires. Values are means ± standard errors.

	N	S	K	Ca	Mg	Mn	Fe
Light fire							
Before fire	2.9 ± 0.0	0.3 ± 0.0	17.1 ± 1.4	23.4 ± 0.6	6.8 ± 0.4	1.1 ± 0.1	6.1 ± 0.2
After fire:							
4 d	2.4 ± 0.0	0.3 ± 0.0	19.5 ± 0.4	23.6 ± 1.0	7.0 ± 0.5	1.1 ± 0.0	6.8 ± 0.4
59 d	3.1 ± 0.0	0.3 ± 0.0	17.9 ± 0.3	20.7 ± 0.3	6.1 ± 0.5	1.2 ± 0.0	5.6 ± 0.8
180 d	2.9 ± 0.0	0.2 ± 0.0	19.5 ± 0.8	27.3 ± 2.3	7.4 ± 0.4	1.1 ± 0.0	7.2 ± 1.2
353 d	2.6 ± 0.0	0.2 ± 0.0	19.2 ± 0.2	23.1 ± 0.5	6.5 ± 0.1	1.1 ± 0.0	19.4 ± 0.0
Intense fire							
Before fire	2.6 ± 0.0	0.3 ± 0.0	19.3 ± 2.0	24.3 ± 1.7	6.8 ± 0.2	1.0 ± 0.0	6.1 ± 0.4
After fire:							
4 d	2.1 ± 0.0	0.3 ± 0.0	20.2 ± 0.6	23.6 ± 1.2	6.6 ± 1.0	1.0 ± 0.0	7.0 ± 0.9
59 d	1.6 ± 0.0	0.4 ± 0.0	20.7 ± 0.6	26.1 ± 0.5	7.5 ± 0.1	1.1 ± 0.0	6.2 ± 0.6
180 d	2.1 ± 0.0	0.2 ± 0.0	19.4 ± 0.8	24.5 ± 0.3	6.8 ± 0.3	1.0 ± 0.0	6.3 ± 0.7
353 d	1.9 ± 0.0	0.1 ± 0.0	23.1 ± 1.9	21.1 ± 1.3	7.4 ± 1.7	1.1 ± 0.0	15.9 ± 0.4

The soil exposed to wildfire showed higher values of organic carbon content ($P < 0.005$), relative to the unburned control area (Figure 1), and this explained observed increases of N, S and Fe content. The mean increase of organic carbon content was 84% over the whole period considered. This increase was probably due to the input of organic matter from the burned canopy. In fact, the effect of fire on carbon pool changed according to the fire type, as also emphasized by Johnson [12], who observed a reduction of soil carbon after intense wildfire but no change or even an increase after a prescribed fire.

Apparently, in this study, the wildfire did not significantly affect the microbial biomass and fungal mycelia (Figure 2).

In our study, an increase in microbial activity was observed after all fire types. Compared with the controls, burned soils exhibited higher values of respiration ($P < 0.005$ for intense fire; $P < 0.01$ for wildfire; Figure 3) and qCO_2 ($P < 0.005$ for both light and intense experimental fires; $P = 0.01$ for wildfire; Figure 4). The mean increase of respiration over the whole period after burning was 29%, 246% and 53% for light, intense and wildfire, respectively. The mean increase of qCO_2 over the whole period after burning was 141%, 334% and 23% for light, intense and wildfire, respectively. Moreover, after experimental fires, the soils showed higher values of CEM ($P < 0.005$), with a mean increase over the whole period of 315% and 334% for light and intense fires, respectively (Figure 3). In contrast, CEM did not appear to be influenced by wildfire (Figure 3). The higher soil microbial activity after wildfire is possibly related mainly to the input of easily decomposable substrate, such as animals and microorganisms or parts of plants destroyed by fire. On the contrary, the increased microbial activity after experimental fires is probably due mainly to stress conditions.

In the area burned by wildfire, CO_2 emissions exhibited a mean increase of 37% ($P < 0.05$), compared with the control area, during 7 months following the fire (Figure 5). Our results are similar to those reported by Poth et al. [13], who also found higher CO_2 emissions after a fire in an experiment performed in the savanna. Other studies suggesting an increase of microbial activity for several metabolic pathways include: higher emissions of N_2O [13, 14], NO [13] and higher uptake of CH_4 [13] from burned soils.

Table 2

Nutrient content (mg g^{-1} dw) in unburned soils (control) and in soils exposed to wildfire measured at different times (days) after fire. Values are means ± standard errors.

	N	S	K	Ca	Mg	Mn	Fe
Control							
103 d	2.1 ± 0.0	0.03 ± 0.01	21.7 ± 0.1	69.2 ± 1.2	4.9 ± 0.0	0.8 ± 0.0	10.6 ± 0.4
110 d	2.6 ± 0.0	0.07 ± 0.01	20.5 ± 0.1	68.5 ± 1.6	5.4 ± 0.2	0.8 ± 0.0	12.0 ± 0.2
116 d	1.7 ± 0.0	0.01 ± 0.01	20.7 ± 0.7	54.0 ± 5.1	5.3 ± 0.1	0.8 ± 0.0	11.9 ± 0.4
123 d	1.9 ± 0.0	0.01 ± 0.00	19.9 ± 0.2	52.2 ± 1.5	5.7 ± 0.2	0.8 ± 0.0	12.1 ± 0.7
131 d	1.7 ± 0.3	0.01 ± 0.00	20.7 ± 0.4	73.1 ± 1.5	6.5 ± 0.1	0.9 ± 0.0	12.3 ± 0.2
167 d	1.7 ± 0.3	0.01 ± 0.00	23.1 ± 0.3	70.8 ± 3.7	7.4 ± 0.0	0.8 ± 0.0	13.5 ± 0.5
181 d	1.3 ± 0.2	0.05 ± 0.00	21.3 ± 0.2	68.6 ± 0.2	6.1 ± 0.1	0.8 ± 0.0	10.9 ± 0.2
187 d	1.8 ± 0.0	0.01 ± 0.01	20.8 ± 0.0	67.6 ± 2.1	7.5 ± 0.2	0.8 ± 0.0	12.9 ± 0.5
196 d	2.2 ± 0.1	0.03 ± 0.01	20.8 ± 0.6	67.4 ± 0.6	6.3 ± 0.0	0.8 ± 0.0	12.1 ± 0.1
209 d	2.4 ± 0.0	0.02 ± 0.00	21.1 ± 0.1	64.9 ± 1.0	6.7 ± 0.1	0.8 ± 0.0	11.9 ± 0.1
221 d	1.8 ± 0.0	0.00 ± 0.00	21.3 ± 0.1	73.8 ± 1.9	6.7 ± 0.4	0.8 ± 0.0	11.8 ± 0.0
439 d	1.5 ± 0.0	n.d.	22.0 ± 0.3	80.3 ± 1.6	6.3 ± 0.3	0.8 ± 0.0	12.4 ± 0.6
482 d	1.5 ± 0.0	0.00 ± 0.00	21.5 ± 0.3	67.4 ± 0.8	5.9 ± 0.3	0.8 ± 0.0	11.1 ± 0.4
Burned							
103 d	2.8 ± 0.0	0.03 ± 0.01	20.3 ± 0.2	64.4 ± 1.4	6.0 ± 0.1	0.9 ± 0.0	13.1 ± 0.2
110 d	3.1 ± 0.1	0.05 ± 0.01	19.6 ± 0.3	61.2 ± 3.9	5.6 ± 0.1	0.8 ± 0.0	12.4 ± 0.0
116 d	3.9 ± 0.0	0.08 ± 0.02	20.5 ± 0.4	49.2 ± 0.5	5.8 ± 0.2	0.8 ± 0.0	11.9 ± 0.5
123 d	5.0 ± 0.1	0.09 ± 0.00	19.9 ± 0.5	59.7 ± 2.7	6.4 ± 0.1	0.9 ± 0.0	13.7 ± 0.3
131 d	3.3 ± 0.0	0.04 ± 0.00	20.7 ± 0.2	52.0 ± 0.2	6.4 ± 0.1	0.7 ± 0.0	12.1 ± 0.1
167 d	2.8 ± 0.0	0.06 ± 0.01	20.1 ± 0.2	62.9 ± 1.0	7.4 ± 0.1	0.9 ± 0.0	15.8 ± 0.3
181 d	1.4 ± 0.0	0.13 ± 0.01	23.6 ± 1.7	57.4 ± 2.7	6.6 ± 0.1	0.9 ± 0.0	13.6 ± 0.7
187 d	2.5 ± 0.0	0.05 ± 0.02	21.0 ± 0.4	65.5 ± 1.0	6.7 ± 0.1	0.8 ± 0.0	13.6 ± 0.1
196 d	5.3 ± 0.1	0.12 ± 0.00	19.3 ± 0.1	62.1 ± 1.3	7.4 ± 0.2	0.8 ± 0.0	13.3 ± 0.1
209 d	3.3 ± 0.0	0.02 ± 0.01	21.8 ± 0.4	57.4 ± 2.7	6.6 ± 0.2	0.8 ± 0.0	12.6 ± 0.4
221 d	5.2 ± 0.1	0.04 ± 0.02	20.0 ± 0.3	60.1 ± 1.0	7.0 ± 0.1	0.8 ± 0.0	13.3 ± 0.3
439 d	3.0 ± 0.1	n.d.	20.9 ± 0.1	54.9 ± 3.0	6.8 ± 0.2	0.8 ± 0.0	13.5 ± 0.2
482 d	1.8 ± 0.0	0.00 ± 0.00	21.3 ± 0.2	66.4 ± 3.1	6.1 ± 0.2	0.8 ± 0.0	13.1 ± 0.6

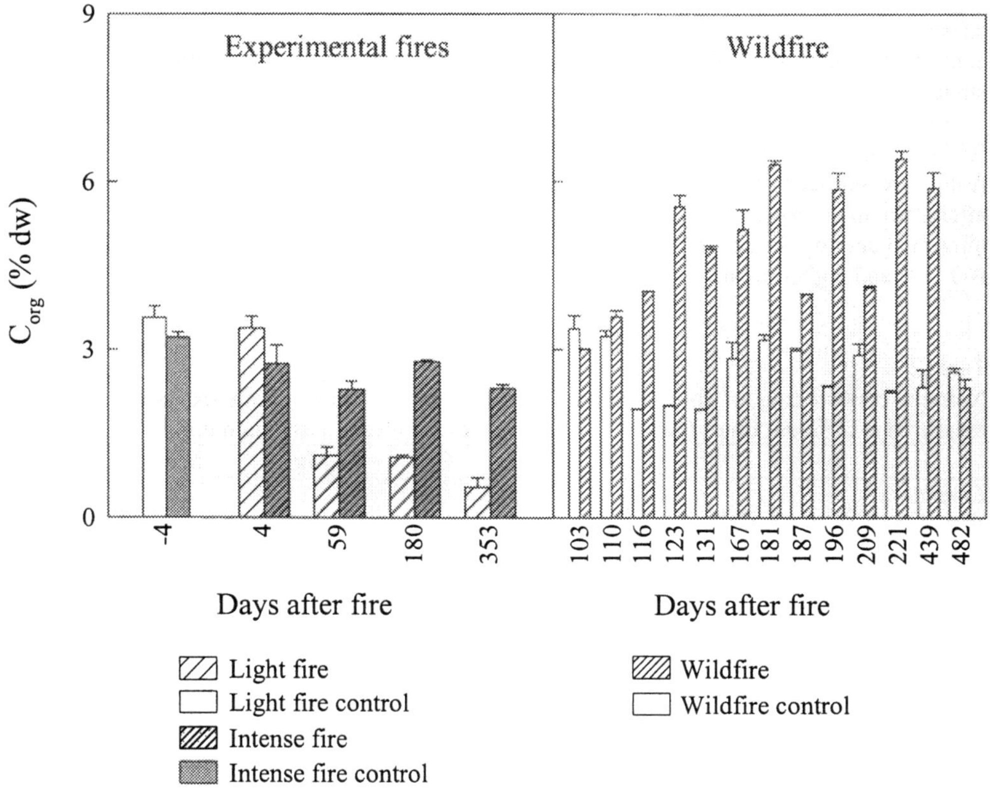

Figure 1. Changes in organic carbon content (C_{org}) in soils exposed to light or intense experimental fire and to wildfire. The bars on the columns indicate the standard error.

4. CONCLUSIONS

It is known that fire affects soil organic matter content and microbial biomass and activity [3, 4]. However, the induced changes may differ with fire type. Moreover, because of vegetation burning, soil temperature and humidity undergo marked diurnal and seasonal fluctuations determining unfavorable conditions for microorganisms.

Our study demonstrates that wildfire and experimental fires produce contrasting effects. Indeed, after wildfire an increase was observed in organic matter in the soil due to the input of parts of plants and other organisms killed by fire; this was likely responsible also for the concomitant increase of N, S and Fe in the soil. No effect on total microbial biomass and fungal mycelia was found. By contrast, after the experimental fires, which developed on the soil surface and were more destructive, a decrease was observed in the organic matter pool, N content, total microbial biomass and fungal mycelia.

Although total microbial biomass and fungal mycelia were not affected by wildfire, they were decreased by experimental fires; both types of fire induced an increase of soil

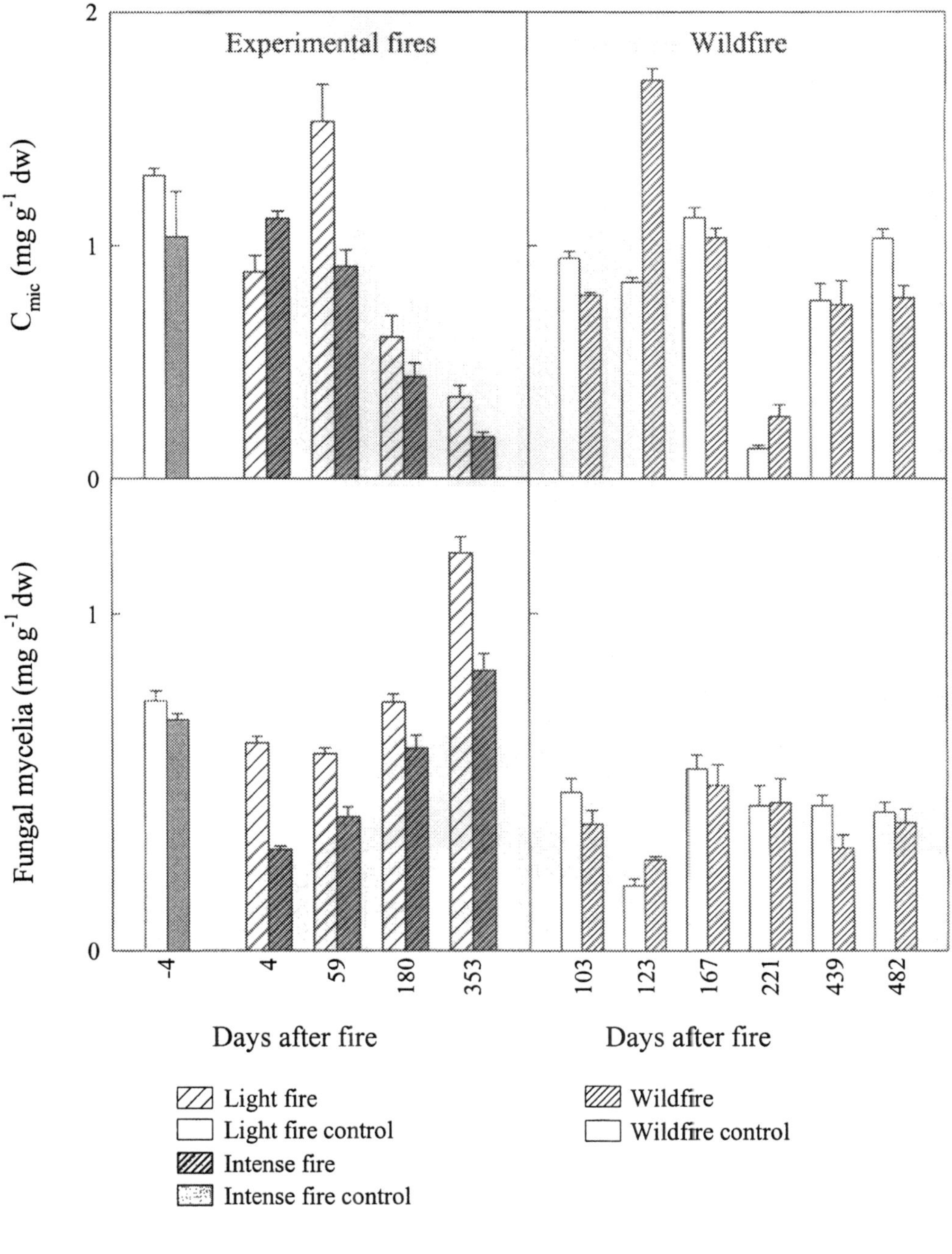

Figure 2. Changes in microbial biomass (C_{mic}) and fungal mycelia in soils burned by light and intense experimental fires and wildfire. The bars on the columns indicate the standard error.

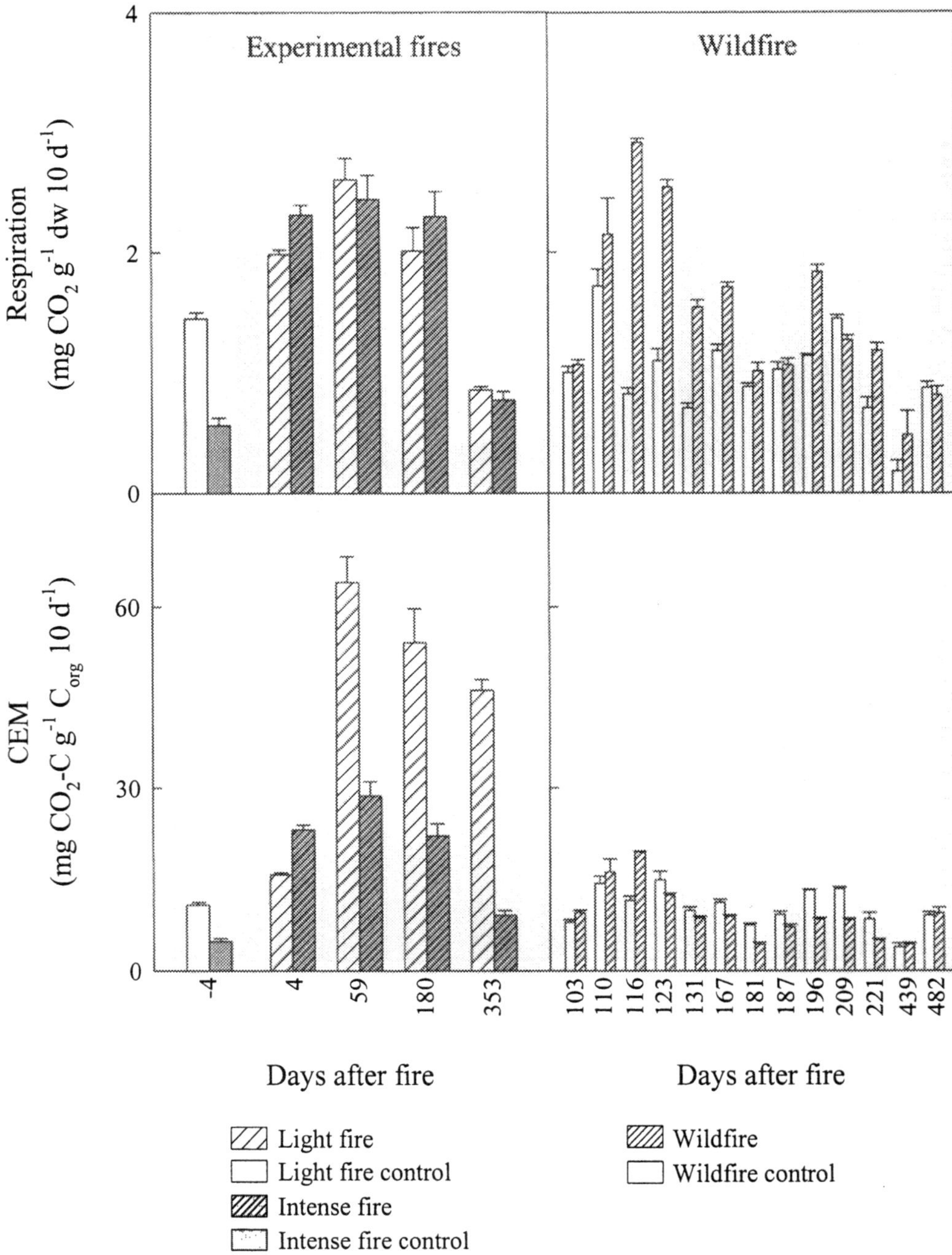

Figure 3. Changes in respiration and in coefficient of endogenous mineralization (CEM) in soils burned by light and intense fires and wildfire. The bars on the columns indicate the standard error.

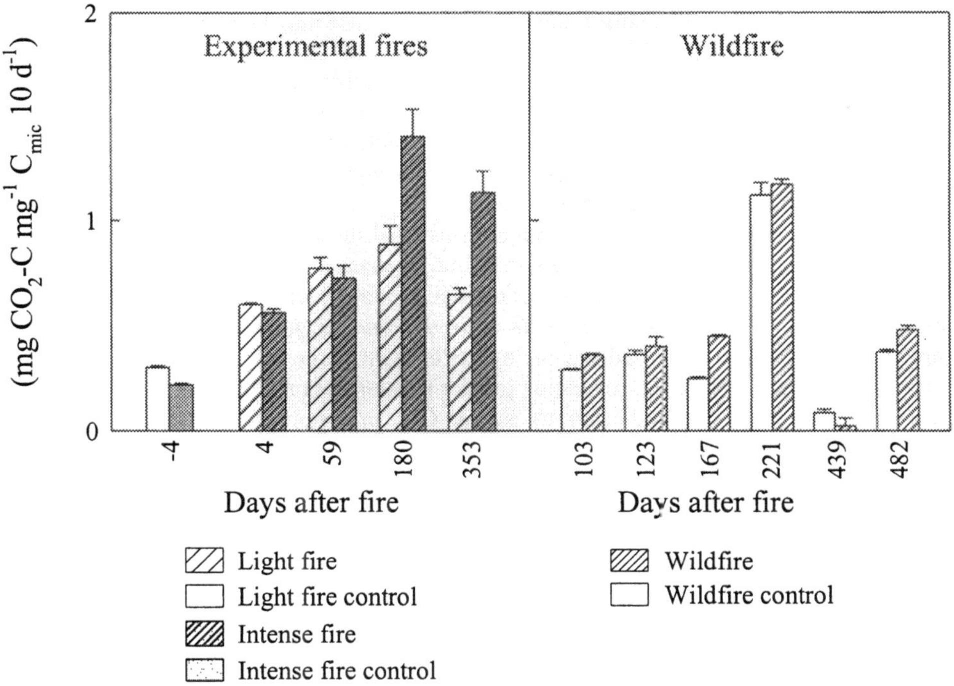

Figure 4. Changes in metabolic quotient (qCO_2) in soils burned by light and intense experimental fires and wildfire. The bars on the columns indicate the standard error.

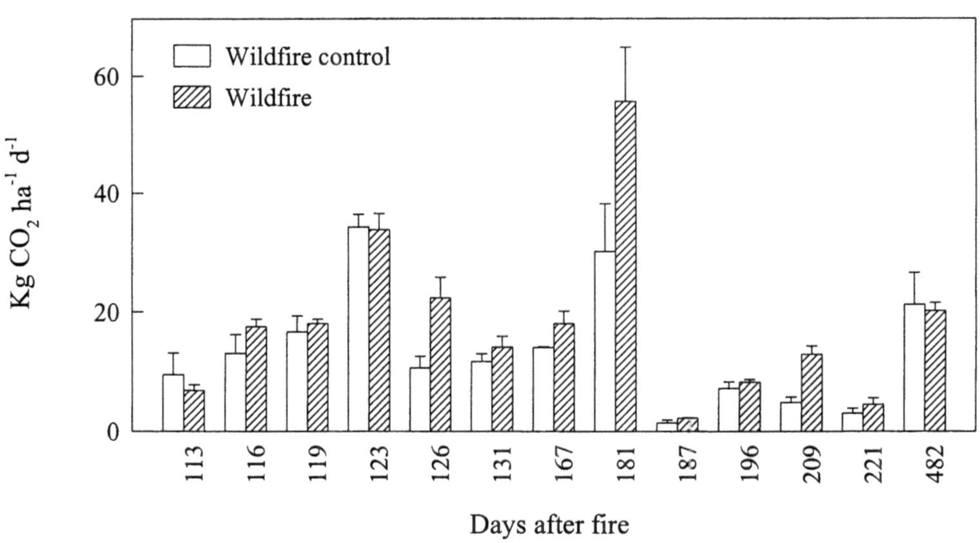

Figure 5. Changes in CO_2 emissions from soil exposed to wildfire. The bars on the columns indicate the standard error.

respiration. In the case of wildfire, the respiration increase may be dependent on the increase of organic matter input into the soil. The increase of respiration after light and intense experimental fires may be explained mainly by unfavorable soil conditions as indicated by the relevant increase (141%-334% as compared to the control) of the metabolic quotient (qCO_2). It is well know that qCO_2 is higher under unfavorable than under favorable conditions [15, 16]. The higher microbial activity also resulted in a higher mineralization rate of soil carbon (CEM) and a depletion of soil carbon content.

Even in the case of wildfire, microorganisms used more energy for maintenance than for growth, as indicated by the unchanged microbial biomass and fungal mycelia and by the 23% increase of qCO_2 as compared to unburned soil. Odum observed that "*repairing damage by disturbances requires diverting energy from growth and production to maintenance*" [17]. So only a low fraction of the killed organic matter brought to the soil by fire was immobilized in microbial biomass, while the remaining greater fraction was released into the atmosphere by microorganisms, as also indicated by higher CO_2 emission from soil in burned than in unburned areas.

Our results allow us to conclude that fire contributes to CO_2 emissions in the atmosphere, not only during biomass burning, but also in several months after fire occurrence. In the case of wildfire, the increase of CO_2 emissions from burned soils depends mainly on the increase in soil microbial activity due to the increase of organic matter in the soil and to some degree on stress conditions; in the case of experimental fires, the increase of CO_2 emissions is due to stress conditions.

ACKNOWLEDGMENTS

Financial support for this research was provided by the Ministero dell'Università e della Ricerca Scientifica e Tecnologica of Italy.

REFERENCES

1. Sparling, G.P., Ross, D.J., 1993. Biochemical methods to estimate soil microbial biomass: current developments and applications. *In*: Mulongoy, K., Merckx, R. (Eds.), Soil organic matter dynamics and sustainability of tropical agriculture. Wiley-Sayce, Leuven, pp. 17-21.
2. Alvarez, C.R., Alvarez, R., Grigera, M.S., Lavado, R.S., 1998. Associations between organic matter fractions and the active soil microbial biomass. Soil Biol. Biochem. 30, 767-773.
3. Pietikainen, J., Fritze, H., 1993. Microbial biomass and activity in the humus layer following burning: short term effects of two different fires. Can. J. Forest Res. 23, 1275-1285.
4. Carballas, M., Acea, M.J., Cabaneiro, A., Trasar, C., Villar, M.C., Diaz-Ravina, M., Fernandez, I., Prieto, A., Saa, A., Vazquez, F.J., Zehner, R., Carballas, T., 1993. Organic matter, nitrogen, phosphorus and microbial population evolution in forest humiferous acid soils after wildfires. *In*: Trabaud, L., Prodron, R. (Eds.), Fire in Mediterranean Ecosystems. Ecosystems Research Report N.5. Commission of the European Communities, Brussels, pp. 379-385.

5. di Castri, F., Goodall, D.W., Specht, R.L., 1981. Mediterranean-Type shrublands. Ecosystems of the World 11. Elsevier, Amsterdam.
6. Esposito, A., Mazzoleni, S., Strumia, S., 1999. Post-fire bryophyte dynamics in Mediterranean vegetation. J. Veg. Sci. 10, 261-268.
7. FAO, 1998. World reference base for soil resources. World Soil Resources. Report n. 84, Rome.
8. Mazzoleni, S., Esposito, A., 1993. Vegetative regrowth after fire and cutting of Mediterranean macchia species. *In*: Trabaud, L., Prodon, R. (Eds.), Fire in Mediterranean Ecosystems. Ecosystems Research Report N.5. Commission of the European Communities, Brussels, pp. 87-99.
9. Jenkinson, D.S., Powlson, D.S., 1976. The effect of biocidal treatments on metabolism in soil. I Fumigation with chloroform. Soil Biol. Biochem. 8, 167-177.
10. Olson, F.C.W., 1950. Quantitative estimates of filamentous algae. T. Am. Microsc. Soc. 69, 272-279.
11. Froment, A., 1972. Soil respiration in a mixed oak forest. Oikos 23, 273-277.
12. Johnson, D.W., 1992. Effects of forest management on soil carbon storage. Water Air Soil Poll. 63, 83-120.
13. Poth, M., Anderson, I.C., Miranda, H.S., Miranda, A.C., Riggan, P.J., 1995. The magnitude and persistence of soil NO, N_2O, CH_4 and CO_2 fluxes from burned tropical savanna in Brazil. Global Biogeochem. Cy. 9, 503-513.
14. Fierro, A.R., Rutigliano, F.A., De Pascale, R.A., Bartoli, G., Virzo De Santo, A., 1996. Soil emissions of N_2O and CO_2 from a burned pine-wood. *In*: Virzo De Santo, A., Alfani, A., Carrada, G.C., Rutigliano F.A. (Eds.), Ecologia. Atti del VII Congresso Nazionale della Società Italiana di Ecologia, pp. 63-66.
15. Anderson, T.H., Domsch, K.H., 1993. The metabolic quotient for CO_2 (qCO_2) as a specific activity parameter to assess the effects of environmental conditions, such as pH, on the microbial biomass of forest soils. Soil Biol. Biochem. 25, 393-395.
16. Wardle, D.A., Ghani, A., 1995. A critique of the microbial metabolic quotient (qCO_2) as a bioindicator of disturbance and ecosystem development. Soil Biol. Biochem. 27, 1601-1610.
17. Odum, E., 1985. Trends expected in stressed ecosystems. Bioscience 35, 419-422.

Developments in Soil Science, Volume 28B
Editors: A. Violante, P.M. Huang, J.-M. Bollag and L. Gianfreda

SOIL DEVELOPMENT AND MICROBIAL COMMUNITY ALONG AN ALTITUDINAL TRANSECT IN TRENTINO MOUNTAINS

M. Castaldini[a], A. Mirabella[a,c], G. Sartori[b], A. Fabiani[a], F. Santomassimo[a] and N. Miclaus[a]

[a]Istituto Sperimentale per lo Studio e la Difesa del Suolo, P.zza D'Azeglio 30, Firenze, Italy

[b]Museo Tridentino di Scienze Naturali, Via Calepina 14, Trento

[c]Corresponding author: Tel.: +39-055-2491232; Fax +39-055-241485.
E-mail: mirabella@dada.it

Three soils, collected at different altitudes in the Genoa Valley in Trentino (Italy), were studied to relate their degree of development to the eubacterial population structure. Organic chelating compounds have led to the formation of smectitic minerals in the surface horizon of the two Haplic Podzols developed at 1830 (TSG2) and 1660 (TSG3) m above sea level (a.s.l.); this process was not evident in the Spodo-Dystric Cambisol at 940 m a.s.l. (TSG5). Soils at the higher altitudes were characterized by higher acidity, exchangeable Al and CEC, when compared to the soil at 940 m a.s.l.. The ARDRA (amplified rDNA restriction analysis) and DGGE (denaturing gradient gel electrophoresis) analysis showed differences in the eubacterial population among the examined horizons of the three soils. Members of α and γ subgroups of the Proteobacteria were found to be more represented in soil at 940 m a.s.l. in hybridization experiments. Moreover, *Nitrosomonas* sp. were the prevalent ammonia oxidizers in TSG2 and TSG3 soils at the higher altitudes, while *Nitrosospira* sp. were in TSG5 soil. Soil pH, exchangeable Al and complexing organic acids seemed to be important factors affecting the microbial community in the examined soils.

1. INTRODUCTION

In forest soil genesis studies, the status of the eubacterial population rarely has been taken into account and little is known about how soil weathering factors may influence the composition of the bacterial consortium.

Microorganisms, by decomposing organic matter, strongly influence the cycle of nutrients in forest ecosystems, while physical and chemical properties of soils affect the structure of soil microflora. To ensure long-term sustainability of forest ecosystems, it is necessary to study not only the vegetation and the soil chemical and physical properties, but also the microbial state of the soil community, which is extremely sensitive to any climatic or anthropic management change. Many studies have taken into account, in different forest types, the relationship between soil properties and microbial communities using classical microbial techniques [1, 2, 3]. In any case, the use of these methods has limitations due to the nonculturability of the largest portion of soil bacteria [4, 5].

Recent advances in molecular techniques to investigate soil microflora by means of amplified rDNA restriction analysis (ARDRA) [6] and denaturing gradient gel electrophoresis (DGGE) [7] allow the description of the whole eubacterial population, intending to overcome limitations of microbial culture techniques; in fact, the RNA genes are useful for such studies since they are present in all bacteria and highly reliable for diagnostic purposes. The advantage of these methods is that fluctuations in the microbial community can be detected by studying the distinct banding patterns resulting from the soil microbiota consortium, and they have already proven to be highly informative in agroecosystems [8, 9]. In addition, amplified rDNA fragments can be hybridized with taxonomic-specific probes [10].

In this study, three soils developing along an altitudinal transect were investigated to relate their degree of weathering, derived from chemical and mineralogical analyses, to the microbial community structure, with a deeper insight on autotrophic ammonia oxidizers, which play a fundamental role as bioindicators of stress and disturbance [11].

2. MATERIALS AND METHODS

2.1. Soil sampling and classification

The characteristics of the sampling sites are shown in Table 1. The soil profiles TSG2 and TSG3, in the altitude range 1830-1660 m a.s.l., are Haplic Podzols [12] and thus have shown well-expressed spodic features; the soil profile TSG5, developing at 940 m a.s.l., was classified as a Spodo-Dystric Cambisol [13]. Soils were sampled, sieved at 2 mm and subjected to physical and chemical analyses. Mineralogical and microbial analyses were carried out on the two upper mineral surface horizons of each soil profile, which was representative of the site considered.

Table 1. Characteristics of the study site.

Profile	Locality	Elevation (m)	Location (°N)	Parent material	Dominant Vegetation	Form of Humus	FAO Legend
TSG2	Malga S. Giuliano	1830	40	granodiorite	*Larix decidua*	Mor	Haplic Podzol
TSG3	Malga Campo	1660	115	granite glacial colluvium	*Picetum* mixed to *Larix*	Dysmoder	Haplic Podzol
TSG5	Cascate Nardis	940	20	granodiorite	*Picea abies* and *Abies alba*	Mor	Spodo-Dystric Cambisol

2.2. Physical and chemical analyses

Particle-size distribution was determined by the pipette method after dispersion in deionized water. Organic matter and total nitrogen content were estimated, respectively, by the Walkley-Black and Kjeldahl methods; pH was measured by potentiometry in a soil:solution (H_2O and KCl) ratio of 1:2.5; CEC and exchangeable bases were determined with a 1 N NH_4OAc extraction, and element concentrations were analyzed by AAS [14]. The exchangeable acidity was obtained by the KCl method; amorphous solid phases were extracted by oxalate at pH 3 (Fe_o and Al_o) [15].

2.3. Clay minerals analyses

Clay particles (< 2 μm) were separated from the fine earth fraction by sedimentation in water with the addition of Calgon as a dispersing agent. X-ray patterns of soil clays were obtained from samples oriented on glass slides with a 3 KW Rigaku D/MAX III C diffractometer, equipped with a horizontal goniometer, a curved beam graphite monochromator and Cu radiation. Slides were step-scanned from 2 to 15° 2θ with 0.02° 2θ steps and 4" counting time. The following treatments were performed: Ethylene glycol solvation, Mg saturation and K saturation, followed by heating for two hours at 335°C and 550°C.

2.4. PCR conditions of 16S and 23S rDNA of the whole eubacterial community

Twenty-five ng of DNA directly extracted from soil [9] were amplified with 16S and 23S universal eubacterial primers (Table 2). The 50 μl reaction contained 250 μM each primer, 250 μM dNTPs, 1,5 mM $MgCl_2$, 1X buffer and 2,5 U Polytaq (Polymed Biotechnology Division, Florence, Italy). 16S rDNA was amplified in a 25 cycles touch down PCR with 30 sec denaturation at 95°C, 30 sec annealing at temperature decreasing five grades from 60°C to 50°C every five cycles, 4 min elongation at 72°C and a final elongation of 10 min at the same temperature. 23S rDNA was amplified in a 30 cycles touch down PCR with 45 sec denaturation at 95°C, 30 sec annealing at temperature decreasing one grade each two cycles from 60°C to 56°C and 1 min 45 sec elongation at 72°C with a final 5 min elongation step at 72°C. PCR products were checked on 0.8% agarose TAE gel.

For DGGE analysis, the V6 to V8 region of eubacterial 16S rDNA was amplified with universal primers (Table 2), as described by Felske et al. [16]. PCR products were checked on 1.2% agarose TAE gel.

Table 2. Primers for PCR amplification of 16S and 23S rDNA genes.

Name-Position (*E.coli*)	Specificity
16S P0f (7-27)	Eubacteria [17]
16S P6r (1514-1495)	Eubacteria [17]
23S P441f (441-470)	Eubacteria [18]
23S M2436r (2436-2417)	Eubacteria [18]
16S P141f (141-158)	Ammonia oxidizers [8]
16S P1314f (1314-1290)	Ammonia oxidizers [8]
16S GC U968f (968-983)	Eubacteria [16]
16S L1401r (1401-1385)	Eubacteria [16]

2.5. PCR conditions for ammonia oxidizers

PCR reaction mix for 50 μl consisted of 5 μl PCR buffer (10x), 1.5 mM MgCl2, 200 μM of each dNTP, 0.2 pmol each primer (Table 2), 40 ng of soil DNA, 2.6 U of Taq. A touch-down program was used: 1 min denaturation at 94°C, 1 min at annealing temperature lowering 2°C every two cycles from 64°C to 54°C for the remaining 30 cycles, 3 min extenuation at 72°C and a final 10 min extension at 72°C. PCR products were checked on 0.8% agarose TAE gel.

2.6. ARDRA analysis

The ARDRA analysis was performed with *Alu*I on 400 ng of 16S and 23S amplified rDNA of the eubacterial community and with *Taq*I on 16S rDNA of ammonia oxidizers population; after 3 hours digestion at 37°C (65°C for *Taq*I digestion), bands were separated by electrophoresis on a 3% agarose gel at 5V/cm for 3 hours in 1x TAE and stained with ethidium bromide.

2.7. DGGE analysis

The DGGE analysis of the V6–V8 amplicons was performed on a 6% polyacrylamide gel (40% acrylamide-bis 37,5:1) with a 42% to 58% denaturing gradient of 7 M urea and 40% formamide at 60V for 17 hours in 1x TAE at 60°C in a BIORAD D-gene system. The gel was stained with SYBR® Green I. (FMC BioProducts).

2.8. Slot-blot hybridization

This analysis was performed with the digoxygenin 3'-end labelled (Roche) probes specific for subgroups of proteobacteria and ammonia oxidizers (Table 3), according to Manz et al. [10] and Stephen et al. [19], respectively.

Table 3. Probes for hybridization of rDNA genes.

Name/Specificity - Position (*E.coli*)
α - proteobacteria 16S (19-35) [10]
β - proteobacteria 23S (1027-1043) [10]
γ - proteobacteria 23S (1027-1043) [10]
δ - proteobacteria 16S (385-402) [5]
Nmo 254a *Nitrosomonas* sp. 16S (236-254) [19]
Nsp 436 *Nitrosospira* sp. 16S (418-436) [19]

3. RESULTS

The climate along the study area differs greatly with elevation. At the nearby station of Pinzolo at 776 m a.s.l., mean annual temperature was 8.75°C, ranging from -1.19°C in January to 19.01°C in July, and mean annual precipitation was 1097 mm. At the nearby station of Malga Bissina at 1792 m a.s.l., mean annual temperature was 5.46°C, ranging from -3.01°C in February to 12.45°C in July, and mean annual precipitation was 1436 mm. Soils developing in the range 1830-1660 m a.s.l. did not show moisture deficit, while a brief summer moisture deficit should be expected to affect the soil TSG5 developing at 940 m a.s.l.

Results of physical and chemical analyses are reported in Table 4.

Texture varied from loam to sandy loam in the selected soils. The lowest pH values were found in the surface horizons of profiles TSG2 and TSG3, with values around 4 in water and 3 in KCl.

In profiles TSG2 and TSG3, organic matter eluviates from the surface horizons toward the lower ones; this process is paralleled by an increase of Al_{ox} and Fe_{ox}, according to the theory of formation of Spodosols that suggests the formation of Al and Fe complexes with organic matter and their downward migration in the soil profile [20].

Profile TSG5 differed from the other two soils because it did not show an increase of organic carbon and Fe_{ox} in the lower horizons and Al_{ox} increased only slightly with depth. Cation exchange capacity was higher in the two soils developing at the altitude range 1830-1660 m a.s.l.. Al was the most represented cation in the soil exchange complex, and the soil at the lowest altitude showed a much higher percentage of base saturation.

Table 4. Physical and chemical properties of the selected soils.

Horizon/ Depth		Sand	Silt	Clay	pH	pH	C	N	Fe_{ox}	Al_{ox}	Ca	Mg	K	Na	Al	H	CEC	Base sat.
Cm		%	%	%	H_2O	KCl	g/kg	g/kg	g/kg	g/kg	cmol/kg							%
Profile TSG2																		
E	10-12	45.1	40.5	14.4	4.1	3.4	77	3.8	0.74	0.33	0.3	0.4	0.2	0.0	7.8	1.5	10.2	10
Bhs	12-17	34.4	46.7	18.9	4.9	4.1	101	4.7	1.88	2.20	0.3	0.2	0.1	0.4	7.3	0.0	8.4	12
Bs	17-38	43.5	31.3	25.2	4.8	4.4	92	3.3	0.78	3.06	0.4	0.1	0.1	0.1	4.1	0.0	4.8	15
BC	38-50+	49.3	48.8	1.9	5.1	4.6	40	1.3	0.30	2.28	0.1	0.0	0.0	0.0	1.8	0.0	2.0	9
Profile TSG3																		
AE	10-16	32.0	49.5	18.5	3.8	3.1	115	6.9	0.50	0.29	0.5	0.7	0.3	0.0	10.1	1.8	13.4	11
Bhs	20-55	27.7	54.6	17.7	4.6	3.9	111	6.0	1.93	1.72	0.0	0.2	0.1	0.1	9.9	0.7	11.1	3
Bs	55-95	38.9	45.7	15.4	4.9	4.3	138	4.8	1.10	3.12	0.1	0.1	0.0	0.1	5.9	0.0	6.1	4
2A	95-115	37.9	48.6	13.5	5.1	4.3	123	3.8	1.95	2.91	0.6	0.1	0.0	0.1	4.6	0.0	5.4	14
Profile TSG5																		
AE	8-16	35.3	41.1	23.6	5.0	4.3	98	4.9	0.84	1.50	4.9	0.7	0.3	0.1	2.9	0.0	8.9	67
Bs1	16-50	47.5	38.0	14.5	5.2	4.5	52	2.5	0.86	1.83	1.7	0.1	0.1	0.2	1.9	0.0	3.9	52
Bs2	50-105	46.4	45.4	8.2	5.3	4.6	34	1.2	0.75	1.76	1.2	0.1	0.1	0.2	0.8	0.0	2.5	67
BC	105-125	58.9	32.4	8.7	5.6	4.8	22	1.0	0.42	1.83	1.4	0.1	0.1	0.3	0.5	0.0	2.4	78

X-ray patterns of the clay samples of the two surface horizons of the three soils are given in Figure 1 (a–f).

The surface horizons of profiles TSG2 and TSG3, respectively horizons E and AE, showed a low charge expandable mineral that could be identified by the peak around 1.65 nm after ethylene glycol solvation, which collapsed to 1.4 nm following Mg saturation (Figures 1a and c). This mineral could not be detected in the lower Bhs horizon of profile TSG3 (Figure 1d), but a component with a less pronounced charge, evidenced by a shoulder toward low angles in the ethylene solvated pattern, was still present in the Bhs horizon of profile TSG2 (Figure 1b). The soil profile TSG5 did not exhibit this mineral in either of the surface horizons (Figures 1e and f). Together with the low charge expandable mineral, vermiculite, evidenced by the peak at 1.4 nm that collapsed to 1.0 nm following K saturation, mica, distinguished by the peak at 1.0 nm, and kaolinite, whose peak at 0.72 nm disappeared after the 550°C heat treatment, were also identified in the surface horizons of profiles TSG2 and TSG3.

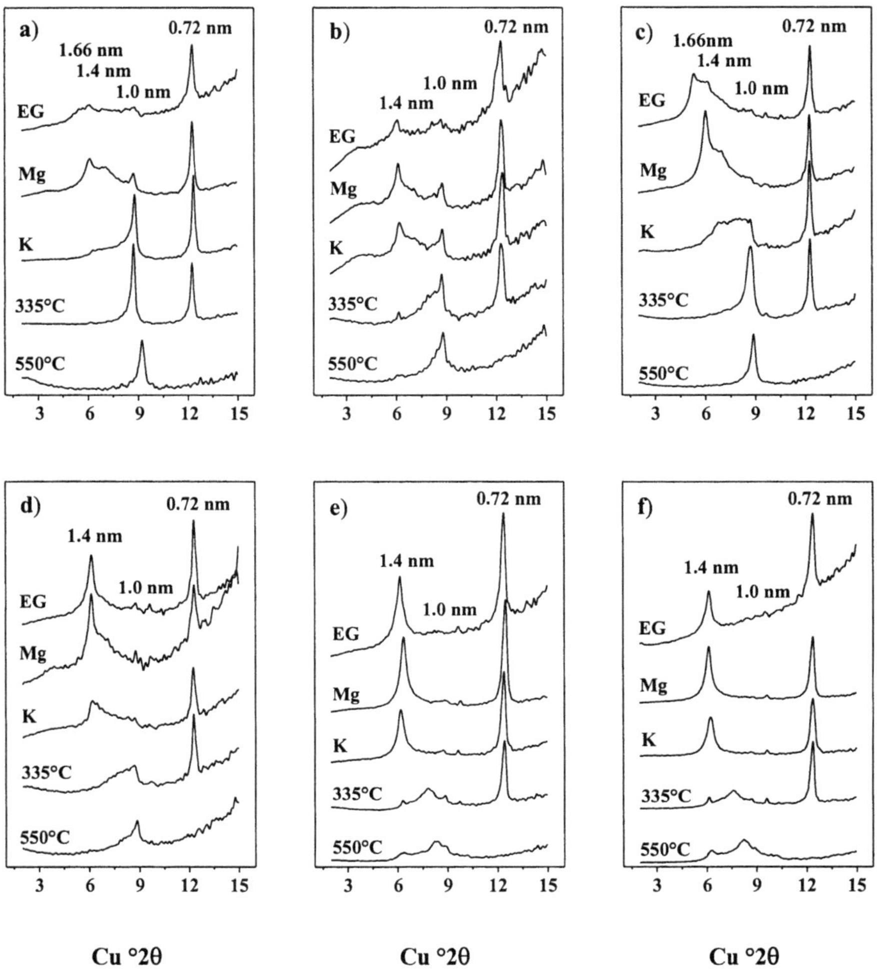

Figure 1. X-ray patterns relative to the clay samples of the following soil horizons: a) E and b) Bhs of profile TSG2; c) AE and d) Bhs of profile TSG3; e) A and f) Bs1 of profile TSG5.

Furthermore, an interstratified mica/vermiculite mineral, shown by the peak around 1.3 nm in the Mg-saturated samples, which shifted to 1.0 nm after K saturation, was detected in the E horizon of profile TSG2 (Figure 1a). This mineral was also present in the Bhs horizon of profile TSG2 and in both the surface horizons of profile TSG3, but vermiculite was hydroxy-interlayered, as shown by the collapse to 1.0 nm of the peak around 1.3 nm after the heat treatment at 335°C (Figures 1b– d).

The Bhs horizons of soil profiles TSG2 and TSG3 contained vermiculite, which was almost completely hydroxy-interlayered, as shown by the peak at 1.4 nm, that resisted collapse after K saturation and shifted to 1.0 nm after heating at 335°C. Traces of chlorite were also present, as indicated by the peak at 1.4 nm in the 550°C pattern. (Figures 1b and d).

The main feature of the two surface horizons of profile TSG5 was the presence of vermiculite that was hydroxy-interlayered, as shown by the collapse to 1.0 nm of the peak at 1.4 nm following K saturation and heating at 335°C (Figures 1e and f). Furthermore, chlorite could also be found in both of the examined A horizons, together with mica (1.0 nm) and kaolinite (0.72 nm), while the interstratified mica/vermiculite mineral was absent.

DNA restriction analysis with AluI of 16S rDNA (Figure 2a) showed a difference in the richness and evenness of the eubacterial community related respectively to the presence of a band of about 700 bp in the patterns of the upper horizons of the three soils (■) and to a different intensity of two close bands at about 600 bp in all horizons (●). Differences in the banding patterns were also detected by the ARDRA analysis on 23S rDNA (Figure 2b) between 400–500 (■) and 200–300 bp (●), though not strictly related to the soil horizons.

The DGGE pointed out that the patterns for the soil developing at the lowest altitude (Figure 3 lane 5, 6) were different considering the region with the most intense bands (■). All horizons showed different patterns in the lower portion of fingerprintings (Figure 3 ●); some similarities were detectable between the patterns of the two Bhs horizons.

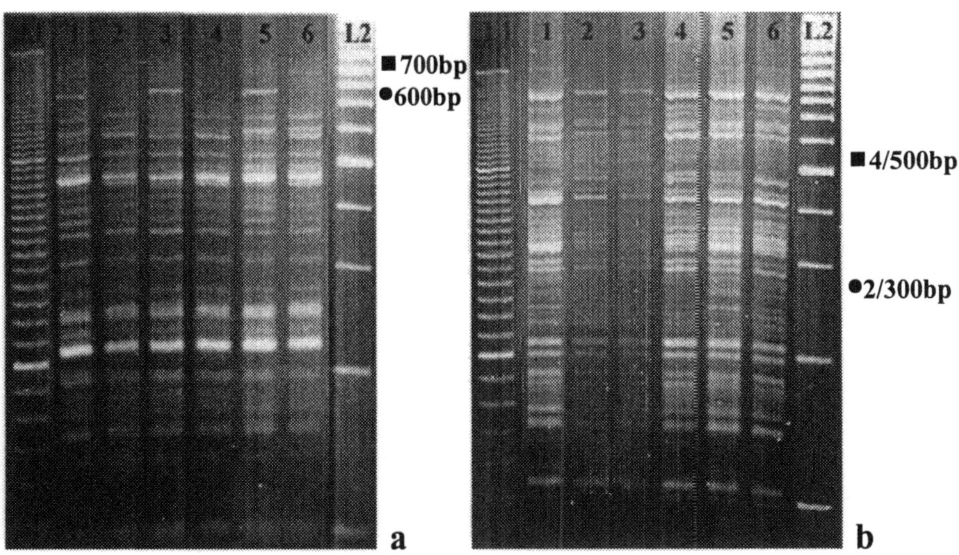

Figure 2. ARDRA patterns obtained with AluI on 16S rDNAs (a) and 23S rDNAs (b) of eubacterial community of forest soil samples: 1,2 TSG2, horizons E and Bhs; 3,4 TSG3, horizons AE and Bhs; 5,6 TSG5, horizons AE and Bsl. L1: 20 bp ladder; L2: 100 bp ladder.

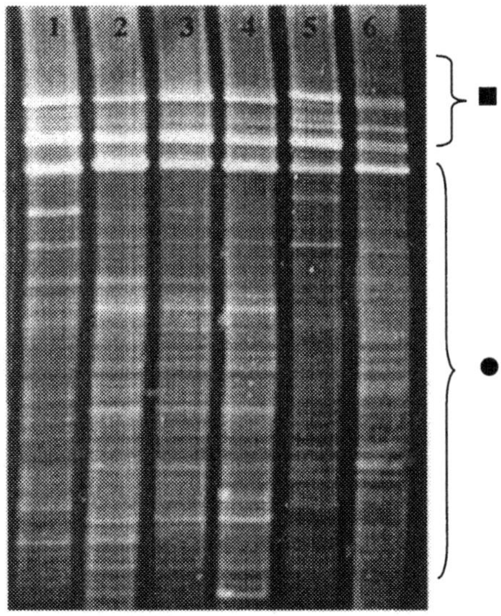

Figure 3. DGGE patterns of 16S rDNAs of eubacterial community of forest samples: 1,2 TSG2, horizons E and Bhs; 3,4 TSG3, horizons AE and Bhs; 5,6 TSG5, horizons AE and Bsl. L1: 20 bp ladder; L2: 100 bp ladder.

The hybridizations with probes specific for α and γ proteobacteria (Figures 4 a and b) showed an increasing signal intensity with the decrease of altitude; the hybridization with β- and δ-specific probes gave comparable signals in all the examined horizons (data not shown).

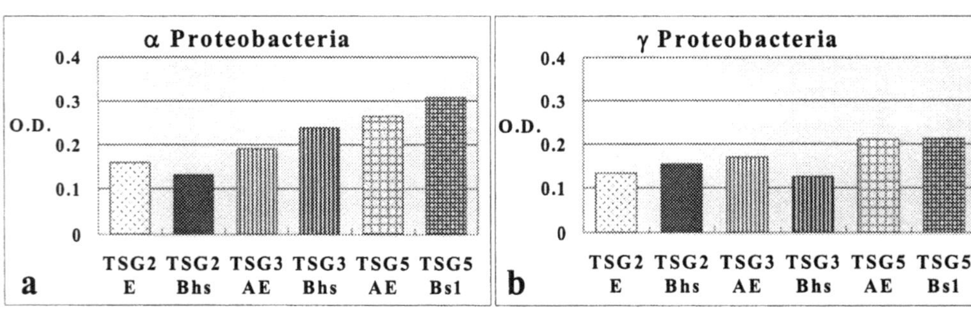

Figure 4. Optical density values of hybridization signals of 16S amplified rDNAs from the two horizons of each profile (a) with α proteobacteria-specific probe and (b) with γ proteobacteria-specific probe.

The detection of ammonia-oxidizing bacteria, performed with ARDRA and subsequent hybridization, indicated that in both the horizons of the soils at the two highest sites *Nitrosomonas* spp. was the prevailing genus (Figure 5a), while in the lowest site *Nitrosospira* spp. was dominant (Figure 5b).

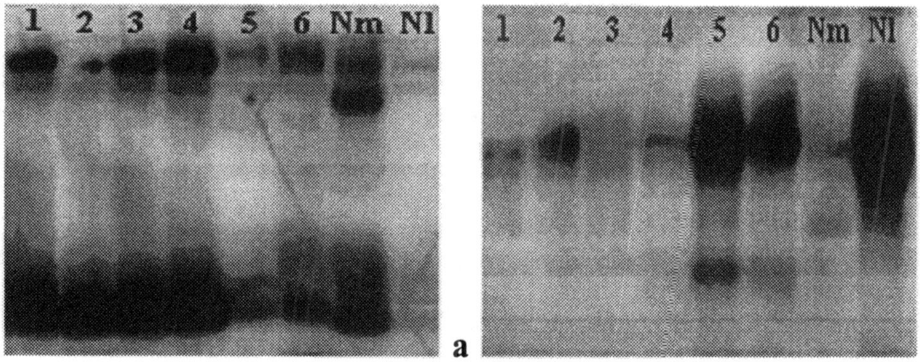

Figure 5. a) Hybridization with *Nitrosomonas* sp.-specific probe Nmo 254a of amplified 16S rDNAs of ammonia oxidizers digested with *HinfI*: 1,2 TSG2, horizon E and Bhs; 3,4 TSG3, horizon AE and Bhs; 5,6 TSG5, horizon AE and Bs1. Nm: *Nitrosomonas eouropaea* ATCC 25978; Nl: *Nitrosospira multiformis* ATCC 25196. b) Hybridization with *Nitrosospira* sp.-specific probe Nsp 436 of amplified 16S rDNAs of ammonia oxidizers digested with *HinfI*: 1,2 TSG2, horizon E and Bhs; 3,4 TSG3, horizon AE and Bhs; 5,6 TSG5, horizon AE and Bs1. Nm: *Nitrosomonas eouropaea* ATCC 25978; Nl: *Nitrosospira multiformis* ATCC 25196.

4. DISCUSSION

Chemical and mineralogical analyses showed a clear difference between the soils developing in the altitude range 1830-1660 m a.s.l. and the soil developing at the lower altitude (940 m a.s.l.). In particular, the different climatic conditions at the higher altitudes, characterized by higher rainfalls in the form of snow in winter and by lower temperatures, has led to the formation of eluvial surface horizons. These conditions have promoted the appearance of a low charge expandable mineral formed from chlorite through removal of hydroxy-interlayers and reduction of layer charge, with the extraction of the interlayer polymers being enhanced by organic complexing agents [21]. This process was not evidenced in soil TSG5, developing at 940 m a.s.l.; in fact, this soil still contains in the surface horizon a higher amount of hydroxy-interlayered 2:1 clay minerals and a small amount of chlorite.

The process indicated in the E and AE horizons of profiles TSG2 and TSG3 is enhanced by the presence of low molecular weight aliphatic and aromatic acids, humic acids or fulvic acids [22] and causes a decrease of pH and a high amount of Al in the exchange complex. Al also migrates in the lower horizons of these two soils, as indicated by the increase of the concentration of Al extracted by oxalate (Al_{ox}) with increasing depth. In deeper horizons, this element may be still chelated by organic matter or present in the form of amorphous short-order phases.

The different genetic processes also affected the diversity of the eubacterial population in the considered soil horizons, as shown by the electrophoretic profiles of ARDRA and DGGE analysis. The 16S rDNA profiles reflect the individual species present in the eubacterial community and therefore the fingerprintings consist of a combination of some more intense bands, several lower intensity ones and an additional number of very weak bands, the last ones often producing a smear. That is why, notwithstanding the high resolution of these molecular techniques, it is not possible to infer the exact number of different species present in soil, as it is estimated that thousands of different bacterial genomes are present in one gram of soil [4].

As reported in grassland soils [23], the soil microbial communities usually contain a few prominent species with many individuals and a lot of species of low abundance. In particular, the 16S rDNA analysis performed on our samples has evidenced a different main microflora structure between the two examined Podzols, developing at a higher altitude, and the Cambisol at 940 m a.s.l.; as regards the less represented eubacterial population, differences in the degree of richness and evenness were detected between the two horizons examined within every soil, even though a certain degree of similarity was detected between the lower Bhs horizons of the podzolic soils.

These findings have been supported by results of the hybridizations with probes specific for the main subgroups of Proteobacteria. In particular, a more intense signal with the α and γ subgroup-specific probes was found in the two examined horizons of the Cambic soil.

When considering the composition of a single bacterial population, such as autotrophic ammonia oxidizers, a different distribution of the two main genera *Nitrosomonas* and *Nitrosospira* was detected and related to soil type and altitude and, therefore, to climate. These two microorganisms play a key role in nitrogen cycling and have been found in acid forest soils [24, 25]. The predominance of *Nitrosomonas* in both the surface horizons of the two Podzols and of *Nitrosospira* in the examined horizons of soil at 940 m a.s.l. may be due to the occurrence of two types of chemolithotrophic nitrification [25]. The environment of the two Podzols is characterized by lower pH values, a much higher content of exchangeable Al and a lower base saturation when compared with the Cambisol. Furthermore, the process of formation of smectitic components in the surface horizons of the Podzols indicates the presence of low molecular weight organic acids and fulvic acids that represent a more readily degradable substrate and may favor the more acid-tolerant chemolithotrophic bacteria. This is in agreement with the fact that all four genes involved in the ammonia oxidation pathway have been found to diverge between the *Nitrosomonas* and the *Nitrosospira* group [26].

5. CONCLUSIONS

This study has related the different characters of soils developing in different climatic conditions to the structure of the eubacterial community, which was found to be influenced by the chemical and mineralogical processes.

Data obtained with molecular techniques allowed a broader description to be made of the changes of soil microflora among the studied soils and constitute a preliminary step in better understanding the relationship between soil genetic processes and the soil bacterial community.

REFERENCES

1. Das, P.K., Nath, S., Banerjee, K., 1991. Distribution of microorganisms in soils under different forest cover at different altitudes. Indian Agric. 35, 217-223.
2. Gregori, E., Miclaus, N., Sasso, A., 1991. Studio di una faggeta dell'Appenino Pistoiese: decomposizione della lettiera fogliare confinata al suolo II. Popolazioni microbiche e attività enzimatiche. Ann. Ist. Sper. Studio Difesa Suolo XX, 95-130.

3. Vinther, F.P., Eiland, F., Lind, A.M., Elsgaard, L., 1999. Microbial biomass and numbers of denitrifiers related to macropore channels in agricultural and forest soils. Soil Biol. Biochem. 31, 603-611.
4. Torsvik, K.V., Goksoyr, J., Daae, F.L., 1990. High diversity in the DNA of soil bacteria. Appl. Environ. Microbiol. 56, 782-787.
5. Amann, R.I., Ludwig, W., Schleifer, K.H., 1995. Phylogenetic identification and in situ detection of individual microbial cells without cultivation. Microbiol. Rev. 59, 143-169.
6. Massol-Deya, A.A., Odelson, D.A., Hickey, R.F., Tiedje, J.M., 1995. Bacterial community fingerprinting of amplified 16S and 16-23S ribosomal DNA gene sequences and restriction endonucleases analysis (ARDRA). Molecular Microbial Ecology Manual. Kluwer Academic Publishers 3.3.2, pp. 1-8.
7. Muyzer, G., Brinkoff, T., Nübel, U., Santegoeds, C., Schäfer, H., Wawer, C., 1998. Denaturing gradient gel electrophoresis (DGGE) in microbial ecology. Molecular Microbial Ecology Manual. Kluwer Academic Publishers, 3.4.4., pp. 1-27.
8. Smit, E., Leeflang, P., Wernars, K., 1997. Detection of shifts in microbial community structure and diversity in soil caused by copper contamination using amplified ribosomal DNA restriction analysis. FEMS Microbiol. Ecol. 23, 249-261.
9. Miclaus, N., Castaldini, M., Ceccherini, M.T., Piovanelli, C., Bazzicalupo, M., 1998. Effetti della fertilizzazione con reflui zootecnici suini sulle popolazioni microbiche del suolo. Agric. Ric. 173, 23-36.
10. Manz, W., Amann, R., Ludwig, W., Wagner, M., Schleifer, K.-H., 1992. Phylogenetic oligodeoxynucleotide probes for the major subclass of Proteobacteria: problems and solutions. Syst. Appl. Microbiol. 15, 593-600.
11. Ceccherini, M.T., Castaldini, M., Piovanelli, C., Hastings, R.C., McCarthy, A.J., Bazzicalupo, M., Miclaus, N., 1998. Effects of swine manure fertilization on autotrophic ammonia oxidizing bacteria in soil. Appl. Soil Ecol. 7, 149-157.
12. FAO – UNESCO, 1990. Soil Map of the World - Revised Legend. Rome, Italy.
13. Mirabella, A., Sartori, G., 1998. The effect of climate on the mineralogical properties of soils from the Val Genova Valley (Trentino, Italy). Fres. Environ. Bull. 7, 478-483.
14. Thomas, G.W., 1982. Exchangeable cations. *In:* Page, A.L. (Ed.), Methods of Soil Analysis, Part 2. Chemical and Microbiological Properties, Agronomy Monograph no. 9, Soil Sci. Soc. Am., Madison, WI, pp. 159-165.
15. Schwertmann, U., 1964. Differenzierung der eisenoxide des bodens durch extraction mit ammoniumoxalat lösung. Pflanzenernähr. Düng Bodenk. 105, 195-202.
16. Felske, A., Wolterink, A., Van Lis, R., Akkermans, A.D.L., 1998. Phylogeny of the main bacterial 16S rRNA sequences in Drentse A grassland soils (The Netherlands). Appl. Environ. Microbiol. 64, 871-879.
17. Lane, D.J., 1991. 16S/23S rRNA sequencing. *In:* Stackebrandt, E., Goodfellow, M. (Eds.), Nucleic Acid Techniques in Bacterial Systematics. John Wiley & Sons, Chichester, UK, pp.115-175.
18. Campbell, K.D., East, A.K., Thompson, D.E., Collins, M.D., 1993. Studies on the large subunit ribosomal RNA genes and intergenic spacer regions of non-proteolytic *Clostridium botulinum* types B, E and F. Res. Microbiol. 144, 171-180.
19. Stephen, J.R., Kowalchuk, G.A., Bruns, M.A.V., McCaig, A.E., Phillips, C.J., Embley, T.M., Prosser, J.I., 1998. Analysis of β-subgroup proteobacterial ammonia oxidizer populations in soil by denaturing gradient gel electrophoresis analysis and hierarchical phylogenetic probing. Appl. Environ. Microbiol. 64, 2958-2965.

20. De Coninck, F., 1980. Major mechanisms in formation of spodic horizons. Geoderma 24, 101-128.
21. Carnicelli, S., Mirabella, A., Cecchini, G., Sanesi, G., 1998. Weathering of chlorite to a low charge expandable mineral in a Spodosol on the Apennine Mountains-Italy. Clays Clay Miner. 45, 28-41.
22. Lundström, U.S., van Breemen, N., Bain, D., 2000. The podzolization process. A review. Geoderma 94, 91-107.
23. Felske, A., Akkermans, A.D.L., Spatial homogeneity of abundant bacterial 16S rRNA molecules in grassland soils. Microb. Ecol. 36, 31-36.
24. Pennington, P.I., Ellis, R.C., Autotrophic and heterotrophic nitrification in acidic forest and native grassland soils. Soil Biol. Biochem. 25, 1399-1408.
25. De Boer, V., Klien Gunnewiek, P.J.A., Troelstra, S.R., Laanbroek, H.J., 1989. Two types of chemolithotrophic nitrification in acid heathland humus. Plant Soil 119, 229-235.
26. Bruns, M.A., Fries, M.R., Tiedje, J.M., Paul, E.A., 1998. Functional gene hybridization patterns of terrestrial ammonia-oxidizing bacteria. Microb. Ecol. 36, 293-302.

Developments in Soil Science, Volume 28B
Editors: A. Violante, P.M. Huang, J.-M. Bollag and L. Gianfreda

SOIL MICROBIAL ACTIVITY IN THREE EVERGREEN OAK (*QUERCUS ILEX*) WOODS IN A MEDITERRANEAN AREA

S. Papa, E. Curcio, A. Lombardi, P. D'Oriano and A. Fioretto

Dipartimento di Scienze della Vita, Seconda Università di Napoli,
Via Vivaldi 45, 81100 Caserta, Italy

Soil respiration and enzyme activities (cellulase, xylanase, invertase, protease and trealase) were studied in three evergreen oak woods in Southern Italy (Campania) in the Nature Reserves of Castel Volturno, Caserta (sea level), of Tirone – Alto Vesuvio, Mt. Vesuvius, Naples (500 m a.s.l.) and in the WWF "Oasi Bosco di S.Silvestro", Caserta (200 m a.s.l.). The experimental sites differed in chemical and physical characteristics and microclimatic conditions. The microbial activity was tested every three months for one year. Microbial respiration showed seasonal variations, with minimum levels in summer and maximum levels in wet seasons, and this pattern closely followed seasonal changes in moisture content. In absolute terms, the highest levels of microbial respiration were found in the Vesuvius soil (7.45μmol CO_2/g d.w./h to 23.09μmol CO_2/g d.w./h), which also had the highest organic C content, whereas the lowest levels were found in the Castel Volturno soil (1.76μmol CO_2/g d.w./h to 8.14μmol CO_2/g d.w./h). Seasonal variations in enzyme activities were also found, with good correlation with soil moisture content. No significant differences in enzyme activity were found among the three studied woods.

1. INTRODUCTION

Soil microorganisms are important for soil fertility and primary production because of their roles in organic matter decomposition and nutrient cycling. The soil contains a large variety of microbial taxa with a wide diversity of metabolic abilities [1].

Ever since the development of systematic studies in soil microbiology there has been interest in assessing the activity of either the total soil microbial community or of specific groups of microorganisms. Measuring enzyme activities in soil can indicate not only the presence or absence of a certain function, but may also quantify the potential of the microbial community to realize this function [2].

The release of extracellular enzymes is dependent on the activity of microorganisms which, in turn, is influenced by temperature, moisture, pH and the quality of substrate [3, 4].

Numerous studies have been performed on microbial activities, notably on enzyme activities in temperate forest soils [5-11], but no information is available for Mediterranean ecosystems where the main factor limiting productivity is seasonal water shortage.

In this study we compare soil microbial respiration and soil microbial enzyme activities (cellulase, xylanase, invertase, protease and trealase) in three natural evergreen oak woods in Campania, with the aim to evaluate whether differences in chemical and physical soil characteristics and in microclimatic conditions promote differences in the biological activity of microbial communities.

2. MATERIALS AND METHODS

2.1. Site description

This study was carried out in areas (3000 m^2) selected in three evergreen oak woods in the Campania region (Southern Italy) located in the Nature Reserve of Castel Volturno, Caserta (at sea level), in the Nature Reserve of Tirone-Alto Vesuvio, Mount Vesuvius, Naples (500 m a.s.l.) and in the WWF Oasis "Bosco di S.Silvestro", Caserta (200 m a.s.l.).

The Nature Reserve of Castel Volturno is located to the south of the Volturno estuary. The topography is characterized by dunes of alluvial deposits with an elevation of 6-9 m above sea level, loose siliceous – calcareous sand of marine origin and pyroclastic products. The experimental plot is located about 300 m from the sea, with a West exposure.

The Nature Reserve of Tirone Alto Vesuvio is located on the slope of the Mount Vesuvius. The selected plot is in "Piana delle Ginestre", at 400-450 m above sea level, with a South South-West exposure, and is gently sloping. This place was affected by the eruption in 1865-68. The soil, evolved from lava, is characterized by surface outcrops of volcanic rock.

The WWF "Oasi Bosco di S.Silvestro" is situated 200 m above sea level with an East exposure. The slope is interrupted by terracing. The soil often presents calcareous formations.

The climate is typically mediterranean with mild, wet winters and hot, dry summers although microclimatic differences, linked to different exposure and altitude, occur in the woods.

2.2. Samples processing

The soils were collected from a depth of 0-5 cm, at three-month intervals for one year, at nine sites each chosen randomly within a 3000 m^2 area.

The soil was pooled into three groups with the same ranges of microbial respiration, measured *in situ*. Each sample was sieved (<2mm) and divided into subsamples to determine soil water content, microbial respiration and enzyme activity measurements. The subsamples used for enzyme activity measurements were stored at –80°C until analysis.

2.3. Soil respiration

Soil respiration was measured as CO_2 evolution from soil at field moisture level. Approximately 5 g of soil was incubated in airtight jars for 2 d at 25°C in total darkness. The CO_2 released was absorbed in NaOH solution (0.1 M) and its amount determined by two-phase titration with HCl (0.025 M) [12]. The CO_2 output from soil was expressed in μmoles g^{-1} dry soil d^{-1}. All measures were performed in triplicate for each soil sample. The water content of the soils at sampling was determined by oven drying of subsamples at 75°C to a constant weight.

2.4. Enzyme assays

Xylanase, CM-cellulase and invertase activities were determined according to Schinner and Von Mersi [13], protease activity according to Ladd and Butler [14], trealase activity according to Kiem and Kandeler [15].

All enzyme assays were performed in triplicate for each soil sample.

2.5. Chemical analyses

The chemical composition of the three soils was determined as follows: the oven-dried soil subsamples were ground to fine powder by a Fritsch Pulverisette (type 00.502, Oberstein, Germany) equipped with an agate pocket and ball mill. The analyses of each sample were carried out in triplicate.

Carbon and nitrogen content were determined by combustion in an Elemental Analyser NA 1500 (Carlo Erba Strumentazione, Milan, Italy).

2.6. pH measurements

The pH of the soil was determined by shaking an aliquot of soil in distilled water for 10 min (10 g of dry soil in 25 ml of water). The suspension was left to stand for 10 min, and the supernatant was used to measure the pH with an electronic pH meter (HI 8424, HANNA Instruments, Sarmeola di Rubano PD, Italy).

The potential pH was determined by shaking an aliquot of soil in 1N KCl solution and then proceeding as already described.

Measurements of each sample were made in triplicate.

2.7. Statistics

The significance of differences among the soils was tested by two-way analysis of variance (ANOVA) followed by Tukey test. Correlations were determined using the simple Pearson correlation coefficient.

3. RESULTS AND DISCUSSION

3.1. Soil properties

The chemical and physical characteristics of the soils in the three evergreen oak woods are presented in Table 1.

The Vesuvius soil is shallow and in many points lava rock appears on the surface. In the surface layer, 0-5 cm, gravel is notable, but coarse and fine sand are dominant. The clay and silt component is about 14.3%.

The Castel Volturno soil is much deeper than the Vesuvius soil. No gravel exists in the layer 0-5 cm while coarse sand is abundant. Unlike the Vesuvius soil, the clay and silt component is practically absent.

The Bosco di S.Silvestro soil is also shallow, with calcareous rock showing on the surface and stones distributed throughout the soil profile. In the 0-5 cm layer, fine sand represents an important component, followed in order of decreasing abundance by the clay and silt component and course sand.

The highest water holding capacity was found in the Bosco di S.Silvestro soil, correlating to the high content of silt, clay and fine sand, followed by the Vesuvius soil and finally by Castel Volturno soil, even though this soil presented a low value of C_{org} (%).

Table 1
Soil characteristics of the three oak woods.

		Mount Vesuvius	Castel Volturno	Bosco di S. Silvestro
Substrate		Volcanic	Sandy (siliceous-calcareous)	Calcareous
Fraction*	Designation	Distribution of particle sizes (mass %)		
1 cm - 2mm	Gravel	4.5	0	3.0
2 - 0.2 mm	Coarse sand	41.2	76.8	17.8
0.2 - 0.02 mm	Fine sand	40.0	22.8	51.4
< 0.02 mm	Clay and silt	14.3	0.4	27.8
Water holding capacity (H_2O g/100g d.w.)		71.7	62	156.7
pH		6.5	8.0	7.0
pH potential		5.0	6.8	6.4
C_{org} (%)		21.7	13.8	13.5
N (%)		1.47	0.51	0.5
C/N		14.8	27.1	26.0

* Gravel > 1 cm was removed before fraction separation.

3.2. Microbial activity

Figure 1 shows the respiration rates and water contents of the soils at each sampling time.

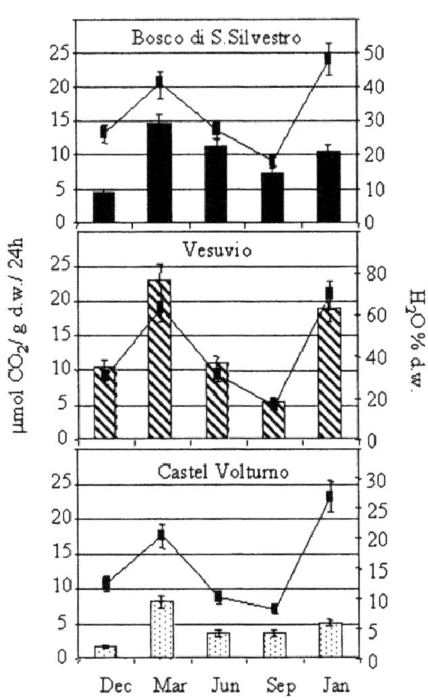

Figure 1. Soil respiration rates at the field moisture (□) and soil water content at sampling (—). Values are means ± SD of 3 measurements with 3 replicates.

Microbial respiration showed pronounced seasonal variations with the lowest levels in summer. The patterns in microbial respiration matched the seasonal changes in moisture content, and significant positive correlations were found between these parameters in the three soils (Figure 2). This suggests that climatic conditions and in particular the summer drought strongly influence microbial activity.

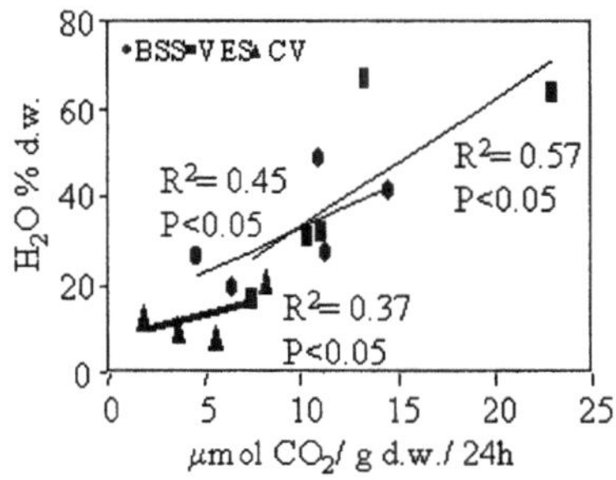

Figure 2. Correlation between respiration rate and water content for all three sites of incubation.

By comparing the three woods, the highest levels of microbial respiration were found in the Vesuvian soil (5.45 μmol CO_2/g d.w./h in the summer to 23.09 μmol CO_2/g d.w./h at the end of winter), which also had the highest organic C content (Table 1), while the lowest levels were found in the Castel Volturno soil (1.76 μmol CO_2/g d.w./h in the summer to 8.14 μmol CO_2/g d.w./h at the end of winter).

Significant differences were found between the oak forest soil at Mt. Vesuvius and the soil at Castel Volturno ($P < 0.05$).

These respiration values were higher than those measured in the same layer of soil in the *Abies alba* forest at Mount Taburno, the *Pinus laricio* forest at Sila and the *Pinus pinea* forest at Vesuvius [16].

3.3. Enzyme activities

Cellulase, xylanase, protease and trealase activities showed seasonal variations with a maximum in wet seasons and a minimum in dry seasons (Figure 3). Less pronounced seasonal variations were observed for invertase activity.

No difference in enzyme activities was observed among the three soils studied.

The seasonal patterns of the invertase and xylanase activities, as observed at the three sites described in the present study, contradict previous reports by Ross et al. [17] and Kandeler & Eder [18] which reported to no seasonal variation for these enzyme activities.

The patterns of cellulase and xylanase activities matched microbial respiration (Figure 4) and a positive and significant correlation was found between these enzyme activities and microbial respiration as well as water content of the soil. This suggests that enzyme activities are closely correlated to microbial activity.

These results are in agreement with those of Ohtonen et al. [19], who found that cellulase activity correlated with basal respiration in Scots pine stands in Oulu, Northern Finland. In contrast, the invertase, protease and trealase activities were correlated only with moisture content (Figure5).

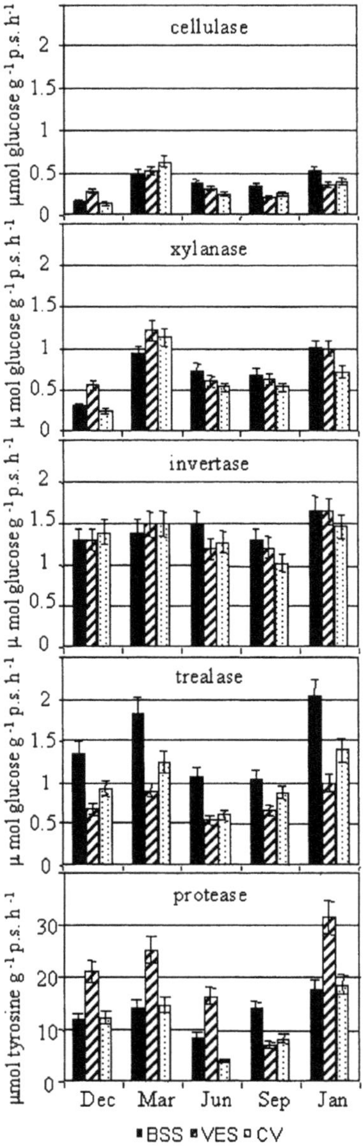

Figure 3. Cellulase, xylanase, invertase, trealase and protease activities of soils. Values are means ± SD of 3 measurements with 3 replications.

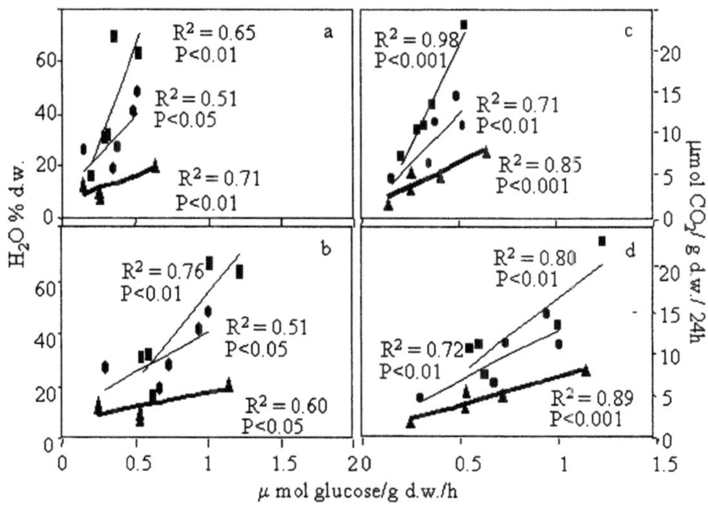

Figure 4. Correlation between water content and cellulase (a) and xylanase activities (b) for all three sites of incubation. Correlation between respiration rate and cellulase (c) and xylanase activities (d) was also reported.

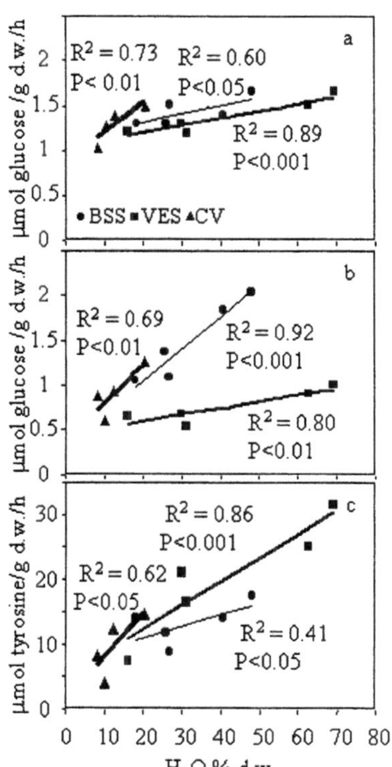

Figure 5. Correlation between water content and invertase (a), trealase (b) and protease (c) activities for all three sites of incubation.

4. CONCLUSIONS

The study in these mediterranean sites has shown that:

- the environmental conditions of the three sites, and particularly the different pedologic characteristics, did not seem to influence the microbial communities. In fact, there was a similar trend for all three woods;
- the climatic conditions strongly influenced microbial respiration and enzymatic activities, with a strong seasonal variation being evident. Both microbial respiration and enzymatic activities were high during the moist winter and low in the dry summer season.

ACKNOWLEDGEMENTS

This research was supported by MURST (Ministero dell'Università e Ricerca Scientifica of Italy) and by ModMedIII project contract n. ENV4CT970680 (European Community). We thank Dr. N. Costantino, who permitted us to work in the Natural Reserve of Castel Volturno, and Tirone Alto Vesuvio, and Dr. F. Paolella, who permitted us to work in the Oasi WWF Bosco di S.Silvestro.

REFERENCES

1. Parkinson, D., Coleman, D.C., 1991. Microbial communities, activity and biomass. Agric. Ecosyst. Environ. 34, 3-31.
2. Nannipieri, P., Kandeler E., Ruggiero, P., 1999. Enzyme activities as indicators of soil functionality. *In:* Dick, R.P. (Ed), Proceedings of the International Conference "Enzymes in the Environment: Ecology Activity and Applications" Granada, Spain, July 12-15, pp. 66.
3. Linkins, A.E., Melillo, J.M., Sinsabaugh, R.L., 1984. Factors affecting cellulase activity in terrestrial and aquatic ecosystems. *In:* Klug, M.J., Reddy, C.A. (Eds.), Current Perspectives in Microbial Ecology. American Society for Microbiology, Washington, DC, pp. 572-579.
4. Sinsabaugh, R.L., Linkins, A.E., 1988. Adsorption of cellulase components by leaf litter. Soil Biol. Biochem. 20, 927-932.
5. Burns, R.G., (Ed.) 1978. Soil Enzymes. Academic Press, London.
6. Ross, D.J., Speir, T.W., Kettles, H.A., Mackay, A.D., 1995. Soil microbial biomass, C and N mineralization, and enzyme activities in a hill pasture – influence of season and slow-release P and S fertilizer. Soil Biol. Biochem. 27, 1434-1443.
7. Moorhead, D.L., Linkins, A.E., Everett, K.R., 1996. Road dust alters extracellular enzyme activities in tussock tundra soils, Alaska, USA. Artic Alp. Res. 28, 346-351.
8. Bergstrom, D.W., Monreal, C.M., King, D.J., 1998. Sensitivity of soil enzyme activities to conservation practices. Soil Sci. Soc. Amer. J. 62, 1286-1295.
9. Chander, K., Goyal, S., Nandal, D.P., Kapoor, K.K., 1998. Soil organic matter, microbial biomass and enzyme activities in a tropical agroforestry system. Biol. Fert. Soils 27, 168-172.

10. Staddon, W.J., Duchesne, L.C., Trevors, J.T., 1998. Acid phosphatase and arylsulphatase activities in soils from a jack pine (*Pinus banksiana* Lamb.) ecosystem after clear-cutting, prescribed burning, and scarification. Biol. Fertil. Soils 22, 1-4.
11. Salam, A.K., Katayama, A, Kimura, M., 1998. Activities of some soil enzymes in different land use systems after deforestation in hilly areas of West Lampung, South Sumatra, Indonesia. Soil Sci. Plant Nutr. 44, 93-103.
12. Froment, A. 1972. Soil respiration in a mixed oak forest. Oikos 23, 273-277.
13. Schinner, F., Von Mersi, W., 1990. Xylanase, CM- Cellulase and Invertase activity in soil: an improved method. Soil Biol. Biochem. 22, 511-515.
14. Ladd, J.N., Butler, J.H.A., 1972. Short-term assays of soil proteolytic enzyme activities using proteins and dipeptide derivatives as substrates. Soil Biol. Biochem. 4, 19-30.
15. Kiem, R., Kandeler, E., 1997. A simple method for the determination of trehalase activity in soils. Microbiol. Res. 152, 19-25.
16. Virzo De Santo, A., Berg, B., Rutigliano, F.A., Alfani, A., Fioretto, A., 1993. Factors regulating early-stage decomposition of needle litters in five different coniferous forests. Soil Biol. Biochem. 25, 1423-1433.
17. Ross, D.J., Speir, T.W., Kettles, H.A., Tate, K.R., Mackay, A.D., 1995. Soil microbial biomass, C and N mineralization, and enzyme activities in a hill pasture – influence of grazing management. Austr. J. Soil Res. 33, 943-959.
18. Kandeler, E., Eder, G., 1993. Effect of cattle slurry in grassland on microbial biomass and on activities of various enzymes. Biol. Fert. Soils 16, 249-254.
19. Ohtonen, R., Lahdesmaki, P., Markkola, A.M., 1994. Cellulase activity in forest humus along an industrial pollution gradient in Oulu, Northern Finland. Soil Biol. Biochem. 26, 97-101.

Developments in Soil Science, Volume 28B
Editors: A. Violante, P.M. Huang, J.-M. Bollag and L. Gianfreda

A METHOD TO RECOVER *SALMONELLA* FROM COMPOST BY IMS-PCR

M. Civilini

Dipartimento di Scienze degli Alimenti, Università degli Studi di Udine,
via Marangoni, 97, 33100 Udine, Italy
E-mail: Marcello.Civilini@DSA.Uniud.it

In this work, a procedure was optimized to recover *Salmonella typhimurium* from a vegetable compost after its artificial contamination. A combination of a bacteriological method, immunomagnetic separation (IMS), and polymerase chain reaction (PCR) techniques allowed a reduction in the detection time to 30 hours, while maintaining high specificity. The lower threshold of direct amplification from extracts of contaminated compost was 10^8 salmonellae g^{-1}. To improve the sensitivity, a combination of shortened preenrichment and enrichment procedures were optimized and the growth of *S. typhimurium* evaluated. Immunomagnetic separation using anti-*Salmonella* Dynabeads permitted the recovery of 30 salmonellae per 50 g of compost, the same value obtained with the traditional microbial method, which takes two days longer. Alternative purification methods to reduce organic compound inhibition of the PCR reaction mixture did not improve *Salmonella* detection under a threshold of 10^5 salmonellae g^{-1}.

1. INTRODUCTION

The use of PCR to detect pathogens in many kinds of samples, including compost, soil and slurries, seems to be limited by a variety of inhibitors, including humic and fulvic acids and phenolic compounds [1, 2]. In compost, the concentration of these inhibitors depends on the source of the materials used for composting and the "maturity" of the compost [3]. Because *Salmonella* must be absent before a compost can be marketed, this study aimed to improve the sensitivity and accelerate the detection of *Salmonella* in compost. To reduce the variables involved, a constant product from a food and vegetable waste processing and composting plant was artificially contaminated with *S. typhimurium*.

Of the different methods available for recovering *Salmonella*, a combination of a traditional microbial method together with the IMS–PCR procedure was optimized and the sensitivity evaluated.

2. MATERIAL AND METHODS

2.1 Bacterial strains

The positive control was *Salmonella typhimurium* II 505 from the National Institute of Public Health and Environmental Hygiene (RIVM), Bilthoven, The Netherlands. *Escherichia coli* and *Hafnia alvei* (University of Udine culture collection) were isolated from vegetables and used as negative controls.

2.2. Compost

The compost samples used in these experiments were obtained from Caviro S.p.A., Faenza, Italy [4].

2.3. DNA extraction from compost

To determine the sensitivity of the PCR method to compost inhibitors, one gram of sterilized compost (sterilized three times in three days at 121°C for 20 min) was artificially inoculated with 200 µl of suspensions of *S. typhimurium* at levels of approximately of 10, 10^2, 10^3, 10^4, 10^5, 10^6, 10^7, and 10^8 cells g^{-1}. Two negative controls were obtained by not adding the inoculum to one sample and adding 200 µl of 10^8 cells g^{-1} of *E. coli* to another. The tests were repeated in triplicate, and inoculum levels were confirmed using standard plate counts on agar (PCA) (Oxoid–Unipath S.p.A., Garbagnate Milanese, Milano, Italy). 20 ml of a 20% solution of sodium hexametaphoshate was added to each test tube. The tubes were then placed on a rotary shaker and mixed for 15 minutes at 100 rpm. The tubes were centrifuged at 100 x *g* for 10 min. The supernatant was then recovered and centrifuged at 12,000 x *g* for 20 min. Twenty ml of phosphate buffer saline (PBS) and 3 g of Diaion SA20A (Supelco-Bellefonte, PA, USA) were added to the pellet obtained. The samples were placed on a rotary shaker and mixed for 10 min at 100 rpm, followed by centrifugation for 10 min at 100 x *g* (15°C). The supernatants were recovered and centrifuged at 12,000 x *g* for 15 min (15°C).

The cell pellets were resuspended with 567 µl of TE buffer (10 m*M* Tris-Cl, pH 8; 0.1 m*M* EDTA) and transferred to 1.5 ml microcentrifuge tubes. Then 30 µl of a 10% sodium dodecyl sulphate solution (SDS) and 60 µg of proteinase K were added, and the samples were incubated statically for 1 h in a 37°C water bath. Next, 700 µl of phenol:chloroform:isoamyl alcohol (25:24:1) were added to the samples, mixed, and centrifuged at 15,000 x *g* for 5 min at 4°C; this procedure was repeated twice more. The upper aqueous phases were placed in clean tubes, and isopropanol was added to each sample at a ratio of 0.6. After centrifugation at 15,000 x *g* for 10 min (4°C), the pellets were washed with cold 70% ethanol, air-dried and then resuspended in 100 µl of TE buffer.

2.4. Purification of DNA extracts

The DNA extracts were purified by gel filtration chromatography with a Sephadex G-200 (Pharmacia Biotech, Uppsala, Sweden) column [5]. About 100 µl of DNA solution in TE buffer were harvested from the spun column.

2.5. Preenrichment of compost samples

Portions of unsterilized compost samples weighing between 10 and 100 g were mixed with buffered peptone water (BPW) (Peptone, 10 g l^{-1}; NaCl, 5 g l^{-1}; Na_2HPO_4 $2H_2O$, 4.5 g l^{-1}; KH_2PO_4, 1.5 g l^{-1}) always in a proportion of 1/10. No *Salmonella* had been detected previously in other portions of the same samples.

After mixing, two samples were used as references, and other duplicate repetitions were inoculated with different concentrations of "stressed" *S. typhimurium* and incubated at 37°C. The reproducibility of the inoculum was standardized by growing *S. typhimurium* on Bacto Brain Heart Infusion (BHI) (DIFCO Laboratories, Detroit, Michigan, USA) at 37°C for 12 h. Next, this culture was heated at 60°C for 2 min to damage the cells, simulating composting.

The total viable count of the culture and the preenrichment immediately after inoculation and after 12 and 24 h of incubation was obtained on standard plate count agar (PCA) (Oxoid-Unipath).

At the same time, duplicate 0.1 ml samples of the preenrichment were plated on the *Salmonella* isolation medium bismuth sulphite agar (BSA) (Oxoid-Unipath). The predominant colony types were subcultured each time and provisionally identified by appropriate tests.

2.6. Enrichment on selective media

After the preenrichment phase, two types of selective media were compared: Rappaport-Vassiliadis enrichment (RV) broth (Oxoid-Unipath) and selenite cystine (SC) broth (Oxoid-Unipath).

One hundred microlitres of preenrichment BPW or pellet obtained from 10 ml of the same cultured media were used to inoculate 9.9 ml of RV. Tubes were incubated in a water bath at 42°C for 12 to 24 hours.

One millilitre of preenrichment BPW was used to inoculate 9 ml of SC. Tubes were incubated in a water bath at 37°C for 12 and/or 24 hours.

The *Salmonella* isolation procedure was carried out by streaking a loop or plating 5 to 50 µl of the enrichment broth on BSA plates. All cultures were incubated at 37°C for 24 h. Presumptive identification of selected colonies was made on the basis of morphology, Gram stain reaction, production of catalase and oxidase, and the triple sugar iron agar test (TSI) (Oxoid-Unipath). The identity of representative strains of the main groups of bacteria that were present after 12 and 24 h was confirmed with the API 20E system (API System, Montalieu Vercieu, France).

2.7. Immunoseparation

Immunoseparation by anti-*Salmonella* Dynabeads (Dynal, Oslo, Norway) was performed as described by Soumet et al. [6].

Different washing steps were performed to purify the nonselective and selective enrichment broths before separating the paramagnetic bead-bacterium complex with magnetic particle concentrator, MPC-E (Dynal). 1 ml of each culture was centrifuged at 12000 x *g* for 20 min at 15°C, followed by combinations of washing with TE buffer alone and with 2% sodium hexametaphosphate (HMP) at pH 8.5 [7].

2.8. Conditions for PCR amplification

The amplifications were performed on 5 µl of samples, except where stated otherwise. 100 µl of PCR reaction mixture consisted of 10 mmol l^{-1} Tris HCl (pH 8.3), 50 mmol l^{-1} KCl, 2.5 mmol l^{-1} $MgCl_2$, 200 µmol l^{-1} of each 2'-deoxynucleoside 5'-triphosphate, dNTP (Boehringer Mannheim Italia S.p.A., Milano, Italy) and 3 µmol l^{-1} of each primer. Two sets of primers were used: SHIMA-L and SHIMA-R [8], which are known to provide sufficient nucleotide mismatches with the *himA* gene in *E. coli* and other enteric bacteria so that *Salmonella* spp. could be detected specifically. ST11 and ST15 [9] were chosen to be specific for the genus *Salmonella*.

The amplifications were carried out using a Gene Amp PCR System 2400, Perkin Elmer thermal cycler (Perkin-Elmer Italia S.p.A., Padova, Italy). The optimum conditions for amplification using the SHIMA primers were as follows: a hot start of 5 min at 94°C, followed by 30 cycles of 15 s at 94°C, 1 min at 55°C, 30 s at 72°C, and final extension at 72°C for 7 min. Using ST primers and maintaining the same hot start and final extension steps, the 30 amplification cycles were performed with DNA denaturation at 94°C for 30 s, annealing at 57°C for 30 s, and extension at 72°C for 1 min.

The PCR products were subjected to electrophoresis at 10 V cm^{-1} in 2% agarose in 0.5x TBE buffer containing ethidium bromide (0.5 $\mu g\ ml^{-1}$).

2.9. Hybridization

For Southern blotting with $\alpha^{32}P$-labeled probe, the DNA was transferred after agarose gel electrophoresis to a nylon membrane (Hybond-N, Amersham) and hybridized with the end-labeled SHIMA-P probe [8] using standard procedures. Autoradiography was performed by exposing the blots overnight to X-ray sensitive film (Kodak) at -80°C.

3. RESULTS AND DISCUSSION

3.1. Recovery of *Salmonella* directly from compost

Preliminary tests were performed to evaluate humic and fulvic acid inhibition during PCR amplification. Purification steps with spun columns reduced DNA concentrations in the extract solutions, and its amplification showed no improvement in comparison with amplification of nonpurified compost extracts. The lower threshold of *S. typhimurium* recovery by PCR amplification was determined with artificial contamination of compost at appropriate dilutions. PCR amplification of DNA extracts without purification was positive only when the concentration of *S. typhimurium* was greater than 10^8 cfu g^{-1} (Figure1a). After Southern hybridization with SHIMA-P probe, the threshold detection levels were 10^7 cfu g^{-1} and 10^6 cfu g^{-1}, respectively after 7 (Figure1b) and 72 hours of exposition.

3.2. Recovery of *Salmonella* after preenrichment phase

To increase the number of salmonellae for PCR analysis and resuscitate the sublethally injured salmonellae [10], traditional preenrichment with a nonselective medium (BPW) was added to compost samples.

Repetitions of viable counts showed that thermal treatment of the inoculum to stress the cells (reproducing the cell damage during the composting process) reduced their values by about 1 *log* unit. After 12 and 24 h of incubation on BPW, the behavior of the inoculated *S. typhimurium* and the total count were as represented in Figure 2.

Several combinations of centrifugation (1,000 x *g*, 10 min, 10°C) and washing with TE buffer and sodium hexametaphoshate (TE 1x, 2x, 3x or TE 2x + HMP 1x) were applied to aliquots of preenrichment cultures (about 8 x 10^6 *S. typhimurium* ml^{-1}) before IMS treatments and PCR amplifications. Results allowed the optimization of the method to recover *S. typhimurium* from the preenrichment broth: compost (50 g) containing added BPW (450 ml) was incubated for 18 h at 37°C; an aliquot of supernatant (1 ml) was used for IMS treatment, and then 5 μl of the harvested solution were used for the PCR experiments. The sensitivity of this method with SHIMA primers was 8 x 10^5 cfu ml^{-1} (Figure 1c). Using ST primers for PCR amplification, a lower sensitivity by about 1 *log* unit was obtained.

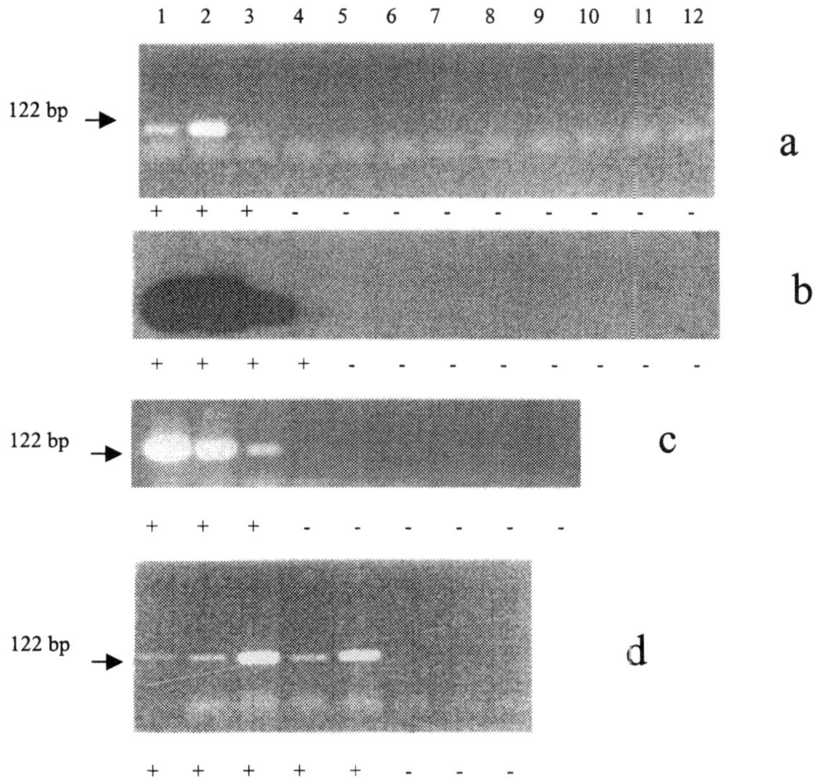

Figure 1. a) Gel electrophoresis of PCR amplification to recover *S. typhimurium* directly from compost samples. Lane 1, marker DNA 122 bp; lane 2, pure culture of *S. typhimurium* on BHI; lanes 3-10, *S. typhimurium* in compost from 10^8 to 10 cfu g^{-1} respectively; lane 11, compost not inoculated; lane 12, reagents control. b) Southern blot of the above gel: hybridization with the end-labeled SHIMA-P probe after 7 hours of exposition. c) Gel electrophoresis of PCR amplification of the preenrichment samples. Lane 1, pure culture of *S. typhimurium* on BHI $8*10^6$ cfu ml^{-1}; lanes 2-6, *S. typhimurium* in pre-enrichment respectively from $8*10^6$ to $8*10^2$ cfu ml^{-1}; lane 7, preenrichment not inoculated; lane 8, pure culture of *E. coli,* 10^8 cfu ml^{-1}; lane 9, reagents control. d) Gel electrophoresis of PCR amplification. Samples of different selective enrichment broth after preenrichment phase. Lane 1, marker DNA 122 bp; lanes 2-7, recovery of *S. typhimurium* after 12 h of selective enrichment of 54 cfu g^{-1} on SC, 54 cfu g^{-1} on RV, 5.4 cfu g^{-1} on SC, 5.4 cfu g^{-1} on RV, 0.54 cfu g^{-1} on SC, 0.54 cfu g^{-1} on RV respectively; lane 8, reagents control.

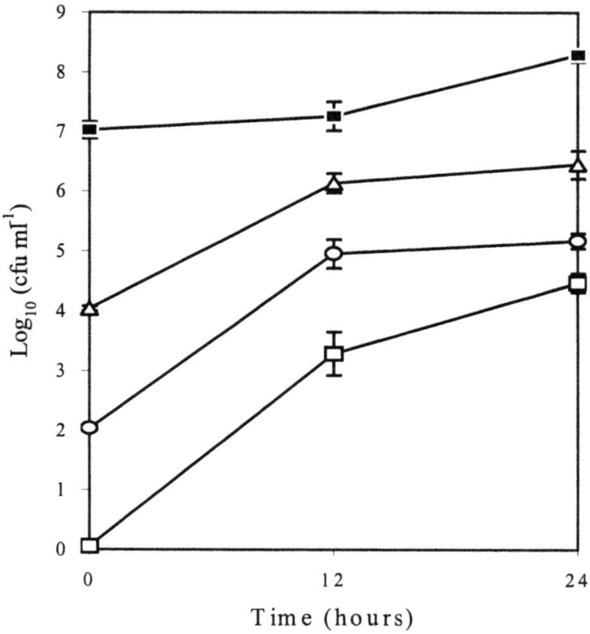

Figure 2. Microbial behavior during pre-enrichment phase with different starting inocula. ■ Mesophilic total count; □ 10 *Salmonellae* g^{-1} compost; ○ 1,000 *Salmonellae* g^{-1} compost; △ 100,000 *Salmonellae* g^{-1} compost.

3.3. Recovery of *Salmonella* after preenrichment and selective enrichment phases

A series of experiments were programmed to combine preenrichment cultures with selective enrichment. The purpose was to determine the lower threshold of *S. typhimurium* recovery and to gain time over the traditional microbial method.

Observing the optimum ratio, after 12 and 24 h, preenrichment cultures were inoculated into RV and SC. At the same time, the detection of *S. typhimurium* was verified by PCR amplification and correlated to its concentration. After 12 h of pre-enrichment and 24 h of incubation on RV,

enrichment was sufficient to obtain higher sensitivity (Table 1A). Reducing the enrichment phase to 12 h, no decrease in sensitivity was obtained (Table 1B). The lower limit detection for *S. typhimurium* after 12 h of BPW incubation and 12 h of RV incubation was about 1.1 – 1.2 cfu g^{-1}.

The comparison between the two selective media showed that SC was 10=fold less sensitive than RV (Figure1d).

Further improvements to the detection limit were obtained with the method shown in Figure 3, in which two differential centrifugations increased the concentration in RV about 100=fold. This strategy allowed the determination of about 30 cfu 50 g^{-1} of *S. typhimurium*, a value very close to the legal limit.

Table 1
PCR amplifications with SHIMA primers of preenrichment and enrichment samples at different times

Pre-enrichment	Time (h)	*S. typhimurium* (cfu g^{-1})				Enrichment	Time (h)	*S. typhimurium* (cfu g^{-1})			
A		0	1.1×10^5	1.1×10^3	1.1			0	1.1×10^5	1.1×10^3	1.1
BPW	12	-	+	-	-	RV	24	-	+	+	+
BPW	24	-	+	+	-	RV	24	-	+	+	+
B		0	1.2×10^3	12	1.2			0	1.1×10^3	12	1.2
BPW	12	-	+	-	-	RV	12	-	+	+	+
BPW	24	-	+	-	-	RV	24	-	+	+	+

+ = positive; - = negative; BPW = buffered peptone water; RV = Rappaport-Vassiliadis enrichment broth

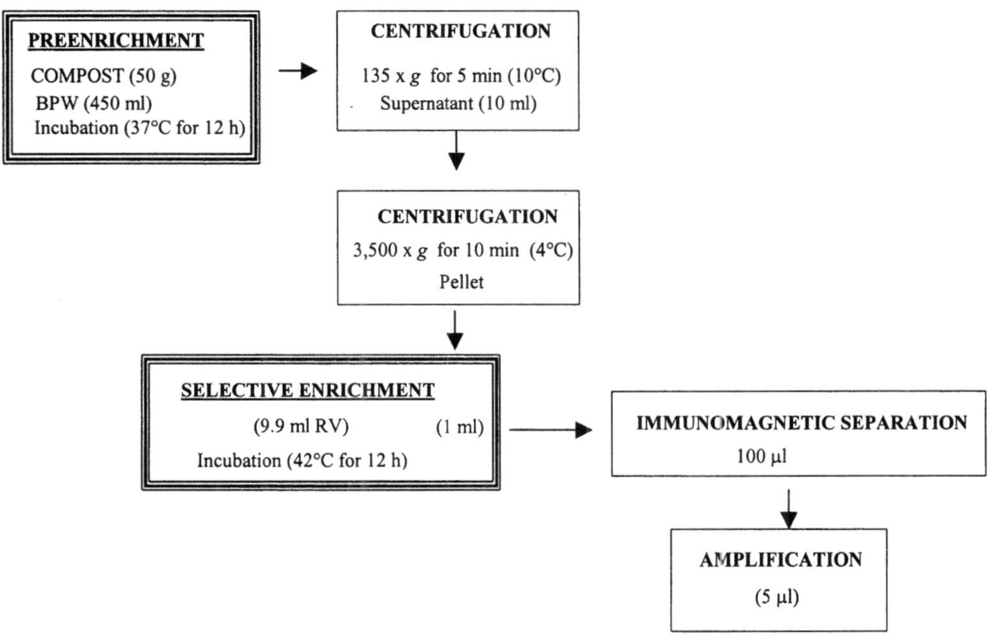

Figure 3. Final procedure to recover *S. typhimurium* in compost

4. CONCLUSIONS

This research showed that, using artificial contamination with *S. typhimurium*, it was possible to detect salmonellae on compost samples with the PCR method with a good degree of sensitivity. The method overcame the inhibition of organic compounds toward PCR amplification, increasing the number of *Salmonella* in preenrichment and enrichment phases and reducing the operational time needed specifically to detect *S. typhimurium* in comparison with microbiological methods.

REFERENCES

1. Tsai, Y.L., Olson B.H., 1992. Detection of low numbers of bacterial cells in soils and sediments by Polymerase Chain Reaction. Appl. Environ. Microbiol. 58, 754-757.
2. Tsai, Y.L., Olson B.H., 1992. Rapid method for separation of bacterial DNA from humic substances in sediment for Polymerase Chain Reaction. Appl. Environ. Microbiol. 58, 2292-2295.
3. Chen, Y., Chefetz, B., Hadar, Y., 1996. Formation and properties of humic substance originating from composts. *In:* de Bertoldi, M., Sequi, P., Lemmes, B., Papi, T. (Eds.), Science of Composting. Blackie Academic & Professional, Glasgow, pp. 382-393.
4. Papi, T., Marani, G., Mannironi, R., de Bertoldi, M., 1996. A new composting plant realized within the THERMIE program of the European Commission. *In*: de Bertoldi, M., Sequi, P., Lemmes, B., Papi, T. (Eds.), Science of Composting. Blackie Academic & Professional, Glasgow, pp. 162-172.
5. Tsai, Y.L., Palmer, C.J., Sangermano L.R., 1993. Detection of *Escherichia coli* in sewage and sludge by polymerase chain reaction. Appl. Environ. Microbiol. 59, 353-357.
6. Soumet, C., Ermel, G., Rose, N., Rose, V., Drouin, P., Salvat, G., Colin, P., 1999. Evaluation of a multiplex PCR assay for simultaneous identification of *Salmonella* sp., *Salmonella enteritidis* and *Salmonella typhimurium* from environmental swabs of poultry houses. Lett. Appl. Microbiol. 28, 113-117.
7. Holben, W.E., Jansson, J.K., Chelm, B.K., Tiedje, J.M., 1988. DNA probe method for detection of specific microorganisms in the soil bacterial community. Appl. Environ. Microbiol. 54, 703-711.
8. Bej, A.K., Mahbubani, M.H., Boyce, M.J., Atlas, R.M., 1994. Detection of *Salmonella* spp. in oysters by PCR. Appl. Environ. Microbiol. 60, 368-373.
9. Aabo, S., Rasmussen, O.F., Rossen, L., Sorensen, P.D., Olsen, J.E., 1993. *Salmonella* identification by the polymerase chain reaction. Molec. Cell. Probes 7, 171-178.
10. Edel, W., Kampelmaker, E.H., 1973. Comparative studies on the isolation of "sublethally injured" salmonellae in nine European laboratories. Bull. World Health Org. 48, 167-174.

Developments in Soil Science, Volume 28B
Editors: A. Violante, P.M. Huang, J.-M. Bollag and L. Gianfreda

STUDY OF TEMPORAL VARIABILITY IN THE MICROBIAL COMMUNITY OF TWO FOREST SOILS: BIOMASS AND ACTIVITY

C. Gigliotti and A. Farini

Department of Crop Production, University of Milan, Via Celoria 2, 20133 Milan, Italy

Within a wider interdisciplinary project carried out to obtain more detailed information on the forest ecosystem, we have studied the behavior of the microbial population of two different forest soils in Val Masino and Val Gerola (Lombardy - North Italy). We considered microbial biomass, respiratory activity, and ecophysiological parameters with the aim to obtain information on the microbial biomass status. The results showed that the changes in the microflora depended on climatic conditions, and the ecophysiological indexes identified and confirmed these findings. The microflora of Val Masino was more active, registering higher activity values and reacting more quickly to temperature changes than what was found for Val Gerola, where greater forest deterioration was evident.

1. INTRODUCTION

Soil microorganisms play a very important role in soil fertility, not only because of their ability to carry out biochemical transformation, but also due to their importance as a source and sink for mineral nutrients [1].

Numerous investigations on the microbial biomass and its activities in agricultural soils have been carried out. The results of these studies have shown that the microbial biomass activity is influenced by many factors and that it is subjected to seasonal variation. Knowledge about the relationships between microbial biomass and seasonal variation is thus of considerable biological importance. However, there is little information available on the temporal variability of the microbial community in forest soil.

This work is part of a wide-ranging project aimed to obtain more detailed information on the forest ecosystems, by means of a specialized and interdisciplinary study, and on its relationship with environmental stresses. In this context, two areas under permanent observation located in the Regional Forests of Val Masino and Val Gerola different in soil, degrees of crown transparency, age, altitude, and temperature, but both in the same climatic area, were chosen.

The main purpose of our work was to:

- identify the microbial level of the areas, because this has been little investigated so far, through the evaluation of the microbial biomass and respiratory activity;
- determine the possible microflora responses to climatic variation;
- carry out a comparison on the state of the microflora of the two forest soils, with the aim to verify whether the microbial community differed.

2. MATERIALS AND METHODS

2.1. Study sites

We conducted the research on two forest soils in Val Gerola and Val Masino (Valtellina - Northern Italy).

In Val Gerola, the investigation site is in the village of Rasura (Sondrio), in a mountainous area called Alpe Culino (46° 6' 4" N, 9° 32' 13" E) at a height of 1490 m a.s.l. The type of soil is Cambic Podzols (FAO, 1990). The soil pH is 4.1, the total nitrogen (Kjeldahl) is 5.5 g kg^{-1}, the cation exchange capacity is 41.6 cmol(+) kg^{-1}, and percentage base saturation is 2.4%. The upper canopy (tree density: 45 trees 2500 m^{-2}) is a copse of mixed contemporary species, with White spruce (*Abies alba)*, which is the prevalent one, Norway spruce (*Picea abies)*, and Larch (*Larix decidua)*. The average annual rainfall is 1769 mm, and the average annual temperature is 5.8°C, the maximum average, in July, being 15.0°C, and the minimum average, in January, -3.3°C (period 1951 - 1964).

In Val Masino, the investigation site was in the village of Bagni di Masino (Sondrio) (46° 14' 19" N, 9° 35' 40" E) at a height of 1190 m a.s.l., the soil type being Humic Cambisol (FAO, 1990). The soil pH is 4.4, the total nitrogen (Kjeldahl) is 5.2 g kg^{-1}, the cation exchange capacity is 38.8 cmol(+) kg^{-1}, and percentage base saturation is 3.4%. The upper canopy (tree density: 81 trees 2500 m^{-2}), is a mixed wood of conifers, with a prevalence of Norway spruce (*Picea abies)* and, to a lesser extent, Larch *(Larix decidua)* and White spruce (*Abies alba),* with occasional broad-leafed species such as Beech (*Fagus sylvatica*), Birch (*Betula alba),* and Sycamore maple *(Acer pseudoplatanus*). The average annual rainfall is 1,353 mm (period 1951 - 1964). Average temperatures for Val Masino are not available.

The two sites show marked differences in overall damage, evaluated observing both transparency and live crown conditions. The Alpe Culino area evidenced greater damage (34% of crown transparency), Bagni di Masino being healthier in appearance (20% of crown transparency) (Fondazione Lombardia per l'Ambiente, 1996).

2.2. Sample collection and preparation

Since the preliminary results of the chemical parameters (not reported here but determined for soil spatial variability at the two sites) showed a spatial variability of 1-2 m, we chose a small site of about 1.5 m^2 (1.5x1 m) with relatively small spatial variation for the study.

For each site three soil samples were investigated, each composed of five subsamples taken at depths of 0-30 cm (horizon A) using plastic tubes (length: 30 cm, diameter: 20 cm) according to Williams [2]. The subsamples were pooled, placed in polyethylene bags and transported in an ice chest to the laboratory. The soils were then sieved (4 mm) and all visible plant debris and soil fauna were removed manually. The water holding capacity of each soil was determined, and the moisture content adjusted to 60%. The soils were stored moist at 4°C. Subsamples for the determination of chemical parameters were air-dried and sieved (2 and 0.5 mm) before analysis.

The experiment was planned as a two year investigation (1996 and 1997) and the sampling was done every two months in the vegetative and plant growth period. During wintertime there was no sampling due to the inaccessibility of the sites. The sampling dates were: for 1996 - 24 May, 18 July, 01 October, 09 December, and for 1997 - 27 May, 15 July, 27 September, 15 November.

2.3. Analytical methods

Soil organic C was determined by wet oxidation with 0.167 M $K_2Cr_2O_7$ and subsequent titration with 0.1 M $(NH_4)_2Fe(SO)_4.6H_2O$.

Microbial biomass C was determined by the chloroform fumigation-extraction method [3]; 25 g (oven-dry basis) of moist soil sample was fumigated with ethanol-free chloroform for 24 h. After removal of the $CHCl_3$, C was extracted from fumigated and unfumigated samples with 0.5 M K_2SO_4 for 30 min on a rotating shaker, using a soil extraction ratio of 1:4 (w:v). Organic C in the filtered soil extracts was determined by wet oxidation with 0.0417 M $K_2Cr_20_7$ and subsequent titration with 0.1 M $(NH_4)_2Fe(S0_4)_2.6H_20$ on 20-ml extract. Biomass C values were calculated as 2.64 Ec [3], where Ec is the extractable C flush, i.e., the difference between extractable C from fumigated and unfumigated samples.

Respiratory activity was determined according to the Pochon and Tardieux method [4], slightly modified, by placing subsamples of moist soil (corresponding to 50 g oven-dry soil) in 1.0 liter jars equipped with a 50-ml beaker containing 10 ml 0.25 M KOH to trap the evolved CO_2. The CO_2 trapped in the KOH solution was measured by back-titration with 0.25 M HCl to end-point pH values of 8.00 and 4.10, with a Radiometer DTS 800 multi-titration system (Radiometer, Copenhagen, Denmark). The CO_2 determination was carried out after 10 days of soil incubation in the dark at 25°C.

2.4. Calculation of ecophysiological parameters

The data of microbial biomass, respiratory activity, and organic carbon were used to calculate two ratios: microbial biomass C to soil organic C and respiratory C to microbial biomass C per unit time (respiratory quotient - qCO_2).

2.5. Statistical analysis

All the results were the average of five replicates and were expressed on an air-dry soil basis. The results of the three soil samples taken for each site, averaged and transformed according to the equation $y = \text{arcs in} \sqrt{p} \cdot n^{-1}$, were subjected to Pearson correlation analysis with Statgraphics (Statistical Graphics Corporation, Rockville, MD, USA).

3. RESULTS AND DISCUSSION

In 1996, the microbial biomass of the soils of the two valleys increased from May to July (Figure 1). This trend was due to both the organic carbon content (Figure 2) and the climatic conditions. Initially, the biomass used the carbon present in the soil, that in fact diminished, and, subsequently, carbon supply by vegetal residues. In the summer period, both the temperature and the rainfall were favorable for the development of the microbial biomass (Figures 3, 4]. In November, the biomass decreased, as did the organic C and the temperature.

Confirmation of a relationship between organic C and biomass comes from the MBC:OC ratio.

In Val Gerola, the trend of the MBC:OC ratio (Figure 5) was similar to that of the microbial biomass. Therefore, from May to July there was, for organic carbon parity, an increase in the biomass fraction of the organic matter in the soil as the nutritional and environmental conditions were such as to put the biomass in favorable growing conditions. The dependence of the MBC:OC ratio on climatic conditions was shown by Insam [5]. Studies of this author demonstrated the high correlation between the MBC:OC ratio and the many environmental variables, particularly the temperature. The reduction in the index in November shows how the biomass had a reduced and less assimilable substrate, therefore diminishing accordingly, also, with temperature.

Compared with Val Gerola, the MBC:OC ratio in Val Masino had a more constant trend, and this revealed that the microbial biomass fluctuated less, its dimensions remaining at a more constant level. The milder temperatures and higher content of nutritious elements in this valley (unpublished data) were most likely responsible for this constancy.

For all of 1996, the trend in the respiratory activity in Val Gerola followed that of the biomass, in accordance with the fact that in normal conditions, an increase in biomass leads to a corresponding increase in respiration (Figure 6) [6], a relationship that was not revealed in the soils of Val Masino. Therefore, it would seem that the respiration and biomass growth are independent of each other, but this is not so, as can be deduced from the respiratory quotient.

In Val Gerola, the microbial biomass and respiratory activity were coincident, determining a certain constancy in such a ratio (Figure 7); it can therefore be believed that the energetic efficiency of the metabolic processes of the microflora did not, in 1996, experience important variations and that this ratio value could be the base level of activity of the microflora in those environmental conditions.

In Val Masino, despite there being no coincidence between biomass trend and the trend of respiration, the metabolic quotient reveals that the July sampling differed in terms of better substrate use; the biomass uses less carbon (as CO_2) during growth. Also, the November sampling was significant: there was no correspondence between biomass reduction and reduced respiration, which had the same values as in September. The value of qCO_2 for this sample was the highest for the year (0.059). The November microbial biomass presented the same MBC:OC as in July, but its maintenance required more work (it burned more substrate). All this is in accordance with the findings of Chander and Brookes [7, 8], Nannipieri et al. [9], and Anderson and Domsch [10], according to whom adverse environmental conditions modify the metabolism of the living in the sense that there is a greater energy consumption of nutrients, not for the production of new biomass, but to maintain the activity of the natural defense mechanisms that allow the organism to survive in less than optimal conditions. All this leads to a greater production of CO_2 per biomass unit and, as a consequence, lower energy efficiency in transforming nutrient substrate into new biomass.

In 1997 in both valleys, the microbial biomass in the soil sampled in May was higher than in the May samples of the previous year (Figure 1). In our opinion this result can, for Val Gerola, be attributed to the higher organic C content (Figure 2) and for Val Masino to the higher temperatures. Although the biomass in May 1997 was higher than in May 1996, the MBC:OC ratio confirms real biomass growth per unit of carbon only in Val Masino. For Val Gerola, the value of the ratio in May 1997 (Figure 5) is near that of May 1996, while for Val Masino it is much higher than in May 1996. Furthermore, for Val Gerola, this value is in line with the other values, unlike for Val Masino. Therefore, it can be concluded that in May 1997 the biomass, compared with 1996, grew more than the level that can be considered the base level for that valley.

The respiratory activity was higher in May 1997 (Figure 6), in apparent accordance with the microbial biomass. But the RAC/MBC (Figure 7) ratio was high, suggesting that the values of respiratory activity for that period were too high with respect to the increase in biomass. Hence, the biomass in the May sampling used more C as CO_2 for growth than in other months, especially in Val Gerola.

Also in 1997, in both valleys, the microbial biomass increased from May to June, while the respiratory activity and the organic carbon decreased. Furthermore, during the same period, the climate in the two sites was favorable for microflora growth. The good functioning of the biomass is confirmed by both the increase in the MBC:OC ratio and the reduction in the respiratory quotient. The high MBC:OC ratio reached in Val Masino, to which the lowest respiratory quotient corresponds, shows a true increase in the microbial population. Therefore, it can be hypothesized that in this period, particular conditions prevailed (climatic and nutritional) that allowed the biomass an optimal distribution of energy for growth and maintenance.

In Val Masino, following this period, the biomass decreased until September, probably due to the quality of the organic C present. It can be supposed that the biomass was using a more complex substrate and was no longer able to maintain its mass, causing it to enter a crisis phase that led to a real reduction in the microbial population. This crisis is confirmed by the considerable reduction in the MBC:OC, which returns to values that can be considered normal for that soil in those environmental conditions. Also in Val Gerola, the biomass decreases in the same period, as does the organic C content, but this decrease is within the norm because the MBC:OC ratio is in line with the other values. The crisis undergone by the biomass in the two valleys was also confirmed by the increase in the respiratory quotient.

After this period, the biomass still increased, from September to November, due to both the arrival of new organic matter in the soil (that the microflora used with moderate efficiency) and to the relatively favorable climatic conditions. This resumption was confirmed by the slight increase in the MBC:OC ratio and the reduction of the respiratory quotient.

The relationship we found between biomass, respiratory activity, and the indexes (MBC:OC ratio and qCO_2) was not always statistically significant (Table 1), but this is most probably because a real field situation was investigated.

Table 1

Matrix of linear correlation coefficients and significance levels between organic C (OC), respiratory activity (RAC), microbial biomass (MBC), the ratio of respiratory activity C to soil organic C (RAC:OC), and the ratio of microbial biomass C to soil organic C (MBC:OC).

	OC	RAC	MBC	RAC:MBC	MBC:OC
			VAL GEROLA		
OC	-				
RAC	0.7138**	-			
MBC	0.5662	0.5998	-		
RAC:MBC	0.1173	0.3789	-0.5121	-	
MBC:OC	-0.3405	0.0125	0.5794	-0.665*	-
			VAL MASINO		
OC	-				
RAC	0.2949	-			
MBC	0.1569	0.1074	-		
RAC:MBC	0.0199	0.5741	-0.7455**	-	
MBC:OC	-0.3524	-0.0636	0.8686***	-0.7260**	-

(*$P \leq 0.1$; ** $P \leq 0.05$; *** $P \leq 0.01$)

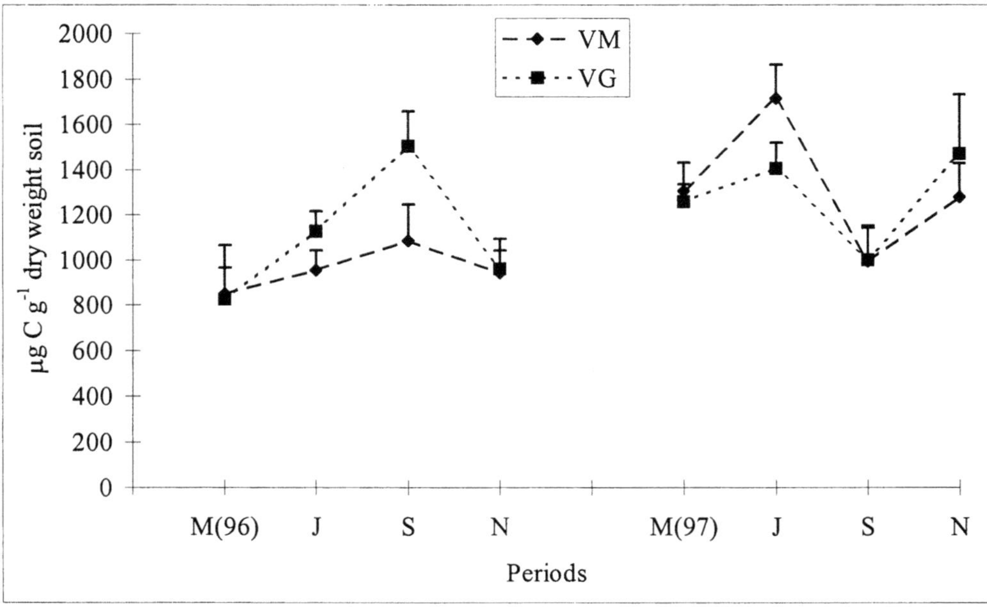

Figure 1. Microbial biomass. Error bars indicate the standard deviation of 15 replicates.
VG = Val Gerola; VM = Val Masino.
M = May; J = July; S = September; N = November.

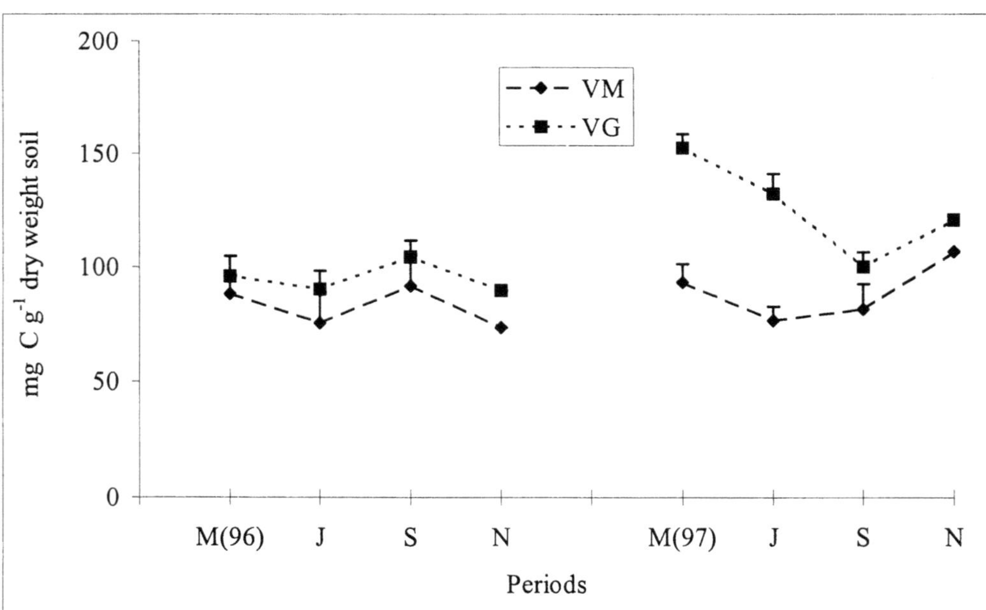

Figure 2. Organic carbon. Error bars indicate the standard deviation of 15 replicates.

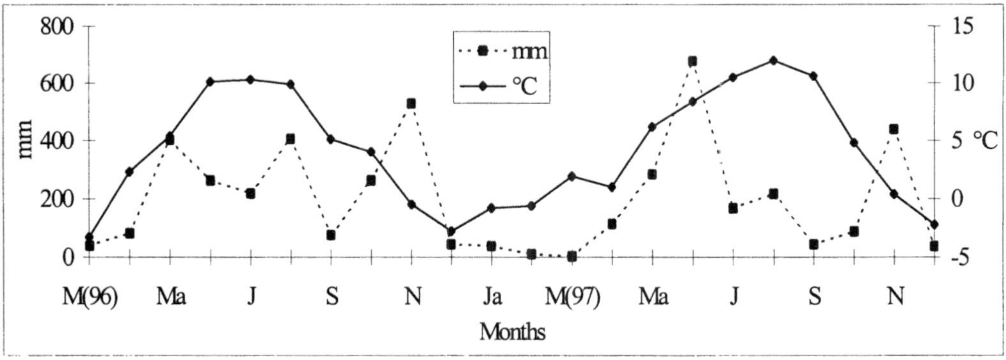

Figure 3. Val Gerola: temperature and precipitation (average values).
M = March; Ma = May; J = July; S = September; N = November; Ja = Junnary.

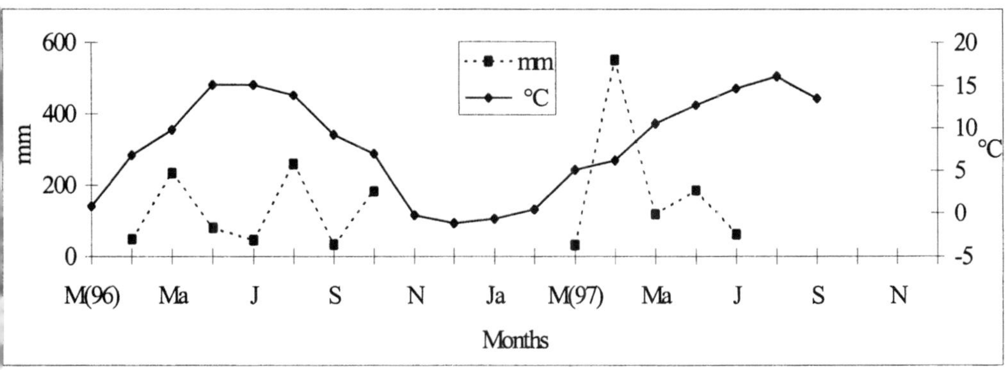

Figure 4. Val Masino: temperature and precipitation (average values)

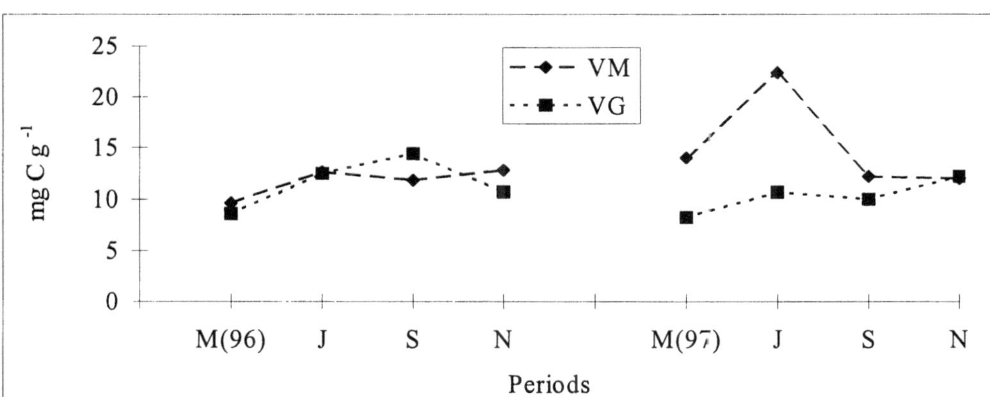

Figure 5. Ratio of microbial biomass C to soil organic C (MBC : 0C).
VG = Val Gerola; VM = Val Masino.
M = May, J = July, S = September, N = November.

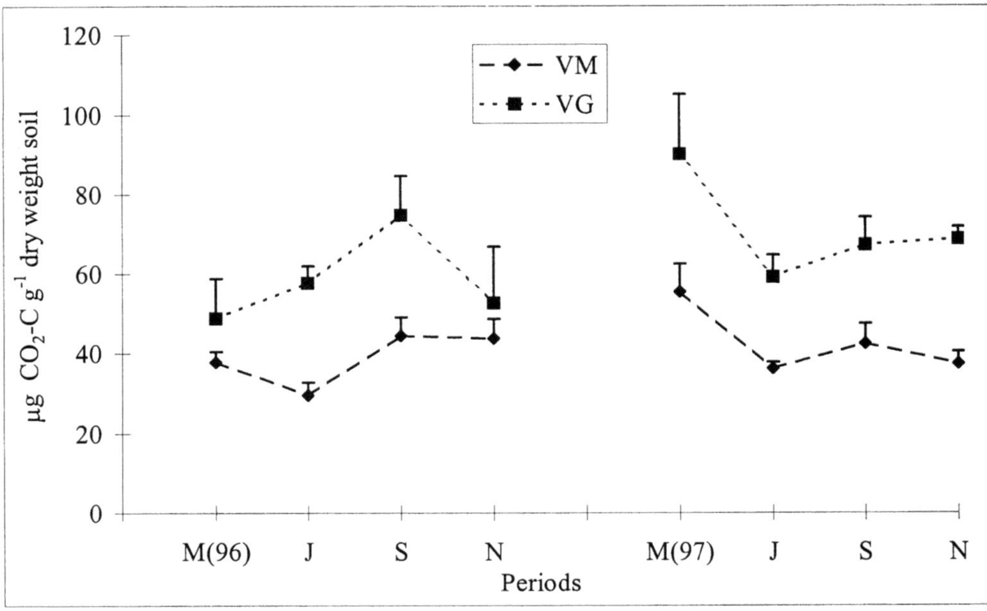

Figure 6. Respiratory activity. Error bars indicate the standard deviation of 15 replicates.
VG = Val Gerola; VM = Val Masino.
M = May, J = July, S = September, N = November.

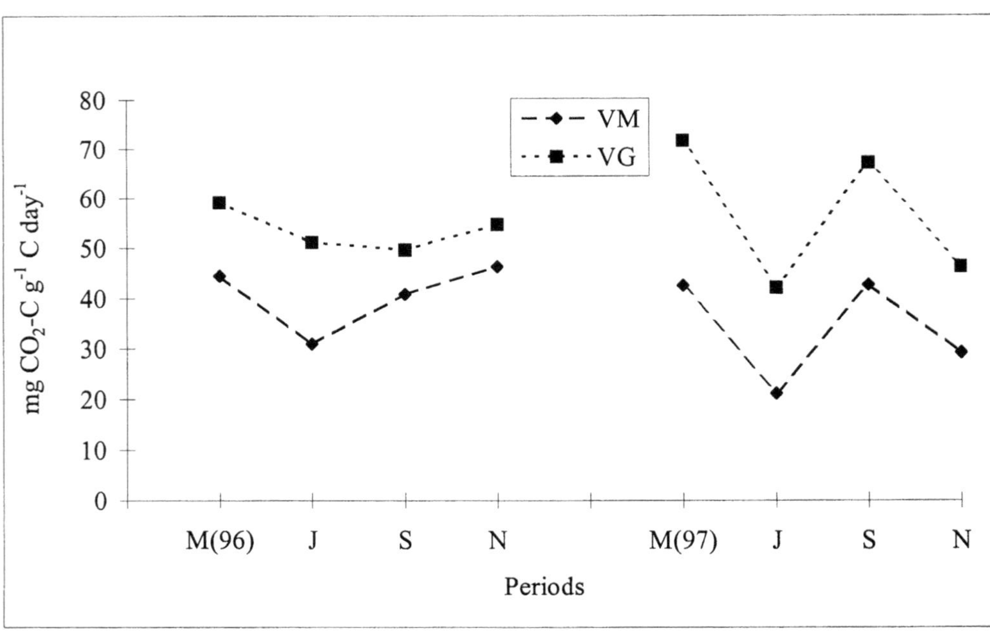

Figure 7. Ratio of respiratory C to microbial biomass C per unit time (RAC : MBC).

4. CONCLUSIONS

This experiment confirms that the biomass, if not disturbed, establishes a balance with its environment and undergoes changes with the latter. Moreover, the results obtained do not enable the establishment of any close relationship between the low activity levels and marked biomass stability (regarded as minor activity in the microflora of the Val Gerola) and the forest deterioration that could be due to environmental pollution. It is more likely that this situation can be attributed to causes, all of which are present in this valley, such as: low availability of nutrients, lower temperatures, and the old age of the forest.

The ecophysiological indexes identify and confirm these findings. Nevertheless, as indicated in the literature, the indexes still do not appear to provide an appropriate diagnostic system for soil alterations because no base value for reference purposes has yet been determined.

Because a real field situation was investigated, where the soil is subjected to variable environmental conditions, our findings could be useful for further field studies on biomass behavior.

REFERENCES

1. Jenkinson, D.S., Ladd, J.N., 1981. Microbial biomass in soil: measurement and turnover. Soil Biol. Biochem. 5, 415-471.
2. Williams, B.L., 1992. Nitrogen dynamics in humus and soil beneath Sitka spruce (*Picea sitchensis*) planted in pure stands and in mixture with Scots pine (*Pinus sylvestris* L.). Plant Soil. 144, 77-84.
3. Vance, E.D., Brookes, P.C., Jenkinson, D.S., 1987. An extraction method for measuring microbial biomass C. Soil Biol. Biochem. 19, 703-707.
4. Pochon, J., Tardieux, P., 1962. Techniques d'analyse en microbiologie du sol. De La Tourelle. St Maundè. Seine.
5. Insam, H., 1990. Are the soil microbial biomass and basal respiration governed by the climatic regime? Soil Biol. Biochem. 22, 525-532.
6. Anderson, T.H., Domsch, K.H., 1985. Determination of ecophysiological maintenance carbon requirements of soil microorganism in a dormant state. Biol. Fertil. Soils. 1, 81-89.
7. Chander, K. Brookes, P.C., 1991. Plant imput of carbon to metal-contaminated soil and effects on the soil microbial biomass. Soil Biol. Biochem. 23, 1169-1177.
8. Chander, K. Brookes, P.C., 1991. Microbial biomass dynamics during the decomposition of glucose and maize in metal-contaminated and non-contaminated soils. Soil Biol. Biochem. 23, 917-925.
9. Nannipieri, P., Grego, S., Ceccanti, B., 1990. Ecological significance of the biological activity in soil. *In*: Stotzky, G., Bollag, J-M. (Eds.). Soil Biochemistry, Vol. 6. Marcel Dekker. New York, pp. 295-355.
10. Anderson, T.H., Domsch, K.H., 1993. The metabolic quotient for CO_2 (qCO_2) as a specific activity parameter to assess the effects of environmental conditions, such as pH, on the microbial biomass of forest soils. Soil Biol. Biochem. 25, 393-395.

Developments in Soil Science, Volume 28B
Editors: A. Violante, P.M. Huang, J.-M. Bollag and L. Gianfreda

EFFECT OF Cu^{2+} AND Hg^{2+} ON GENE TRANSFER BY TRANSFORMATION AND TRANSDUCTION IN *BACILLUS SUBTILIS*

M. Franchi, L. Rinaldi and E. Gallori*

Department of Animal Biology and Genetics, University of Florence, Via Romana 17, 50125 Florence, Italy, E-mail: gallori@dbag.unifi.it

The effects of Cu^{2+} and Hg^{2+} on gene transfer *via* transduction by phage PBS1, free or adsorbed on the clay minerals montmorillonite (M) and kaolinite (K), and *via* transformation by chromosomal DNA, free or adsorbed on the clays, were studied in the soil bacterium *Bacillus subtilis*. The metal cations inhibited (minimal inhibiting concentration, MIC) the growth of *B. subtilis* at $2.5 \cdot 10^{-4}$ M for Cu^{2+} and at $7.5 \cdot 10^{-7}$ M for Hg^{2+}. In the presence of inhibiting concentrations of the metal cations, the phage titers decreased by about one order of magnitude for free phage, but they remained essentially constant for the clay-associated phage. This indicates that the survival of the phage in the presence of heavy metals was increased by its adsorption on clay particles. Transduction frequencies by free and clay-associated phages were similar when the metal cations were present only during the infection process, but there was transduction only by clay-associated phages when the metals were also present on the selection plates. Only DNA bound on clay was capable of transforming competent cells when the metal cations were present at a concentration higher than the MIC during the transformation process. The frequency of transformation remained essentially constant for free and clay-bound DNA when the metal cations were also present on the selection plates at a concentration lower than the MIC.

1. INTRODUCTION

It has been widely demonstrated that soil fertility is primarily the result of the activity of the soil microbiota (bacteria, fungi, etc.), the most active component of the soil. Because soil is the major sink for environmental pollutants, *i.e.*, heavy metals, released in the biosphere, soil microorganisms are inevitably exposed to the toxic effects of these compounds [1].

Abiotic stress caused by the addition of heavy metals (in inorganic and organic forms) affects the growth, morphology, and metabolism of microorganisms in soils through functional disturbance, protein denaturation, or destruction of the integrity of cell membranes [2]. Consequently, soil pollution by heavy metals can reduce the size and activity of the microbial biomass [3].

The interactions between soil microorganisms and heavy metals have been studied mainly to

—

determine the toxic effects of metals on microbial populations and communities, their activities, and the development of resistant microorganisms. However, little is known about the effect of metal ions on bacterial gene transfer, even though a) the gene exchange network (conjugation, transformation, and transduction) is the most important driving force of bacterial dynamics and evolution [4, 5], and b) the genetic material (transforming DNA molecule, transducing virus particles) can persist for a long time in natural soils as a result of its interaction with soil components, mainly clay minerals [6, 7].

The purpose of this work was to investigate the effect of two heavy metals (Cu^{2+} and Hg^{2+}) on gene transfer *via* transduction (both free and clay-bound PBS1 bacteriophage) and transformation (both free and clay-bound DNA) among different strains of the soil bacterium *B. subtilis*.

2. METHODS

2.1. Microorganisms and media

The strains of *B. subtilis* used were FB300 (*arg* F4 *met* A29 *thi* B4 Rfm^r) [7], a spontaneous mutant resistant to rifampicin (Rfm) isolated from *B. subtilis* 1A125 (*arg* F4 *met* A29 *thi* B4) (BGSC), BD170 (*thr*-5 *trp* C2), and BD1512 (*his* A1 *met* B5 *leu* A8 *Cm^r*) resistant to chloramphenicol (Cm) (D. Dubnau). Phage PBS1 [8] is a pseudotemperate phage of *B. subtilis* capable of generalized transduction [9].

The strains of *B. subtilis* were grown in antibiotic medium no. 3 (Oxoid) on tryptose blood agar base (TBAB; Difco), or in minimal medium (MM) (Davis and Mingioli) [10] containing the required nutritional factors at a final concentration of 25 $\mu g\ ml^{-1}$. When required, Rfm and Cm were added to the media to a final concentration of 25 $\mu g\ ml^{-1}$ and 10 $\mu g\ ml^{-1}$, respectively. The strains were maintained on slants of Nutrient Agar (NA, 1.5%, w v^{-1}; Difco). The phenotypes of the strains were verified regularly.

The pH of all the media and solutions utilized was verified regularly and, eventually, adjusted to neutrality.

2.2. Preparation of homoionic clays

The < 2μm fractions of montmorillonite (M), a 2:1 (tetrahedral Si:octahedral Al) layer swelling clay mineral (Crook County, Wyoming), and kaolinite (K), a 1:1 (tetrahedral Si:octahedral Al) layer non-swelling clay mineral (Zettlitz, Germany), were made homoionic to Ca^{2+}, as described by Fusi *et al.* [11]. The two clays differed in their cation exchange capacity (CEC) (M: 76.4 cmol kg^{-1}; K: 5.8 cmol kg^{-1}) and specific surface area (SSA) (M total=800 $m^2\ g^{-1}$, M external=78 $m^2\ g^{-1}$; K total=15 $m^2\ g^{-1}$).

2.3. Preparation of clay-phage complexes

Transducing lysates of PBS1 were prepared from *B. subtilis* BD170, as described by Vettori et al. [7].

Clay-phage complexes were prepared by mixing appropriate amounts of clay and phage in 1 ml of double distilled water (ddH_2O), according to the method described by Vettori et al. [7]. The unbound/loosely bound phages were removed from the clay pellets by washing with ddH_2O, as described by Vettori et al. [7].

2.4. Preparation of clay-DNA complexes

Chromosomal DNA from donor strain *B. subtilis* BD170 was prepared as described by Khanna and Stotzky [12].

The clay-DNA complexes were prepared by reacting chromosomal DNA (50 μg) with 100 μl of a suspension (20 mg ml^{-1}) of the clay (M or K) in a total volume of 1 ml of deionized distilled water (ddH_2O), as described by Gallori et al. [13].

2.5. Titer of free and clay-adsorbed phage

Free lysate prepared from *B. subtilis* BD170 was allowed to infect *B. subtilis* FB300, as described by Vettori et al. [7].

The titration of clay-adsorbed phage was performed by resuspending the pellets of clay-phage complexes in Y medium [8] and using the suspension obtained to infect *B. subtilis* FB300, as described [7].

2.6. Transduction by free and clay-adsorbed phage

Recipient cells of *B. subtilis* FB300 and phage were mixed to give a multiplicity of infection (MOI) of about one and then incubated, as described by Vettori et al. [7]. The cells were plated on the appropriate selective media, and bacterial cells not treated with phage were plated on the selective media as controls, as previously described [7]. The transduction frequency was expressed as the number of Met^+ transductants/number of total viable cells.

The transduction by clay-adsorbed phage was performed by resuspending the pellets of clay-phage complexes in Y medium containing sufficient colony-forming units (cfu) of *B. subtilis* FB300 to yield an MOI of 1.0. Transduction frequency was determined as for free phage.

2.7. Transformation by free and clay-adsorbed DNA

Recipient cells of *B. subtilis* strain BD 1512 were made competent by the method described by Khanna and Stotzky [12].

Transformation by free DNA was performed by incubation of competent cells with saturating levels of chromosomal DNA (2 to 10 μg ml^{-1}), as described by Gallori et al. [13]. The cells were plated on the appropriate selective media, and bacterial cells not treated with DNA were plated on the selective media as controls, as previously described [13]. The transformation frequency was expressed as the number of His^+ transformants/number of total viable cells.

The transformation by clay-adsorbed DNA was performed by resuspending the pellets of tightly bound clay-DNA complexes (2.0 mg of montmorillonite containing 8 to 20 μg of chromosomal DNA) in 1 ml of competent cells and rotating the suspensions at 70 rpm at 37°C for 30 min. The cells were plated on the appropriate selective media, as described by Gallori et al. [13].

2.8. Statistics

All experiments were performed in triplicate. Statistical analysis included standard errors of the means (± S.E.M.) and Student's *t*-test ($P < 0.05$ is considered significant). Data reported in the tables are expressed as means ± S.E.M.

3. RESULTS AND DISCUSSION

3.1. Effect of metal cations on bacterial growth

The minimal inhibiting concentration (MIC) of Cu^{2+} on *B. subtilis* FB300 and *B. subtilis* BD1512 was $2.5\cdot 10^{-4}$ M, and the MIC of Hg^{2+} was $7.5\cdot 10^{-7}$ M.

3.2. Effect of metal cations on transduction

3.2.1. Survival of PBS1 phage adsorbed on M and K in the presence of Cu^{2+} and Hg^{2+}

To evaluate the effect of the heavy metal cations on bacteriophage survival, free and clay-associated lysates of PBS1 were incubated for 1 h in the presence of different cation concentrations: $[Cu^{2+}]$: 0, 10^{-4} and 10^{-3} M; $[Hg^{2+}]$: 0, $5x10^{-7}$, $7.5x10^{-7}$ and 10^{-6} M, and then titrated.

In the presence of inhibiting concentrations of Cu^{2+} (10^{-3} M) and Hg^{2+} (10^{-6} M), the titers decreased by about one order of magnitude for free phage, but they remained essentially constant for the clay-associated phage (Table 1). These results indicate that environmental pollutants, such as heavy metals, negatively affect bacteriophage activity. However, the survival of the phage in the presence of metal cations was increased by its adsorption on clay minerals.

Table 1
PBS1 phage survival in the presence of metal cations (titers expressed as pfu/ml x 10^{-8})

	Cations						
	Cu^{2+}			Hg^{2+}			
	0	10^{-4} M	10^{-3} M	0	$5\cdot 10^{-7}$ M	$7.5\cdot 10^{-7}$ M	10^{-6} M
Free	5.1±0.4	2.0±0.5	0.3±0.03	22±7	24±5	5±0.6	1.3±0.3
K-adsorbed	0.3±0.05	0.2±0.03	0.5±0.06	1.3±0.6	1.1±0.5	2.1±0.4	1.3±0.6
M-adsorbed	0.3±0.04	0.4±0.06	0.1±0.05	3.1±0.4	1.4±0.4	1.5±0.4	1.4±0.4

3.2.2. Transducing ability of PBS1 phage adsorbed on M and K in the presence of Cu^{2+}

The effect of Cu^{2+} on transduction was evaluated by transducing *B. subtilis* in the presence of different concentrations of Cu^{2+}: 0, 10^{-4}, $2.5x10^{-4}$, $5x10^{-4}$, $7.5x10^{-4}$ and 10^{-3} M.

To discriminate between the effects of the metal cation on the mechanism of transduction and on cell growth, transduction experiments were carried out in two different conditions: a) the metal cation was present during the infection process (30 min); b) the metal cation was present during the infection process (30 min) and also in the plates of selection.

When the metal cation was present only during the infection process (Table 2), the frequency of transduction, for free and clay-associated phage, remained essentially constant (about 10^{-5}) for all concentrations assayed. The titers of both viable cells and transductants decreased proportionally when the concentration of Cu^{2+} increased. These results indicate that Cu^{2+} directly affected bacterial growth and the infection process, but did not directly affect the mechanism of transduction.

Table 2
Transducing ability of phage PBS1 in the presence of Cu^{2+} during the infection process (30 min)

$[Cu^{2+}]$ (M)	PBS1 a	PBS1 b	PBS1 c	M-PBS1 a	M-PBS1 b	M-PBS1 c	K-PBS1 a	K-PBS1 b	K-PBS1 c
0	35±7	24±7	1.4±0.7	35±7	24±6	14±6	32±6	24±6	1.4±.6
$1\cdot 10^{-4}$	14±2	2.8±0.8	4.9±2.3	20±6	28±8	0.7±0.4	52±10	19±4	2.7±1.1
$2.5\cdot 10^{-4}$	13±3	1.6±0.3	8.3±3.5	39±8	21±5	1.8±.3	45±9	14±4	3.2±1.6
$5\cdot 10^{-4}$	6.1±1.2	1.9±0.4	3.2±1.4	20±4	17±4	1.2±.5	59±12	13±5	4.5±2.1
$7.5\cdot 10^{-4}$	4.5±.9	4.5±1.3	1.0±0.5	27±4	24±4	1.2±.4	24±5	15±3	1.6±.7
$1\cdot 10^{-3}$	0.25±.04	0.15±0.03	1.6±0.6	11±3	27±7	0.4±0.2	14±2	19±4	0.8±0.3

a. Number of transductants (cfu/ml) x 10^{-2}
b. Number of viable cells (cfu/ml) x 10^{-7}
c. Transduction frequency (number of transductants/number of total viable cells) for *met* (x 10^{5})

When the metal cation was also present in the plates of selection (Table 3), a protective action by clay minerals was observed. In the presence of inhibiting concentrations of Cu^{2+}, there was no transduction for free phage, but gene transfer occurred with clay-adsorbed phage, probably as a result of the adsorption/binding of the metal cation on clay particles. Colonies grown on plates containing inhibiting concentrations of Cu^{2+} exhibited an unusual morphology, being smaller and translucent.

There was no growth on the selective medium of bacteria not exposed to the phage.

Table 3
Transducing ability of phage PBS1 in the presence of Cu^{2+} during the infection process and also in the selection plates

$[Cu^{2+}]$ (M)	PBS1 a	PBS1 b	PBS1 c	M-PBS1 a	M-PBS1 b	M-PBS1 c	K-PBS1 a	K-PBS1 b	K-PBS1 c
0	13±2	3.3±0.6	3.9±1.4	35±8	24±5	1.4±0.6	23±6	2.3±0.5	1.0±0.4
$1\cdot 10^{-4}$	11±3	4.6±1.1	2.3±1.3	34±8	13±3	2.6±1.2	6.2±2	4.9±1	1.3±0.6
$2.5\cdot 10^{-4}$	N.D.	0.9±0.5	N.D.	2.1±0.5	8.0±2	2.7±1.3	1.3±0.4	1.5±.3	0.9±0.5
$5\cdot 10^{-4}$	N.D.	3.9±2	N.D.	1.1±.2	4.7±1	2.3±1.2	12±4	5.2±1	2.4±1.3
$7.5\cdot 10^{-4}$	N.D.	N.D.	N.D.	0.15±0.1	0.5±0.2	0.3±.15	2.0±0.6	6.5±1.5	3.3±1.8
$1\cdot 10^{-3}$	N.D.	N.D.	N.D.	N.D.	1.0±0.5	N.D.	N.D.	1.0±.4	N.D.

a. Number of transductants (cfu/ml) x 10^{-2}
b. Number of viable cells (cfu/ml) x 10^{-7}
c. Transduction frequency (number of transductants/number of total viable cells) for *met* (x 10^{5})

3.2.3. Transducing ability of PBS1 phage adsorbed on M and K in the presence of Hg^{2+}

The effect of Hg^{2+} on the transduction process was evaluated by transducing *B. subtilis* in the presence of different concentrations of Hg^{2+}: 0, $5x10^{-7}$, $7.5x10^{-7}$ and 10^{-6} M).

Transduction experiments were carried out in the presence of the metal cation only during the infection process (30 min) (Table 4) Both the titers of viable cells and the frequency of transduction, for free and clay-associated phage, remained essentially constant for all the concentrations assayed. It is possible that 30 min of exposure to the metal cation was not sufficient to have a toxic effect.

There was no growth on the selective medium of bacteria not exposed to the phage.

Table 4
Transducing ability of phage PBS1 in the presence of Hg^{2+} during the infection process

	PBS1			M-PBS1			K-PBS1		
$[Hg^{2+}]$ (M)	a	b	c	a	b	c	a	b	c
0	1.6±0.3	1.4±0.2	1.1±.4	11±3	11±2	1.0±0.4	12±2	27±5	0.5±0.2
$5\cdot 10^{-7}$	7.0±2	7.5±1	0.9±0.4	15±3	15±1	1.0±0.3	4.8±10	44±12	0.1±0.05
$7.5\cdot 10^{-7}$	10±2	5.3±1	2.0±0.8	16±3	17±2	0.9±0.3	6.8±2	33±7	0.2±0.1
$1\cdot 10^{-6}$	5.7±2	6.7±1.5	0.9±0.5	14±3	12±3	1.0±0.5	17±3	27±4	0.6±0.2

a. Number of transductants (cfu/ml) x 10^{-2}
b. Number of viable cells (cfu/ml) x 10^{-7}
c. Transduction frequency (number of transductants/number of total viable cells) for *met* (x 10^{5})

3.3. Effect of Cu^{2+} on transformation

The effect of Cu^{2+} on transformation was evaluated by transforming *B. subtilis* in the presence of different concentrations of Cu^{2+}: 0, $7.5x10^{-5}$, 10^{-4}, $2.5x10^{-4}$ and $5x10^{-4}$ M. To discriminate between the effects of the metal cation on the mechanism of transformation and on cell growth, transformation experiments were carried out in two different conditions: a) the metal cation was present during the transformation process (30 min); b) the metal cation was present during the transformation process (30 min) and also in the plates of selection.

When the metal cation was present only during the transformation process (Table 5), the frequency of transformation for free DNA remained essentially constant (about 10^{-5}) for concentrations of Cu^{2+} lower than the MIC. In contrast, the frequency of transformation decreased by about one order of magnitude (about 10^{-6}) when $[Cu^{2+}]$ = MIC, and there was no transformation in the presence of inhibiting concentrations ($2.5x10^{-4}$ M) of the cation. These results indicate that the metal cation directly affects the transformation process, reducing its efficiency.

The transformation frequency remained essentially constant with clay-bound DNA (about 10^{-6}) for all concentrations assayed, probably as a result of the adsorption and binding of the metal cation on clay particles [14, 15].

Table 5

Transformation of *B. subtilis* BD1512 by chromosomal DNA in the presence of Cu^{2+} during the transformation process

	DNA			M- DNA			K- DNA		
$[Cu^{2+}]$ (M)	a	b	c	a	b	c	a	b	c
0	78±20	60±12	1.3±.6	32±6	65±13	0.5±0.2	14±4	65±15	0.2±0.1
$7.5 \cdot 10^{-5}$	97±24	29±5	3.4±1.5	2.2±0.3	32±6	0.7±0.3	6.5±2	46±13	0.14±0.06
$1 \cdot 10^{-4}$	81±21	35±8	2.3±1.1	4.4±0.9	16±2	0.3±0.1	3.5±0.6	31±7	0.1±0.04
$2.5 \cdot 10^{-4}$	0.05±0.02	0.6±0.2	0.08±0.03	41±7	52±10	0.8±0.3	45±9	59±9	0.8±0.03
$5 \cdot 10^{-4}$	N.D.	N.D.	N.D.	0.15±0.05	0.3±0.08	0.5±0.3	0.15±0.07	0.4+0.2	0.4±0.2

a. Number of transformants (cfu/ml) x 10^{-2}
b. Number of viable cells (cfu/ml) x 10^{-7}
c. Transformation frequency (number of transformants/number of total viable cells) for *his* (x 10^{5})

Table 6

Transformation of *B. subtilis* BD1512 by chromosomal DNA in the presence of Cu^{2+} during the transformation process and also in the selection plates

	DNA			M- DNA			K- DNA		
$[Cu^{2+}]$ (M)	a	b	c	a	b	c	a	b	c
0	77±15	63±11	1.3±0.5	32±5	65±16	0.5±0.2	14±	64±12	0.2±0.1
$7.5 \cdot 10^{-5}$	0.1±0.02	0.1±0.02	1.0±0.4	1.1±0.3	5.1±1	0.2±0.1	0.7±0.2	1.9±0.3	0.4±0.2
$1.0\ 10^{-4}$	0.15±0.02	0.1±0.02	1.5±0.6	0.7±0.15	0.8±0.2	0.9±0.4	0.6±0.1	0.9±0.2	0.6±0.2
$2.5 \cdot 10^{-4}$	N.D.	0.09	N.D.	0.15±0.03	0.4±0.1	0.4±0.2	0.15±0.06	0.6±0.2	0.3±0.2
$5 \cdot 10^{-4}$	N.D.	N.D.	N.D.	N.D.	N.D.	N.D.	N.D.	N.D.	N.D.

a. Number of transformants (cfu/ml) x 10^{-2}
b. Number of viable cells (cfu/ml) x 10^{-7}
c. Transformation frequency (number of transformants/number of total viable cells) for *his* (x 10^{5})

When the metal cation was also present in the selection plates (Table 6), the frequency of transformation remained essentially constant for clay-bound DNA (about 10^{-6}) up to $[Cu^{2+}]$ = MIC; transformation with free DNA was observed only for $[Cu^{2+}]$ lower than MIC The titers of viable cells decreased by about one order of magnitude for free and clay-bound DNA at $[Cu^{2+}] = 7.5x10^{-5}$ M.

The presence of the metal cation in the selection plates negatively affected cell growth; thus it was difficult to get significant results for concentrations of Cu^{2+} higher than MIC.
There was no growth on the selective medium of bacteria not exposed to the DNA.

Results obtained, even if preliminary, indicated that heavy metal pollutants can affect the genetic function of soil microorganisms. Such information could be useful in monitoring the biological quality of a soil.

REFERENCES

1. Leita, L., De Nobili, M., Muhlbachova, G., Mondini, C., Marchil, L., Zerbi, G., 1995. Bioavailability and effects of heavy metals on soil microbial biomass survival during laboratory incubation. Biol. Fertil. Soils 19, 103-108.
2. Collins, Y.E. Stotzky, G., 1996. Changes in the surface charge bacteria caused by heavy metals do not affect survival. Can. J. Microbiol. 42, 621-627.
3. Brookes, P.C., McGrath, S.P., 1984. Effects of metal toxicity on the size of the soil microbial biomass. J. Soil Sci. 35, 341-347.
4. Cohan, F.M., 1994. Genetic exchange and evolutionary divergence in prokariotes. Trends Ecol. Evol. 9, 175-180.
5. Yin, X., Stotzky, G. 1997. Gene transfer among bacteria in natural environments. Adv. Appl. Microbiol. 45, 153-212.
6. Stotzky, G., Gallori, E., Khanna, M., 1996. Transformation in soil. *In*: Akkermans, A.D.L., van Elsas, Y.D., de Brujin, F.J. (Eds.), Molecular Microbial Ecology Manual. Kluwer Academy Publishers, Dordrecht (NL), 5, 1.2, pp.1-28.
7. Vettori, C., Stotzky, G., Yoder, M., Gallori, E., 1999. Interaction between bacteriophage PBS1 and clay minerals and transduction of *Bacillus subtilis* by clay-phage complexes. Environ. Microbiol. 1, 347-355.
8. Takahashi, I., 1963. Transducing phages for *Bacillus subtilis*. J. Gen. Microbiol. 31, 211-217.
9. Hemphill, H.E., 1993. Transduction in Gram-positive bacteria. *In*: Streips, U.N, Yasbin, R.E (Eds.), Modern Microbial Genetics. Wiley-Liss, New York, pp. 271-275.
10. Davis, B.D., Mingioli, E.S., 1950. Mutants of *Escherichia coli* requiring methionine or vitamin B12. J. Bacteriol. 60, 17-28.
11. Fusi, P., Ristori, G., Calamai, L. Stotzky, G., 1989. Adsorption and binding of protein on "clean" (homoionic) and "dirty" (coated with oxyhydroxides) montmorillonite, illite, and kaolinite. Soil Biol. Biochem. 21, 911-920.
12. Khanna, M., Stotzky, G., 1992. Transformation of *Bacillus subtilis* by DNA bound on montmorillonite and effect of Dnase on the transforming ability of bound DNA. Appl. Environ. Microbiol. 58, 1930-1939.
13. Gallori, E., Bazzicalupo, M., Dal Canto, L., Fani, R., Nannipieri, P., Vettori, C., Stotzky, G., 1994. Transformation of *Bacillus subtilis* by DNA bound on clay in non-sterile soil.

FEMS Microbiol. Ecol. 15, 119-126.
14. Stotzky, G., 1986. Influence of soil mineral colloids on metabolic processes, growth, adhesion and ecology of microbes and viruses. *In*: Huang, P.M., Schinitzer, M. (Eds.), Interactions of Soil Minerals with Natural Organics and Microbes. SSSA, Madison, WI, pp. 305-428.
15. Jackson, T.A., 1995. Effects of clay minerals, oxyhydroxides, and humic matter on microbial communities of soil, sediment, and water. *In*: Huang, P.M., Berthelin, J., Bollag, J.M., McGill, W.B., Page, A.L. (Eds.), Environmental Impact of Soil Component Interactions. CRC Press, Boca Raton, FL. pp. 165-200.

Developments in Soil Science, Volume 28B
Editors: A. Violante, P.M. Huang, J.-M. Bollag and L. Gianfreda

CADMIUM AND ZINC TOXICITY TO SOIL MICROBIAL BIOMASS AND ACTIVITY

G. Renella[a], P.C. Brookes[b] and P. Nannipieri[a]

[a]Department of Soil Science and Plant Nutrition, University of Florence, P.[le] delle Cascine, 28, 50144- Florence, Italy

[b]Soil Science Department, IACR-Rothamsted, Harpenden, AL5 2JQ, Hertfordshire, United Kingdom

The effects of sewage sludge selectively enriched with zinc at the maximum permitted level and cadmium at concentration four times the current EU limits on microbial biomass and activity were tested using soils under different management in a laboratory incubation experiment. The microbial biomass and activity and the heavy metal mobility were measured over a period of six months. The results showed that neither zinc at the maximum permitted level nor the cadmium at concentrations as high as four times the current EU limits may have major adverse effects on the microbial biomass and respiration of soils.

1. INTRODUCTION

In 1986, the European Union (EU) produced a directive containing mandatory limits to prevent the accumulation of heavy metals in agricultural soils above safe limits. Particularly, the limit for cadmium is kept very low (Table 1). It is considered to be very toxic because it is not required for any known biological process, it is very mobile due to its low affinity for the soil colloids [2] and might be easily transferred to crops [16].

Table 1
Maximum permitted concentrations of some heavy metals in agricultural soils, according to the current EU regulation [9]

Total concentration (mg kg^{-1} soil)	
element	*range*
Cadmium	1-3
Chromium	100-150
Copper	50-140
Lead	50-300
Nickel	30-75
Zinc	150-300

Over the past decade or so, there has been increasing evidence that heavy metals at or around the current EU permitted limits may have adverse effects upon the soil microbial ecosystem, while crop growth may apparently be unaltered. Thus, soils from the Woburn Garden Market Experiments at Rothamsted (UK), which last received metal-contaminated sewage sludge more than thirty years ago, now contain only about half as much total microbial biomass (measured as biomass C or ATP) as similar plots that received farmyard manure or inorganic fertilizers (NPK) over the same period. The microbial populations of the high metal plots still hold a higher specific respiration rate [3,4] which, according to the theoretical biology, is an index of stress [18]. Brookes et al [5] also reported a strong decrease of microbial acetylene reduction in the same soils indicating a decrease of the nitrogenase activity; McGrath et al [15] showed that in sludge-treated soils the N_2 fixation activity was ineffective though the nodulation of the clover roots occurred.

Since the 1960's, all plots have received NPK only and now the maximum metal concentrations in the previously sludged plots are around the maximum current EU limits, except for cadmium, which may be more than three times the limit in some plots [3]. Heavy metals are present in the human-derived sewage sludge because household and industrial wastes are currently collected by the same routes, and once in soils, they persist and are concentrated.

We set up an experiment to determine whether the microbiological properties of the high metal plots depend on the high cadmium concentration. We also investigated the occurrence of synergistic adverse effects of cadmium and zinc when their concentrations exceed and meet the current EU limits, respectively.

2. MATERIALS AND METHODS

The experiment was carried out using Rothamsted soils (Aquic Paleudalf, USDA 1992), 27% clay, $pH_{(H2O)}$ 5.2, under arable and grassland management. Soils were both amended with cadmium- or zinc-enriched sludge (12 and 300 mg kg^{-1} of metals, respectively) equivalent to a sludge application rate of 40 t ha^{-1}. The metal enrichment of the sludge was made as follows: Moist sludge cake was homogenized by hand and portions equivalent to 100 g of dry matter were placed in plastic containers and allowed to dry in fan-equipped cupboards. When 20% of water was lost, the sludge were treated with the same amount of water lost or cadmium and zinc (as sulphates) solutions. Single and bimetallic solutions were used to reach the desired metal concentrations in soils. The treated sludges were then mixed by hand and allowed to dry for two weeks until a constant weight was reached. To achieve better incorporation in the soils and rapid mineralization, all the sludges were dried and milled to <1 mm after drying. Unamended soils and soils amended with unspiked sludge served as controls. The elemental composition of the sewage sludge used is in Table 2.

The soils (120 g o.d. equivalent) were incubated in 1.1 L dark glass bottles with separate vials of water and 1 M NaOH (25 ml each). Cumulative CO_2 evolved was measured over 24 weeks by automated titration with standardized 1M HCl, according to Tinsley et al [20]. The vials of NaOH were replaced at 7 d intervals. Soil microbial biomass C was measured by fumigation extraction [21], and the cadmium and zinc mobility were determined using 1 M NH_4NO_3 [19].

Table 2.
Elemental analysis of the Banbury sewage sludge used in the experiment. The total elemental content was determined using aqua regia digestion of the dry powdered sludge, according to McGrath and Cunliffe [13] and are expressed as mg Kg^{-1} of sludge. The organic C and the total N were reported as percent of dry matter.

Organic C	Total N	Cd	Cr	Cu	Pb	Zn
48	20	0.5	36.5	46.7	73.5	17.0

The LSD (Tukey's test, $P < 0.05$) was calculated for the biomass C values to assess the statistical significance of difference of the means (n = 3).

3. RESULTS AND DISCUSSION

Neither cadmium at four times the current EU limit nor zinc at the maximum permitted level decreased the total soils microbial biomass (Figure 1) and respiration (Figure 2) of the arable and grassland soils over the period of 6 months.

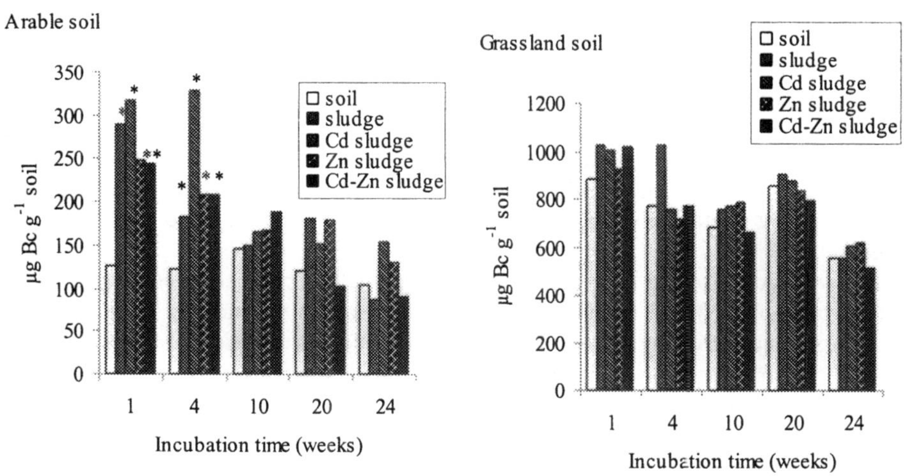

Figure 1. Microbial biomass C of the Rothamsted arable and grassland soils. * = $P<0.05$

The sludge incorporation induced a significant microbial biomass growth only in the arable soil; the increase was not significant in the grassland, where the biomass content was already very high (Figure 1). Then the biomass declined, in all soils, to values not significantly different from those of the respective controls. The trends displayed by the soil respiration were consistent with those of the biomass for all soils and treatments.

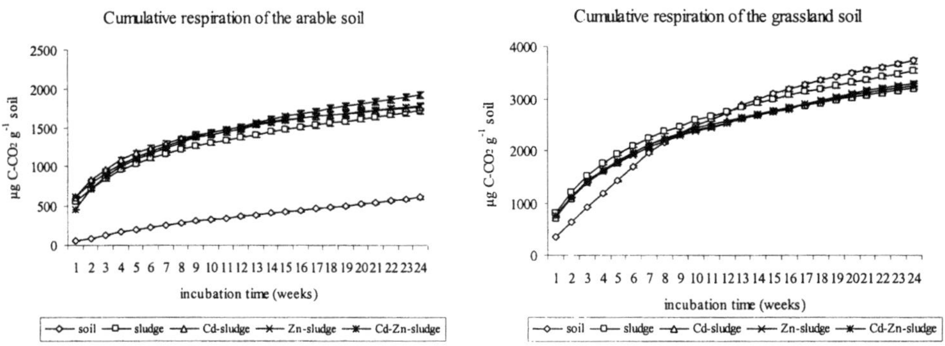

Figure 2. Cumulative respiration of the Rothamsted arable and grassland soils. The error bar is the standard error (n = 3) of the mean.

The cadmium and zinc mobility increased with time, and their mobility was different in the two soils, being generally more mobile in the arable soil (Figure 3). This was probably due to the lower organic matter content of the arable soil because it is known that soil organic matter can reduce heavy metal mobility by binding them [1]. After 24 weeks, about 10% of the total cadmium and 12% of the total zinc were NH_4NO_3-soluble in the arable soil while 3% of the total cadmium and 6% of the total zinc were NH_4NO_3-soluble in the grassland soil, approaching a plateau concentration in both the soils.

Moreover, no particular differences in heavy metal mobility were observed depending on the single- or bimetallic treatment, although it is well known that heavy metals compete for the adsorption sites when added to soils simultaneously [8] before they undergo to adsorption. Possibly the heavy metal releasing over time was due to the sludge organic matrix decomposition.

Anyway, we would exclude later effects of the metals added because an equilibrium (i.e. a minimum metal mobility) had surely been reached after about six months since the sludge addition unless dramatic changes such as rapid acidification or addition of chelating agents occurs.

The results presented suggest that cadmium, even at concentration as high as four times the current EU limit might be not toxic to soil microganisms or at least no more toxic than zinc at or a little above the current EU limits.

Chaudri et al. [7], using soils from other long-term experiments located at Braunschweig, Germany, demonstrated that the zinc concentrations of those soils was responsible for the decline in rhizobia numbers, enumerated by the MPN technique, although zinc total concentration did not exceed the maximum EU limit.

In our opinion the total content of heavy metals and the plant uptake cannot be the only criteria to be taken into account for establishing heavy metals limits in soils. For example Dahlin et al. [10], by comparing soils with different total heavy metals contents, reported adverse effects in the less polluted soils even if their total heavy metal concentrations were below the current EU limits. McGrath et al. [17] demonstrated that heavy metal toxicity, as measured by using biosensors, is better correlated with the free ion concentration rather than the total soil concentration.

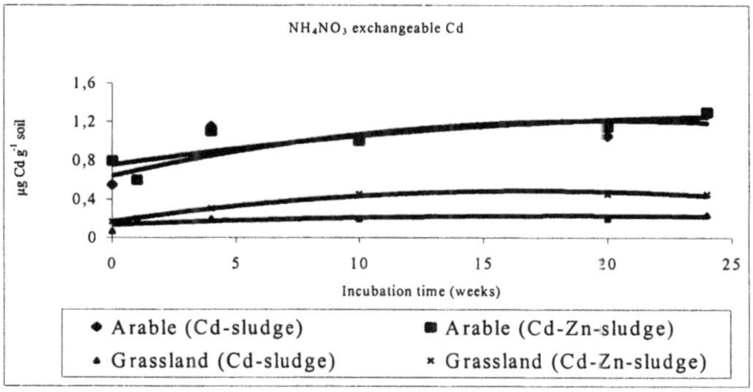

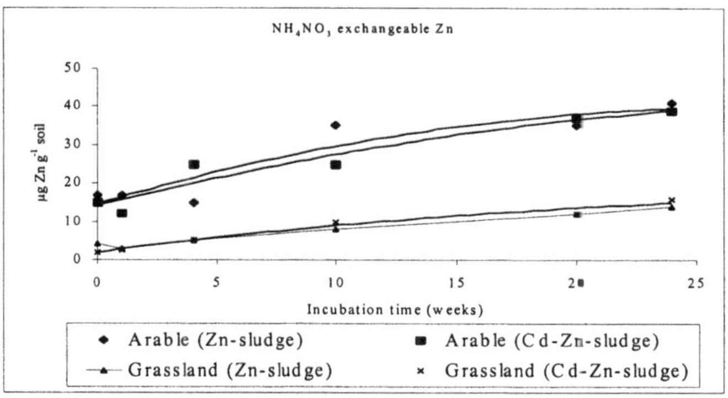

Figure 3. Cadmium and zinc mobility in the Rothamsted soils amended with Cd-, Zn-, and Cd+Zn-enriched sludge. The error bar is the standard error of the mean (n = 3).

It has been also reported that heavy metal toxicity to microorganisms might be enhanced by the synergistic action of different pollutants. It seems that microorganisms become more sensitive to some chemicals when exposed to other analogous compounds [11]. Chander and Brookes [6] described such an additive effects of heavy metals in soils from the Gleadthorpe experiments (UK). Soil amendments with single-metal (Cu, Ni, or Zn) polluted sludge had only small effects on the microbial biomass, but in plots receiving bimetallic sludge (Ni+Zn or Cu+Zn) the microbial biomass content was reduced by half compared with the reference soil.

Indeed, polluted soils usually contain more than a single heavy metal. However, such an additive effect of cadmium and zinc was not observed here and more work is needed for clarifying the mechanisms behind this phenomenon.

4. CONCLUSIONS

The use of human-derived sewage sludge as a fertilizer is a common agricultural practice and in some European countries 50% of this material produced is spread on farmlands. The sludge production is continuously increasing and, because sea disposal in the EU is illegal since 1998 and the other feasible disposal practices (e.g. incineration) are expensive and release gaseous and solid pollutants, an increase in the amounts of sludge used as fertilizer is likely.

The results presented suggest that cadmium eventually present in this material may not cause decrease of microbial biomass or alter its activity even if its content exceeds the current EU limit. Conversely, there is evidence that other heavy metals, such as copper, nickel, and zinc can cause such effects at metal concentrations at or a little above the current EU limits.

However, a better understanding of the synergistic interactions between heavy metals will enable us to use the responses of the soil microbial biomass as a reliable index of the environmental impact of the incorporation into soils of heavy metal-contaminated wastes *sensu* Brookes [3].

ACKNOWLEDGMENTS

Authors wish to thank Mrs. Carol Grace for her help in the sample preparation and Mr. A. Crossland for the elemental analysis of the soils. IACR-Rothamsted receives grant-aided support from the United Kingdom Biotechnology and Biological Science Research Council (BBSRC). G. Renella was funded by the University of Florence, Italy.

REFERENCES

1. Adriano, D.C. 1986. Trace Elements in the Terrestrial Ecosystems. Springer-Verlag, New York.
2. Alloway, B.J., 1995. Cadmium. *In:* Alloway, B.J. (Ed.), Heavy Metals in Soils. Blackie A & P, London, UK, pp. 122-151.
3. Brookes, P.C., 1995. Use of microbial parameters in monitoring soil pollution by heavy metals. Biol. Fertil. Soils 19, 269-279.
4. Brookes, P. C., McGrath, S. P., 1984. Effects of metal toxicity on the size of the soil microbial biomass. J. Soil Sci. 35, 341-346.
5. Brookes, P.C., McGrath, S.P., Heijnen, C, 1986. Metal residues in soils previously treated with sewage sludge and their effects on growth and nitrogen fixation by blue-green algae. Soil Biol. Biochem. 18, 345-353.
6. Chander, K., Brookes, P.C., 1993. Residual effects of zinc, copper and nickel in sewage sludge on soil microbial biomass in a sandy loam. Soil Biol. Biochem. 25, 1231-1239.
7. Chaudri, A.M., McGrath, S.P., Giller, K.E., Sauerbeck, D.R., 1993. Enumeration of indigenous *Rhizobium leguminosarum* biovar *trifolii* in soils previously treated with metal-contaminated sewage sludge. Soil Biol. Biochem. 25, 301-309.
8. Christensen, T.H., 1987b. Cadmium sorption at low concentration. VI. A model forf zinc competition. Water Air Soil Poll. 34, 304-314.
9. Commission of the European Communities, 1986. Council directive on the protection of the environment, and in particular the soil, when sewage sludge is used in agriculture. Off. J. Eur. Comm. L81 (Annex 1A), 6-12.
10. Dahlin, S., Witter, E., Martenson, A., Turner, A., Baath, E., 1997. Where's the limit? Microbiological properties of agricultural soils at low levels of metal contamination. Soil Biol. Biochem. 29, 1405-1415.

11. Gadd, G.M., Griffiths, A.J., 1978. Micro-organisms and heavy metal toxicity. Microb. Ecol. 4, 303-317.
12. Giller, K.E., McGrath, S.P. Hirsh, P.R., 1989. Absence of nitrogen fixation in clover grown on soil subject to long-term contamination with heavy metals is due to survival of only ineffective *Rhizobium*. Soil Biol. Biochem. 21, 841-848.
13. Giller, K.E., Witter, E., McGrath, S.P., 1998. Toxicity of heavy metals to micro-organisms and microbial processes in agricultural soils: A review. Soil Biol. Biochem. 30, 1389-1414.
14. McGrath, S.P., Cunliffe, C.H., 1985. A simplified method for extraction of the heavy metals Fe, Zn, Cu, Ni, Cd, Pb, Cr, Co and Mn from soils and sewage sludge. J. Sci. Food Agric., 36, 794-798.
15. McGrath, S.P., Brookes, P.C., Giller, K.E. 1988. Effects of potentially toxic metals in soils derived from past applications of sewage sludge on nitrogen fixation by *Trifolium repens* L. Soil Biol. Biochem. 20, 415-424.
16. McGrath, S.P., Chaudri, A.M, Giller, K.E, 1995. Long-term effects of metals on sewage sludge on soils, microorganisms and plants. J. Ind. Microbiol. 14, 94-104.
17. McGrath, S.P., Knight, B., Killham, K., Preston, S., Paton, G.I., 1999. Assessment of the toxicity of metals in soils amended with sewage sludge using a chemical speciation technique and a *lux*-based biosensor. Env. Toxicol. Chem. 18, 659-663.
18. Pirt, S.J., 1975. Principles of Microbe and Cell Cultivation. Blackwell Scientific Publication, Oxford.
19. Prueβ, A., 1998. Action values for mobile (NH_4NO_3-extractable) trace elements in soils based on the German Nation Standard DIN 19730.
20. Tinsley, J., Taylor, T.J. Moore, J.H., 1951. The determination of carbon dioxide derived from carbonates in agricultural and biological materials. Analyst 76, 300-310.
21. Vance, E.D., Brookes, P.C., Jenkinson, D.S. 1987. An extraction method for measuring microbial biomass C. Soil Biol. Biochem. 19, 703-707.

Developments in Soil Science, Volume 28B
Editors: A. Violante, P.M. Huang, J.-M. Bollag and L. Gianfreda

IMPACT OF TILLAGE PRACTICES ON SEASONAL NITRATE DYNAMICS, LEACHING AND MICROBIAL ACTIVITY IN A CROP ROTATION

R. Papini, C. Gamba, C. Piovanelli and G. Brandi

Istituto Sperimentale per Studio e la Difesa del Suolo,
Piazza D'Azeglio 30, 50121 Florence, Italy
Fax: +39 055 241485 E-mail: r.papini@dada.it

A field study was conducted on a loam typic Udifluvent soil to determine the effects of tillage on N-NO_3^- and N-NH_4 dynamics during the growing season, on soil profile residual N-NO_3^-, on organic carbon, on respiratory activity, on mineralization index, and on microbial biomass. Conventional tillage was compared with shallow tillage in a three-year crop rotation (maize-wheat-bean). Soil samples were taken periodically for a three-year period for nitrate and ammonium determination and in the sixth year after the introduction of alternative tillage practices, for OC and microbial activity. Results showed that the plots under minimum tillage maintained significantly higher N-NO_3^- amounts both in 0-20 and 0-60 cm depth during the growing season. The N-NH_4 content instead proved not to be influenced by tillage practices. The residual N-NO_3^- content measured on the soil profile in the fall was not influenced by the tillage systems. Moreover, the minimum tillage plots showed high respiratory activity, mineralization index, and microbial biomass.

1. INTRODUCTION

Reduced tillage leads to many changes in the soil's physical properties, which can influence the turnover of nitrogen and the leaching of nitrate in groundwater. Indeed the decrease of soil disturbance leads to modifications in the soil structure [1] that affect the water-holding capacity and the movement of air and water in the soil [2, 3]. The effect of reduced tillage on nitrification remains somewhat obscure; however, both nitrification and leaching are known to be influenced by soil moisture and structure. However, it is important to remember that the consequences of conservation tillage on water and chemical movement depend on a number of site-specific environmental factors, including drainage, slope, soil texture, porosity and aggregation and, most importantly, the weather [4]. For this reason, with respect to conventional tillage, reduced tillage has in some cases increased leaching [5]; in others, it has reduced leaching [6]; and in still other cases, no significant differences were found between different tillage practices [7].

It is a wellknown fact that tillage enhances mineralization of organic matter in the soil by exposing organic matter previously not accessible to microbial attack. Instead, the reduced ploughing depth with shallow tillage or minimum tillage leads to the accumulation of soil organic matter near the surface. This generally means that in no-tillage, soil organic matter tends to concentrate in the upper few centimeters [8], whereas in conventional tillage, the distribution of organic matter is more uniform throughout the ploughed layers [9, 10].

Different management regimes also modify microorganism populations, and these changes are caused by the differences in rooting with effects of the environment, such as different yields and NO_X emission [11].

To assess the effects of disturbance on soil quality, there are bioindicators that are sensitive to perturbation, such as the soil microbial biomass, which could be used as an indicator of the early changes in soil organic matter brought about by management practices, such as tillage and straw incorporation [12, 13], the CO_2 emitted during the basal respiration, and the mineralization index.

The aim of this study was to determine the effect of different tillage practices on nitrate dynamics during the growing season and on residual nitrogen level in the soil over a three year period and to measure the changes in the organic carbon content and in same microbial activities after six years of reduced tillage on a typical crop rotation in a Mediterranean region.

2. MATERIALS AND METHODS

2.1. Experimental design

The study was conducted at the farm belonging to the Research Institute for Soil Study and Conservation near Scarperia, Florence (Central Italy) on a loam soil typic Udifluvent [14]. This area has a temperate climate with a dry summer and an average annual rainfall of 1024 mm. The soil composition was: skeleton 5%, sand 45%, silt 38%, clay 17%, pH 7.8, organic matter 1.8% and total N 0.12%.

Six experimental plots (10x50 m), that had been cultivated with maize since 1970 were laid out in three blocks and treated with two different tillage practices starting in 1994: conventional tillage (CT) and minimum tillage (MT).

Conventional tillage consists of a moldboard ploughing to a depth of 40 cm and a passage with a seed harrow to prepare for planting; minimum tillage involves passing with a disk harrow twice to a depth of 10 cm.

A three-year rotation maize-wheat-bean was applied as split-plot treatment, dividing each plot into three subplots starting to 1996, as shown in Figure 1.

Sub plot 1		Sub plot 2		Sub plot 3	
MT	*CT*	*MT*	*CT*	*MT*	*CT*
M	B	M	M	M	W
W	M	B	W	B	B
B	W	W	B	W	M

Figure 1. Experimental layout in 1996. *MT* = minimum tillage; *CT* = conventional tillage. M = maize, W = wheat, B = bean.

Maize (*Zea mays* L.) was usually sown at the end of April and received 60 kg N ha^{-1} as 8-24-24 and 170 kg N ha^{-1} as urea at sowing; it was ridged and fertilized with 50 kg N ha^{-1} as ammonium nitrate in June and harvested at the end of October. Winter wheat (*Triticum L.*) was sown in November; it received 60 kg N ha^{-1} as 8-24-24 at sowing and 60 kg N ha^{-1} as ammonium nitrate as topdressing at the beginning of wheat tillering and was harvested in July. Bean (*Vicia faba L.*) was sown in November and harvested in July and received no fertilization.

Crop residues were incorporated into the soil on maize and bean plots, and straw was removed on wheat plots. Wheat plots were tilled during the fall, bean plots the subsequent spring, and maize plots after the harvest.

2.2. Sampling procedure

Samples for nitrate and exchangeable ammonium measurement were taken from March 1996 to March 1999. To determine the seasonal N-NO_3^- and N-NH_4 dynamics, soil samples were taken each month during the growing season to a depth of 60 cm, in 20 cm increments. For estimating nitrate losses during the winter, soil samples were collected in November and in the following March in increments of 20 cm to a depth of 100 cm. At each sampling time, soil samples were taken at three different locations of the experimental plots, set in plastic bags, and frozen until the time of analysis.

In June 1998, three soil samples were randomly collected from each plot to determine the microbial activities and the organic carbon content at 0-20 cm and 20-40 cm depth. The soil was sieved at 2 mm and kept at 5°C.

2.3. Sample analysis

N-NO_3^- and N-NH_4 was extracted by shaking 20 g of soil in 100 ml of 2M KCl solution for 1 h. Soil moisture was determined by the gravimetric method. Nitrate was determined by the cadmium-reduction method and subsequent reaction by sulphanilamide [15]; ammonium was determined with reaction by sodium salicylate, sodium nitroprusside and sodium hypochlorite at pH 13 [16] using the Technicon Analyzer II system. All results were expressed on a dry-soil basis.

The respiratory activity was determined using Edwards' method [17]. The latter was modified as follows: the soil samples were divided into three subsamples of 100 g, adjusted to 50% of water-holding capacity (WHC), and incubated at 25°C in 1 l flasks with soda lime pellets. The soda lime was substituted twice in 10 days. The evolved CO_2, as CO_2-C, was measured by the gravimetric method and expressed as mg CO_2 kg^{-1} dry soil d^{-1}.

Microbial biomass C (MBC) was estimated with the fumigation-incubation (FI) method, using the following equation: $Bc = Fc / kc$ [18] where Fc is CO_2-C (evolved from fumigated soil during the 10 days incubation period) minus CO_2-C (evolved from nonfumigated soil during the same incubation period) and kc is 0.45 [19]. The biomass was expressed in mg C kg^{-1} of dry soil.

The organic carbon content (OC) was evaluated by oxidation with the method described by Yeomans and Bremner [20] using a Mettler automatic titrator.

The mineralization index was calculated from the ratio between the respiratory CO_2-C of 7 days and OC [21].

2.4. Statistical analysis

The analyses of variance were run on soil N-NO_3^-, N-NH_4, and H_2O data using split-plot design, when the factor A applied at the unit is the tillage system (tillage) and the factor B applied at the subunits is the crop with which the rotation starts (starting crop). The factor A was considered a fixed factor and the factor B was considered a random one [22].

Samples were taken from the plots over a three-year period, so that each subunit would be subject to the complete rotation. The differences that can be seen in the B factors are therefore due to the variations which in the parameters under consideration are caused by the various weather conditions occurring over the three years in question. The statistical analyses of the periodic sampling were carried out first of all separately for the 0-20 cm layer, then for the complete depth of sampling surveyed (0-60 cm).

The surface layer was considered separately so as to evaluate better the effects of the tillage in the layer that is most affected by microbial activity responsible for the mineralization of the nitrogen.

The statistical analyses of the nitrogen data were done after logarithmic transformation [23].

The microbiological data were analyzed with the same system. The factor B represents the effect of the particular crop being grown in the year of sampling.

Means were separated by Duncan's multiple range test.

2.5. Rainfall

The total amounts of rainfall during each of the three years of sampling were 1240, 850, and 1044 mm respectively. Regarding the monthly distribution of the rain (Figure 2), it can be noted that in 1996-97 and 1998-99, precipitations occurred principally in April, September, October and during the winter, whereas in 1997-98, there was a higher rainfall during the summer (August) and very little precipitation in September and October.

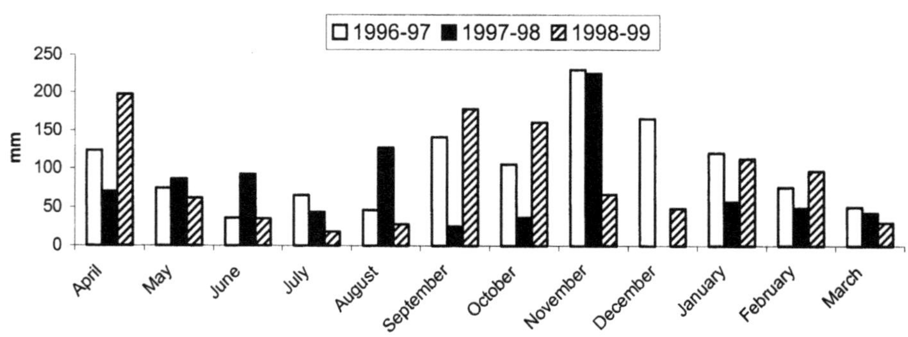

Figure 2. Monthly precipitation during the study period.

3. RESULTS

3.1. N-NO_3^- and N-NH_4 dynamics

The tillage systems influenced the average N-NO_3^- content in the soil during the growing season in 0-20 cm and 0-60 cm soil depths (Table 1). In particular MT plots showed a higher content of N-NO_3^-.

Taking into account the $N\text{-}NO_3^-$ dynamic during the three-year sampling period, the MT plots showed higher values than CT in the majority of sampling times, but the difference became greater as the $N\text{-}NO_3^-$ content in soil increased (Figure 3).

The starting crop did not influence the $N\text{-}NO_3^-$ amount in the 0-20 cm depth, but in the 0-60 cm layer, the highest values of $N\text{-}NO_3^-$ were in the plots starting with the wheat crop and ending with maize. This fact also demonstrated an interaction between crop and climatic condition on $N\text{-}NO_3^-$ dynamic during the year, especially for the maize crop. In fact, as we can see in figure 4, the $N\text{-}NO_3^-$ content was quite different in maize crop during the three examined years: in the last year, the values were higher than in the preceding years due to the lower rainfall during the growing cycle (Figure 2).

Table 1
Analysis of variance and Duncan's multiple range test for $N\text{-}NO_3$, $N\text{-}NH_4$ and H_2O content.

		N-NO₃		N-NH₄		H₂O %	
	df	0-20 cm	0-60 cm	0-20 cm	0-60 cm	0-20 cm	0-60 cm
Tillage (A)	1	0.201	0.137	0.55	0.041	1.86	9.16
		**	***	ns	ns	ns	***
Blocks (R)	2	0.042	0.012	0.25	0.046	0.61	0.34
		ns	**	ns	ns	ns	*
Error 1 (AxR)	2	0.009	0.001	0.95	0.099	0.230	0.029
Stating crop (B)	2	0.001	0.058	2.57	1.248	7.69	6.84
		ns	**	**	***	***	***
Interaction (AxB)	2	0.004	0.009	0.24	0.099	0.05	0.53
		ns	ns	ns	ns	ns	ns
Error 2	8	0.032	0.07	0.48	0.079	0.04	0.24
CT		14.12 b	9.77 b	1.92 a	1.67 a	16.95 a	17.60 a
MT		17.29 a	12.16 a	2.27 a	1.98 a	16.31 a	16.17 b
M-W-B		12.68 a	10.98 b	1.69 b	1.83 ab	17.48 a	17.68 a
W-B-M		22.58 a	13.00 a	2.91 a	2.3 a	17.05 a	17.38 a
B-M-W		11.80 a	8.68 c	1.73 b	1.37 b	15.35 b	15.67 b

* $P < 0.1$; ** $P < 0.05$; *** $P < 0.01$.

During the wheat and bean crops the soil $N\text{-}NO_3^-$ content was far lower than maize starting in June in each year. Moreover the $N\text{-}NO_3^-$ dynamics in the three examined years were quite similar for these crops (Figure 4).

Tillage did not show a significant influence on $N\text{-}NH_4$ content in 0-20 and 0-60 cm depth, but the $N\text{-}NH_4$ was influenced by the crop from which the rotation started, with higher values in the plots starting with wheat.

The H_2O content was not influenced by the tillage in the 0-20 cm depth. However in the whole sampling depth the H_2O content in CT plots was higher. The effect of the starting crop was significant at 0-20 and 0-60 cm depth.

Crop production in the three-year study period, was not significantly affected by the type of tillage operation. The average productions were: 3.7 t ha^{-1} in CT and 4.5 in MT for maize; 4.9 t ha^{-1} in CT and 4.1 in MT for wheat; 2.5 t ha^{-1}in CT and 2.8 in MT for bean.

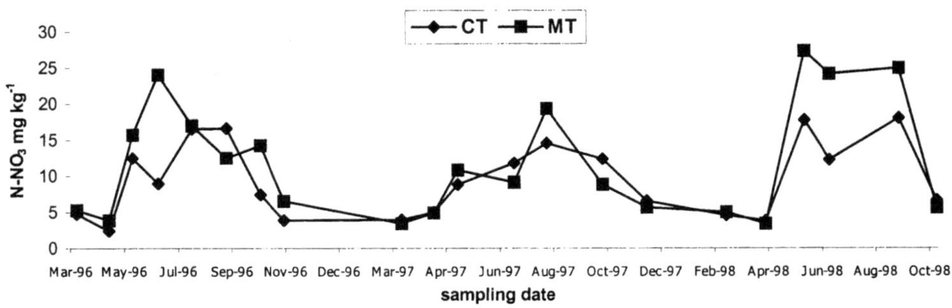

Figure 3. Seasonal N-NO_3^- dynamics in 0-60 cm depth during 1996-1998 in CT and MT.

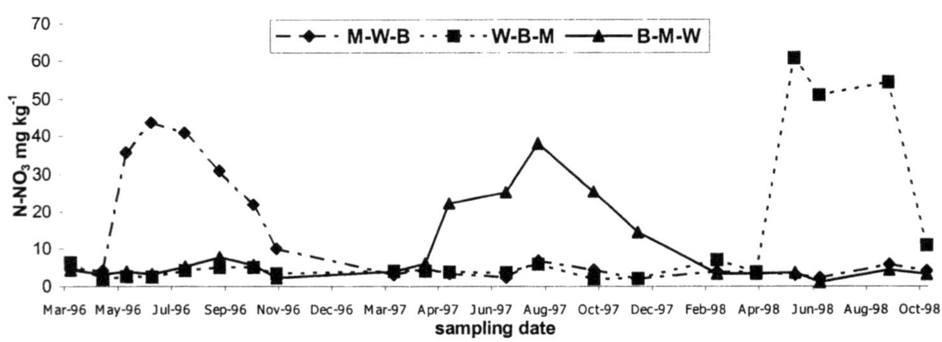

Figure 4. Seasonal N-NO_3^- dynamics in 0-60 cm depth during 1996-1998 for the different crop sequences

3.2. Residual N-NO_3^- in soil profile

Even if the samples at 1 m depth were taken in November, to have a good estimation of the N-NO_3^- moving down the soil profile, we also took into account the nitrate present at 0-60 cm depth in October, when autumnal rainfall starts. In fact, in this month, the plots with wheat and beans were fallow and the maize did not absorb N-NO_3^- from soil either. For this reason, N-NO_3^- present in the soil in this month was susceptible to leaching.

The tillage system did not show any significant effect on the N-NO_3^- content in the profile in October, November, or March (Table 2). The starting crops instead showed a significant effect on residual N-NO_3^- in November and March. The high significance of the interaction between the tillage and starting crop in October and November underlines that tillage influences N-NO_3^- content in a different way depending on the climatic condition and the crop.

The H_2O content in soil was instead significantly influenced by tillage system at each sampling date: in particular, CT plots showed higher H_2O content. For this parameter, the starting crop was also important, even if there was a significance also for the interaction tillage x starting crop.

Table 2
Analysis of variance and Duncan's multiple range test for $N\text{-}NO_3$ and H_2O content soil profile in October, November, and March.

		$N\text{-}NO_3$			H_2O %		
	df	Oct.	Nov.	March	Oct.	Nov.	March
Tillage (A)	1	0.259	0.002	0.001	7.53	9.78	18.13
		ns	ns	ns	*	***	**
Blocks (R)	2	0.103	0.004	0.019	0.324	0.078	0.385
		ns	ns	ns	ns	ns	ns
Error 1 (AxR)	2	0.050	0.008	0.0 11	0.643	0.054	0.289
Starting crop (B)	2	0.060	0.248	0.270	4.66	6.68	4.06
		ns	***	**	***	***	ns
Interaction (AxB)	2	0.423	0.250	0.030	0.991 *	2.609	2.17
		***	***	ns		**	ns
Error 2	8	0.040	0.013	0.042	0.301	0.495	0.874
CT		14.8a	5.3a	3.7a	16.6a	19.7a	19.5a
MT		19.2a	6.1a	3.4a	15.3b	18.2b	17.5b
M-W-B		12.5b	5.1b	2.6b	16.6a	19.9a	19.1a
W-B-M		24.4a	6.3a	4.3a	14.9b	19.3a	19.2a
B-M-W		13.2b	5.8ab	3.9a	16.3a	17.8b	17.6a

* $P < 0.1$; ** $P < 0.05$; *** $P < 0.01$.

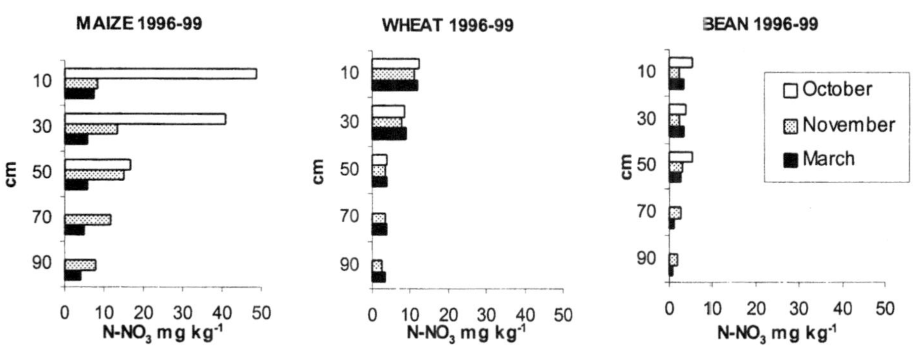

Figure 5. $N\text{-}NO_3^-$ content in soil profile in October, November, and March in the different crops.

If we consider the $N\text{-}NO_3^-$ content at the various depths in October, November, and March in the three year period, we can observe that in all the crops the greatest movement of $N\text{-}NO_3^-$ in the soil profile occurred from October to November (Figure 5). In November, the deeper layers turned out to be richer in $N\text{-}NO_3^-$ than the upper ones in most cases, indicating that $N\text{-}NO_3^-$ moved down in the soil profile as soon as autumn rainfall arrived.

This means, especially in the case of wheat and bean, which leave the soil bare early on, that any nitrate losses due to leaching are to be expected in autumn. This is very evident in the bean plots which are not ploughed or fertilized until the following spring. Indeed, in November, the nitrate content in this crop was always lower than in October, whereas in March it was possible to have a higher content than the value measured in November.

The maize plots showed the most evident $N\text{-}NO_3^-$ losses by leaching during fall and winter. In fact, the enrichment in deeper layers that was evident in November disappeared in the following March, indicating that the $N\text{-}NO_3^-$ movement had descended beyond the examined depth. This is also supported by the fact that in this soil very little loss due to denitrification was measured (unpublished data).

3.3. Microbiological activities

The use of different soil tillage systems strongly affected the soil microbiological activities (respiration, biomass, and C mineralization) and the total carbon content in the 0-20 cm layer, but in the deep layers, there were no differences between the two tillage systems (Table 3). In the CT, there was a more uniform distribution of the organic matter in the soil profile, and there were no differences between 0-20 and 20-40 cm layers (Figure 6). In the MT system, the organic matter is more abundant in the surface than in the plough layer.

Table 3
Analysis of variance and Duncan's multiple range test for total organic carbon and soil microbiological activities.

		O. C.		Soil resp.		Biomass C		C min.	
	df	0-20 cm	20-40 cm	0-20 cm	20-40 cm	0-20 cm	20-40 cm	0-20 cm	20-40 cm
Tillage (A)	1	0.046 **	0.003 ns	15.17 **	1.17 ns	3808 **	6.36 ns	0.414 *	0.353 *
Blocks (R)	2	0.003 ns	0.005 ns	0.38 ns	0.38 ns	127.4 ns	64.17 ns	0.029 ns	0.096 ns
Error 1 (AxR)	2	0.001	0.001	0.25	0.25	151.8	45.20	0.037	0.030
Crop (B)	2	0.004 **	0.010 ***	2.97 ***	0.11 ns	1637 ***	1416 ***	0.026 ns	0.265 ***
Interaction (AxB)	2	0.008 ***	0.008 ***	11.99 ***	1.95 ns	1111 ***	68.11 ns	0.279 ***	0.065 ns
Error 2	8	0.001	0.001	0.17	0.76	34.27	42.84	0.027	0.021

* $P < 0.1$; ** $P < 0.05$; *** $P < 0.01$.

The microbial biomass in the superficial layer was stimulated by the presence of crop residue left on the soil by preceding crops. The accumulation of crop residue in the superficial layer, that is characteristic of MT, causes intense microbial growth in 0-20 cm depth. In CT, the value of the microbial biomass was far inferior and above all presented no difference between 0-20 and 20-40 cm layers, both subjected to tillage.

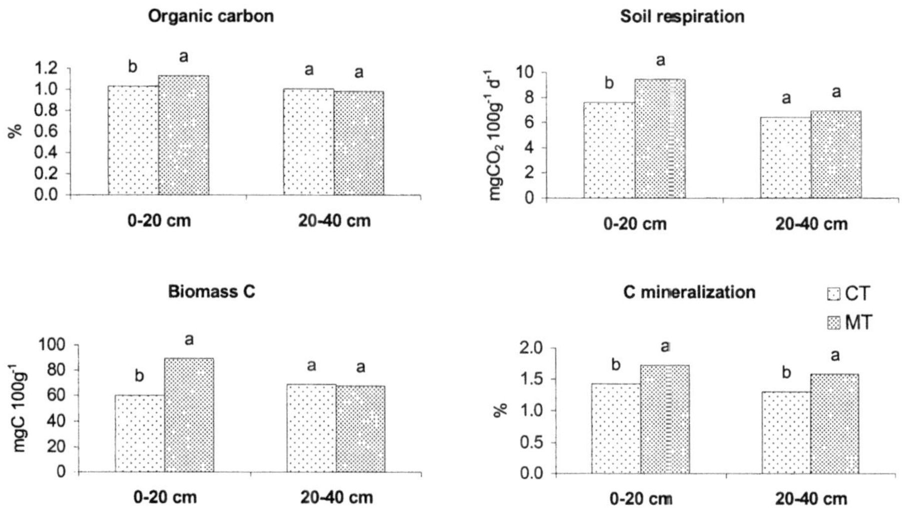

Figure 6. Soil organic carbon and soil microbial activities in the CT and MT.

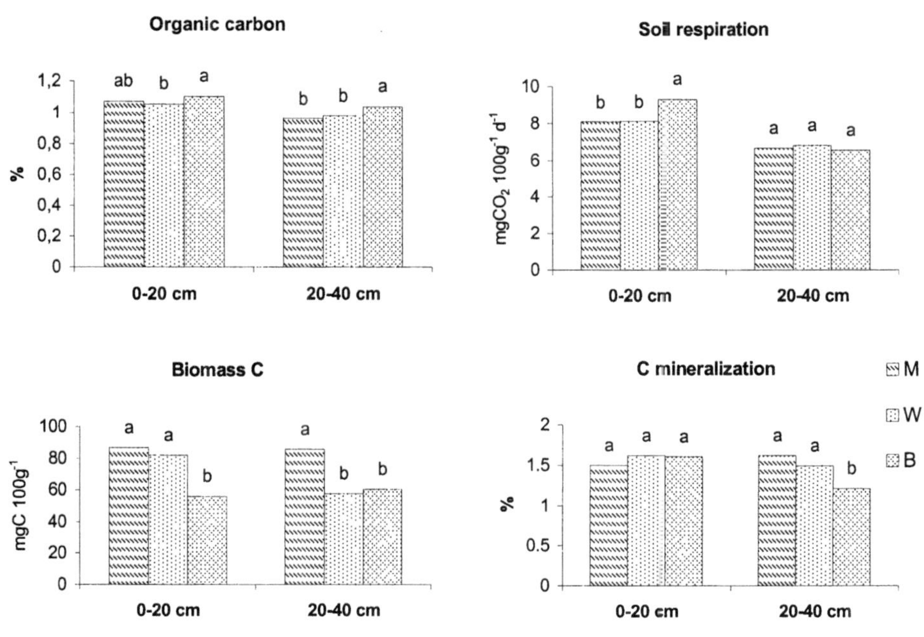

Figure 7. Soil organic carbon and soil microbial activities in the different crops.

A similar situation was found for the soil respiration activity and the carbon mineralization, which were higher in MT than CT in the 0-20 cm layer.

The crops (Figure 7) proved to have less influence than tillage on the soil respiration and carbon mineralization activities, with the exception of higher respiration in the case of bean crop in the 0-20 cm layer. The interaction tillage x crop was, however, highly significant, which means that the tillage factor was in all cases more important than the differences between the individual crops.

The organic carbon content was dependent on the residual crop left on the soil by the preceding crop. This also influenced the soil microbial biomass; in fact, the biomass was higher in the maize and wheat crops, in which soil microorganisms can make use of the crop residues left respectively by the bean and maize crops. The soil biomass of maize crop was high also in the deep layer, but in the wheat crop, it was drastically reduced.

4. DISCUSSION

The higher $N\text{-}NO_3^-$ quantity in the 0-60 cm layer found with MT can be attributed to the influence that the different type of tillage has on the soil structure and consequently on the water-holding capacity and the movement of air and water in the soil [2]. Various authors have found that reduced tillage leads to an increase in macroporosity with respect to conventional tillage, and this allows a greater water infiltration capacity [24, 25] combined with a reduction in surface runoff [4). Moreover, an enhancement of macroporosity and hydraulic conductivity in the upper 10 cm with MT was found by Pagliai et al. [26] in an experiment on maize single-crop farming carried out in the same environment as this study , and this enhancement is confirmed by the lower moisture content found in the reduced tillage in all the sampling carried out in this study. Furthermore, a greater quantity of nitrate in the soil during the growing of wheat and barley was found on a loam soil by Riley [27], and in maize by Papini et al. [28] in a two-year test on maize single-crop farming in the same environment.

The MT reduced alteration of the structure and increased the macroporosity; this guarantees better water and air circulation in the upper soil layer. From our results, it was evident that this causes a more intense microflora activity, as has been shown by the larger quantity of biomass and the greater respiratory and mineralization activity of the carbon found in the MT plots with every crop. Indeed, the increase in mineralization and organic matter content found in the surface layer of the plots under MT was the reason for the greater quantity of nitrate in this treatment. This could also explain the greater differences between the two tillage systems in the periods in which mineralization is more intense.

In particular, the MT favored the distribution of crop residues only in the surface layer with increment of organic carbon, as reported by Etana et al., [29] and Yang ad Wander [8] and enhanced the biomass C content and respiration activity. The low organic carbon in the deeper layer was rapidly mineralized by microflora, which were generally high as in the surface layers. CT caused the uniform distribution of the crop residues throughout the plough layer and the oxygenation of the depth layer, as reported by Campbell et al. [10]. In these conditions, there were no great differences in the soil biomass and its activity in the 0-40 cm depth.

The $N\text{-}NO_3^-$ content and distribution in the soil was greatly influenced by the meteorological trend. In maize plots, the $N\text{-}NO_3^-$ content during the growing season was very different in the three years examined. The low rainfall that occurred during the summer in the last year influenced the nitrogen uptake, provoking a big reduction of maize yield (data not shown) and a great amount of $N\text{-}NO_3^-$ in the soil.

The crop did not seem to influence the microbial activities, except for the effects of the residues left on the soil after the maize and fava bean crop. This amount of fresh organic matter, combined with aeration of soil, generally produced an increment of biomass C. The wheat crop did not leave any residues in the soil; in this way an impoverishment of biomass C in the following year (bean crop) was possible.

Different tillage methods did not influence the quantity of nitrates in the profile at the end of the growing time and, as a consequence, the loss of these ions for leaching in the autumn-winter period. In the same environment, on the other hand, with a single crop of maize [28], a greater nitrate loss resulted because of leaching with reduced tilling. Very probably the effect of alternating the crops, such as wheat or beans with maize, which leaves low quantities of nitrate in the soil, mitigates the differences between the tilling methods found in the single crops.

5. CONCLUSIONS

The results highlight the influence of the tillage method on the quantity of nitrate present in the soil during the growing cycle and after the harvest. The greatest nitrate concentration, which was found with minimum tillage, can be attributed to the lesser degree of alteration of the soil structure that emerges in this treatment, in which more favorable conditions both for nitrifying activity and mineralization of organic matter are created. Six years after set up the experiment, we can note variations in the content and distribution of organic matter and microbiological activity in the soil profile due to the different tillage methods. Minimum tillage favors the accumulation of organic matter in the surface layer and a more intensive microbial activity in the surface layer with respect to conventional tillage. CT on the other hand tends to homogenize the microbial activity in the 0-40 cm profile, thus favoring the conservation of organic matter in the deeper layer thanks to a continuous carbon supply.

These results are evidence of how the use of minimum tillage, which neither causes major nitrogen losses because of leaching nor reduces production, but does improve the microbiological characteristics of the soil, is to be considered in study conditions a practice capable of reducing the environmental impact of agriculture.

REFERENCES

1. Pagliai, M., Raglione, M., Panini, T., Maletta, M., La Marca, 1995. The structure of two alluvial soils in Italy after 10 years of conventional and minimum tillage. Soil Till. Res. 34, 209-223.
2. Patruno, A., Cavazza, L., Pistelli, E., 1986. Variazione del contenuto in azoto nitrico del terreno. Influenza della lavorazione, della coltura e del decorso meteorico. Riv. Agron, 20, 204-217.
3. Varshney, P, Kanwar, R.S,. Baker, J.L., Anderson, C.E., 1993. Tillage and nitrogen management effects on nitrate-nitrogen in the soil profile. Am. Soc. Agric. Eng. 36, 783-789.

4. Shipitalo, M.J., Dick, W.A., Edwards, W.M., 2000 Conservation tillage and macropore factors that affect water movement and the fate of chemicals. Soil Till. Res. 53, 167-183.
5. Tyler, D.D., Thomas, G.W.,1977. Lysimeter measurements of nitrate and chloride losses from soil under conventional and no-tillage corn. J. Environ. Qual. 6, 63-66.
6. Goss, M.J, Colbourne, P., Harris, G.L., Howse, K.R.,1998. Leaching of nitrogen under autumn-sown crops and the effect of tillage. *In:* Nitrogen Efficiency in Agricultural Soils. Jenkinson, D.S., Smith, K.A. (Eds.), Elsevier Applied Science, Barking, UK, pp. 269-282.
7. Groffman, P.M., Hendrix, P.F., Crossley, D.A., 1987. Nitrogen dynamics in conventional and no-tillage agroecosystems with inorganic fertilizer or legume nitrogen inputs. Plant Soil 9, 315-332.
8. Yang, X.M., Wander, M.M., 1993. Tillage effects on soil organic carbon distribution and storage in a silt loam soil in Illinois. Soil Till. Res. 52, 1-9.
9. Kern, J.J., Johnson M.G., 1993. Conservation tillage impacts on national and atmospheric carbon levels. Soil Sci. Soc. Am. J. 57, 200-210.
10. Campbell, C.A., McConkey, B.G., Zentner, R.P., Selles, F., Curtin, D., 1996. Tillage and crop rotation effects on soil organic C and N in a coarse-textured Typic Haploboroll in southwestern Saskatchewan. Soil Till. Res. 37, 3-14.
11. Whalley, W.R. E., Dumitru, A.R., Dexter, T., 1995. Biological effects of soil compaction. Soil Till. Res. 35, 53-68.
12. Powlson, D.S., Brookes, P.C., Christensen, B.T., 1987. Measurement of microbial biomass provides an early indication of changes in total soil organic matter due to the straw incorporation. Soil Biol. Biochem. 19, 159-164.
13. Smith, J.L., Halvorson, J.J., Bolton, H. Jr, 1995. Determination and use of a corrected control factor in the chloroform fumigation method of estimating soil microbiol biomass. Biol. Fertil. Soils 19, 287-291.
14. Soil Survey Staff, 1996. Keys to Soil Taxonomy. 7th Edition. Pocahontas Press Inc., Blacksburg, VA, USA.
15. Keeney, D. R., Nelson, D. W., 1982. Nitrogen-inorganic forms. *In*: Madison, A. L. (Ed.), Methods of Soil Analysis. Part 2, WI, USA, pp. 643-698.
16. Buurman P., Van Lagen, B., Velthorst, E.J., 1996. Manual for Soil and Water Analysis. Backhuys Publishers Laiden.
17. Edwards, N.T., 1982. The use of soda-lime for measuring respiration rates in terrestrial systems. Pedobiologia 23, 321 330.
18. Jenkinson, D.S., Powlson, D.S., 1976. The effects of biocidal treatment on metabolism in soil. V. A method for measuring soil biomass. Soil Biol. Biochem. 8, 209-213.
19. Jenkinson, D.S., 1988. The determination of microbial biomass carbon and nitrogen in soil. *In*: Wilson, J. R. (Ed.), Advances in Nitrogen Cycling in Agricultural Ecosystems. CAB International, Wallingford, pp. 368-386.
20. Yeomans, J.C., Bremner, J.M., 1988. A rapid and precise method for routine determination of organic carbon in soil. Commun. Soil Sci. Plant Anal. 19, 1467-1476.
21. Dommergues, Y., 1960. La notion de coefficent de mineralisation du carbone dans les sols. Agron. Trop. 15, 54-60.
22. Camusi, A., Moller, F., Ottaviano, E., Sari Gorla, M,1986. Metodi Statistici per la Sperimentazione Biologica. Zanichelli, Bologna, Italy.
23. White, R.E., Haigh, R.A., McDuff, J.H., 1987.Frequency distribution and spatially dependent variability of ammonium and nitrate concentration in soil under grazed and ungrazed grassland. Fertil. Res. 11, 193-208.

24. Kanwar, R.S., Baker, J.L., Laflen, J.M., 1985. Nitrate movement through the soil profile in relation to tillage system and fertilizer application method. Transactions of the ASAE 28, 1802-1807.
25. Mφller Hansen, E., Djurhuus, J., 1997. Nitrate leaching as influenced by soil tillage and catch crop. Soil Till. Res. 41, 203-219.
26. Pagliai, M., Rousseva, S., Vignozzi, N., Piovanelli, C., Pellegrini, S., Miclaus, N., 1998. Tillage impact on soil quality. I. Soil porosity and related physical properties. It. J. Agron. 2, 11-20.
27. Riley, H. C. F., 1998. Soil mineral-N and N-fertilizer requirements of spring cereals in two long-term tillage trials on loam soil in southeast Norway. Soil Till. Res. 48, 265-274.
28. Papini, R., Piovanelli, C., Brandi, G., Miclaus, N., 1998. Tillage effects on seasonal nitrate dynamics, leaching and microbial activity in continuous maize production. Fres. Envir. Bull. 7, 472-477.
29. Etana, A., Håkansson, I., Zagal, E., Buas, S., 1999. Effects of tillage depth on organic carbon content and physical properties in five Swedish soils. Scil Till. Res. 52,129-139.

Developments in Soil Science, Volume 28B
Editors: A. Violante, P.M. Huang, J.-M. Bollag and L. Gianfreda

CHARACTERIZATION OF ENZYMATIC OR ABIOTIC IMMOBILIZATION OF XENOBIOTICS IN SOIL

J.-M. Bollag, M. Strynar, M.-Y. Ahn and J. Dec

Laboratory of Soil Biochemistry, Center for Bioremediation and Detoxification,
The Pennsylvania State University, University Park, Pennsylvania

Immobilization processes occurring in soil are of great environmental importance because they may lead to a considerable reduction in the bioavailability of xenobiotics. Xenobiotics can be bound to soil constituents by physical forces or chemical reactions of varying strength. From an environmental point of view, chemical interactions are preferred, because they frequently lead to the formation of strong covalent bonds that are difficult to disrupt by microbial activity or other means. Chemical bonds may be formed through oxidative coupling reactions, which are catalyzed either biologically by polyphenol oxidases and peroxidases, or abiotically by certain clays and metal oxides. These naturally occurring processes are believed to result in the detoxification of xenobiotics and have long been suggested as a means of decontamination. The ability of enzymes to mediate the immobilization processes has been demonstrated in a variety of model studies, in which xenobiotic compounds were incubated with humic monomers or natural humic acids in the presence of various phenoloxidases. The incubations were followed by determining the extent of xenobiotic binding to humic material, and--whenever possible--by identifying the resulting complexes. The pioneering research on xenobiotic immobilization is mostly based on the application of ^{14}C-labeled xenobiotics and radiocounting. Several recent studies have demonstrated that further progress in the evaluation of binding can be achieved by applying ^{13}C- or ^{15}N-labeled xenobiotics combined with ^{13}C- and ^{15}N-NMR spectroscopy. The basis of the NMR approach was that any binding-related modification in the original arrangement of the labeled atoms automatically induced changes in the position of the corresponding signals in the NMR spectra. The delocalization of the signals indicated whether or not covalent binding had occurred and, if so, the type of covalent bond that had been formed.

1. INTRODUCTION

It is well known that large amounts of organic xenobiotics that reach the terrestrial system become sequestered or bound to soil. The environmental significance of these phenomena was well documented when it was shown that between 20 and 70% of applied pesticides or their degradation products resisted extraction from soil by water and organic solvents [1]. Neither laboratory experiments nor practical experience have detected significant negative or toxic impact of bound xenobiotics on the environment, and consequently, the immobilization of pollutants in the soil has been recognized as a promising decontamination mechanism [2, 3].

Xenobiotics interact with soil colloids through several mechanisms, and a number of reviews describe the various possibilities [4, 5]. Adsorption occurs primarily as a consequence of the attraction between the solid surface of soil and the soluble or vapor phase of the xenobiotic. Organic compounds adsorb to soil colloids via several mechanisms, including van der Waals forces, charge-transfer complexation, hydrogen bonding, and hydrophobic interactions [6]. The nature and strength of adsorption largely depend on the chemical class or structure of the molecule. Adsorption is a reversible process and the released chemicals are available to interact with the biota. However, there is abundant evidence suggesting that with longer exposure to soil, adsorbed residues become more resistant to extraction and degradation [1, 7]. This resistance may result from a gradual sequestration or slow incorporation of the pollutant into humus.

The most persistent complexes result from the covalent binding of xenobiotics to humic material. These complexes, often referred to as "bound residues", are highly resistant to acid and base hydrolysis, thermal treatment, and microbial degradation [8-10]. Our studies on the mode of covalent binding mostly involved xenobiotic phenols and anilines. As chemical analogues of naturally occurring humic acid precursors, xenobiotic phenols and anilines can be readily incorporated into humus during the humification process. Oxidative coupling is one of the most important chemical reactions leading to such incorporation. The oxidative coupling involving either phenols or anilines is a free radical reaction. The resonance-stabilized free radicals, which are formed upon the loss of an electron and a proton from the molecule of phenol, couple to soil organic matter via the atoms hosting the unpaired electron. Phenolic reactants are mostly linked through C-C and C-O bonds; aromatic amines form C-N and N-N linkages [11].

The decontamination potential, through covalent bonds with soil organic matter, reduces the compounds' bioavailability, and therefore reduce the toxicity of the pollutants. Binding also restricts the movement of xenobiotics by leaching, thus preventing the contamination of aquatic environments. As indicated by many studies, the incorporation of xenobiotics into humus, and therefore the decontamination effect, can be enhanced by the addition of biological or abiotic catalysts, or by altering the physicochemical conditions of the soil environment.

2. OXIDATIVE COUPLING REACTIONS

Oxidative coupling can be mediated by a number of biological and abiotic catalysts, including microbial or plant-derived enzymes, inorganic chemicals (e.g., ferric chloride, cupric hydroxide) and clay minerals [12, 13]. However, in the natural environment, it is often difficult to determine which catalyst is responsible for a transformation reaction.

Extracellular oxidoreductases capable of catalyzing the coupling of aromatic compounds are produced by a variety of soil microorganisms. The enzymes are classified as either peroxidases or polyphenol oxidases. All peroxidases contain an iron porphyrin ring and require the presence of peroxides (e.g., hydrogen peroxide) for activity. Horseradish peroxidase (HRP) has been shown to catalyze the polymerization of a wide range of phenolic and anilinic compounds, and its use in the detoxification of industrial wastewater has been thoroughly examined [14-16].

The products formed during the cross-coupling of pollutants and humic substances are highly heterogeneous and are therefore difficult to identify. In order to elucidate the mechanism by which pesticides are incorporated into humus and to identify the products formed, our laboratory has studied the binding process through the use of relatively simple model systems. Generally, different fungal laccases were incubated with various chlorinated phenols or aromatic amines in the presence of specific phenolic humic constituents. The hybrid products formed were separated by thin-layer chromatography or high-performance liquid chromatography and characterized by mass and nuclear magnetic resonance spectroscopy. For instance, 2,4-dichlorophenol (2,4-DCP), a degradation product of the herbicide 2,4-D, was shown to cross-couple with humus-derived compounds, such as orcinol, syringic acid, vanillic acid, and vanillin [17]. In further studies, phenols containing one to five chlorines (4-chlorophenol, 2,4- and 2,6-dichlorophenol, 2,3,6- and 2,4,5-trichlorophenol, 2,3,5,6-tetrachlorophenol, and pentachlorophenol) and various chloroanilines (4-chloroaniline, 3,4-dichloroaniline, and 2,4,5-trichloroaniline) were cross-coupled with syringic acid or guaiacol by the laccase from the fungus *Rhizoctonia praticola* [18]. Two types of cross-coupling products were formed: (1) quinonoid oligomers consisting of chlorophenols linked by ether bonds to orthoquinoline products of syringic acid; and (2) phenolic oligomers consisting of chlorophenols bound by ether linkages to decarboxylated products of syringic acid. The anilines were mostly coupled via imine linkages to quinone derivatives of the humus constituents.

In another study, the incorporation of ^{14}C-labeled phenols into humus was examined under conditions that approximated the natural soil habitat [19]. The research involved incubation of ^{14}C-labeled 2,4-DCP with humic or fulvic acid isolated from stream water in the presence of various oxidoreductases. Upon dialysis of the reaction mixtures, about 65% of the initial radioactivity was firmly incorporated into fulvic acid. The coupling occurred over a wide pH and temperature range.

3. EFFECT OF BIOTIC AND ABIOTIC CATALYSTS

A number of investigations have been performed to compare the efficiency of biotic and abiotic agents in the transformation of phenolic substances and to compare the chemical structures of the resulting reaction products. For instance, guaiacol was incubated with various oxidants in aqueous, buffered solutions [20]. Under the reaction conditions used (small amounts of oxidants and short incubation times), only the initial low molecular weight products were formed. With higher oxidant concentration, or longer reaction times, additional higher molecular-weight compounds were generated. However, the insight into the initial structures was sufficient to draw conclusions regarding the effect of various oxidation agents [20].

The first seven products eluted from the HPLC column following the peroxidase-catalyzed oxidation of guaiacol as the only substrate were structurally identified using mass spectrometry and proton nuclear magnetic resonance spectroscopy (Figure 1). They included five dimers (compounds B through F) and two trimers (compounds G and H) with quinoid and phenolic structural features. The same seven oligomers were produced using other inorganic

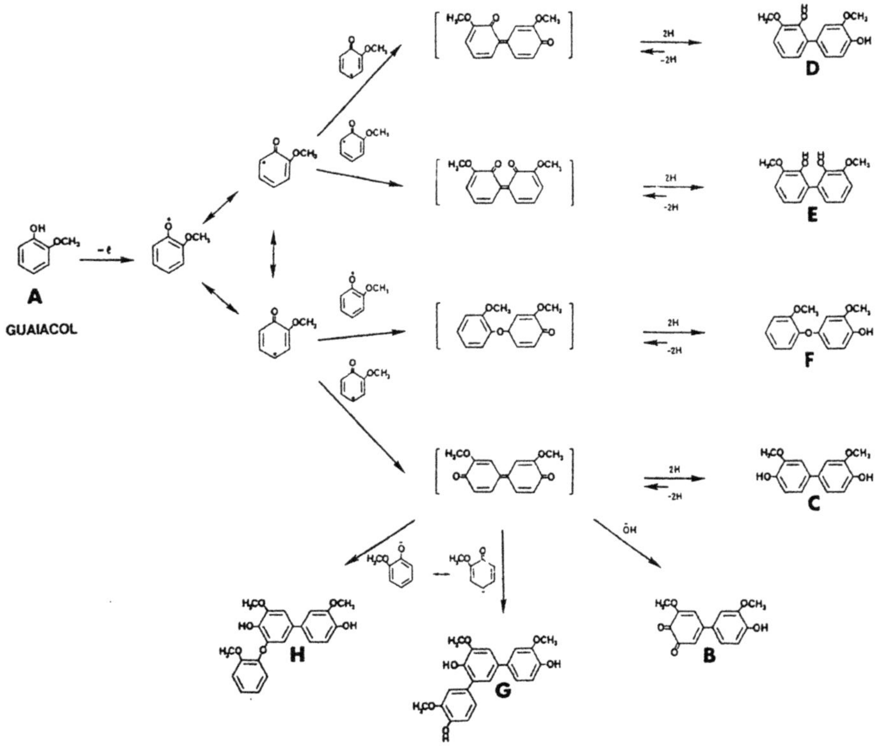

Figure 1. Oxidative coupling products of guaiacol

and biological agents occurring in soil [e.g., manganese dioxide, clay (bentonite), tyrosinase, and the laccases of *Trametes versicolor* and of *Rhizoctonia praticola*]. Figure 2 compares the chromatograms of six oxidation reactions. The chromatograms of the control samples using boiled enzymes showed only the presence of unreacted guaiacol. The qualitative similarity of the product mixtures for the six reaction solutions is evident. Note, however, that the products obtained in the clay-mediated oxidation of guaiacol do not include compound B, a dimer. It is likely that this dimer, which is an orange-colored quinone, was adsorbed to, or reacted with clay, as was suggested by the orange color of the clay residue. Nonetheless, the other oligomers (C through H) were detected in the clay reaction solution.

Although the same seven guaiacol-derived products were formed in six different reaction solutions, the amounts of the respective oligomers varied (Figure 2). The differences were most likely a result of different reaction conditions such as pH of the reaction mixture or rate of the oxidation. Regardless of the amount of each oligomer, the fact remains that the same products were formed, and this is a direct result of guaiacol being in an environment where its oxidation was favored by the loss of an electron from the oxygen atom.

In separate experiments, the same inorganic and biological oxidation agents were examined for their ability to mediate the co-oligomerization of guaiacol and 4-chloroaniline [21]. The same five co-oligomers were found in the reaction mixtures regardless of the nature

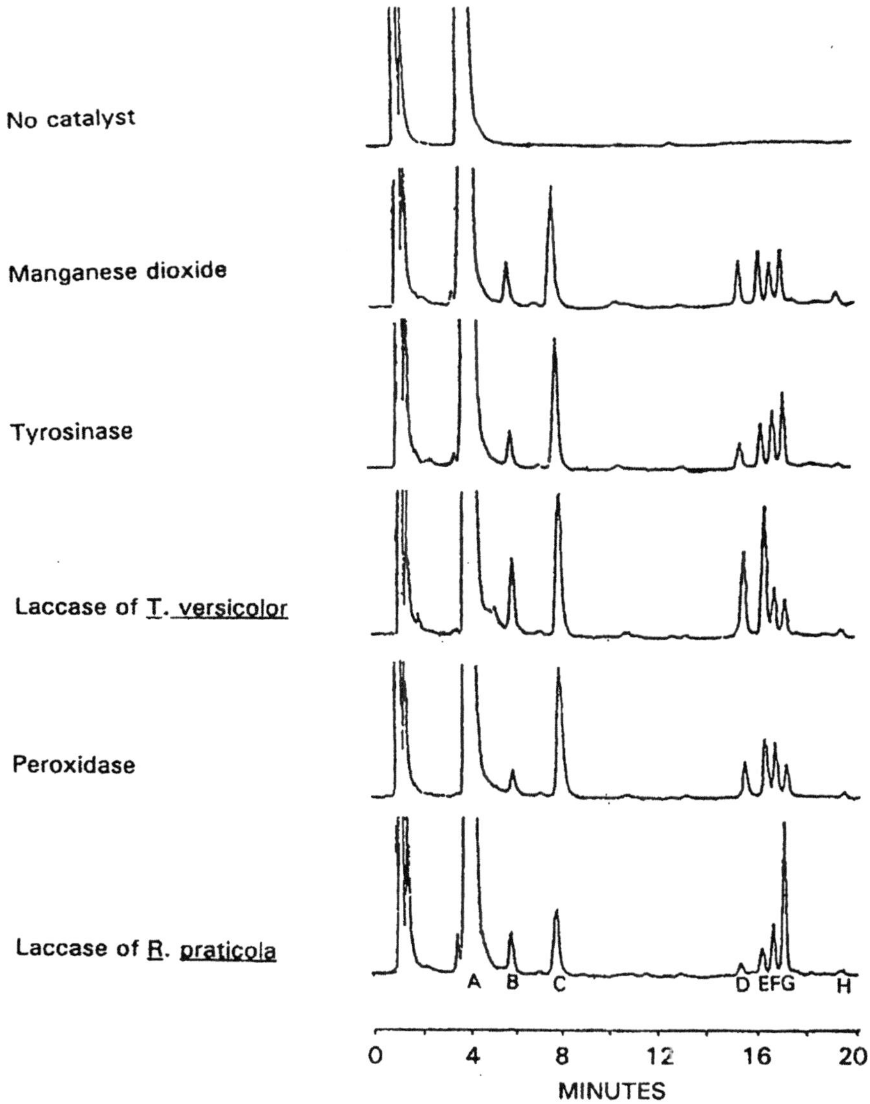

Figure 2. HPLC chromatograms of reaction solutions containing guaiacol and various oxidative catalysts

of the agents (manganese dioxide, peroxidase, *Trametes versicolor* laccase, *Rhizoctonia praticola* laccase and tyrosinase) used to catalyze the oxidations.

An important research topic was the elucidation of the role of biotic and abiotic agents in the transformation of aromatic compounds [22]. In particular, it was essential to determine whether abiotic factors behave as true catalysts during transformation reactions. As indicated by measuring oxygen uptake and analyzing substrate disappearance, abiotic and biotic agents behaved similarly in their capacity to oxidize various phenolic compounds, but the mechanisms were different. For instance, although the transformation of 2,6-dimethoxyphenol occurred in both abiotically and biotically catalyzed reactions, oxygen uptake was demonstrated only in enzyme-catalyzed reactions. These results suggest that in enzymatically catalyzed reactions, oxygen functions as the electron acceptor, whereas in birnessite-catalyzed reactions, the electron acceptor is manganese oxide. Pal et al. [22] concluded that enzymes as homogeneous catalysts appeared to be more effective oxidative agents than abiotic agents that are heterogeneous catalysts, but because of their abundance in nature, abiotic agents play a significant role in the transformation of phenolic compounds.

In the case of birnessite, the Mn ion serving as an electron acceptor is thereby reduced. The oxidative coupling reaction may involve the formation of free-radical intermediates and the same electron transfer mechanism as experienced with enzymes. Although catalysts are frequently defined as substances that accelerate chemical reactions without themselves undergoing change, this definition is too restrictive. The properties of all real catalysts are prone to change with use [23]. In fact, many substances classified as catalysts are destroyed either as a result of the process that gives them their catalytic activity or because of their subsequent combination with the products. From a practical point of view, a catalyst is a substance that changes the rate of a desired reaction, regardless of the fate of the catalyst itself [24].

Incubation of chlorinated phenols and anilines with oxidoreductive catalysts (peroxidase, laccase, tyrosinase, and birnessite) in the presence of humic acid led to oligomerization of the xenobiotics or to their binding to organic matter [25]. The effect of humic acid on the overall transformation depended on the type of catalyst, the chemical structure of the xenobiotic, and the concentration and source of the humic acid. In experiments with ^{14}C-labeled phenols and anilines, for instance, 4-chlorophenol was mainly bound to humic acid and formed few oligomers, whereas 4-chloroaniline was largely subject to oligomerization, with relatively little binding to humic acid.

Naidja et al. [26] characterized the products of catechol oxidation, catalyzed by either the enzyme tyrosinase or the mineral birnessite (δ-MnO_2). Oxidation of catechol led to the formation of two types of catechol-melanin: a) green-colored polymers formed in the presence of δ-MnO_2, and b) brown-colored polymers formed in the presence of tyrosinase. The reaction products formed in the presence of tyrosinase or birnessite consisted of various heterogeneous polycondensates and fragments having a wide range of molecular weights (MWs) as indicated in mass spectra (Figure 3). These species may be divided into two groups: (1) organic components with MWs less than 600 and (2) organic components with MWs of 600 to 900, which were more abundant in the tyrosinase-catechol system than in the birnessite-catechol supernatant. In summary, the oxidation of catechol catalyzed by δ-MnO_2 produced polycondensates with lower degrees of aromatic ring condensation and lower molecular weights than those produced by tyrosinase.

Ruggiero et al. [27] stated in their review on 'Soil as a catalytic system' that the determination of the rates and extent of abiotic and biotic transformations is a prerequisite for using soil in the elimination of toxic compounds that have accumulated as a result of intended and unintended human activities. If one is interested in applying catalysts to control

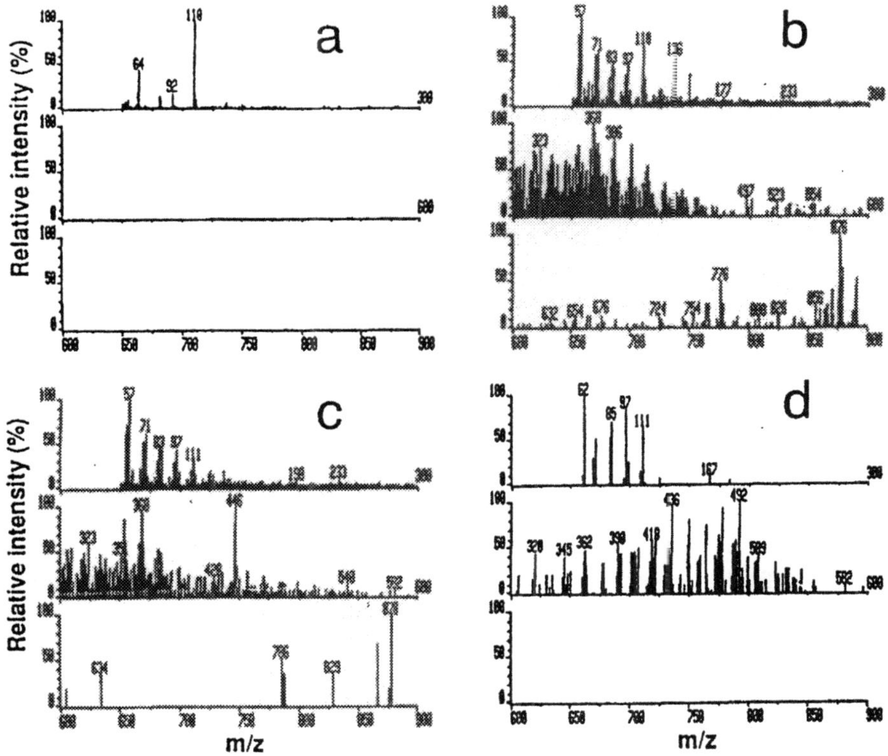

Figure 3. Mass spectra of the reaction products formed from catechol in the presence of tyrosinase and birnessite

pollution in the environment, future research must address the question of how to apply enzymes or abiotic catalysts (oxidizing agents) in natural habitats. It is very likely that in nature, biotic and abiotic catalysts interact in the transformation of natural and anthropogenic compounds. Consequently, considering the use of catalysts for environmental cleanup purposes, it will be of particular interest to study the simultaneous activities of various catalysts and to evaluate their combined effect on the oxidation of xenobiotics in the soil.

4. APPLICATION OF NMR TO DETERMINE PHYSICAL OR CHEMICAL BINDING IN THE SOIL

The formation of hybrid products during model reactions with humus components strongly supports the assumption that substituted phenols and anilines may bind covalently to humus. To provide evidence for this hypothesis, studies have recently been conducted using ^{13}C-NMR spectroscopy, which is a non-destructive technique allowing direct observation of the ^{13}C label and its associations. Using the NMR approach, Hatcher et al. [28] investigated the enzymatic incorporation of ^{13}C-labeled 2,4-dichlorophenol into humic acid and found that the ^{13}C label generated sharp distinct signals in specific regions of the spectrum. From the position of the signals in the spectra, it was concluded that 2,4-dichlorophenol was covalently bound to humic acid through ester, phenolic ether, and carbon-carbon linkages.

Thorne et al. [29] used ^{15}N-NMR spectroscopy to demonstrate covalent binding of ^{15}N-labeled aniline to humic acid when the two components were dissolved in water and stirred for 5 days at pH 6. The changes in the chemical shifts of the ^{15}N atom indicated that binding was due to nucleophilic addition reactions of aniline with the quinone or carbonyl groups typically present in humic substances. The labeled chemical was incorporated in the form of anilinohydroquinone, anilinoquinone, anilide, imine and heterocyclic nitrogen. The latter comprised over 50% of the bound amines.

^{15}N-NMR spectroscopy was used in recent studies [30] to demonstrate the bound nature of ^{15}N-trinitrotoluene (TNT) in humic acid or humin isolated from a soil/compost system designed to promote binding. NMR spectra indicated that nitro groups of TNT had been reduced to amino groups that were subsequently involved in the formation of covalent bonds with soil organic matter. Mass balance of added ^{14}C-TNT indicated that 82.5% of the originally added TNT was unextractable after an anaerobic/aerobic incubation. Analysis of the extractable fraction revealed that the remaining ^{14}C-TNT was strongly associated with soluble humic material and no free ^{14}C-TNT was present.

The NMR technique helped obtain evidence for covalent binding of secondary amines through mechanisms that do not involve the nitrogen atom. In the study of Haider et al. [31], for instance, the fungicide anilazine [4,6-dichloro-*N*-(2-chloro-phenyl)-2,3,5-triazine-2-amine] labeled with ^{14}C and ^{13}C in the triazine ring was found to be immobilized in soil through ligand exchange. The immobilization occurred within a few days (about 80% of the initial radioactivity) under both sterile and non-sterile conditions, indicating the abiotic character of the binding reaction. The NMR spectra of humic acid extracted from soil together with the bound fungicide revealed that the chlorine substituents located at the C-4 and C-6 positions were removed from the triazine ring and replaced by the oxygen-containing functional groups of soil organic matter. This exchange resulted in the formation of strong ether and ester linkages between the dehalogenated molecule of anilazine and the humic matrix.

^{13}C-NMR spectroscopy was also used to evaluate soil-bound residues of the fungicide cyprodinil [4-cyclopropyl-6-methyl-2-phenylaminopyrimidine], which was labeled with ^{13}C and ^{14}C in either the phenyl or the pyrimidyl ring [32, 33]. The transformation of cyprodinil by aerobic microorganisms was a prerequisite to binding; negligible immobilization of the fungicide took place under anaerobic or sterile conditions. The rates of binding determined for the phenyl- and pyrimidyl-labeled compound were essentially the same. However, significant differences occurred in the rates of $^{14}CO_2$ evolution upon the incubation of ^{14}C-cyprodinil in soil (a 3-fold difference in favor of the phenyl label) and in the rates of plant uptake (a 4-fold difference in favor of the pyrimidyl label). The differences suggested significant changes in the chemical structure of the studied compound, including cleavage between the phenyl and pyrimidyl rings. The possibility of cleavage and separate binding of both aromatic moieties to soil was confirmed by the results of ^{13}C-NMR analysis of the humic acid fraction dissolved in 1% NaOD [32, 33].

5. CONCLUSIONS

The presence of immobilized xenobiotics in soil will continue to be the source of environmental concern. A frequently asked question is whether bound organic compounds are safely "locked" in the soil matrix or rather serve as a pool of toxic residues that may be released in the future. Immobilization is presently believed to be the dominant cause of declining bioavailability of xenobiotics as they age in soil. In general, however, information on the effects of xenobiotic binding is scarce, and more experimental data are needed to provide regulatory agencies with the factual background for risk reassessment.

REFERENCES

1. Calderbank, A., 1989. The occurrence and significance of bound residues in soil. Rev. Environ. Contam. Toxicol. 108, 71-103.
2. Bollag, J.-M., 1992. Decontaminating soil with enzymes Environ. Sci. Technol. 26, 1876-1881.
3. Alexander, M., 2000. Aging, bioavailability, and overestimation of risk from environmental pollutants. Environ. Sci. Technol. 34, 4259-4265.
4. Pignatello, J.J., 1989. Sorption dynamics of organic compounds in soils and sediments. *In:* Sawhney, B.L., Brown, K. (Eds.), Reactions and Movement of Organic Chemicals in Soil. SSSA Spec. Publ. 22, Madison, WI, pp. 45-80.
5. Koskinen, W.C., Harper, S.S., 1990. The retention process: Mechanisms. *In:* Cheng, H.H. (Ed.), Pesticides in the Soil Environment. SSSA Book Series, Madison, WI, pp. 51-77.
6. Khan, S.U., 1978. Kinetics of hydrolysis of atrazine in aqueous fulvic acid solution. Pestic. Sci. 9, 39-43.
7. Hatzinger, P.B., Alexander, M., 1995. Effect of aging of chemicals in soil on their biodegradability and extractability. Environ. Sci. Technol. 29, 537-545.
8. Helling, C.S., Krivonak, A.E., 1978. Physicochemical characteristics of bound dinitroaniline herbicides in soils. J. Agric. Food Chem. 26, 1156-1163.
9. Katan, J., Fuhremann, T.W., Lichtenstein, E.P., 1976. Binding of ^{14}C-parathion in soil: A reassessment of pesticide persistence. Science 193, 891-894.
10. Roberts, T.R., Klein, W., Still, G.G., Kearney, P.C., Drescher, N., Desmoras, J., Essen, H.O., Aaharonson, N., Vonk, J. 1984. Non-extractable pesticide residues in soils and plants. Pure Appl. Chem. 56, 945-956.
11. Sjoblad, R.D., Bollag, J.-M., 1981. Oxidative coupling of aromatic compounds by enzymes from soil microorganisms. *In:* Paul, E.A., Ladd, J.N. (Eds.), Soil Biochemistry. Vol. 5. Marcel Dekker Inc., NY, pp. 113-152.
12. Bollag, J.-M., Minard, R.D., Liu, S.-Y., 1983. Cross-linkage between anilines and phenolic humus constituents. Environ. Sci. Technol. 17, 72-80.
13. Wang, T.S.C., Huang, P.M., Chou, C.-H., Chen. J.H., 1986. The role of soil minerals in the abiotic polymerization of phenolic compounds and formation of humic substances. *In:* Huang, P.M., Schnitzer, M. (Eds.), Interactions of Soil Minerals with Natural Organics and Microbes. SSSA Special Publication No. 17, Soil Science Society of America, Madison, WI, pp. 251-281.

14. Klibanov, A.M., Alberti, B.N., Morris, E.D., Felshin, L.M., 1980. Enzymatic removal of toxic phenols and anilines from waste waters. J. Appl. Biochem. 2, 414-421.
15. Maloney, S.E., Maule, A., Smith, A.R.W., 1993. Purification and preliminary characterization of permethrinase from a pyrethroid-transforming strain of *Bacillus cereus*. Appl. Environ. Microbiol. 59, 2007-2013.
16. Dec, J., Bollag, J.-M., 1994. Dehalogenation of chlorinated phenols during oxidative coupling. Environ. Sci. Technol. 28, 484-490.
17. Bollag, J.-M., Liu, S.-Y., Minard, R.D., 1980. Cross-coupling of phenolic humus constituents and 2,4-dichlorophenol. Soil. Sci. Soc. Am. J. 44, 52-56.
18. Bollag, J.-M., Liu, S.-Y., 1985. Copolymerization of halogenated phenols and syringic acid. Pestic. Biochem. Physiol. 23, 261-272.
19. Sarkar, J.M., Malcolm, R.L., Bollag, J.-M., 1988. Enzymatic coupling of 2,4-dichlorophenol to stream fulvic acid in the presence of oxidoreductases. Soil Sci. Soc. Am. J. 52, 688-694.
20. Simmons, K.E., Minard, R.D., Bollag, J.-M., 1988. Oxidative coupling and polymerization of guaiacol, a lignin derivative. Soil Sci. Soc. Am. J. 52, 1356-1360.
21. Simmons, K.E., Minard, R.D., Bollag, J.-M., 1989. Oxidative co-oligomerization of guaiacol and 4-chloroaniline. Environ. Sci. Technol. 23, 115-121.
22. Pal, S., Bollag, J.-M., Huang, P. M., 1994. Role of abiotic and biotic catalysts in the transformation of phenolic compounds through oxidative coupling reactions. Soil Biol. Biochem. 26, 813-820.
23. Twigg, M.V., 1989. Catalyst Handbook. 2nd ed. Wolfe Publishing Ltd., London.
24. Moore, J.W., Pearson, R.G., 1988. Kinetics and Mechanism. Third edition. John Wiley & Sons, New York.
25. Park, J.-W., Dec, J., Kim, J.-E., Bollag, J.-M., 1999. Effect of humic constituents on the transformation of chlorinated phenols and anilines in the presence of oxidoreductive enzymes or birnessite. Environ. Sci. Technol. 33, 2028-2034.
26. Naidja, A., Huang, P.M. Bollag, J.-M., 1998. Comparison of reaction products from the transformation of catechol as catalyzed by birnessite or tyrosinase. Soil Sci. Soc. Amer. J. 62, 188-195.
27. Ruggiero, P., Dec, J., Bollag, J.-M., 1996. Soil as a catalytic system. In: Stotzky, G. Bollag, J.-M. (Eds.), Soil Biochemistry. Vol. 9. Marcel Dekker, NY, pp. 79-122.
28. Hatcher, P.G., Bortiatynski, J.M., Minard, R.D., Dec, J., Bollag, J.-M., 1993. Use of high-resolution ^{13}C NMR to examine the enzymatic covalent binding of ^{13}C-labeled 2,4-dichlorophenol to humic substances. Environ. Sci. Technol. 27, 2098-2103.
29. Thorne, K.A., Pettigrew, P.J., Goldenberg, W.S., Weber, E.J., 1996. Covalent binding of aniline to humic substances. 2. ^{15}N NMR studies of nucleophilic addition reactions. Environ. Sci. Technol. 26, 107-116.
30. Strynar, M., Dec, J., Bollag, J.-M., 2001. Anaerobic/aerobic composting of soil contaminated with 2,4,6-trinitrotoluene. (in preparation).
31. Haider, K., Spiteller, M., Wais, A., Fild, M., 1993. Evaluation of the binding mechanism of anilazine and its metabolites in soil organic matter. Intern. J. Environ. Anal. Chem. 53, 125-137.

32. Dec, J., Haider, K., Benesi, A., Rangaswamy, V., Schäffer, A., Plücken, U., Bollag, J.-M., 1997. Analysis of soil-bound residues of the ^{13}C-labeled fungicide cyprodinil by NMR spectroscopy. Environ. Sci. Technol. 45, 514-520.
33. Dec, J., Haider, K., Rangaswamy, V., Schäffer, A., Fernandes, E., Bollag, J.-M., 1997. Formation of soil-bound residues of cyprodinil and their plant uptake. J. Agric. Food Chem. 45, 514-520.

Developments in Soil Science, Volume 28B
Editors: A. Violante, P.M. Huang, J.-M. Bollag and L. Gianfreda

ENZYMES IN SOIL: PROPERTIES, BEHAVIOR AND POTENTIAL APPLICATIONS

L. Gianfreda, M.A. Rao, F. Sannino, F. Saccomandi and A. Violante

Dipartimento di Scienze Chimico-Agrarie, Università di Napoli Federico II
Via Università 100, 80055 Portici, Italy

Soil enzymatic proteins are organic components unique to the habitat in which they function. As such, they show peculiar properties and behavior and have a potential for environmental purposes. Different features of enzyme activities in soil have been addressed by using enzymatic systems that simulate those naturally occurring in the soil environment. Synthetic enzyme-soil colloid complexes, prepared under laboratory conditions, have been utilized. In particular, three enzymes (invertase, phosphatase and urease, which have a key role in carbon-, phosphorus- and nitrogen-cycles), as well as pure clays (montmorillonite), dirty clays (i.e., montmorillonite covered by different amounts of OH-Al species), and tannic acid, have been used as representative of inorganic and organic soil components. The influence of metal (Al, Fe and Mn) species on the interactions between enzymes and tannic acid has been investigated as well. Synthetic enzyme complexes showed catalytic and stability properties that depended on the matrix and the bonding mechanism involved in their formation. Usually, a reduced activity and a kinetic behavior conforming to Michaelis-Menten kinetics were observed. Various responses in terms of thermal and proteolytic stability were observed. The influence of clays, humic substances and pesticides on the same enzymes, both in free and/or immobilized forms, was studied. In addition, the effect of pesticides on the activity of invertase, phosphatase and urease in natural soils was investigated. Various responses were found. The state of the enzyme (i.e., if free or complexes with clay, organo and/or organo-mineral matrices) strongly affected the response to the presence of a pesticide. Similarities between results obtained with synthetic systems and whole soils allow some behavior to be hypothesized. Finally, the potentiality of a laccase, an oxidoreductive enzyme, to detoxify a phenol-polluted system has been addressed. The enzyme showed a satisfying efficiency towards different phenolic molecules when acting in the presence of complex system (phenol-soil colloids suspensions).

1. INTRODUCTION

Soil is a particular and peculiar ecosystem in which a high abundance and diversity of biota coexist and develop several biochemical reactions. The intense biological and biochemical activity occurring in soil led Quastel [1] to claim that soil may be considered a biological entity, i.e., a living tissue.

All biochemical transformations occurring in soil are dependent on or related to the presence of enzymes. In soil, enzymes are present as a composite of various intracellular and extracellular enzymatic components, derived from animal, microbial and vegetal sources [2, 3]. Their existence in soil may occur under different "states" or categories that considerably influence their performance and properties. Several categories have been identified and different classifications of enzymes in soils have been provided by Kiss et al. [4], Skujins [5] and Burns [6]. A common category, present in all classifications, is the so-called "immobilized enzymes", which includes enzymatic proteins bound to inorganic and organic soil colloids and shows particular kinetic and stability properties.

Natural events and, more markedly, anthropogenic activities (e.g., industrial activities, release of toxic compounds, uncontrolled utilization of natural resources), may have a drastic impact on the environment and the quality of life. Among the ecosystems, soil can be considered one of the more affected by this impact.

Since soil enzymatic activities are capable of transforming both endogenous and non-endogenous inorganic and organic soil components, they may represent a direct measurement of the biological potential of a soil. Therefore, the determination of soil enzymatic activity changes, induced by natural and anthropogenic events [7], may provide an useful tool to assess soil status, its quality and productivity [8]. Indications on the inherent and endogenous ability of a soil to oppose and reduce adverse effects of polluting compounds might be also obtained.

Enzymes in soil may be also involved in the transformation and removal of xenobiotic substances [9]. Intracellular and extracellular hydrolases, such as esterases, acylamidase, and phosphatases, are able to catalyze the hydrolysis of several pollutants, such as pesticides, into less toxic products [10]. Removal of other pollutants of a phenolic nature, such as phenols and aromatic amines, may be accomplished by phenoloxidases that can convert these chemicals to water-insoluble polymers or bind them to soil organic matter [9, 10]. This capability renders phenoloxidases potential agents in the decontamination of polluted soils.

The aim of this chapter is to give a general picture of investigations performed by our research group on some aspects of enzymes in soil, namely a) enzymes in soil, as an example of "naturally immobilized enzymes", b) enzymes in soil, as affected by natural and anthropogenic factors, and c) enzymes as potential detoxifying agents of polluted systems, including soils, with specific reference to phenolic pollutants.

2. ENZYMES IN SOIL AS AN EXAMPLE OF "NATURALLY IMMOBILIZED ENZYMES"

Soil may be considered a harsh habitat for the survival of enzymatic proteins because several potential denaturating factors (e.g., microbial and enzymatic degradation, chemical-mediated denaturation) are usually present. Soil is, however, able to display a catalytic and enzymatic activity, even in the absence of living organisms. These two apparently contradictory situations have led soil enzymologists to assume that the majority of enzymatic activities in soil exist as stable complexes with inorganic and organic soil colloids. As reported by Gianfreda and Bollag [7], these enzymes can be regarded as "naturally immobilized enzymes" and as such, they exhibit altered catalytic, kinetic and stability properties. Therefore, enzymatic activities present in soil may be considered unique to the habitat, i.e., soil, in which they function and they may be labelled "soil enzymes."

Several findings have demonstrated that clay minerals, organic matter and organo-mineral complexes are involved in the formation of immobilized enzymes in soil [2, 11-15].

Two basic strategies commonly have been used to study soil enzymes and to understand the relationships between enzymes in soil and their clay or humic supports. The first strategy is basically founded on the isolation, purification, and characterization of active enzymatic fractions from soils. Although various experimental procedures and several biochemical techniques were used for the extraction, separation, and purification of these enzymes, very few successful examples are reported in the literature [16].

The second strategy, mainly used in our investigations, is to study the properties and the behavior of synthetic complexes, prepared *in vitro* by simulating interactions between enzymes and the main soil components (e.g., clays, humic substances and organo-mineral complexes). Different chemical and physical mechanisms involved in the complexation process are also considered.

2.1. Clay-enzyme complexes

Numerous studies support the concept that clays are significant elements in the formation of immobilized enzymes in soil [12, 13, 15]. Moreover, as demonstrated by Stotzky [17], the binding of proteins on clay minerals decreases to various extents the availability of these macromolecules as energy sources for microorganisms.

In several soils, such as those characterized by acidic pHs, and sediments, short-range–ordered Al or Fe precipitation products, as well as "dirty clays," i.e., clay minerals coated partially or totally by polymeric species or non-crystalline oxides of aluminum and iron, are particularly abundant, and they, rather than "pure clays," play an important role in the adsorption and surface-catalyzed processes involving enzymatic proteins.

Our studies dealt with the adsorption of invertase (I), urease (U) and phosphatase (P) on pure and dirty clays [18-20]. The three enzymes are particularly important because they are key enzymes in the carbon-, nitrogen-, and phosphorus-cycles.

According to several reports [12, 13, 21], our main results demonstrated that the amount of enzymes adsorbed, as well as the resulting properties of the formed complexes, were strictly influenced by the nature and properties of both enzymes and clays. Our studies were particularly devoted to examining the influence of the coating of a pure montmorillonite with different amounts of OH-Al species on the adsorption process. The affinity of enzymes for the phyllosilicates, as well as the activity of the adsorbed enzymes, was strongly influenced by the partial coating of montmorillonite with hydrolytic products of aluminum [18-20].

Table 1 reports the values of X_m and K constants, as defined in the Langmuir equation referred to invertase and urease adsorption isotherms on a montmorillonite (M), a chlorite-like complex, an Al(OH)x-montmorillonite complex (AM) loading 18 mmol of Al per gram of montmorillonite, and a non-crystalline Al hydroxide (A). Both invertase and urease were adsorbed on montmorillonite with a higher binding energy and adsorptive intensity as compared to AM and A.

A negative effect by OH-Al species coated on M surfaces was also observed on the amount of urease adsorbed on the clays [19]. Indeed, by increasing the Al species loadings on montmorillonite surfaces from 3 to 18 meq g^{-1} clay, the percentage of adsorbed urease decreased from 82 to 69% [19].

Similar results were obtained by Naidja et al. [22] when tyrosinase was adsorbed on a montmorillonite with different levels of hydroxyaluminum coatings. The authors found that hydroxy-Al coatings onto montmorillonite decreased the ability of this latter to adsorb the enzyme. The presence of polymeric oxyhydroxides of Fe (III) on montmorillonite, illite and kaolinite did not influence strongly the extent of β-lactoglobulin adsorption, whereas lower

Table 1
Langmuir X_m and K values and correlation coefficients for the adsorption of invertase (pH 4.65) and urease (pH 7.00) by the clay minerals (M=montmorillonite, AM=$Al(OH)_x$-montmorillonite complex loading 18 mmol of Al per gram of M, AL=aluminum hydroxide)

Clay minerals	Invertase			Urease		
	X_m	K	r^2 *	X_m	K	r^2 *
M	6.9	0.28	0.96	19.8	168	0.94
AM	23.5	0.57	0.97	32.5	13	0.87
AL	--	--	--	9.0	17	0.82

*As determined by a computed linear regression analysis.

levels of catalase were usually loaded by the dirty clays as compared to that adsorbed onto the corresponding Ca-saturated phyllosilicates [23].

A completely different behavior was observed with phosphatase. It was not possible to describe classical equilibrium adsorption isotherms because a complete adsorption of the proteins, in terms of mg, took place on all clays except on A [20].

Many explanations have been proposed to account for the different capabilities of clay minerals to adsorb proteins [21]. Usually, correlations between surface (external and internal) area, cation exchange capacity (CEC) or surface charge density of the clays and amounts of proteins adsorbed have been attempted. Our results [18, 19] did not support a significant correlation between the surface area of the clay minerals and the amount of invertase and urease adsorbed (Table 1). Furthermore, since no interstratification of both enzymes in the interlamellar spaces of montmorillonite was revealed by X-ray diffraction analysis [18, 19], it was not possible to ascribe the greatest amounts of the enzymes adsorbed onto M to its larger specific area. By contrast, a detectable intercalation was observed with acid phosphatase adsorbed on Na-saturated montmorillonite [20]. The gradual shift to higher d-spacing values of tyrosinase-Ca-montmorillonite complex observed with increasing amounts of enzyme indicates that tyrosinase molecules were interstratified into the interlamellar species of montmorillonite [22] as well.

The different behavior observed with tyrosinase/acid phosphatase and invertase/urease could be ascribed to too small amounts of both invertase and urease used in adsorption experiments [18, 19]. The different size of protein molecules involved in the phenomenon has also to be considered. Acid phosphatase and tyrosinase are small proteins (100 and 120 kDa, respectively) compared to invertase (270 kDa) and urease (480 kDa). Therefore, their interstratification into the layers of the phyllosilicate might be much easier. When the montmorillonite interlayers are partially preoccupied by hydroxy-Al ions, a steric hindrance to the intercalation of enzyme molecules may occur. Usually, at a coating level > 3.0 mol Al or Fe kg^{-1} clay, hydroxy-Al or Fe ions constitute a stable blocking material that occupies the interlamellar spaces in the montmorillonite and prevents any intercalation of macromolecules [24]. This is confirmed by the progressive decrease of the interlayering of acid phosphatase in Al(OH)x montmorillonite complexes that is observed when increasing the amount of Al species [20]. In particular, no significant changes in the d-spacing of $Al(OH)_x$-montmorillonite complexes (loading 5 and 6 mmol g^{-1} Al montmorillonite) were measured by X-ray diffraction analysis after adsorption of tyrosinase and acid phosphatase [20, 22]. Similarly, catalase and β-lactoglobulin were unable to penetrate the spaces between the spheres and rods of the polyhydroxides of Fe on the surface of the clays [23].

Usually, adsorption of enzyme on clay mineral surfaces produces a decrease, even substantial, of enzyme activity. An increased stability of enzyme molecules to proteolytic attack and thermal denaturation usually derives [3, 15]. The presence on clay mineral surfaces of Fe, Al and Mn oxides and hydroxides may modify the interactions occurring between enzymes and adsorbent supports; a subsequent effect on the activity of the adsorbed enzyme may result [18, 20, 25, 26].

The presence of OH-Al species on montmorillonite surfaces resulted in lower activity levels for invertase and urease, whereas a higher activity was measured with phosphatase [18-20]. This effect was much more evident when urease and phosphatase were adsorbed on Al-montmorillonite complexes, loading increasing quantities of OH-Al species. The higher the amount of Al species on M surfaces, the lower the activity of urease and the higher that of phosphatase (Table 2).

Table 2
Residual activity (%) of urease and phosphatase on montmorillonite (M) and $Al(OH)_x$-montmorillonite complexes (AM) containing different amounts of aluminum (3, 9 and 18 m-equiv Al g^{-1} montmorillonite)

	Residual activity (%) *	
Clay	Urease	Phosphatase
M	71	18
AM_3	62	21
AM_9	62	32
AM_{18}	48	54

*As a percentage of the specific activity of the free enzymes.

This behavior could be ascribed to several factors that include properties of both proteins and matrix, and environmental conditions under which the adsorption process occurs. Among these factors, the type of the adsorbing matrix plays a predominant role. Although it is not easy to define precisely what kind of interaction occurs between enzyme and clay and how enzymatic molecules may change upon adsorption, it seems clear that the enzymatic active site may be involved in the phenomenon and, as consequence, its native structural conformation may be modified.

The different response of adsorbed urease and phosphatase to the presence of OH-Al species on M surfaces was attributed to a) different arrangements held by the proteins on clays, and b) different kinds of electrostatic interactions occurring between the enzyme molecules and the binding sites on clays, differently charged at the pHs at which adsorption studies were carried out [19, 20].

2.2. Organo- and organo-mineral-enzyme complexes

In soil, associations between enzymes and organo-mineral complexes are expected to be more common than clay- or humus-enzyme complexes. Indeed, several soil constituents (e.g., pure clays, phenolic compounds, inorganic components such as Fe, Mn and Al oxides and hydroxides, primary minerals and OH-Al polymers) may interact with each other, giving rise to stable organo-mineral complexes. Enzymatic proteins, present in the environment, may be involved in one or more than one of these complexes or may successively interact with them.

In these processes, monomers, polymers and non-crystalline products of aluminum and iron, and manganese compounds play an important role. These latter may facilitate the interaction between organic molecules and clay interfaces and produce organo-mineral complexes, showing peculiar physico-chemical properties and reactivity toward nutrients, organics and pollutants [24]. Furthermore, they may accelerate the abiotic polymerization of simple and complex aromatic compounds (e.g., mono- and polyphenols, starting humic-like precursors), favoring the formation of polymeric aggregates [12, 27].

To better simulate the interactions possibly occurring in soil between enzymes and soil components, our studies have been extended to investigate the formation and the properties of enzymatic complexes obtained by interaction of enzyme molecules with organic and organo-mineral components [20, 26, 28, 29].

The studies still have been performed with the enzymes (invertase, urease and phosphatase) previously used in clay-protein investigations. Montmorillonite and tannic acid (T) were used as representative of pure clay and humus-like constituents, respectively. The influence of soluble and insoluble species of Fe, Mn and Al on the formation and catalytic properties of enzymatic complexes was examined as well.

The formation of active tannate-enzyme complexes was influenced by both the contact time and the ratio between tannic acid and enzyme. For example, tannic acid suddenly and tightly interacted with urease molecules to form soluble urease-tannate complexes that were less active at higher tannic acid/urease (T/U) ratios [29]. By increasing the contact time, a further decay of activities and the appearance of solid material in the mixture suggested that enzyme-tannic acid complexes with lower apparent activity were formed.

A similar behavior, both in terms of residual activity of enzymatic complexes and formation of insoluble materials, was observed with phosphatase. In this case, the phenomenon was much more complicated. Two different trends were observed between 0 and 2 h and greater than 2 h incubation time (Figure 1). In the first time interval (0-2 h), a more pronounced decay of activity was observed; after this time, a leveling off to a long plateau was detected. This suggests that a complex mechanism of two steps with different kinetics is involved in the interaction between phosphatase and tannic acid [20]. In both steps, the higher the T/P ratio, the more pronounced was the phenomenon (Figure 1). The complex behavior was explained by assuming that in the first time interval (0-2 h), a higher number of free enzyme molecules were removed from the solution and immobilized with tannic acid with a progressive diminution of the activity. Then, a lower number of molecules were involved and a much smaller decrease was observed.

Interaction between invertase and tannic acid occurred at a lower rate than in the case of phosphatase and urease [28]. In fact, in the first time of incubation (0-2 h), no marked decreases of activity were measured whatever the initial T/I ratio. By increasing incubation time and the ratio between tannic acid and invertase (up to 50 T/I w:w), a progressive reduction of enzymatic activity was measured. The browning of the solution with time and the formation of visible precipitate formed particularly in the presence of higher amounts of

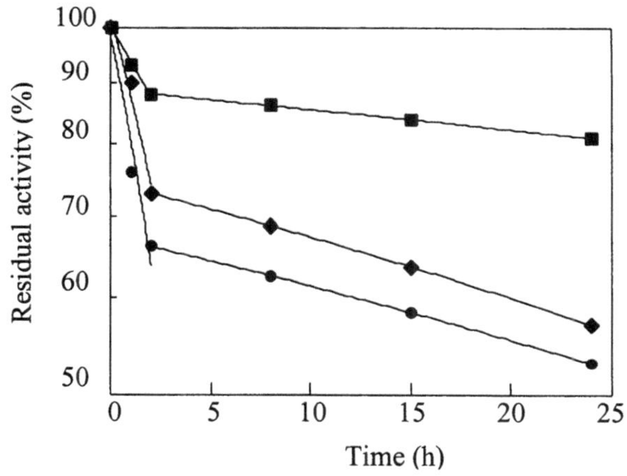

Figure 1. Influence of the incubation time and amount of tannic acid (T) on the activity of acid phosphatase (P). [■,T/P (w:w) = 5; ◆, T/P = 10; ●, T/P = 20]. As modified from [20].

tannic acid (> 25 mg) suggest the formation of a different population of soluble and insoluble invertase-tannic acid complexes.

These results demonstrate that tannic acid gave rise to stable interactions with the three enzymes. However, the association of phosphatase with tannic acid produced complexes with higher activity levels than with urease and invertase. A higher number of phosphatase than invertase and/or urease molecules were removed from solution and immobilized in tannic acid-enzyme complexes as well.

These results could be explained by hypothesizing that two phenomena occur during interaction between tannic acid and enzymatic molecules. Tannic acid may have either interacted directly with enzyme molecules through one or more sites present on the surface of proteins [30] (see below), or it may have undergone an oxidative polymerization process, involving enzyme molecules [31]. The formation of organo-enzyme complexes showing different physical, chemical, and catalytic features may have resulted.

When the complexation phenomenon between enzymes and tannic acid occurred in the presence of aluminum, iron and/or manganese species, a higher number of enzymatic molecules were involved in the complexes and showed greater activity levels [20, 26, 28, 29].

Table 3 reports the activities of phosphatase- and urease-tannic acid complexes obtained in the presence of OH-Al species and Fe and/or Mn, both as soluble (ions) and insoluble (oxides) species. Detectable increases of residual activity were measured for both the enzymes in the presence of OH-Al species and Fe_2O_3. Significantly higher activities were shown also by tannic acid-invertase complexes when OH-Al species were involved [28].

The presence of Fe and Mn ions positively changed the activity of urease-tannic acid complexes (increases up to 46 and 24%, respectively), but no similar effects were observed with acid phosphatase.

The beneficial influence displayed by OH-Al species on the activity of enzyme-tannic acid-OH-Al-complexes can be explained by various phenomena that can all occur in the presence of OH-Al species. Direct interactions between tannic acid and OH-Al species may

Table 3
Residual activities (%) of enzyme-tannic acid complexes formed in the presence or in the absence of OH-Al species, Fe and Mn ion and oxides

	Residual activity (%) *	
	Phosphatase	Urease
Tannic acid	33	19
Tannic acid + Al(OH)	48	64
Tannic acid + Fe^{3+}	33	46
Tannic acid + Mn^{2+}	36	24
Tannic acid + Fe_2O_3	58	60
Tannic acid + MnO_2	33	26

* Percentages of the initial enzymatic units added.

favor an enhanced precipitation of soluble tannic acid-enzyme complexes; OH-Al polymers may promote the formation of Al-precipitation products with different sorptive properties for enzyme molecules; OH-Al-tannate complexes may have sites with different bonding properties; and, finally, OH-Al polymers may facilitate the possible flocculation of more active tannate-enzyme complexes.

The different effects shown by Fe and Mn oxides on phosphatase- and urease-complexes could be accounted for by different phenomena promoted by the two oxides. Both Fe and Mn oxides are able to promote the abiotic polymerization of phenolic compounds, Mn oxide being a stronger polymerizing catalyst [27, 32, 33]. Moreover, the oxides may favor the adsorption of organic molecules such as phenolic compounds and proteins.

MnO_2 was probably most effective in polymerizing tannate molecules, and promoted the formation of insoluble and large polymers, which may have entrapped many less active phosphatase and urease molecules. Conversely, Fe oxide may have favored the adsorption of enzyme molecules and/ or small tannate-enzyme complexes on its particles, rather than entrapping the enzyme in large tannate polymers. Exposed adsorbed enzyme molecules would have given higher activities.

A general further increase of activity of enzymatic insoluble complexes was observed when tannic acid, enzymes, soluble and insoluble metal species, and montmorillonite were simultaneously present and interacted with each other.

For example, the activities of urease- and phosphatase-Fe^{3+}-tannate–montmorillonite complexes were 40 and 84% higher than those of complexes obtained in the absence of montmorillonite, respectively. The capability of both Fe ions and/or OH-Al species for acting as bridges between tannic acid molecules and montmorillonite surfaces [24] probably facilitated the attachment of active soluble enzyme-tannic acid complexes on clay surfaces.

2.3. Catalytic behavior

Immobilized enzymes usually display an altered catalytic behavior due to the "heterogeneous nature" achieved by catalysis brought about by a "static enzyme" [34-38].

Several findings have demonstrated that enzymes adsorbed to clay minerals or entrapped in organo and/or organo-mineral complexes usually show reduced residual activities and altered kinetic parameters. They usually have lower V_{max} and higher K_m values than their free counterparts. Furthermore, shifts in the apparent pH optimum, as well as an increased stability toward physical, chemical, and biological denaturating agents, are observed [2, 11-15, 34, 37].

Different factors may contribute to affect the behavior of an enzyme upon immobilization on solid supports, namely: i) possible conformational changes in enzyme structure; ii) steric limitations on the penetration of substrate into the less accessible enzymatic active site; iii) limitations to the diffusion of both substrates and products from/to the bulk solution to/from the immobilized enzyme; and iv) microenvironmental effects due to the presence of negative or positive charges on the surface of the immobilizing support, leading to changes in the local pH (i.e., the pH at the support interface) [35, 38 and references therein].

Points i), ii) and iii) are responsible of the changed values of the kinetic parameters (i.e., K_m and V_{max}), while point iv) may account for the shifts in the apparent pH optimum usually shown by immobilized enzymes.

Furthermore, the prevention of temperature-promoted conformational changes of enzyme molecules by either fixation of these latter to solid supports or their embedding in tridimensional structures justifies the thermostability of enzymes adsorbed and/or entrapped on/in inorganic and organic matrices. A protection against the attack by proteases, large proteic molecules, may also derive as the result of restricted access of these latter to immobilized enzyme molecules.

As previously reported, the results achieved in our studies with invertase, urease, and phosphatase immobilized on clays and/or tannic acid and organo-mineral complexes were generally in agreement with the "immobilization-decline activity" rule. In all cases, the three enzymes showed Michaelis-Menten kinetics [18, 19, 26, 29, 39].

Both invertase and phosphatase adsorbed on pure and dirty clays displayed reduced V_{max} values and increased K_m values as compared to free enzymes. Lower V_{max} and K_m values, i.e., an increased affinity for the substrate, were, instead, measured for urease adsorbed on all clays (Table 4). Although unusual, similar findings were obtained for other immobilized enzymes [40, 41]. A reduced affinity (higher K_m) was also detected for phosphatase and urease immobilized on tannic acid (Table 4). The presence of Fe ions and OH-Al species during the complexation of the enzymes with tannic acid influenced the structural characteristics of the resulting complexes, whose catalytic properties approximated those of the free enzymes (Table 4).

In agreement with the presence of a microenvironmental effect by the support on the activity of immobilized enzymes [35], pH-activity profiles of phosphatase-OH-Al-montmorillonite complex showed a shift toward the acidic region. An opposite shift toward more alkaline pHs was displayed by urease- and phosphatase-tannic acid complexes.

In the first case, phosphatase was adsorbed on a positively charged carrier, such as montmorillonite covered by O-Al species. Consequently, a locally higher pH (more alkaline) due to an accumulation of hydroxyl ions in the microenvironment at the support interface occurred and surrounded the immobilized enzyme. A more acidic pH in the bulk solution was then necessary to achieve the local optimum pH for the immobilized enzyme.

An opposite situation was present when the enzymes were immobilized on tannate molecules that are negatively charged. In this case, the microenvironment was more acid (lower pH) than the bulk solution and a more alkaline pH was required.

Table 4
Michaelis-Menten kinetic constants V_{max} and K_m of phosphatase and urease immobilized on clays and tannic acid, in the presence and absence of Fe^{3+} ions and O-Al species

Complexes	Urease		Phosphatase	
	V_{max} (%)	K_m(mmol L^{-1})	V_{max} (%)	K_m (mmol L^{-1})
Free- Enzyme (E)	100	24	100	0.3
E-M *	71	8	20	0.4
E-AM_3*	64	14	58	0.6
E-AL*	15	18	45	0.7
E-T	19	37	33	0.6
E-T- Fe^{3+}	45	25	33	1.0
E-T-Al(OH)	64	24	48	0.3

*see Table 1.

The stability to both thermal and proteolytic denaturation of the three enzymes was variably affected by the immobilization on the inorganic, organic and organo-mineral matrices.

Invertase and phosphatase increased their thermal stability after adsorption on pure and dirty clays, whereas a reduced sensitivity to temperature was observed for urease. Clay-enzyme complexes showed a general low stability to proteolysis. No substantial improvements of thermal and proteolytic stabilities were measured when the enzymes interacted with tannic acid and inorganic species.

Some exceptions to this behavior were, however, found. As compared to the free enzyme, the complexes phosphatase-tannate-OH-Al species showed a gain in activity of about 45 and 55% after thermal deactivation at 60°C and 2 h exposure to a protease (proteinase K), respectively. A similar beneficial effect on the stability of phosphatase-tannate complex was observed when Fe ions were involved in the complex.

A general conclusion can be drawn from these results. When invertase, urease and phosphatase were adsorbed on clay surfaces or interacted with tannic acid in the presence of OH-Al or iron species, an improvement of their catalytic features was achieved, thereby confirming the beneficial effect of these species on enzyme function.

3. ENZYMES IN SOIL AS AFFECTED BY NATURAL AND ANTHROPOGENIC FACTORS

The production, activity, catalytic behavior and persistence of enzymes in soil may be affected by natural and anthropogenic factors [7 and references therein]. Natural factors include a) seasonal and geographical (i.e., seasonal changes, geographical location, in situ distribution), and b) physico-chemical (i.e., content of organic carbon and nitrogen, moisture and salinity, humus, clays minerals, biomass) properties of soil. In addition to natural factors, several anthropogenic causes may induce significant changes in the activity of enzymes in soil. Some agricultural practices (e.g., use of vegetative cover, amendments and managements, application of fertilizers and pesticides) usually designed to affect soil productivity may adversely or positively affect enzyme activities in soil.

3.1. Effect of clay minerals and humic compounds

Inhibitory effects induced by clay minerals on enzyme activity have been widely studied [15 and references therein]. Inhibition studies performed with pure and dirty clays, used in the adsorption studies above described (see Section 2.1.), have demonstrated that clays displayed various inhibitory effects on the activity of invertase, urease, and phosphatase. The results were generally in agreement with those usually reported in the literature and exhaustively reviewed by Kiss et al. [15].

As described in detail by Rao et al. [20], classical inhibition tests were performed by measuring the kinetics of phosphatase in the presence of increasing amounts of clays and by evaluating the variations of the kinetic parameters K_m and V_{max}. Table 5 reports the effects of two different amounts of clay minerals on the activity of acid phosphatase [20]. Results reported in Table 5 clearly indicate that clays inhibited acid phosphatase activity. The higher the amount of clay, the lower the enzymatic activity. Furthermore, the inhibition was depressed or partially removed when montmorillonite was covered by Al(OH)x species. The inhibitory effect decreased from the highest value (>70%) measured in the presence of montmorillonite to about a half-value (35%) when the clay was covered by the greater amount (18 meq of Al per gram of clay) of OH-Al species (Table 5).

The higher K_m values observed in the presence of clays indicate that a lower affinity for the substrate arises. Similar experiments carried out with invertase [18] demonstrated that the addition of increasing amounts of clays decreased the V_{max} values of invertase without affecting the K_m values, which remained unchanged at any quantity of clay added. A detailed kinetic analysis was performed by using various kinetic models [42]. The behavior of kinetic parameters of invertase was explained by a partial non-competitive model, whereas no well-defined inhibitory mechanism was demonstrated for phosphatase. A mixed inhibition effect might, however, be suggested by the simultaneous decrease of V_{max} and increase of K_m values (Table 5).

A direct inhibitory effect on enzyme activity by humus compounds and related substances is also well documented by several reports [14, 43-46].

The activity of invertase, urease and phosphatase was inhibited to various extents in the presence of tannic acid. Many findings have demonstrated that tannins may inhibit enzymatic proteins as the result of direct and indirect effects on the enzymatically mediated reactions [7, 14, 47].

Table 5
Effect of clay minerals (M, A, AM_3, AM_9, AM_{18}) on acid phosphatase activity

Clay	Amount (mg)	K_m (mmol L^{-1})	V_{max} (%)
Free Phosphatase (P)	--	0.63	100
M	40	0.60	27
M	80	1.04	27
AL	40	1.07	97
AL	80	1.66	66
AM_3	40	1.80	98
AM_3	80	1.64	51
AM_9	40	1.14	67
AM_9	80	2.51	68
AM_{18}	40	2.06	84
AM_{18}	80	2.06	65

As modified from [20].

Table 6 reports tannic acid concentration that caused a 50% reduction of the enzymatic activity of invertase, urease and phosphatase, as determined at saturating substrate concentration [48]. Tannic acid showed the strongest inhibition effect on acid phosphates, whereas invertase offered the greatest resistance to the inhibitor [48].

The different response of the three enzymes to the influence of tannic acid (Table 6) could be partly explained by the different nature of enzymatic proteins involved. Invertase is a glycoprotein (with more than 50% of glucidic residues) and may offer a higher resistance to complexation phenomenon by phenolic compounds [49]. By contrast, urease, lacking any carbohydrate part, is much more sensitive to inhibition by tannic acid.

Table 6
Tannic acid concentrations which gave 50% inhibition of invertase, urease, and acid phosphatase activities

Enzyme	Tannic acid (mM)
Invertase	$4.8\ 10^{-1}$
Urease	$5.0\ 10^{-2}$
Acid Phosphatase	$2.0\ 10^{-2}$

From [48].

The role of the carbohydrate chain in the stabilization of glycoenzymes is well known [16, 50]. Although this glycosid part is not necessary for enzyme function, it can be very important for enzyme action because it contributes to the stability of the active proteic component against proteolytic activity, microbial denaturation and oxidizing agents. A glycosid vesicle enveloping the amino acid chain probably forms. The stabilizing effect of enzymes by carbohydrates is supported by the fact that the majority of enzymes extracted from soil (considered true free enzymes) are present in soils strongly associated with a carbohydrate portion, chemically bound to the peptidic part through the R groups of specific amino acids [50].

The highest inhibition by tannic acid toward acid phosphatase (Table 6), a glycoprotein with a 17% carbohydrate content, indicates a greater affinity of tannic acid for the enzyme and seems to contrast what was affirmed above. A great affinity between tannic acid and acid phosphatase was also demonstrated by Hsu et al. [51], who used tannic acid in the purification of acid phosphatase from potato. The content of carbohydrate residues in the phosphatase molecule was probably too low to efficiently prevent the interaction with phenolic compounds.

3.2. Effect of pesticides

Pesticides have been and are widely used in agriculture. As a consequence, they have been found in numerous natural systems, and soil is probably the ecosystem mostly exposed to their influence. Once in soil, pesticides, which are involved in several processes such as retention, transformation and transport, may affect soil biochemical and microbial processes [52]. Because soil enzymes have an important role in soil fertility and nutrient cycling, several studies have been devoted to investigating the effect of pesticides on enzyme activities in soil [53 and references therein].

Most of these studies have been devoted to evaluating short- and long-term effects of pesticides on dehydrogenase, which is active only in viable cells, and then it is a measurement of the active biomass in soil. Other enzymes widely studied are phosphatase and urease. A timely review, providing an exhaustive examination of studies performed with pesticides and several enzymes in soil, has been published by Shaffer [53]. One of the most important conclusions drawn by the author was that contradictory results (i.e., activation, inhibition or no-effect) are often reported about the action of a given pesticide on the activity of a certain enzyme in soil.

These confusing results might be partly due to the extreme complexity of enzymes in soil. As reported above, enzymes in soil exist as the combination of several enzymatic categories or "states," every one having its "own" catalytic features and characteristics. As a consequence, each enzymatic state may respond in a different way and with a different intensity to the action of a given factor.

To overcome some of these drawbacks, we have addressed the study of pesticide effects on enzyme activities in soil with a different methodological approach, making use of simplified systems.

The effects of four pesticides, three herbicides (atrazine, paraquat, glyphosate) and one insecticide (carbaryl), on the activity of invertase, urease and phosphatase was studied under the following "states":

a) free enzymes, which should simulate the fractions of enzymes free in soil solution, if any;
b) synthetic clay-, organo- and organo-clay-enzyme complexes, which should simulate enzyme-soil colloid associations; and
c) enzyme in whole soils, which represent natural systems.

The synthetic enzymatic complexes used in our studies were enzyme-montmorillonite (I-M, U-M, P-M), enzyme-tannic acid (I-T, U-T, P-T), and enzyme-Al(OH)x-tannic acid-montmorillonite complexes (I-ATM, U-ATM, P-ATM). All of them were previously studied in terms of catalytic and stability properties and have been previously discussed (see Sections 2.1, 2.2 and 2.3).

Tables 7 and 8 summarize the results obtained with free and immobilized enzymes exposed to a given amount of each pesticide. The experiments were performed under laboratory conditions and are described in detail in Gianfreda et al. [54-56] and Sannino and Gianfreda [57]. It was not possible to determine the influence of paraquat on phosphatase activity because the pesticide caused a complete interference with the analytical determination of *p*-nitrophenol, produced by phosphatase action on *p*-nitrophenylphosphate, used as the substrate in the activity assay [54].

Looking at the results reported in Tables 7 and 8, some conclusions can be drawn:

a) the "state" of the enzyme affected the response of the enzymatic activity to the presence of a pesticide; and
b) no generalizations can be made, being enzyme or pesticide specific.

For instance, the activity of free invertase was markedly activated by glyphosate and paraquat, whereas the performance of urease appears to be uninfluenced by both pesticides (Table 7). An activation effect by the two pesticides and by paraquat only was also detected on invertase and urease immobilized on montmorillonite, respectively (Table 8). By contrast, both pesticides did not affect or caused slight decreases in the activity of invertase and urease immobilized on tannic acid. A measurable increment of the activity of U-ATM was instead caused by glyphosate.

The response of phosphatase to the presence of glyphosate was completely different. All the enzymatic complexes were inhibited by the herbicide in a range of 20-40%, as compared to the activity in the absence of pesticide (Table 8).

When considering the influence of the two pesticides atrazine and carbaryl, the effect of methanol, used as solvent, on the activity of enzymes was also significant (Tables 7 and 8). A general inhibition of enzyme activity was measured when methanol was present in the activity assay and a partial removal of this effect occurred with atrazine and/or carbaryl. The extent of activity decrease by methanol, as well as the partial recovery of activity in the presence of atrazine and carbaryl, varied with both the involved enzyme and its state. For example, carbaryl strongly inhibited free invertase, but the inhibitory effect significantly decreased when the enzyme was immobilized on ATM (Tables 7 and 8).

The heterogeneity of the above results clearly indicates the complexity of the systems under investigations. When attempts are made to give some reasonable explanations, a few important considerations should be made.

The enzymatic systems investigated differ substantially from each other, namely:

a) when free enzymes interact with pesticides, the system is homogeneous and direct interactions at the molecular level may occur.
b) when enzymes are present in an immobilized form, the system is heterogeneous and several interactions may be generated: <u>direct interactions</u> of the pesticide with the enzyme molecules, which could have varied their catalytic feature upon immobilization, but also <u>indirect effects</u>, deriving from the interactions between pesticide and inorganic and organic supports. Competition phenomena between immobilized enzyme and pesticide molecules could occur, which could result in a possible release of free enzymatic molecules from matrices.

The marked increase in urease activity of the U-M complex with paraquat (Table 8) was explained as due to an indirect effect of the pesticide rather than to a direct effect on catalysis by the enzyme. The adsorption of paraquat, present as cation, on external and internal montmorillonite surfaces was considered as a possible cause of release of free urease molecules from specific sites, with a partial recovery of activity, lost during the immobilization process [55].

Complex responses were obtained when the effects of the four pesticides were studied in natural systems, i.e., whole soils [55-57]. Contrasting results (increases, decreases and no effects) with glyphosate and paraquat were measured for the activity of invertase in eleven soils sampled in different sites of Campania (Italy) and characterized by different physico-chemical properties and initial invertase activity levels [56].

Table 7
Effect of atrazine, carbaryl, glyphosate and paraquat on the activity of free enzymes

	Relative activity (%)				
Enzymes	Glyphosate	Paraquat	Atrazine	Carbaryl	Methanol
Invertase	1.80	1.23	0.71	0.53	0.64
Urease	0.93	0.96	0.58	0.53	0.64
Phosphatase	1.05	---	0.61	0 76	0.74

As modified from [57].

Table 8
Effect of atrazine, carbaryl, glyphosate and paraquat on the activity of immobilized enzymes

	Relative activity (%)				
Immobilized enzymes	Glyphosate	Paraquat	Atrazine	Carbaryl	Methanol
Invertase-M	1.10	1.10	0.62	0.67	0.51
Urease-M	1.03	1.79	0.54	0.60	0.57
Phosphatase-M	0.61	---	0.87	0.86	0.87
Invertase-T	0.98	0.85	0.54	0.61	0.43
Urease-T	0.99	0.95	0.67	0.62	0.73
Phosphatase-T	0.79	---	0.85	0.80	0.85
Invertase-ATM	0.93	0.92	0.55	0.82	0.43
Urease-ATM	1.20	1.03	0.58	0.65	0.63
Phosphatase-ATM	0.59	---	0.70	0.69	0.71

As modified from [57].

When investigations were extended to 22 additional soils, still belonging to the same sampling area, no univocal responses were observed in all soils [57]. For example, activity increments ranging from +4% to as much as 204% were measured for soil invertase in the presence of glyphosate but also decreases by an average of 30% were observed in other soils.

A similar behavior was observed with soil urease, which was studied also in soils from different sites of Italy. In this case, increments, decrements and no effects were measured for soil urease, in the presence of paraquat and glyphosate, as well [55].

A complex opposite behavior was observed with phosphatase. Glyphosate inhibited on average by 50% the activity of the enzyme in all soils [57].

Contradictory results were also observed when enzymes in whole soils were exposed to methanol, atrazine and/or carbaryl. Methanol exhibited variable inhibitory effects (from 5 to 80%) on invertase activity in all soils and the presence of atrazine further contributed to depressing the activity of the enzyme [57].

A complete different behavior, except for very few cases, was shown by soil urease and phosphatase in the presence of methanol and methanol-pesticide solutions. In all soils tested (about 33), urease activity was strongly enhanced by methanol and the addition of carbaryl and/or atrazine further increased the activity of the enzyme in several soils. A significant activation effect (ranging from 3 to 547%) by methanol was also observed for soil phosphatase, whereas magnification or depression of the methanol effect was measured with atrazine and carbaryl [57].

To explain our results we addressed the problem with two different approaches:

a) we attempted to find reliable and predictable relationships, if any, between observed effects and physico-chemical properties of soil. In this case sample and multiple regression analyses were used;
b) we compared the results measured in whole soils to those obtained with the synthetic enzymatic systems (Table 8).

Although several physical properties, generally known as strictly related to enzyme activities, were considered, no reliable relationships between soil properties and pesticide effects were found for all enzymes.

Conversely, some satisfying conclusions were reached by comparing the results obtained with the synthetic enzymatic systems and whole soils. For example, a similar behavior (i.e., decrease of activity) was observed when phosphatase, free, immobilized on solid supports or in whole soils, was exposed to glyphosate. A decrease of activity was also caused by methanol for all invertase categories (i.e., free, immobilized on solid supports and in whole soils).

By contrast, the increases of soil urease and phosphatase activities, observed in the presence of methanol, are not supported by the results obtained with the model enzymatic systems (Table 7 and 8). While the inhibitory effect of methanol on free and immobilized enzymes is more predictable because it is supported by much experimental evidence on enzyme behavior (it is widely reported in the literature that organic solvents inhibit the activity of free and immobilized enzymes [58 and references therein]), the marked increase in urease and phosphatase activity of soils by the solvent is more difficult to explain.

Methanol probably promoted an increase in cell wall permeability as well as a cellular lysis of microbial cells present in soil [59]. An increase of accessibility of the substrate molecules to intracellular enzymes, with a subsequent increase of activity, may have resulted [55].

Some of the above considerations could appear as simple speculations and in our opinion further reliable experimental evidence should validate these suppositions. However, our methodological approach has been an alternative attempt to address this difficult topic.

Direct, unequivocal evidence of the interactions between pesticides and all the possible forms of enzymes existing in a soil have yet to be presented and is a challenge for future research activities.

4. ENZYMES AS DETOXIFYING AGENTS OF POLLUTED SOILS

Several natural and anthropogenic activities have contributed to increase the pollution of the environment. Consequently, the atmosphere, the hydrosphere and many ecosystems have become polluted by a large variety of xenobiotic compounds. Soils and groundwater are natural and preferred sinks for contamination, and their pollution represents an important concern for human and environmental health. Recently, a growing interest has been dedicated to the search for alternative methodologies to clean-up polluted environments.

Bioremediation is considered the use of biologic agents to transform pollutants into innocuous products. The main biological agents of bioremediation are microorganisms, plants and enzymes. A large number of bacteria and fungi are able to transform, partly or completely, many xenobiotic organic compounds. A complete mineralization of pollutants to simple nontoxic compounds may occur and this is often the desired outcome in bioremediation. The use of microorganisms to remediate a polluted environment may, however, present some limitations, such as adsorption of microbial cells on solid particles, diffusion constraints, competition effects by indigenous microorganisms, and inhibition by toxic compounds.

A possible strategy for overcoming these drawbacks is the use of enzymes as free or immobilized forms. A review aimed at examining the potential of isolated enzymes for the detoxification of organic pollutants has been recently published by Gianfreda and Bollag [60]. Basically, three classes of enzymes were considered by the authors, i.e., hydrolases, dehalogenases, and oxidoreductases, including enzymes highly efficient in the transformation of several pollutants.

Among oxidoreductases, laccases, belonging to the group of phenoloxidases, are able to oxidize phenolic and non-phenolic, aromatic compounds through oxidative coupling reactions that lead to the formation of polymeric products of increasing complexity.

Laccases are multi-copper proteins and represent a very interesting group of ubiquitous, oxidoreductive enzymes that seem to offer great potential for the decontamination of terrestrial and aquatic environments [61]. They are widespread in plants and fungi [62], and their physiological function is mainly connected with lignin transformation [63-65].

Recently, Gianfreda et al. [61] and Xu [66] have reviewed several aspects of these enzymes. Gianfreda et al. [61] have dedicated attention to the current knowledge of origins, distribution, production, molecular and kinetic properties of laccases. The basic and applied areas of the enzymes have been also considered. Recent progress in properties, enzymology, and applications of laccases has been reported by Xu [66].

One of the most interesting aspects of these enzymes is their capability to transform relatively recalcitrant compounds if additional, so-called enhancing co-substrates or proper redox mediators are present [61 and references therein, 63, 67, 68].

The catalytic mechanism of laccase action has been thoroughly reviewed by Yarapolov et al. [68]. Laccases reduce oxygen to water and, typically, oxidize a phenolic substrate. Usually, the reaction is characterized by a two-step mechanism, involving initially the formation of a free radical from an organic substrate and the subsequential reduction of O_2. The organic substrate (which may be represented by various compounds, including *o*, *p*-diphenols, aminophenols, polyphenols, polyamines, lignin, some inorganic ions and aryl diamines) is oxidized [69, 70], and molecular oxygen is reduced directly to water by a four-electron mechanism without involvement of any mediator [68]. Organic pollutants are transformed by laccase into highly reactive radicals, which may spontaneously polymerize, giving rise to less soluble, high-molecular-weight compounds [71], or which may interact with and become incorporated into organic soil colloids, thus being removed from the ecosystem [72].

We have dedicated our attention to evaluating and quantifying the performance of a fungal laccase in the presence of phenolic compounds of different origins [63, 73]. The enzyme was isolated and purified in high yield as a constitutive form from the fungus *Cerrena unicolor* [63]. Kinetic studies demonstrated that the enzyme has catalytic features (i.e., kinetic parameters, pH- and temperature-profiles, sensitivity to deactivation agents) almost similar to those observed with laccases from different origins [63].

Table 9 reports the removal of some substances of aromatic character, which behave as pollutants of natural environments. In particular we have considered some humus-like precursors - resorcinol, pyrogallol and gallic acid - and phenolic compounds-derivatives of common pesticides such as 2,4-dichlorophenol (2,4-DCP), the most common intermediate of the herbicide 2,4-D, *p*-tyrosol (and its *m*- and *o*-isomers), hydroxytyrosol, methylcatechol, and catechol, which are present at high concentration in olive oil mill waste waters. Catechol may also be considered a humic precursor or a usual product of the microbial degradation of several phenoxy compounds.

The enzyme readily transformed all the phenolic compounds, thus confirming its broad substrate specificity [63, 73] According to findings reported in the literature [61 and references therein], the oxidizing capability of laccase from *C. unicolor* depended either on the position or the nature of the substituents on the aromatic ring, as well as on the number of OH-groups present. For instance, meta-substituted isomers (i.e., resorcinol and *m*-tyrosol) were less oxidized as compared to their corresponding *o*- and *p*-isomers (i.e., catechol. *o*- and *p*-tyrosol, respectively). Similarly, compounds typified by three –OHs (i.e., pyrogallol and gallic acid) were more readily oxidized than those having two or one OH groups (in the order listed) (Table 9).

Time course experiments performed on 2,4-DCP and humic precursors (catechol, resorcinol, gallic acid and pyrogallol) demonstrated that the times necessary to achieve 50% transformation of the initial substrate concentration ($T_{50\%}$) conformed to the previous behavior. The greater the number of –OH groups on the aromatic ring, the lower the value of $T_{50\%}$, and the higher the rate of compound oxidation. For example, for 2,4-DCP a $T_{50\%}$ (6.5×10^4 seconds) greater by an order of magnitude was measured as compared to that observed for catechol (2.9×10^3 s), thus indicating the higher suitability of catechol as substrate in laccase assay [73].

An effective and useful utilization of an enzyme for practical applications requires that the final catalyst localization and its possible reusability be known. Thus, the residual activity levels of laccase, after its initial catalytic action on added substrates, were evaluated (Table 9). Standard assay conditions (i.e., ABTS as synthetic substrate, a pH optimum of 3.5 and a temperature of 25°C [63]) were used to measure laccase activity.

All treatments showing a measurable removal of phenolic substrates usually displayed a reduction of laccase activity. Except for the cases of pyrogallol and gallic acid, an inverse relationship between residual laccase activity and phenol removal was observed. Specifically, the higher the substrate removal, the lower the residual activity, with an apparent complementarity between them (Table 9).

Table 9
Substrate removal and residual enzymatic activity (ABTS assay) after 24 h incubation

Substrate	$-(OH)_x$	Substituent	Substrate Removal (%)	Residual Laccase Activity (%)
2,4-DCP	1	-2 Cl	66	34
p-Tyrosol	1	$-CH_2CH_2OH$	73	26
o-Tyrosol	1	$-CH_2CH_2OH$	28	72
m-Tyrosol	1	$-CH_2CH_2OH$	11	88
Catechol	2	-	100	0
Resorcinol	2	-	40	76
Methylcatechol	2	$-CH_3$	76	24
Hydroxytyrosol	2	$-CH_2CH_2OH$	86	18
Pyrogallol	3	-	100	89
Gallic acid	3	-COOH	98	83

From [73].

To explain this apparent correspondence between phenol removal and the extent of the reduction of laccase residual activity, alternative hypotheses were made [73]. The enzyme may have been directly inhibited by phenolic substrates or by polymeric products derived from phenol oxidation. Results discussed in Section 3.1 on the interactions between tannic acid and invertase, urease and phosphatase seem to support this hypothesis. Furthermore, direct inhibitory tests performed with the same phenolic compounds used as substrates for laccase activity demonstrated that all the compounds inhibited the activity of the enzyme. The number of –OH groups and other substituents and the concentration of the compounds affected the extent of inhibition of laccase activity [73]. Investigations performed with peroxidase, another oxidoreductive enzyme capable of oxidizing phenolic compounds, showed that the enzyme was inactivated during the transformation of various phenols [74]. An interaction between the enzyme's active site and phenoxy radicals, formed during the oxidative reaction, was proposed [75-76].

The incorporation and/or the adsorption of laccase molecules into or onto the newly formed polymers was also considered [63]. The incorporation of enzyme molecules into insoluble reaction products may cause a lower (or absence of) accessibility of the substrate to the bound laccase. A lower or absence of enzymatic activity may result.

In experiments performed at increasing 2,4-DCP concentrations [63], we demonstrated that a decrease of laccase activity with increasing 2,4-DCP transformation occurred. For each 2,4-DCP concentration, the residual laccase activity still present in the samples was measured by using ABTS as substrate (Figure 2).

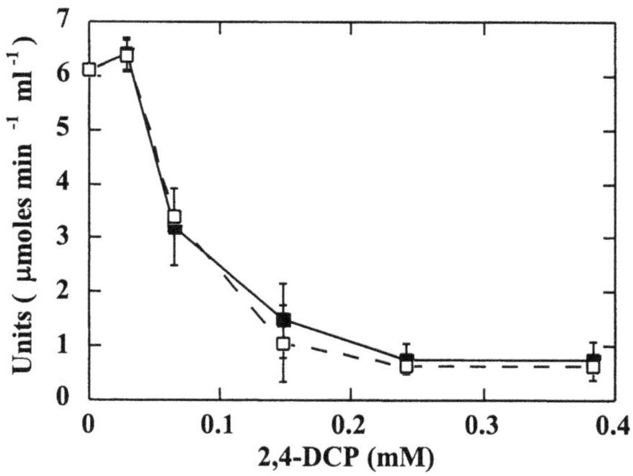

Figure 2. Residual laccase activities of filtered (continuous line) and unfiltered (dashed line) 2,4-DCP–enzyme mixture. The standard deviation ranged from 0.00 to 0.70.

The data reported in Figure 2 show that the higher the 2,4-DCP concentration, the more pronounced was the loss of enzymatic activity. A direct inhibitory effect on laccase activity by 2,4-DCP was ruled out because activity tests performed on samples filtered with filters that specifically adsorb 2,4-DCP gave activity values similar to those obtained without sample filtration (dashed line in Figure 2). Therefore, the results shown in Table 9 and Figure 2 seem to infer that the oxidative transformation of phenolic compounds gave rise to the formation of polymeric products on which laccase molecules may be immobilized and progressively removed from the solution.

No interactions between laccase and organic colloids of aromatic character were demonstrated in investigations performed with laccase and humic acids [77, 78]. The oxidative capability of the enzyme towards 2,4-DCP was evaluated in the presence of two humic acids (HA_1: extracted from lignite, and HA_2: extracted from oxidized coal [79]). The two humic acids, characterized by a high aromaticity, were representative of organic colloids usually present in soil. Experiments in the presence of the chlorite-like complex AM_{18}, representative of an inorganic colloid, were also carried out.

Table 10 summarizes 2,4-DCP transformation and residual laccase activity, measured at two different amounts of laccase and in the presence of the two humic acids and the chlorite-like complex.

When the inorganic colloid was present in the reaction mixture, a remarkable influence, depending on laccase concentration, was detected on the oxidation of 2,4-DCP and resulted in a dramatic decrease of the residual enzymatic activity. At the lowest laccase concentration

(10 mg ml^{-1}), a complete adsorption phenomenon of enzymatic molecules on clay surfaces predominated with a concomitant loss of activity of the adsorbed enzyme (Sannino et al, unpublished results). No transformation of 2,4-DCP took place. By increasing the laccase concentration by four times (40 mg ml^{-1}), the enzyme was only partially adsorbed and the complete oxidation of 2,4-DCP by the enzyme remaining in the solution occurred. By contrast, a slightly increased removal of 2,4-DCP and no effect on the residual enzymatic activities were detected with humic substances.

When both AM_{18} and humic acids were simultaneously present in the mixture, the effect of AM_{18} presence predominated over that of humic acids, and results similar to those obtained with AM_{18} alone were observed (Table 10).

Table 10
2,4-DCP removal and laccase residual activity in the presence of inorganic and organic soil colloids

	2,4-DCP removal (%)		Laccase residual activity (%)	
	Laccase (mg ml^{-1})			
	10	40	10	40
L	56	100	61 (100)*	50 (100)*
L+AM_{18}	0	93	0 (0)	31 (23)
L+HA_1	59	100	65 (90)	75 (100)
L+HA_2	57	100	64 (97)	52 (100)
L+AM_{18}+HA_1	8	95	0 (0)	55 (50)
L+AM_{18}+HA_2	17	94	0 (0)	48 (55)

* Values in parentheses are the residual enzymatic activities of control samples without 2,4-DCP.

The results obtained with laccase and clay minerals are in accordance with those previously described in Section 2.1 and further demonstrate that adsorption of enzymes on clays usually results in a significant decrease of activity of the adsorbed enzyme.

A completely different behavior was observed when laccase was bound to a clay support by chemical binding. Ruggiero et al. [80] and Gianfreda and Bollag [81] showed that a laccase from *Trametes versicolor*, upon immobilization on a montmorillonite or a kaolinite by a covalent immobilization method, retained high residual activity and elevated stability to thermal denaturation, storage and proteolytic attack. Furthermore, the authors evaluated the capability of free and immobilized laccase to detoxify soils polluted by 2,4-dichlorophenol. Investigations were performed under laboratory conditions with soils differing in their chemical and physical properties. Both free and immobilized laccase displayed catalytic activity in soil environments and for repeated incubation cycles. Laccase was able to remove up to 60% of the initial substrate concentration in the presence of soil with a low content of organic matter [81]. By increasing the organic matter content, an inhibitory effect on the activity of free and immobilized enzyme was observed. The inhibition, however, decreased after each incubation cycle, indicating that it was, at least in part, due to soluble soil components easily extracted with substrate solution [81].

Several findings have demonstrated that laccase can be inhibited by various reagents [61, 65 and references therein]). Compounds, including small anions, metal cations, and sulfhydryl reagents, among others, are able to inhibit the activity of laccases from different origins to various extents.

An interesting inhibiting effect on laccase activity was reported by Filazzola et al. [73], who demonstrated that the polymerizing activity of laccase is influenced by some pesticides possibly present in soil. Laccase activity towards 2,4-dichlorophenol and catechol as substrates was measured in the presence of simazine and prometryn, two triazine herbicides, which are not substrates of laccase and may likely be present at phenol-contaminated sites. Both compounds inhibited laccase activity; the extent of the inhibition differed with the nature of the substrate used. Furthermore, higher residual laccase activities were measured when simazine was present in the reaction mixture, thus reflecting an induced decrease of the enzymatic polymerizing activity. The inhibitory behavior of simazine towards laccase was demonstrated also by Ruggiero and Radogna [82] who showed that triazine herbicides are able to reduce the activity of laccase-soil complexes.

The overall results so far discussed indicate that laccase, both free or immobilized on solid supports, might be a suitable enzyme to treat polluted environments. However, a serious concern derives from either the inhibitory effect observed with soil components and herbicides or the inactivation undergone by the enzyme after its catalytic action. Both of these factors might be drawbacks to the possible use of laccase for the removal of phenols from polluted systems.

5. CONCLUSIONS

The investigations briefly examined in this review were performed by using simplified enzymatic systems that should have simulated those naturally occurring in soil.

In particular, the overall results achieved have shown that:

- enzymes associated with matrices resembling inorganic, organic and organo-mineral soil colloids have characteristics and properties that are strongly affected by several factors. They include: the nature and type of immobilizing support; the bonding mechanism between enzyme and matrix; the inherent characteristics of the enzyme molecule; the conformational changes achieved by the native protein upon immobilization; and the microenvironmental effects due to the charge and nature of the support surface.
- the activity and performance of free and immobilized enzymes is affected by the presence of compounds likely present in soil, such as clays, humic materials, and pesticides. The responses are, however, significantly influenced by the "state" of the enzyme. In whole soils, a much more complicated situation arises and results not easily interpretable are obtained. These studies may have also provided useful information for the use of soil enzymes as bioindicators. Reliable indicators of soil health are necessary, when increased levels of crop production and simultaneously the remediation of the adverse effects of environmental pollution are desired. Although soil enzymes have been considered as possible useful indicators of soil health and quality, straightforward

relationships between soil enzymatic levels and soil biological potential have not been easily determined because of the complexity of soil enzymes.

- oxidoreductive enzymes can be a useful tool for the recovery and restoration of polluted systems, including soils. The investigations performed with laccase have demonstrated that for an effective use of this enzyme in environmental clean-up, improvement in the efficiency of the enzymatic process, as well as methods for efficiently prolonging the half-life of the enzyme, must be developed. Furthermore, a thorough understanding of the influence of inhibitor molecules during laccase application *in situ* should be required.

In our opinion, these results may have contributed to clarifying some aspects of the complexity of enzyme activities in soil.

REFERENCES

1. Quastel, J.H., 1946. Soil Metabolism. Royal Institute of Chemistry of Great Britain and Ireland. London.
2. Ladd, J.N., 1985. Soil enzymes. *In:* Vaugham, D., Malcom, R. E. (Eds.), Soil Organic Matter and Biological Activity. Martinus Nijhoff/Dr. W. Junk Publishers, Dordrecht, Netherlands, pp. 175-221.
3. Burns, R.G., 1978. Soil Enzymes, Academic Press, London.
4. Kiss, S., Dragan-Bularda, M., Radulescu, D., 1975. Biological significance of enzymes accumulated in soil. Adv. Agron. 27, 25-87.
5. Skujins, J.J., 1967. Enzymes in soils. *In*: McLaren, A.D., Peterson, G. H. (Eds.), Soil Biochemistry. Vol. 1. Marcel Dekker, NY, pp. 371-414.
6. Burns, R.G., 1982. Enzyme activity in soil: location and a possible role in microbial ecology. Soil Biol. Biochem. 14, 423-427.
7. Gianfreda, L., Bollag, J.-M., 1996. Influence of natural and anthropogenic factors on enzyme activity in soil. *In*: Stotzky, G., Bollag, J.-M. (Eds.), Soil Biochemistry. Vol. 9. Marcel Dekker, NY, pp. 123-194.
8. Dick, R.P., 1997. Soil enzyme activities as integrative indicators of soil health. *In*: Pankhurst, C.E., Doube, B.M., Gupta, V.V.S.R. (Eds.), Biological Indicators of Soil Health. CAB International, pp. 121-156.
9. Bollag, J.-M., 1992. Enzymes catalysing oxidative coupling reactions of pollutants. *In*: Sigel, H., Sigel, A. (Eds.), Metal ions in biological systems. Marcel Dekker, Inc., NY, pp. 206-217.
10. Nannipieri, P., Bollag, J.-M., 1991. Use of enzymes to detoxify pesticide-contaminated soils and waters. J. Envir. Qual. 20, 510-517.
11. McLaren, A.D., 1975. Soil as a system of humus and clay immobilized enzymes. Chem. Scripta 8, 97-99.
12. Burns, R.G., 1986. Interaction of enzymes with soil mineral and organic colloids. *In*: Huang, P.M., Schnitzer, M. (Eds.), Interactions of Soil Minerals with Natural Organics and Microbes. Soil Sci. Soc. Am., Madison, WI, pp. 429-451.
13. Boyd, S.A., Mortland, M.M., 1990. Enzyme interactions with clays and clay-organic matter complexes. *In*: Bollag, J.-M., Stotzky, G. (Eds.), Soil Biochemistry. Vol. 6. Marcel Dekker, NY, pp. 1-28.

14. Ladd, J.N., Butler, J.H.A., 1975. Humus-enzyme systems and synthetic organic polymer-enzyme analogs. *In*: Paul, E.A., McLaren, A.D. (Eds.), Soil Biochemistry. Vol. 3. Marcel Dekker, NY, pp. 143-194.
15. Kiss, S., Dragan-Bularda, M., Pasca, D., 1986. Activity and stability of enzyme molecules following their contact with clay mineral surfaces. Stud. Univ. Babes. Boyai Biol. 2, 3-29.
16. Tabatabai, M. A., Fu. M. 1992. Extraction of enzymes from soils. *In*: Stotzky, G., Bollag, J.-M. (Eds.), Soil Biochemistry. Vol. 7. Marcel Dekker, NY, pp. 197-227.
17. Stotzky, G., 1986. Influence of soil mineral colloids on metabolic processes, growth, adhesion, and ecology and microbes and viruses. *In*: Huang, P.M., Schnitzer, M. (Eds.), Interactions of Soil Minerals with Natural Organics and Microbes. Special Publ. No. 17. Soil Sci. Soc. Am., Madison, WI, pp. 305-428.
18. Gianfreda, L., Rao, M.A., Violante, A., 1991. Invertase (β-fructosidase): effects of montmorillonite, Al-hydroxide and $Al(OH)_x$-montmorillonite complexes on activity and kinetic properties. Soil Biol. Biochem. 23, 581-587.
19. Gianfreda, L., Rao, M.A., Violante, A., 1992. Adsorption, activity and kinetic properties of urease on montmorillonite, aluminum hydroxide and $Al(OH)_x$-montmorillonite complexes. Soil Biol. Biochem. 24, 51-58.
20. Rao, M.A., Palmiero, F., Gianfreda L., Violante, A., 1996. Interaction of acid phosphatase with clays, organic molecules and organo-mineral complexes. Soil Sci. 161, 751-760.
21. Theng, B.K.G., 1979. Formation and properties of clay-polymer complexes. Elsevier, NY.
22. Naidja A., Violante A., Huang, P.M., 1995. Adsorption of tyrosinase onto montmorillonite as influenced by hydroxyaluminum coatings. Clays Clay Min. 43, 647-655.
23. Fusi, P., Ristori, G.G., Calamai, L., Stotzky, G., 1989. Adsorption and binding of protein on "clean" (homoionic) and "dirty" (coated with Fe oxyhydroxides) montmorillonite, illite and kaolinite. Soil Biol. Biochem. 21, 911-920.
24. Violante, A., Gianfreda, L., 2000. Role of biomolecules in the formation and reactivity towards organics and plant nutrients of variable-charge minerals and organo-mineral complexes in soil. *In*: Bollag, J.-M., Stotzky, G. (Eds), Soil Biochemistry. Vol. 10. Marcel Dekker, NY, pp. 207-270.
25. Gianfreda, L., De Cristofaro, A., Rao, M.A., Violante, A., 1995. Kinetic behavior of synthetic organo- and organo-mineral-urease complexes. Soil Sci. Soc. Am. J. 59, 811-815.
26. Rao, M.A., Gianfreda, L. 2000. Properties of acid phosphatase-tannic acid complexes formed in the presence of Fe and Mn. Soil Biol. Biochem. 32, 1921-1926.
27. Huang, P.M., 1990. Role of soil minerals in transformations of natural organics and xenobiotics in soil. *In*: Bollag, J.-M., Stotzky, G. (Eds.), Soil Biochemistry. Vol. 6. Marcel Dekker, NY, pp. 29-115.
28. Gianfreda, L., Rao, M.A., Violante A., 1993. Interactions of invertase with tannic acid, hydroxy-aluminum (OH-Al) species or montmorillonite. Soil Biol. Biochem. 25, 671-677.
29. Gianfreda, L., Rao, M.A., Violante, A., 1995. Formation and activity of urease-tannate complexes affected by aluminum, iron, and manganese. Soil Sci. Soc. Am. J. 59, 805-810.

30. Haslam, E., 1988. Plant polyphenols (syn vegetable tannins) and chemical defense - a reappraisal. J. Chem Ecol. 14, 1789-1805.
31. Hagerman, A.E., 1989. Chemistry of tannin-protein complexation. *In*: Hemingway, R.W., Karchesy, J.J. (Eds.), Chemistry and significance of condensed tannins. Plenum Press, London, pp. 323-333.
32. Shindo, H., Huang, P.M., 1985. The catalytic power of inorganic components in the abiotic synthesis of hydroquinone-derived humic polymers. Appl. Clay Sci. 1, 71-81.
33. Shindo, H., Huang, P.M., 1992. Comparison of the influence of Mn(IV) oxide and tyrosinase on the formation of humic substances in the environment. Sci. Tot. Environ. 117/118, 103-110.
34. Goldstein, L., 1976. Kinetic behaviour of immobilized enzyme systems. *In*: Mosbach, K. (Ed.), Methods in Enzymology. Vol. 44, Academic Press, London, pp. 397-443.
35. Katchalski, E., Silman, I., Goldman, R., 1971. Effect of the microenvironment on the mode of action of immobilized enzymes. Adv. Agron. 27, 25-87.
36. McLaren, A.D., Packer, L., 1970. Some aspects of enzyme reaction in heterogeneous systems. Adv. Enzymol. 33, 245-308.
37. McLaren, A.D. 1978. Kinetics and consecutive reactions of soil enzymes. *In*: Burns, R. G. (Ed.), Soil Enzymes. Academic Press, London, pp. 97-116.
38. Nannipieri, P., Gianfreda, L., 1998. Kinetics of enzyme reactions in soil environments. *In*: Huang, P.M., Senesi, N., Buffle, J. (Eds.), Structure and Surface Reactions of Soil Particles. John Wiley & Sons, NY, pp. 449-479.
39. Rao, M.A., Violante A., Gianfreda, L., 2000. Interaction of acid phosphatase with clays, organic molecules and organo-mineral complexes: kinetics and stability. Soil Biol. Biochem. 32, 1007-1014.
40. Kobaiashi, Y., Aomine, S. 1967. Mechanism of inhibitory effect of allophane and montmorillonite on some enzymes. Soil Sci. Plant Nutr. 13, 189-194.
41. Makboul, H.E., Ottow, J.C.G., 1979. Michaelis constant (K_m) of acid phosphatase as affected by montmorillonite, illite and kaolinite clay minerals. Microb. Ecol. 5, 207-213.
42. Segel, I.H., 1975. Enzyme Kinetics. John Wiley & Sons, NY.
43. Ladd, J.N., Butler, J.H.A., 1969. Inhibitory effect of soil humic compounds on the proteolytic enzyme pronase. Aust. J. Soil Res. 7, 241-251.
44. Ladd, J.N., Butler, J.H.A., 1969. Inhibition and stimulation of proteolytic enzyme activities by soil humic acids. Aust. J. Soil. Res. 7, 253-261.
45. Ladd, J.N., Butler, J.H.A., 1970. The effect of inorganic cations the inhibition and stimulation of protease activity by soil humic acids, Soil Biol. Biochem. 2, 33-40.
46. Ladd, J.N., Butler, J.H.A., 1971. Inhibition by soil humic acids of native and acetylated proteolytic enzymes. Soil Biol. Biochem. 3, 157-60.
47. Müller-Wegener, U., 1988. Interaction of humic substances with biota. *In*: Frimmel, F.H., Christman, R.F. (Eds.), Humic substances and their role in the environment. John Wiley & Sons, NY, pp. 17-192.
48. Rao, M.A., Gianfreda, L., Violante, A., 1998. Interactions between tannic acid and acid phosphatase. Soil Biol. Biochem. 30, 111-112.
49. Strumeyer, D.H., Malin, M.J., 1970. Resistance of extracellular yeast invertase and other glycoproteins to denaturation by tannins. Biochem. J. 1, 88-99.
50. Mayaudon, J. 1986. The role of carbohydrates in the free enzymes in soil. *In*: Fuchsman, C.H. (Ed.), Peat and Water. Elsevier, NY, pp. 263-309.

51. Hsu, R.Y., Cleland, W.W., Anderson, L., 1966. Mechanism of action of the nonspecific phosphomonoesterase from potatoes. Biochemistry 5, 799-807.
52. Bollag, J.-M., Liu, S.-Y., 1990. Biological transformation processes of pesticides. *In*: Cheng, H.H. (Ed.), Pesticides in the Soil Environment: Processes, Impacts, and Modeling. Soil Science Society of America, Book Series, No. 2, Madison, WI, pp. 169-211.
53. Schäffer, A., 1994. Pesticide effects on enzyme activities in the soil ecosystem. *In*: Bollag, J.-M., Stotzky, G. (Eds.), Soil Biochemistry. Vol. 8. Marcel Dekker, NY, pp. 273-340
54. Gianfreda, L., Sannino, F., Violante, A., 1993. Influence of pesticides on the activity and kinetics of invertase, urease and acid phosphatase enzyme. Pest. Sci. 39, 245-250.
55. Gianfreda, L., Sannino, F., Ortega, N., Nannipieri, P., 1994. Activity of free and immobilized urease in soil: effects of pesticides. Soil Biol. Biochem. 26, 777-784.
56. Gianfreda, L., Sannino, F., Violante, A., 1995. Influence of pesticides on the behaviour of free, immobilized and soil invertase. Soil Biol. Biochem. 27, 1201.
57. Sannino, F., Gianfreda, L., 2001. Pesticide influence on soil enzymatic activities Chemosphere 22, 1-9.
58. Gianfreda, L., Scarfi, M. R., 1991. Enzyme stabilization: state of the art. Molec. Cell. Biochem. 199, 97-123.
59. Fenton, D.M., 1982. Solvent treatment for β-D-galactosidase release from yeast cells. Enzyme Microb. Technol. 4, 229-232.
60. Gianfreda, L., Bollag, J.-M., 2002. Isolated enzymes for the transformation and detoxification of organic pollutant. *In*: Burns, R.G., Dick, R. (Eds.), Enzymes in the Environment: Activity, Ecology and Applications. Marcel Dekker, NY, pp. 491-538.
61. Gianfreda, L., Feng, X., Bollag, J.-M., 1999. Laccases: a useful group of oxidoreductive enzymes. Bioremed. J. 3, 1-25.
62. Nerud, F., Misurcova, Z., 1996. Distribution of ligninolytic enzymes in selected white-rot fungi. Folia Microbiologika 41, 264-266.
63. Gianfreda, L., Sannino, F., Filazzola, M. T., Leonowicz, A., 1998. Catalytic behavior and detoxifying of a laccase from the fungal strain *Cerrena unicolor*. J. Molec. Catal. B: Enzymatic 14, 13-23.
64. Youn, H.-D., Hah, Y.-C., Kang, S.-O., 1995. Role of laccase in lignin degradation by white-rot fungi. FEMS Microb. Lett. 132, 183-188.
65. Rodriguez, A., Falcon, M.A., Carnicero, A., Peestelo, F., De la Fuente, G., Trojanowski, J., 1996. Laccase activities of *Penicillium chrysogenum* in relation to lignin degradation. Appl. Microbiol Biotechnol. 45, 399-403.
66. Xu, F., 1998. Recent progress in laccase study: properties, enzymology, production, and applications. *In*: Flickinger, M.C., Drew, S.W. (Eds.), Encyclopedia of Bioprocessing Technology: Fermentation, Biocatalysis, and Bioseparation. Vol. 4. John Wiley & Sons, NY, pp. 1545-1554.
67. Roper, J.-C., Sarkar, J. M., Dec, J., Bollag, J.-M., 1995. Enhanced enzymatic removal of chlorophenols in the presence of co-substrates. Wat. Res. 29, 2720-2724.
68. Yaropolov, A.-I.,. Skorobogat'ko, O.-V, Vartanov, S.-S., Varfolomeyev, S.-D., 1994. Laccase: properties, catalytic mechanism, and applicability. Appl. Biochem. Biotechnol. 49, 257-278.
69. Bollag, J.-M., Shuttleworth, K.L., Anderson, D.H., 1988. Laccase-mediated detoxification of phenolic compounds. Appl. Environ. Microbiol. 54, 3086-3091.

70. Faure, D., Bouillant, M.-L., Jacoud, C., Bally, R., 1996. Phenolic derivatives related to lignin metabolism as substrates for *Azospirillum* laccase activity. Phytochemistry 42, 357-359.
71. Roy-Arcand, L., Archibald, F.S., 1991. Direct dechlorination of chlorophenolic compounds by laccases from *Trametes* (Coriolus) *versicolor*. Enz. Microb. Technol. 13, 194-203.
72. Ruttimann-Johnson C., Lamar, R.-T., 1996. Polymerization of pentachlorophenol and ferulic acid by fungal extracellular lignin-degrading enzymes. Appl. Environ. Microbiol. 62, 3890-3893.
73. Filazzola, M.T., Sannino, F., Rao, M.A., Gianfreda, L., 1999. Effect of various pollutants and soil-like constituents on laccase from *Cerrena unicolor*. J. Environ. Qual. 28, 1929-1938.
74. Klibanov, A. M., Tu, T., Scott, K.P. 1983. Peroxidase-catalysed removal of phenols from coal-conversion waste waters. Science 221, 259-261.
75. Nicell, J.A., Bewtra, J.K., Biswas, N., Taylor, E., 1993. Reactor development for peroxidase catalyzed polymerization and precipitation of phenols from wastewaters. Wat. Res. 11, 1629-1639.
76. Nakamoto, S., Machida, N., 1992. Phenol removal from aqueous solutions by peroxidase-catalysed reaction using additives. Wat. Res. 1, 49-54.
77. Filazzola, M.T., Sannino, F., Gianfreda, L., 1998. The influence of biological agents on the behaviour of pollutants in soil. Fres. Envir. Bull. 7, 484-489.
78. Sannino, F., Filazzola, M.T., Violante, A., Gianfreda, L., 1999. Fate of herbicides as influenced by abiotic and biotic mechanisms. Chemosphere 39, 333-341.
79. Piccolo, A., 1993. Metodi chimici e spettroscopici per lo studio della reattività della sostanza organica dei suoli. *In*: Nannipieri, P. (Ed.), Ciclo della sostanza organica del suolo: aspetti agronomici, chimici, ecologici e selvicolturali. Patron Editore. Bologna, Italy, pp. 245-303.
80. Ruggiero, P., Sarkar, J.M. , Bollag, J.-M. 1989. Detoxification of 2,4-dichlorophenol by a laccase immobilized on soil or clay. Soil Sci. 147, 361-370.
81. Gianfreda, L., Bollag, J.-M., 1994. Effect of soils on the behavior of immobilized enzymes. Soil Sci. Soc. Am. J. 56, 1672-1681.
82. Ruggiero, P., Radogna, V.M., 1988. Humic acids-tyrosinase interactions as a model of soil humic-enzyme complexes. Soil Biol. Biochem. 20, 353-359.

Developments in Soil Science, Volume 28B
Editors: A. Violante, P.M. Huang, J.-M. Bollag and L. Gianfreda

RESPONSE OF ORGANIC MATTER POOLS AND ENZYME ACTIVITIES IN PARTICLE SIZE FRACTIONS TO ORGANIC AMENDMENTS IN A LONG-TERM FIELD EXPERIMENT

M.H. Gerzabek[a], G. Haberhauer[a], E. Kandeler[b], A. Sessitsch[a] and H. Kirchmann[c]

[a]Austrian Research Centers, Division of Life Sciences, A-2444 Seibersdorf, Austria

[b]University Hohenheim, Institute of Soil Science, Emil-Wolff-Straße 27, D-70599 Stuttgart, Germany

[c]Swedish University of Agricultural Sciences, Dept. of Soil Sciences, Box 7014, S-75007 Uppsala, Sweden

Physical soil fractionation may have significant potential to characterize soil organic matter pools of differing stability. This hypothesis was tested in a long-term field experiment on a Eutric Cambisol soil located in Ultuna, central Sweden. In 1956, the soil (0–20 cm depth) contained 15 g kg^{-1} of organic carbon, sand/silt/clay contents of 22%/41%/37% and had a pH of 6.6. Seven treatments, continuous fallow, no nitrogen fertilizers (NoN), $Ca(NO_3)_2$, green manure (green grass), animal manure, peat (sphagnum) and sewage sludge were selected. The application of organic amendments was based on addition of equal amounts of carbon (2000 kg C ha^{-1} yr $^{-1}$). Topsoil (0-20 cm) was sampled in October 1998, dispersed using low-energy sonication and then fractionated by a combination of wet sieving and repeated centrifuging (2000–200 μm, 200–63 μm, 63–2 μm, 2–0.1 μm and < 0.1 μm). The contribution of size fractions to the total organic carbon (OC) present in bulk soil differed significantly between treatments and was correlated with bulk OC contents. Specifically, the silt fraction, which yielded in all treatments the highest amount of OC compared to other fractions, showed increasing OC contributions with increasing bulk OC, as did the two sand-sized fractions. The relative contribution of the two smallest sized, and presumably most stable, OC fractions (2 – 0.1 μm and < 0.1 μm) decreased with increasing bulk OC. The ^{13}C abundances of the latter fractions were less influenced compared to the smaller fractions by OC from the amendments. Cluster analysis of isotopic data of particle separates revealed greatest similarity between green manure and animal manure, $Ca(NO_3)_2$ and NoN treatment, and among fallow, sewage sludge and peat. Xylanase and invertase activities of the bulk soil increased after the long-term addition of green manure, animal manure and sewage sludge. Xylanase activity was mainly located in the sand-sized particles, whereas invertase was associated with the clay and silt fraction. The response of xylanase and invertase to different organic amendments can be easily detected in the sand fraction. 16S rRNA-based analysis showed increasing microbial diversity and abundance with decreasing particle size. Analysis of bacterial diversity and enzyme measurements suggested different metabolic activities in different particle size fractions.

1. INTRODUCTION

Stabilization of organic carbon (OC) in soils is presently a matter of intensive scientific discussion. This is due to the importance of OC both for agronomic factors such as soil fertility and for environmental factors such as terrestrial carbon sequestration and soil conservation. The size of the recalcitrant OC pool in soil is influenced by soil physical, chemical and biological properties. Although single factors such as soil texture, intensity of tillage, or mineral fertilizer and manure applications have been extensively investigated [1, 2, 3, 4], the processes of stabilization of OC are not well understood.

Changes in OC contents of soils due to alternative management or climate change can be observed only after long time periods. An option could be to identify a "sensitive" pool of OC within the bulk soil that responds more quickly than the bulk OC. Physical soil fractionation may have significant potential in this respect because size fractions have proven differences of soil organic matter (SOM) quality and biological activity. Many authors (e.g., 5, 6) point out that SOM contents increase with diminishing particle size, whereas the C-to-N ratio becomes smaller. Agricultural management affects the accumulation of SOM by influencing the amount of plant residues returned to the soil and the rate at which the residues and organic matter decompose. Turnover rates and the role of SOM in soil particle fractions have been studied by various authors [e.g., 7, 8, 9]. In this respect, ^{13}C is a useful tracer for studying the decomposition and incorporation of organic material into more stabile SOM [10, 11]. It has been shown that the ^{13}C content of SOM corresponds closely to the ^{13}C content of the plant material and/or of the organic manure [12] from which it originates.

During decomposition of organic material, complex temporal variations occur in microbial communities. One option to follow decomposition processes is to investigate soil enzymes that mediate the degradation process [13]. Little information is available on the enzymatic response to manure amendments in soil fractions, although recent literature is available on enzymes in particle size fractions influenced by plant litter input and tillage [14, 15].

In the present work, we combined a physical fractionation procedure following low energy sonication, the determination of natural abundances of $^{13}C/^{12}C$, enzyme activities and assessment of bacterial diversities by using 16S rDNA-based methods to evaluate the effect of different amendments on soil organic matter in the Ultuna long-term field experiment located in central Sweden, on a Eutric Cambisol. The main aim of the study was to evaluate the contribution of different particle size fractions to the medium-term storage (decades) of OC and to investigate the response of microbial processes to a range of long-term OC-inputs.

2. MATERIALS AND METHODS

2.1. The Ultuna long-term field experiment

The field experiment is in Uppsala, central Sweden, on a Eutric Cambisol (FAO) with 37% clay and 41% silt. The parent material consists of postglacial clay, with illite as the main clay mineral. A complete documentation of the experiment and compilation of data can be found in Kirchmann et al. (1994) [16]. In 1956, the soil (0–20 cm depth) had 15 g kg^{-1} of organic carbon, 1.7 g kg^{-1} of nitrogen and a pH of 6.6. The experimental design consists of fourteen treatments, laid out with four replicates in a randomized block design. The individual plots (2 m x 2 m) were separated by wooden frames. Seven of the treatments, fallow (continuous bare fallow), NoN (plots did not receive nitrogen fertilizers), $Ca(NO_3)_2$ (80 kg N ha^{-1} yr^{-1}), green manure (green grass), animal manure (well decomposed), peat (sphagnum) and sewage sludge were selected for the study. NoN and $Ca(NO_3)_2$ received root and stubble input only. The application of organic amendments, analyzed before use, was based on equal amounts of ash-free organic matter containing 2000 kg C ha^{-1} yr^{-1}, on average. Organic matter was added in 1956, 1960, 1963 and thereafter every second year by hand. All plots received a dressing of 20 kg P ha^{-1} as superphosphate and 35–38 kg K ha^{-1} as KCl, annually in spring. Cereals (70% of crop rotation), rape crops (25%) and fodder beet (5%) were grown alternately. At harvest, the aboveground portion of the crop was completely removed. Topsoil samples (0–20 cm) were taken in October 1998, one year after the organic matter addition in autumn 1997. Samples were deep frozen. Organic amendments of different years were characterized: OC contents ranged from 304 g kg^{-1} (sewage sludge) to 459 g kg^{-1} (peat), N_t contents from 7.7 g kg^{-1} (peat) to 32.9 g kg^{-1} (sewage sludge); C/N ratios and $\delta^{13}C$ values were 22.6, 22.4, 9.22, 69.6 and -27.25‰, -27.45‰, -25.25‰ and -25.61‰ for green manure, animal manure, sewage sludge and peat, respectively.

2.2. OC and ^{13}C natural abundances in particle fractions

The size fractionation was accomplished with sieved samples (≤ 2 mm) as described in detail in Stemmer et al. (1998)[17]. To minimize destruction of labile particulate organic matter, the soil-water suspension was dispersed using low-energy sonication (0.2 kJ g^{-1} output energy) and then fractionated by a combination of wet sieving (coarse fractions larger than 63 μm) and repeated centrifuging (150 *g*, and 3900 *g*, based on a particle size of 2 and 0.1 μm, respectively). This successively yielded the size fractions 2000–200 μm (coarse-medium sand), 200–63 μm (fine sand), 63–2 μm (silt), 2–0.1 μm (clay) and < 0.1 μm (fine clay) without addition of a flocculent. The fraction samples were freeze-dried and weighed. Fractionation was done in duplicate.

Organic C was determined by dry combustion in an elemental analyzer (Carlo Erba NA 1500). The released CO_2 was separated by a gas-chromatographic column from the other gases and introduced via a gas-splitting interface into an isotope ratio mass spectrometer (Finnigan MAT 251) for measurement of isotopic abundances. The ^{13}C results are expressed in the relative δ scale according to the equation $\delta^{13}C‰ = (R_{sample}/R_{standard} - 1) \times 10^3$, where $R = {}^{13}C/{}^{12}C$; the results are related to the Pee Dee Belemnite (PDB. The estimation of manure-derived organic carbon in particle size fractions is described in detail elsewhere [18].

2.3. Enzyme activities

Xylanase activity was estimated by incubating 0.5 g of moist soil (deep frozen samples thawed over night) or soil fractions with 5.0 ml of a substrate solution (1.7% w/v xylan from oat spelts suspended in 2 M acetate buffer, pH 5.5) and 5.0 ml 2 M acetate buffer (pH 5.5) for 24 h at 50°C. The reducing sugars released were measured colorimetrically by the Prussian blue reaction [19] after reduction of potassium hexacyanoferrate (III) in an alkaline solution. To measure invertase activity, 0.3 g of the bulk soil or fractions were incubated with 5.0 ml of 50 mM sucrose solution and 5.0 ml of 2 M acetate buffer (pH 5.5) for 3 h at 50°C. Reducing sugars were determined as described for xylanase activity. For the determination of urease activity, 0.5 g of the moist bulk soil or fractions were incubated with 1.5 ml of a 79.9 mM urea solution for 2h at 37°C. Released ammonium was extracted with 13.5 ml of 2 M KCl solution and determined colorimetrically by a modified Berthelot reaction [20]. Alkaline phosphomonoesterase (alkaline phosphatase) was determined by a modified disodium phenylphosphate method: 0.3–0.4 g of bulk soil or soil fractions were incubated in 2.0 ml of 0.2 M borate buffer (pH 10.0) and 1.0 ml buffered phenylphosphate solution at 37°C for 3 h; released phenol was estimated by a color reaction [21]. For the determination of arylsulfatase, 0.3–0.4 g of bulk soil or soil fractions were incubated with 1 ml of *p*-nitrophenylsulfate and 4 ml of acetate buffer (0.05 M, pH 5.8) for 1 hour at 37°C [22]. Released *p*-nitrophenol was determined spectrophotometrically at 420 nm.

2.4. DNA isolation

For the isolation of DNA from soils, the protocol described by van Elsas and Smalla [23] was slightly modified. 0.12 g soil and 0.38 g lysozyme were resuspended in 0.75 ml 0.12 M sodium phosphate buffer (pH 8.0) and incubated for 15 min at 37°C. After cooling the samples on ice, 750 mg acid-washed glass beads (Sigma; 0.09-0.13 mm) were added and bead-beating was performed three times for 90 s each at 100% in a Retsch mill with intervals of 30 s. Sodium dodecylsulfate (45 µl of a 20% solution) was added, and the solution was incubated for 15 min at room temperature. Subsequently, an equal to the total volume of solution so far of phenol was added, mixed and centrifuged for 5 min at 10000 *g*. The organic phase was extracted again with 0.12 M sodium phosphate solution; the aqueous phases were pooled and extracted with 1 vol. chloroform. After centrifugation for 5 min at 10000 *g*, 0.1 vol. 5 M potassium acetate was added to the aqueous phase, and humic acids were precipitated for 15 min at room temperature. Then the samples were centrifuged for 5 min at 10000 *g*, and the supernatant was amended with 0.1 vol. 5 M NaCl and 0.7 vol. isopropanol and kept for 30 min at –80°C to precipitate DNA. DNA was obtained by centrifugation for 10 min at 10000 *g*, and the resulting pellet was washed with 70% ethanol, dried and resuspended in 80 µl TE (10 mM Tris/HCL, pH 8.0, 1 mM EDTA). For further purification, spin-columns were prepared containing sepharose CL-6B (Pharmacia) and polyvinylpyrrolidone (20 mg ml^{-1} CL-6B). In general, 2 columns were needed to remove all PCR-inhibiting substances.

2.5. DGGE analysis

Total genomic DNA was used as template to amplify a fragment of about 470 bp comprising the V6 to V8 variable regions of the 16S rRNA gene using the primers U968/GC and 1385R (the number in each primer designation indicates the position of the 5′ end of the primer in *Escherichia coli* 16S rRNA) under previously described conditions [24]. Denaturating gradient gel electrophoresis (DGGE) and silver staining were performed as previously described by Muyzer et al. [25] with denaturating agent concentrations ranging from 20 to 70%.

2.6. T-RFLP analysis

The fluorescently labeled terminal-restriction fragment length polymorphism analysis [26]employed PCR in which one of the two primers used was fluorescently labeled at the 5' end (8f-Fam). Partial 16S rDNA amplification products (primer set 8f-Fam and 518r) were digested with the restriction enzymes *Hha*I and *Hae*III. The reaction mixtures were then directly applied on a polyacrylamide gel and the fluorescently labeled terminal restriction sizes were analyzed using an ABI 373A automated DNA sequencer (PE Applied Biosystems Inc., Faster City, CA, USA). The lengths of the labeled fragments were determined by comparison with the internal Genescan 500 Rox standard (PE Applied Biosystems Inc., Faster City, CA, USA).

2.7. Statistical analysis

Analysis of variance and correlation analyses were accomplished using the program STATISTICA 5.0 and Microsoft EXCEL 97, respectively.

3. RESULTS AND DISCUSSION

3.1. Soil organic matter contents in bulk soil and particle fractions

Forty-two years of different treatments resulted in drastic changes of organic carbon (OC) in bulk soil (Figure 1), ranging from 1.08 (fallow) to 3.2% OC (peat). Plots receiving root and stubble input (NoN: 1.17% OC, Ca-nitrate: 1.42% OC) showed less decrease in OC than the fallow plots. Manure applications resulted in an increase of OC since 1956 (green manure, animal manure, sewage sludge: 1.82, 2.25, 2.58% OC, respectively). Peat, which is less available to microbial turnover due to the unfavorable C/N ratio, accumulated the most OC, followed by sewage sludge, animal manure and green manure. This order directly reflects the stability of the organic input since equal amounts of OC were applied for all manures. The turnover rates of manure-OCs have been discussed previously [12].

The fractionation procedure yielded mass recoveries of 96–100%. Mass recoveries for some particle size fractions, especially fine sand (data not shown), silt and clay, differed among the treatments (Figure 2). Treatments with increasing accumulation of OC were lower in clay-sized and higher in silt-sized particles. This can be explained by an increase of stable microaggregates in the silt fraction, which are not disrupted by the gentle low-sonication dispersion used. Treatments receiving no (fallow) or little (NoN) OC had especially high clay-sized fractions.

The average OC contribution of particle size fractions to total OC decreased in the following order: silt > clay > fine clay > fine sand > coarse + medium sand. The sand fractions were depleted in OC compared to the bulk soil; the finer fractions were enriched in OC. This effect has been shown by several authors [27, 28]. The most important observation was that the contribution of size fractions to the total OC present in bulk soil differed significantly between treatments and was correlated with bulk OC contents (Figure 3). Specifically, the silt fraction, which always yielded the highest amount of OC compared to other fractions, showed increasing OC contributions with increasing bulk OC. Similar and significant trends were observed for both sand-sized fractions. The relative contribution of the two smallest, and presumably, most stable OC fractions (2–0.1 μm and < 0.1 μm) decreased significantly with increasing bulk OC. The largest variation of OC in size fraction between treatments was observed for the fine sand fraction, followed by coarse and medium sand, silt, fine clay and clay. The above results clearly indicate that OC in the different size fractions belongs to pools of different stability. The silt fraction in our case acted as a medium-term OC sink, which might be explained by "physically protected" OC within microaggregates. During the time observed in our experiment, the clay fraction was less important in sequestering OC from the manures.

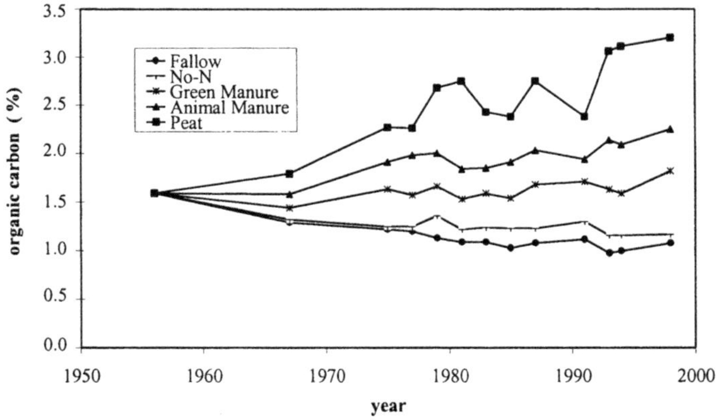

Figure 1. Response of SOC to selected treatments in the Ultuna long-term field experiment over 42 years since its start ([12], amended). (NoN: not nitrogen fertilized, but cropped).

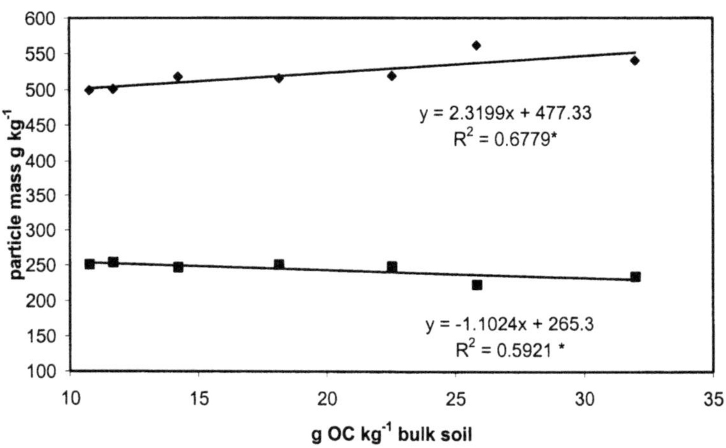

Figure 2. Response of silt-sized (upper line) and clay-sized (lower line) particles to changes in OC contents in the field plots of the Ultuna long-term experiment 42 years after its start (n = 4).

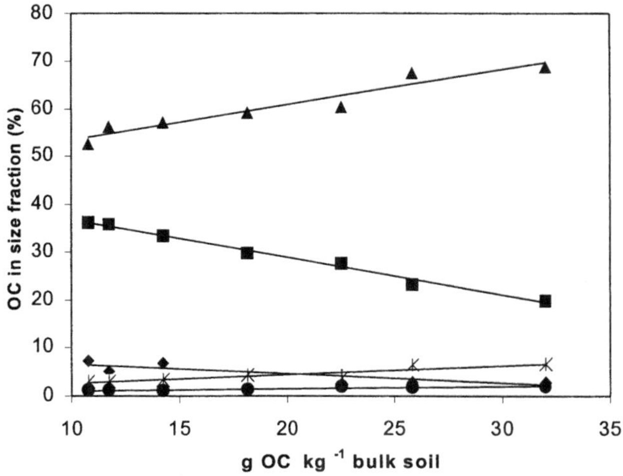

Figure 3. Contribution of OC in particle size fractions from topsoils with different bulk soil OC levels in the Ultuna long-term experiment (redrawn after [18]), n = 4); (● 2000–200 μm; ∗ 200–63 μm; ▲ 63–2 μm; ■ 2–0.1 μm; ◆ < 0.1 μm).

3.2. Carbon natural isotopic abundances

$^{13}C/^{12}C$ natural abundance in bulk soil and particle size fractions was significantly influenced by the different treatments. In 1956, the $\delta^{13}C$ value of untreated topsoil OC was -26.3‰. The fallow treatment showed an increase by 0.5‰ due to the preferential mineralization of the lighter ^{12}C isotope. Compared to the fallow, NoN and Ca-nitrate treatments exhibited more negative delta values (-26.3 and -26.2‰) because of the input of plant residues depleted in $\delta^{13}C$ (-27.1 ± 0.9‰). $\delta^{13}C$ values of bulk samples in manure-treated plots (green manure: -27.0‰, animal manure: -26.9‰, sewage sludge: -26.1‰) were altered according to the isotopic signature of the respective organic materials. Peat OC was not different from soil OC in NoN and Ca-nitrate plots that were used as reference.

Size fractions within treatments often showed larger variations of ^{13}C abundances than bulk samples between treatments. Peat plots exhibited the smallest differences between the lowest and the highest value (0.8), and green manure and NoN plots the highest (2.0). The $\delta^{13}C$ values generally increased from coarse to fine fractions. This phenomenon has also been shown by Stemmer et al. [4] and is probably due to preferential loss of ^{12}C during microbial turnover. Overall ^{12}C losses, thus, were largest for very small organic particles that have been subjected to repeated microbial attack. Cluster analysis of isotopic data of particle separates revealed the greatest similarity between green and animal manure and between $Ca(NO_3)_2$ and NoN treatments (Figure 4). Sewage sludge, peat and fallow exhibited decreasing similarity with green and animal manure treatments. Balance calculations using the isotopic signatures were

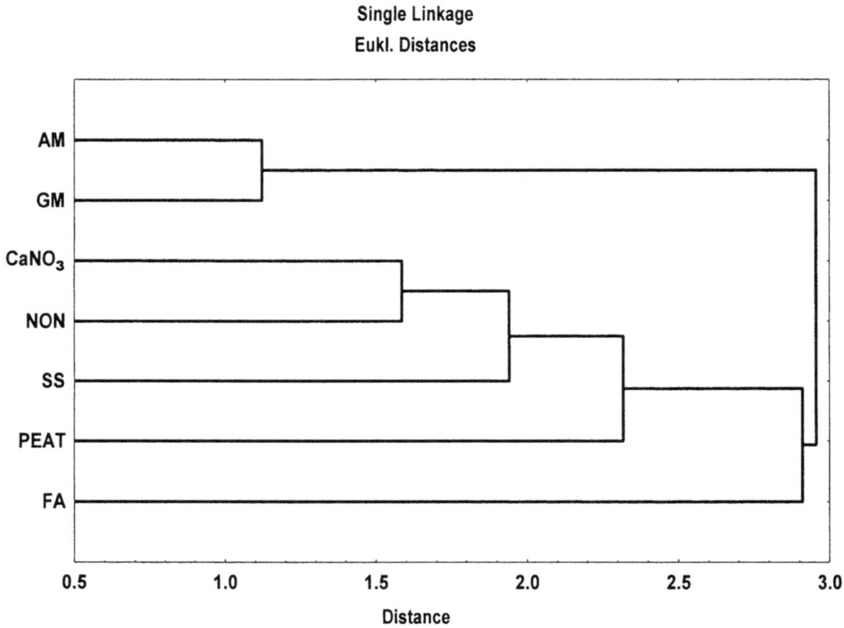

Figure 4. Cluster analysis of ^{13}C natural abundances of OC in particle size fractions of topsoil samples (Ultuna long-term experiment, 4 repl., 7 fractions). AM: animal manure, GM: green manure, NoN: treatment without N-application, SS: sewage sludge, FA: continuous fallow.

limited [18] due to (i) the small differences between investigated carbon sources compared with estimations based on plant carbon originating from different photosynthetic pathways [29] and (ii) the possible isotopic fractionation of different C_{org} pools of the organic amendments and their preferential accumulation in size fractions. Benner et al. [30] already showed significant differences in isotopic composition between plant tissues of varying stability.

3.3. Enzyme activities

Organic amendments significantly influenced the different enzyme activities of bulk soil (Figure 5). Alkaline phosphatase exhibited the largest variation between treatments (maximum divided by minimum = 10.7), followed by xylanase (6.0), urease (4.5), arylsulfatase (4.4) and invertase (2.52). The fallow treatment, in general, exhibited the lowest enzyme activities. At low OC contents, activities increased with the amount of OC present in the topsoil, reaching a maximum primarily in the animal manure treatment. Xylanase activity was equally high in animal manure- and sewage sludge-treated plots. Peat, and sometimes sewage sludge treatments, led to significantly lower enzyme activities. This result again clearly reflects the low microbial digestibility of peat, which is due to already stabilized organic matter and/or an unfavorably high C/N ratio [31]. Statistical analysis showed that there were no statistically significant correlations between enzyme activities and OC contents [32].

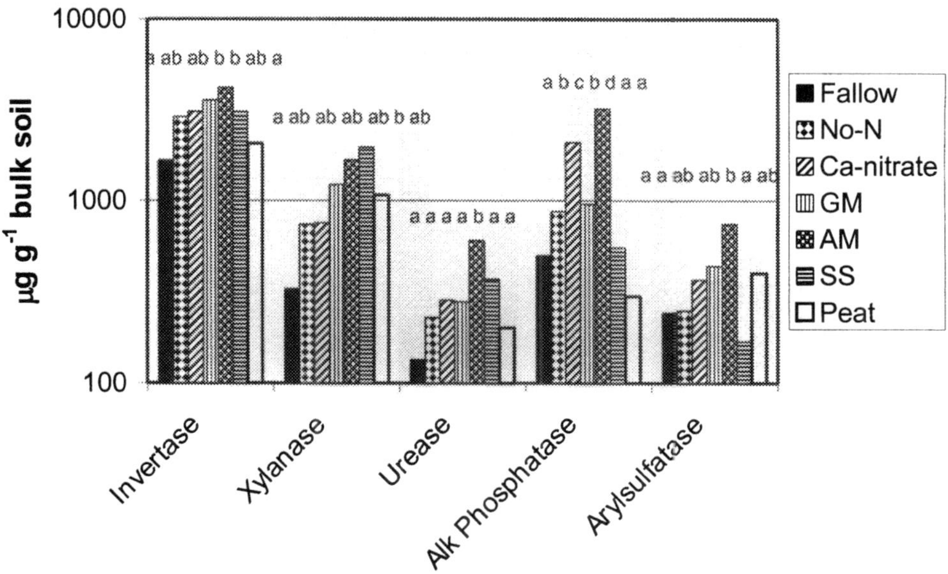

Figure 5. Enzyme activities in bulk soil samples after 42 years of different treatments in the Ultuna long-term field experiment (n = 4). (invertase, xylanase: µg glucose g^{-1} $3h^{-1}$ and $24h^{-1}$; urease: µg N g^{-1} $2h^{-1}$; alkaline phosphatase: µg phenol g^{-1} $3h^{-1}$; arylsulfatase: µg p-nitrophenol g^{-1} h^{-1}). Values for urease and arylsulfatase were multiplied by 10. Letters indicate significant differences between treatments at $p < 0.05$. GM: green manure, AM: animal manure, SS: sewage sludge.

Because biological activity during decomposition processes is highly dependent on the physical availability of organic debris [33], invertase, xylanase and urease (data not shown) activities were additionally analyzed in particle size fractions and related to OC present in the respective fraction. Invertase was mainly associated with silt- and clay-sized particles, whereas xylanase activities were highest in the fine sand fraction (Figure 6).

The distribution of invertase within particle size fractions confirms previous findings of Stemmer et al. and Kandeler et al. [14, 15]. Since the invertase is strongly bound and protected by clay-humus-enzyme complexes, the enzyme may remain stable for long periods of time [34]. It may be concluded that invertase activities in silt- and clay-sized particles represent an enzyme pool quite resistant to decay. Variation of invertase in the fine sand fraction could be due to enzymes newly produced in response to the introduction of organic debris into the soil. The clear activity maximum in the green and animal manure-treated plots again reflected the favorable conditions for microbial OC-turnover provided by these treatments. Xylanase, one extracellular enzyme, was mainly located in the sand- and silt-sized fractions. The highest activity related to OC occurs in the sand-sized particles because the xylanase is mainly bound to particulate organic matter. Cellulases have been shown to be adsorbed to holocellulose and lignin [35)]. Xylanase may be similarly bound.

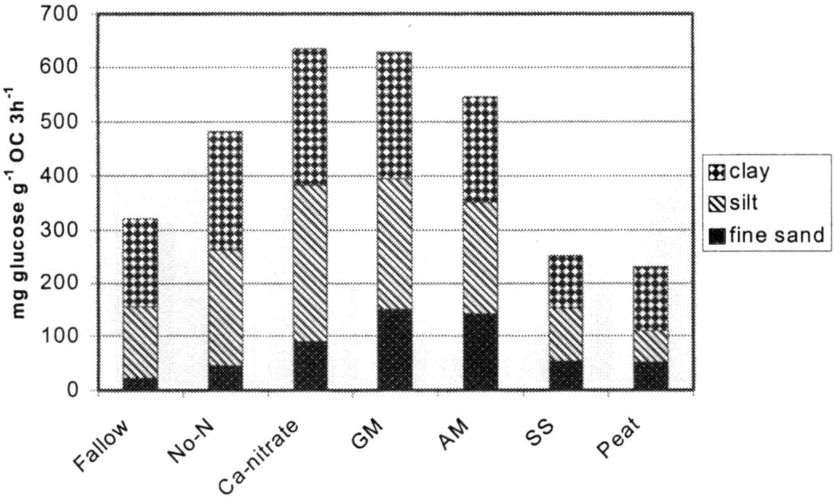

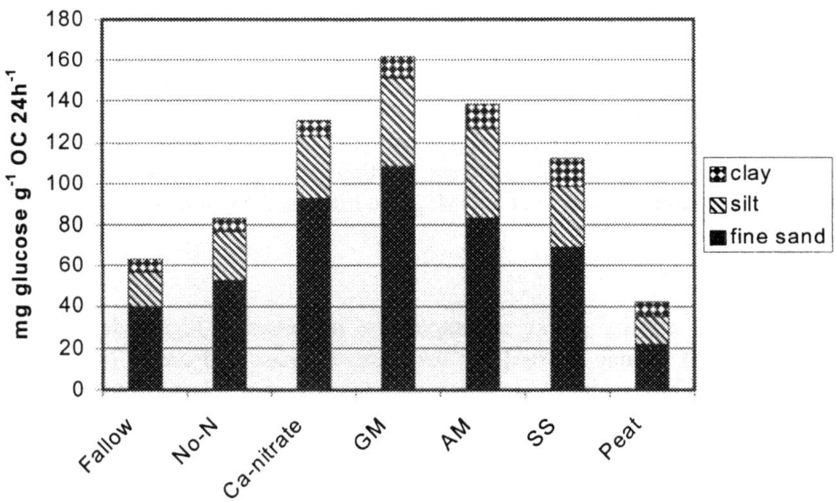

Figure 6. Invertase (top) and xylanase (bottom) activity in particle size fractions related to OC in the fraction (n = 4) (calculated from Kandeler et al. [32]). GM: green manure, AM: animal manure, SS: sewage sludge.

Sewage sludge treatment resulted in a relatively higher xylanase activity compared to invertase activity compared to the other treatments (Figure 5). It has to be taken into account that, in contrast to the other treatments, sewage sludge and peat additions caused significant changes of soil pH. These treatments both resulted in a soil pH of 5.8, whereas the values in all other plots ranged from 6.3 (fallow, green manure) to 6.9 (Ca-nitrate). Since xylanase is mainly produced by fungi, the high xylanase activity of sludge-amended plots may be caused by a shift of the microbial community towards a more fungi-dominated community [36]. The decrease of pH might also explain the lower activities of enzymes other than xylanase in sewage sludge-treated plots compared to green and animal manure-treated plots.

3.4. 16S-rRNA

16S rRNA-based analysis revealed the presence of different microbial populations in the different treatments and particle size fractions. DGGE and T-RFLP analysis showed that the number of visible bands increased with decreasing particle size. At the same time, the intensity of most bands increased (see, e.g., Figure 7).

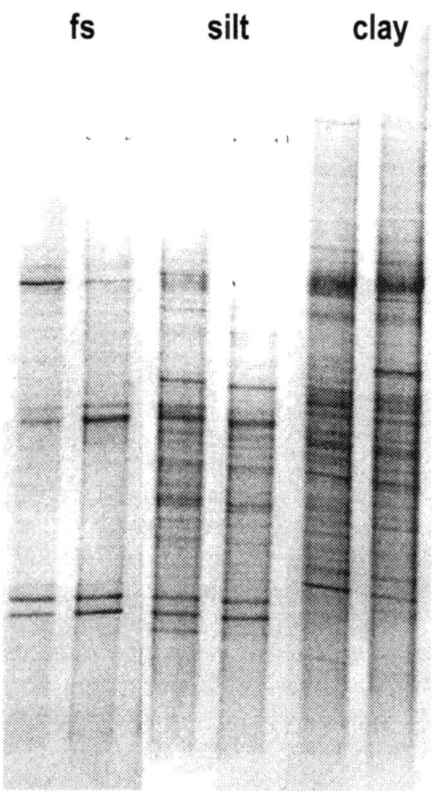

Figure 7. Denaturating gradient gel electrophoresis of the V6 to V8 variable regions of the 16S rRNA gene. Duplicates for fine sand, silt, and clay from peat-treated plots.

This implies that some bacteria present in silt- and clay-sized particles could not be found in the fine sand. However, some bands showed higher intensities in the coarse fractions. Other bacteria seemed to be present in all fractions. The lower diversity of the bacterial community in fine sand-sized particles perhaps makes the fungal origin of xylanase found in that fraction even more likely (compare chapter 3.3.) and supported our general finding of higher biological activities in smaller particle size fractions. Interestingly, differences in bacterial community structures were less pronounced in the fallow treatment. A cluster analysis comparing the fallow and animal manure treatments showed that the bacterial populations found in the sand and the silt fractions of the fallow treatment were similar, whereas the sand fraction of the animal manure treatment was only distantly related with the silt and clay fraction of the same treatment (Figure 8).

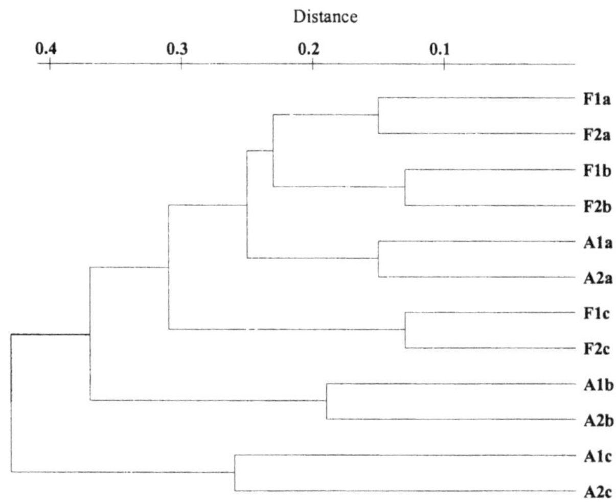

Figure 8. Cluster analysis of fragments obtained in a T-RFLP analysis of sand (a), silt (b) and clay (c) particle size fractions of the two treatments "animal manure" (A) and "fallow" (F). Numbers 1 and 2 indicate replicate samples.

The fallow treatment represents a static system to which few nutrients have been added for 42 years; therefore, this soil may be colonized by well-adapted microorganisms. We can also assume that the bacterial populations found in the fallow treatment resemble most of the initial community structure. In addition, differences in OC and other nutrients between different size fractions may be less pronounced in that treatment. In other treatments, organic material has been degraded, particularly in the coarser fractions, implying a rapid turnover of nutrients. This suggests that microbes respond to the altered nutrient conditions and therefore show lower diversities in the sand fractions. Enzymatic tests and microbial DNA fingerprints indicate that different metabolic activities were induced in different soil particle size fractions.

T-RFLP patterns of total microbial communities of all treatments showed different patterns, although fragments could be identified that were found in all treatments (Table 1). Other fragments were linked to particular treatments, such as a fragment of 290 bp that was mainly found in the sewage sludge treatment. Cluster analysis revealed that the NoN treatment did not cause drastic changes in the bulk soil microflora, whereas the community structures in the animal manure and sewage sludge differed most from the initial composition (data not shown).

Table 1
T-RFLP peaks in various treatments of topsoil bulk samples of the Ultuna long-term experiment

Fragment length (bp)	fallow	noN	$CaNO_3$	peat	green manure	animal manure	sewage sludge
59		■	■	■	■	■	■
66			■	■	■	■	■
75		■	■	■	■	■	
81	■	■	■	■	■	■	■
95	■	■	■		■	■	
194						■	
195	■	■	■	■	■	■	■
196	■	■	■	■	■	■	■
201			■				
204							
207	■	■	■	■	■	■	
209						■	■
216				■		■	
218						■	
221						■	
224						■	
225						■	
290							■
294	■	■	■	■	■	■	■
302							■
305		■	■	■	■	■	■
306					■		■
330				■		■	
413			■				
476	■	■					

4. CONCLUSIONS

- Forty-two years of applications of organic amendments with different stabilities led to distinct differences in OC concentrations in topsoils of the Ultuna long-term organic matter experiment.
- Physical fractionation after low-energy sonication revealed soil OC fractions of significantly different stability and biological activity.
- Silt-sized particles were the most important medium-term sink for OC introduced mainly from organic manures and/or plant residues.
- OC in sand-sized fractions increased most with increasing OC levels of bulk soil. OC levels in sand-sized fractions might be regarded as indicators of OC in bulk soil of the investigated soils.

- The clay-sized fraction, containing more stable OC, was less affected by 42 years of treatments than the other particle size fractions.
- Natural abundance of $^{13}C/^{12}C$ in bulk soil and particle size fractions responded both to application of organic materials and to plant root and stubble input. $\delta^{13}C$ values became more positive with decreasing particle size, indicating the greater "functional" age of the smaller fractions.
- Soil enzyme analysis strongly supported our observations concerning unfavorable conditions for OC turnover in sewage sludge- and peat-treated plots. A sort of "optimum curve" related to increasing OC concentrations in topsoils was observed.
- Invertase was mainly located in silt- and clay-sized particles and xylanase in fine sand. The latter enzyme bound to plant debris seemed to be mainly derived from fungal activity.
- 16S rRNA-based analysis showed increasing bacterial diversity and abundance with decreasing particle size, revealing that the microbial biomass within the clay fraction was predominantly due to soil bacteria.

ACKNOWLEDGMENT

We thank the Austrian Science Foundation (Fonds zur Förderung der wissenschaftlichen Forschung) for financing this bilateral project. A. Sessitsch was financed by an APART fellowship of the Austrian Academy of Sciences.

REFERENCES

1. Golchin, A., Oades, J.M., Skjemstad, J.O., Clarke, P., 1995. Structural and dynamic properties of soil organic matter as reflected by ^{13}C natural abundance, pyrolysis mass spectrometry and solid-state ^{13}C NMR spectrometry in density fractions of an oxisol under forest and pasture. Austr. J. Soil Res. 33, 59–76.
2. Wanniarachchi, S.D., Voroney, R.P., Vyn, T.J., Beyaert, R.P., MacKenzie, A.F., 1999. Tillage effects on the dynamics of total and corn-residue-derived soil organic matter in two southern Ontario soils. Can. J. Soil Sci. 79, 473–480.
3. Aoyama, M., Angers, D.A., N'Dayegamiye, A., Bissonette, N., 1999. Protected organic matter in water-stable aggregates as affected by mineral fertilizer and manure applications. Can. J. Soil Sci. 79, 419–425.
4. Stemmer, M., von Lützow, M., Kandeler, E., Pichlmayer, F., Gerzabek, M.H., 1999. The effect of maize straw placement on mineralisation of C and N in soil particle size fractions. Eur. J. Soil Sci. 50, 73–86.
5. Christensen, B.T., 1986. Straw incorporation and soil organic matter in macro-aggregates and particle size separates. J. Soil Sci. 37, 125–135.
6. Monrozier, L.J., Ladd, J.N., Fitzpatrick, R.W., Foster, R.C., Raupach, M., 1991. Components and microbial biomass content of size fractionations in soils of contrasting aggregation. Geoderma 49, 37–62.
7. Tisdall, J.M., Oades, J.M., 1980. The effect of crop rotation on aggregation in a red-brown earth. Austr. J. Soil Res. 18, 423–433.

8. Schulten, H.-R., Leinweber, P., Sorge, C., 1993. Composition of organic matter in particle-size fractions of an agricultural soil. J. Soil Sci. 44, 677–691.
9. Cambardella, C.A., Elliott, E.T., 1994. Carbon and nitrogen dynamics of soil organic matter fractions from cultivated grassland. Soil Sci. Soc. Am. J. 58, 123–130.
10. Andreux, F., Cerri, C., Vose, P.B., Vitorello, V.A., 1989. Potential of stable isotope, ^{15}N and ^{13}C, methods for determining input and turnover in soils. *In*: Harrison, A.F., Ineson, P., Heal, O.W. (Eds.), Nutrient Cycling in Terrestrial Ecosystems. Field Methods, Application and Interpretation. Elsevier Applied Science, London, pp. 259–275.
11. Piccolo, A., Spaccini, R., Haberhauer, G., Gerzabek, M.H., 1999. Increased sequestration of organic carbon in soil by hydrophobic protection. Naturwissenschaften 86, 496–49.
12. Gerzabek, M.H., Pichlmayer, F., Kirchmann, H., Haberhauer G., 1997. The response of soil organic matter to manure amendments in a long-term experiment in Ultuna, Sweden. Eur. J. Soil Sci. 48, 273–282.
13. Moorhead, D.L., Sinsabaugh, R.L., Linkins, A.E., Reynolds, J.F., 1996. Decomposition processes: modelling approaches and applications. Sci. Tot. Env. 183, 137–149.
14. Stemmer, M., Gerzabek, M.H., Kandeler, E., 1999. Invertase and xylanase activity of bulk soil and particle-size fractions during maize straw decomposition. Soil Biol. Biochem. 31, 9–18.
15. Kandeler, E., Palli, S., Stemmer, M. Gerzabek, M.H., 1999. Tillage changes microbial biomass and enzyme activities in particle-size fractions of a Haplic Chernozem. Soil Biol. Biochem. 31, 1253–1264.
16. Kirchmann, H., Persson, J., Carlgren, K., 1994. The Ultuna long-term soil organic matter experiment, 1956 - 1991. Reports and Dissertation 17, Dep. Soil Sciences, Swedish University of Agricultural Sciences Uppsala.
17. Stemmer, M., Gerzabek, M.H., Kandeler E., 1998. Soil organic matter and enzyme activity in particle size fractions of soils obtained after low-energy sonication. Soil Biol. Biochem. 30, 9-17.
18. Gerzabek, M.H., Haberhauer, G., Kirchmann H., 2001. Soil organic matter pools and ^{13}C natural abundance in particle size fractions of a long-term agricultural field experiment receiving organic amendments. Soil Sci. Soc. Am. J. 65, 352–358.
19. Schinner F., Öhlinger, R., Kandeler, E., Margesin, R., 1996. Methods in Soil Biology. Springer-Verlag, Berlin.
20. Kandeler, E., Gerber, H., 1988. Short-term assay of soil urease activity using colorimetric determination of ammonium. Biol. Fertil. Soils 6, 68–72.
21. Hoffmann, G., 1968. Eine photometrische Methode zur Bestimmung der Phosphatase-Aktivität in Böden. Z. Pflanzenernäh. Bodenk. 118, 161–172.
22. Tabatabai, M.A., Bremner J.M., 1970. Arylsulfatase activity of soils. Soil Sci. Soc. Am. Proc. 34, 225–229.
23. van Elsas, J.D., Smalla, K., 1995. Extraction of microbial community DNA from soils. *In*: de Bruijn, F.J., Akkermans, A.D.L., van Elsas, J.D., (Eds.), Molecular Microbial Ecology Manual 1.3.3. Kluwer Academic Publishers, The Netherlands, pp. 1–11.
24. Felske, A., Engelen, B., Nübel, U., Backhaus, H., 1996. Direct ribosomal isolation from soil to extract bacterial rRNA for community analysis. Appl. Environ. Microbiol. 62, 4162–4167.

25. Muyzer, G., Brinkhoff, T., Nübel, U., Santegoeds, C., Schäfer, H., Wawer, C., 1998. Denaturating gradient gel electrophoresis (DGGE) in microbial ecology. *In*: de Bruijn, F.J., Akkermans, A.D.L., van Elsas, J.D. (Eds.), Molecular Microbial Ecology Manual 3.4.4. Kluwer Academic Publishers, The Netherlands, pp. 1–27.
26. Liu, W.T., Marsh, T.L., Cheng, H., Forney, L.J., 1997. Characterization of microbial diversity by determining terminal restriction fragment length polymorphisms of genes encoding 16S rRNA. Appl. Environ. Microbiol. 63, 4516–4522.
27. Desjardins, T., Andreux, F., Volkoff, B., Cerri, C.C., 1994. Organic carbon and ^{13}C contents in soils and soil size-fractions, and their changes due to deforestation and pasture installation in eastern Amazonia. Geoderma 61, 103–118.
28. Guggenberger, G., Christensen, B.T., Zech, W., 1994. Land-use effects on the composition of organic matter in particle-size separates of soil: I. Lignin and carbohydrate signature. Eur. J. Soil Sci. 45, 449–458.
29. Jastrow, J.D., Boutton, T.W., Miller, R.M., 1996. Carbon dynamics of aggregate-associated organic matter estimated by carbon-13 natural abundance. Soil Sci. Soc. Am. J. 60, 801–807.
30. Benner, R., Fogel, M.L., Sprague, E.K., Hodson R.E., 1987. Depletion of ^{13}C in lignin and its implications for stable carbon isotope studies. Nature (London) 329, 708–710.
31. Gerzabek, M.H., Kirchmann, H., Haberhauer, G., Pichlmayer, F., 1999. The response of soil nitrogen and ^{15}N natural abundance to different amendments in a long-term experiment at Ultuna, Sweden. Agronomie: Agriculture and Environment 19, 457–466.
32. Kandeler, E., Tscherko, D., Sessitsch, A., Kirchmann, H., Gerzabek, M.H., unpublished data.
33. Gaillard, V., Chenu, C., Recous, S., Richard, G., 1999. Carbon, nitrogen and microbial gradients induced by plant residues decomposing in soil. Eur. J. Soil Sci. 50, 567–578.
34. Ginafreda, L., Rao, M.A., Violante, A., 1991. Invertase (β-fructosidase): effects of montmorillonite, Al-hydroxide and Al(OH)x-montmorillonite complex on activity and kinetic properties. Soil Biol. Biochem. 23, 581–587.
35. Sinsabaugh, R.L., Linkins, A.E., 1988. Adsorption of cellulase components by leaf litter. Soil Biol. Biochem. 20, 927–932.
36. Kandeler, E., Tscherko, D., Bruce, K.D., Stemmer, M., Hobbs, P.J., Bardgett, R.D., Amelung, W., 2000. Structure and function of the soil microbial community in microhabitats of a heavy metal polluted soil. Biol. Fertil. Soils. 32, 390-400.

Developments in Soil Science, Volume 28B
Editors: A. Violante, P.M. Huang, J.-M. Bollag and L. Gianfreda

INDIRECT APPROACHES FOR ASSESSING INTRACELLULAR ARYLSULFATASE ACTIVITY IN SOIL

F. Fornasier

Istituto Sperimentale per la Nutrizione delle Piante, S.O.P. di Gorizia.
Via Trieste 23, 34170 Gorizia, Italy.
Fax +39-481-520208. E-mail: f.fornasier@isnp.it

A long-standing question about soil enzymology is differentiating between intracellular and extracellular enzyme activity. To address this, different soil treatments were compared for their effect on arylsulfatase activity. A rapid fumigation (5 min) with liquid diethyl ether increased arylsulfatase activity over chloroform and toluene. Diethyl ether plus 50 mM pH 7.00 tris-HCl buffer further increased total arylsulfatase activity by 27%. Arylsulfatase released from microbial cells was largely adsorbed on soil colloids, as only 3% of total activity was extracted by 50 mM pH 7.00 tris-HCl buffer. However, pre-treatment of soil with a solution of bovine serum albumin (BSA) increased the extraction yield by 11%. Purified arylsulfatase from mollusc *Patella vulgata* was tested as a reference material and was added to the soil to check for arylsulfatase adsorption. Once added to soil, its activity fell by a factor of 7, but about 30% of the enzyme activity added was extracted by tris-HCl buffer. Pre-treatment of soil with BSA did not influence the extraction yield, which suggested that purified arylsulfatase exhibits different behaviour than microbially derived ones. Diethyl ether did not denature purified arylsulfatase when added to the soil.

1. INTRODUCTION

The most important obstacle in interpreting soil enzymatic activity is its partition into intracellular and extracellular fractions [1]. In spite of the extensive research, a method that adequately distinguishes between the two activities has not been found. Methods have been used with chemical agents that unmask the intracellular enzyme activity by either lysing cells or increasing permeability of substrates and products in cells [2]. These agents increase activity over untreated sample, and this difference can be an estimation of intracellular enzymatic activity. The unmasking agents also must not interfere with enzymes, which is not the case for some agents that have been tested [3].

Recently, Klose and Tabatabai [4] evaluated chloroform fumigation as a method for determining arylsulfatase activity of microbial biomass. They found that the toluene widely used in soil enzyme assays was less effective than chloroform in increasing arylsulfatase activity. In addition they found that chloroform fumigation decreased activity of pure arylsulfatases obtained from the molluscs *Helix pomatia* and *Patella vulgata* by 50 and 23%, respectively.

However, for a proper evaluation of intracellular enzymatic activity, other factors need to be taken into account. It is well known that when enzymes are added to soil, they are adsorbed onto soil components, and this usually leads to a considerable decrease in activity, depending on surface characteristics [5]. The same is expected to occur when cells are lysed.

The objective of this research was to investigate: (i) the effectiveness of three unmasking agents on cell disruption; (ii) surface arylsulfatase adsorption and arylsulfatase inhibition after adsorption.

2. MATERIALS AND METHODS

2.1. Soil

A Mollic Hapludalf (USDA) coarse loamy soil under permanent grassland was sampled, sieved (2 mm), and stored moist at 4°C. Selected soil characteristics are reported in Table 1.

Table 1
Physical and chemical characteristics of soil used for the experiment

Sand (2-0.02 mm) $^{(+)}$	673
Silt (0.02-0.002 mm) $^{(+)}$	207
Clay (<0.002 mm) $^{(+)}$	120
Total organic C $^{(+)}$	22
Total N $^{(+)}$	2.3
Microbial biomass C $^{(++)}$	420
pH (water)	6.7

$^{(+)}$ g kg^{-1}. $^{(++)}$ μg C g^{-1} dry soil.

2.2. Enzyme assays

Arylsulfatase activity was measured with a modified method of Tabatabai and Bremner [6]. One g of soil (oven dry basis) was treated with 5 mL of 0.5 M pH 5.2 acetate buffer (optimum pH value for this soil) and p-nitrophenylsulfate as substrate (final concentration 5 mM) in capped polyethylene tubes (10 mL). Soil mixture was incubated for 1 hour under shaking (90 rev min^{-1}) in a Dubnoff bath set at 37 °C. To stop enzymatic reaction, tubes were put in an ice bath and after addition of 1 mL of 0.5 M $CaCl_2$ and 4 mL of 0.5 M NaOH, they were rapidly centrifuged (2500 g for 4 min) [7]. Clear extracts were read at 400nm. Assays were replicated four times.

Extracellular enzymatic activity was assumed to be the activity in absence of any unmasking agent, including toluene.

Three unmasking agents were applied before the assay to expose intracellular arylsulfatase activity:

a) toluene (0.3 mL) was added directly to moist soil [8].
b) chloroform (0.7 mL) was added to cover the soil completely, followed by gentle shaking for 5 min then centrifugation (2500 g for 2 min). Residual chloroform was removed by vacuum in a desiccator for about 90 minutes.
c) diethyl ether was directly added to soil at two different doses, 1 and 2 mL. Tubes were shaken for 5 min then acetate buffer was added and the ether was removed.

Intracellular arylsulfatase activity was calculated by subtracting extracellular activity from the activity in the presence of one of the three unmasking agents.

2.3. Influence of buffer and albumin on adsorption and extractability of arylsulfatase

Arylsulfatase was extracted from unfumigated soil and soil fumigated with 2 mL of diethyl ether by adding 5 mL of 50 mM pH 7.00 tris-HCl to 1 g of soil; tubes were gently shaken for 5 min and then centrifuged at 2500 g for 4 min. The enzyme assay was performed by adding 2 mL of 1.0 M pH 5.2 acetate buffer and 1 mL 25 mM *p*-nitrophenilsulfate (dissolved in 0.5 M acetate buffer) to 2 mL of filtered solution (Whatman GF/C). Residual arylsulfatase activity in centrifuged soil was also measured.

The influence of the buffer during cell lysis on extractability of arylsulfatase was investigated by adding 1 mL of tris-HCl buffer 5 minutes before the addition of 2 mL of diethyl ether.

The adsorption of intracellular arylsulfatase was further investigated using BSA. Soil samples were gently shaken for 5 min with 1 mL of Tris-HCl buffer containing 0.5% or 2% BSA (Sigma A-8022) before treatment with 2 mL of diethyl ether. Arylsulfatase activity in soil extracts as well as that in extracted soil (residual activity) was determined as described above.

2.4. Addition of exogenous arylsulfatase to soil

Arylsulfatase from *Patella vulgata* (Sigma S-8629) was dissolved in water just before use, giving an activity of 365 μM PNP h^{-1} mL^{-1} (measured in 0.5 M pH 5.2 acetate buffer after 30 min incubation).

One mL of 50 mM pH 7.00 tris-HCl buffer was added to soil 5 min before adding 0.2 mL of the arylsulfatase solution (corresponding to 73 μM PNP h^{-1}). Arylsulfatase activity was measured after 5 min. Activity in soil extracts and residual activity in soil were also determined, as described above.

The influence of BSA on activity and adsorption of added arylsulfatase was determined by adding 1 mL of 0.5% BSA 5 min before adding 0.2 mL of the enzyme solution.

The possible denaturation of arylsulfatase from *Patella vulgata* extractable or residual was studied in soil samples treated with 2 mL of diethyl ether after sequential addition of one mL of 0.5% BSA and 0.2 mL of exogenous arylsulfatase.

2.5. Statistical analysis

ANOVA and Student-Newman-Keuls (SNK) tests were performed using the SPSS 9.0 statistical package.

3. RESULTS

The three unmasking agents increased total arylsulfatase activity to a very different extent (Table 2): the chloroform effect was twice that of toluene. The highest effect was observed, however, when 2 mL of ether were added to soil (Table 2).

No extractable arylsulfatase activity was detected in unfumigated soil, whereas the residual enzyme activity of the extracted soil was 1.13 μM PNP g^{-1} soil h^{-1}.

Table 2
Increase of arylsulfatase activity after addition of various chemical agents to soil

Treatment	Total arylsulfatase activity$^{(+)}$	Intracellular arylsulfatase activity$^{(++)}$
	---- μM PNP g^{-1} dry soil h^{-1} ----	
Control (extracellular)	1.25 e	-
Toluene	1.95 d	0.70 d
Chloroform	2.65 b	1.40 b
Ether (1 mL)	2.15 c	0.90 c
Ether (2 mL)	3.05 a	1.80 a

$^{(+)}$ Lower case letters within the same column indicate significant differences at p=0.01 probability level using SNK test.
$^{(++)}$ Calculated as difference between activity in treated soil minus activity in untreated soil.

Even though adsorption of the major part of the enzymes released after treatment with 2 mL of diethyl ether always occurs, addition of any buffer before fumigation was effective in influencing total arylsulfatase activity (Table 3).

Tris-HCl influenced residual activity in soil but at the same time reduced enzyme extractability. BSA significantly increased the extraction of the released enzyme, especially at the highest concentration tested (2%), when about 11% of the total activity was detected in the extract.

Addition of arylsulfatase from *Patella vulgata* to soil caused a dramatic decrease of its activity (Table 4). By contrast, about 30% of its activity was recovered in the extract. BSA pre-treatment affected neither extraction of the enzyme nor the residual activity.

4. DISCUSSION

Chloroform and diethyl ether were used as liquids in order to ensure an immediate lysis so as to reduce the potential for degradation of released intracellular arylsulfatase (fumigated soils retain their protease activity after chloroform fumigation) [9, 10].

Table 3
Effect of addition of tris-HCl buffer and BSA before treatment with 2 mL of diethyl ether on arylsulfatase activity

Treatment	Arylsulfatase activity (−)		
	Extracted (1)	Residual in soil (2)	Total (1+2)
	----------------	μM PNP g^{-1} dry soil h^{-1}	----------------
Control	0.17 b	3.03 b	3.20
TRIS buffer	0.10 c	3.85 a	3.95
BSA 0.5%	0.22 b	3.66 a	3.88
BSA 2%	0.45 a	3.62 a	4.07

(+) Lower case letters within the same column indicate significant differences at p=0.01 probability level using SNK test.

Table 4
Arylsulfatase activity after addition of arylsulfatase from *Patella vulgata* (0.73 μM *p*-nitrophenol (PNP) g^{-1} dry soil) in: a) unfumigated soil with or without pre-treatment of soil with 0.5% BSA or b) soil treated with 2 mL of diethyl ether after BSA pre-treatment

Treatment	Arylsulfatase activity		
	Extracted (+)	Residual in soil (+)	Total
	----------------	μM PNP g-1 dry soil h^{-1}	----------------
a) no cell lysis			
no BSA	-	-	1.35
no BSA, extraction	0.23 a	1.24 a	1.48
BSA	-	-	1.35
BSA, extraction	0.25 a	1.34 a	1.60
b) diethyl ether			
BSA, extraction	0.53	3.62	4.15

(+) Lower case letters within the same column indicate significant differences at p=0.01 probability level using SNK test.

The quantitative effects of toluene and chloroform on arylsulfatase activity are consistent with Klose and Tabatabai [4], who found the latter treatment to be more effective in increasing arylsulfatase activity. It should be noted, however, that they used vapors of chloroform to lyse the cells.

Results obtained in the present work show that diethyl ether is the most effective agent when used at a dose of 2 mL g^{-1} soil. Besides efficient cell membrane disruption, diethyl ether has a low specific gravity and water solubility, which allows for rapid separation.

The presence of 1 mL of buffer (with or without BSA) during plasmolysis caused an average increase of 38% for intracellular activity (i.e. $(3.95\text{-}3.2) * (3.2\text{-}1.25)^{-1} *100$) (Table 3). It is reasonable to hypothesize a more efficient cell lysis in the presence of water or a different form of adsorption on soil colloids. However the low water solubility of diethyl ether and the effect of BSA pre-treatment, which basically did not increase total arylsulfatase activity but increased the amount of extracted activity, support the hypothesis based on the lower adsorption of the enzymes on soil colloids.

To test whether an exogenous arylsulfatase can be used for calibrations, experiments with arylsulfatase from *Patella vulgata* were performed. Klose and Tabatabai (4) used this enzyme as a standard enzyme reference and found that fumigation caused a 23% reduction of activity. Results in our experiment showed that this arylsulfatase when added to unfumigated soil: (a) decreased its activity by a factor of 7; (b) was about 30% extractable; and (c) produced no change in extractability and residual activity in soil after pre-treatment with BSA. It seems therefore that behaviour of arylsulfatase from *Patella vulgata* was different than arylsulfatase(s) released from soil microbes (Table 3). As a consequence this material seems unsuitable for calibrating arylsulfatase adsorption.

Calibration using purified enzymes from soil microbes would be a solution, but commercially available arylsulfatases are obtained from other sources and extraction of enzymes from soil is difficult and time-consuming [11].

Because the treatment with ether decreased neither the amount of extracted exogenous arylsulfatase, nor the residual activity, it seems reasonable to conclude that this procedure does not denature the arylsulfatase added to soil.

5. CONCLUSIONS

Findings of this study showed that intracellular arylsulfatase activity can be estimated by diethyl ether in a simple and rapid procedure.

Determination of extractable and residual activity under various conditions showed that adsorption of released enzyme is an important factor that must be taken into account for a correct estimation of intracellular arylsulfatase activity and that it is possible to reduce enzyme adsorption by pretreating the soil with the protein BSA.

Studies performed with arylsulfatase obtained from *Patella vulgata* showed that this enzyme behaves differently compared to those released from soil microbial cells, particularly with respect to extractability. In addition diethyl ether did not denature this enzyme once added to soil.

Further research is needed in order to better understand factors influencing estimation of intracellular arylsulfatase activity by means of unmasking agents.

ACKNOWLEDGMENTS

I am very indebted to Dr. Liviana Leita, who helped me during the experimental work and manuscript preparation with important suggestions and criticism. Many thanks also to two anonymous referees and the editor for constructive reviews of this paper which did much to improve its content and clarity.

REFERENCES

1. Nannipieri, P., Grego, S., Ceccanti, B., 1990. Ecological significance of the biological activity in soil. *In*: Bollag, J.-M., Stotzky, G. (Eds.), Soil Biochemistry. Vol. 6, Marcell Dekker, Inc. New York, pp. 293-396
2. Skujins, J.J., 1967. Enzymes in soil. *In*: McLaren, A.D., Peterson, G.H. (Eds.), Soil Biochemistry. Vol. 1, Marcell Dekker, Inc. New York, pp. 371-414.
3. Frankenberger, W.T. Jr., Johanson, J.B., 1986. Use of plasmolytic agents and antiseptics in soil enzyme assays. Soil Biol. Biochem. 18, 209-213.
4. Klose, S., Tabatabai, M.A., 1999. Arylsulfatase activity of microbial biomass in soils. Soil Sci. Soc. Am. J. 63, 569-574.
5. Boyd, S.A., Mortland, M.M., 1990. Enzyme interactions with clays and clay-organic matter complexes. *In*: Bollag, J.-M., Stotzky, G. (Eds.), Soil Biochemistry. Vol. 6, Marcell Dekker, Inc. New York, pp.1-28.
6. Tabatabai, M.A., Bremner, J.M., 1970. Arylsulfatase activity in soils. Soil Sci. Soc. Am. Proc. 34, 225-229.
7. Alef, K., Nannipieri, P., Trasar-Cepeda, C., 1995. Phosphatase activity. *In*: Alef, K., Nannipieri, P. (Eds.), Methods in Applied Soil Microbiology and Biochemistry. Academic Press, London, pp. 335-344.
8. Tabatabai, M.A., 1994. Soil enzymes. Arylsulfatase. *In*: Weaver, R.W., Angle, J.S., Bottomley, P.S. (Eds.), Methods of Soil Analysis. Part 2. Microbiological and Biochemical Properties. SSSA, Madison, WI, pp. 814-818.
9. Amato, M., Ladd, J.N., 1988. Assay for microbial biomass based on ninhydrin-reactive nitrogen in extracts of fumigated soils. Soil Biol. Biochem. 20, 107-114.
10. Zelles, L., Palojarvi, A., Kandeler, E., Von Lutzow, M., Winter, K., Bai, Q.Y., 1997. Changes in soil microbial properties and phospholipid fatty acid fractions after chloroform fumigation. Soil Biol. Biochem. 29, 1325-1336.
11. Nannipieri, P., Sequi, P., Fusi, P., 1996. Humus and enzyme activity. *In*: Piccolo, A., (Ed.), Humic Substances in Terrestrial Ecosystems. Elsevier, Amsterdam, pp. 293-328.

Developments in Soil Science, Volume 28B
Editors: A. Violante, P.M. Huang, J.-M. Bollag and L. Gianfreda

CHANGES IN SOIL ORGANIC MATTER, ENZYMATIC ACTIVITIES AND HEAVY METAL AVAILABILITY INDUCED BY APPLICATION OF ORGANIC RESIDUES

P. Burgos, E. Madejón and F. Cabrera

Instituto de Recursos Naturales y Agrobiología (IRNAS) (CSIC)
Avenida Reina Mercedes, 10, P.O. Box 1052, 41080- Sevilla, Spain

A 3-year field experiment on a sandy soil supporting a strawberry crop was carried out to study the effects of the application of three organic residues (AC, olive mill wastewater sludge compost; MWC, municipal solid waste compost; and PW, paper mill waste) on organic matter content, enzymatic activities, and available heavy metals. Significant increases in organic carbon content (total organic carbon (TOC), total extractable carbon (TEC) and humic acid carbon (HAC)) were observed in soils, depending on the nature of organic amendments. The application of the organic residues also caused significant increases in dehydrogenase, phosphatase, ß-glucosidase, urease and BAA-protease activities. This favourable effect on soil biological activity was more noticeable in MWC and PW treatments. Significant positive correlation ($p < 0.01$) between enzymatic activities and total organic carbon was found for all treatments. Available heavy metal contents increased slightly in the soils treated with MWC and PW but did not affect negatively soil enzymatic activities. A discriminant analysis generated two functions (F1 and F2) based on linear contributions of the variables (TOC, TEC, HAC, DH-ase, Phosp, β-glu, Ure, and BAA). F1 is correlated with DH-ase, TOC and BAA and separates treatments C and AC from PW and MWC. F1 underlines the effect of the higher doses of organic matter applied with MWC and PW. F2 is correlated with humic substances (TEC and HAC), and β-glu separates treatments C and PW from AC and MWC. This function shows the difference of the effect of composted (AC and MWC) and fresh residues (PW) on soil properties. Results showed that the application of organic residues to a sandy soil improves its agronomic quality by increasing soil organic matter and enhancing soil enzyme activities.

1. INTRODUCTION

The pressures to sustain modern agricultural systems cause a progressive degradation of soil structure and a continued depletion of soil fertility level as a result of organic matter reduction [1]. The strawberry crop has great socio-economic importance in the Comarca Costa de Huelva (SW Spain), in which sandy soils with a very low organic matter content prevail. Intensive agricultural practices reduce the fertility of these soils. Consequently, these soils need organic matter application to improve and maintain their agronomic quality.

The use of organic residues as a source of organic matter is a common practice to improve physical, chemical and biochemical soil properties; their effect on soil can be observed for a long period after their application [2–7].

Soil enzymatic activities are important biochemical properties in the evaluation of soil fertility because many soil enzymes are involved in the cycle of nutrients [8]. The importance of soil enzymes lies on their participation in the processes involved in organic matter turnover [9]. Enzymatic activities are considered as important markers of soil regeneration through manuring practices [4] because they might indicate the potential soil fertility [1, 9].

On the other hand, the application of organic residues to soil could be one of the most useful and ecological approaches to solve two problems: wastes disposal and soil organic matter reduction. However, the use of certain organic residues can imply some hazards derived from their contents of heavy metals. Successive application of these residues can cause an accumulation of heavy metals in soil. Heavy metals content is a limiting factor for the continual and generalized agronomic utilization of some organic residues. A continuous monitoring of the concentration and availability of heavy metals in soil seems to be a judicious and necessary practice when organic residues rich in heavy metals are used.

This paper deals with the effect of the application as amendment of three organic residues to a sandy soil, over three successive applications on i) the content of organic matter, ii) the contents of available heavy metals (Fe, Mn, Cu, Zn, Cd, Co, Ni and Pb) and iii) the activities of five selected enzymes (dehydrogenase, phosphatase, ß-glucosidase, BAA-protease and urease).

2. MATERIALS AND METHODS

2.1. Soil and organic materials

The soil, a Typic Endoaquent [10], is a very representative soil of the "Comarca Costa de Huelva" (SW Spain), in which sandy soils prevail. The most relevant characteristics of the soil at the beginning of the experiment are in Table 1.

Table 1
Some soil characteristics at the beginning of the experiment

pH (sat. p)	6.1	Available-P (mg kg^{-1})	22.8
EC (1:5) (dS m^{-1})	0.03	Available-K (mg kg^{-1})	116
$CaCO_3$ (g kg^{-1})	< 2.00	Sand (g kg^{-1})	808
TOC (g kg^{-1})	3.6	Silt (g kg^{-1})	36
Kjeldahl-N (g kg^{-1})	0.3	Clay (g kg^{-1})	156

EC: Electrical conductivity.

Three organic materials from different origins were used in the experiment: i) an agricultural compost, made by olive mill wastewater sludge mixed with different agricultural wastes (AC), ii) a municipal solid waste compost (MWC) from the city refuse treatment plant of Villarrasa (Huelva, SW Spain) and iii) a paper mill waste (PW), produced by the paper industry of San Juan del Puerto (Huelva SW, Spain). Table 2 shows the most relevant characteristics of these three materials.

2.2. Field study

The experimental field consisted of thirty-two 45 m^2 plots cropped with strawberries (*Fragaria ananassa*, cv. Camarosa). The plots were amended with the organic residues before planting. The following doses were annually applied for three years: treatment AC, 10000 kg ha^{-1}; treatment MWC, 48000 kg ha^{-1}; and treatment PW, 48000 kg ha^{-1}. The organic residues were incorporated into the first 0–20 cm of soil by ploughing. A treatment C, without organic amendment, was used as control for comparison. In the first two years, plots MWC and PW also received 110 and 220 kg ha^{-1} of urea (46% N), respectively, to avoid N-immobilization because of the high C/N of these materials (Table 2). All plots also received the normal mineral fertigation used by the farmers in the area. The experiment was carried out in a completely randomized design with eight replications per treatment.

Table 2
Some chemical and biochemical characteristics of the organic materials

	AC[a]	SD[b]	MWC[c]	SD[b]	PW[d]	SD[b]
Moisture (g kg^{-1})	198	13.0	368	9.88	769	0.82
TOC (g kg^{-1})	152	9.47	268	19.8	260	11.5
Kjeldahl-N (g kg^{-1})	9.94	0.56	12.6	0.66	2.74	0.62
C/N	15.3	-	21.3	-	94.9	-
Fe (mg kg^{-1})	12669	231	7616	332	3268	1134
Mn (mg kg^{-1})	286	3.01	156	58.2	3734	1016
Cu (mg kg^{-1})	63.5	2.52	168	13.6	1472	480
Zn (mg kg^{-1})	61.4	1.82	366	63.2	1963	336
Cd (mg kg^{-1})	0.14	0.03	0.55	0.03	7.98	2.65
Co (mg kg^{-1})	7.82	0.46	3.23	0.14	146	50.4
Ni (mg kg^{-1})	71.8	2.77	60.9	17.2	91.8	11.6
Pb (mg kg^{-1})	11.3	0.48	188	25.8	32.6	3.45
DH-ase (mg INTF kg^{-1} h^{-1})	47.3	3.87	1.72	0.38	6.18	3.09
Phosp (mg PNP kg^{-1} h^{-1})	2931	633	3914	897	789	123
ß-glu (mg PNP kg^{-1} h^{-1})	316	245	144	94.9	37.1	13.5
Ure (mg NH_3 kg^{-1} h^{-1})	203	22.4	20.4	11.0	7.31	1.00
BAA (mg NH_3 kg^{-1} h^{-1})	15.2	0.94	17.9	4.80	1.80	1.05

[a] AC, agricultural compost. [b] SD, standard deviation. [c] MWC, municipal waste compost. [d] PW, paper waste.
DH-ase: dehydrogenase activity; Phosp: phosphatase activity; β-glu: β-glucosidase; Ure: urease; BAA: BAA-protease.

2.3. Soil sampling and analytical methods

Four soil samples from the surface layer (0–20 cm) were randomly taken from each plot to make a composite sample per plot, eight months after the last soil amendment. Samples were air-dried and ground to 0.5 mm. Wet subsamples were stored until analysis at 4°C for biochemical assays. Soil pH was determined in the saturated soil extract and electrical conductivity (EC) in the 1:5 soil/water extract. Kjeldahl-N, available P and available K were determined by the methods described by Hesse [11].

The total organic carbon content (TOC) was determined by oxidation with potassium dichromate oxidation and titration with ferrous ammonium sulphate [12]. To determine total extractable carbon (TEC), soil samples were extracted with 0.1 M sodium pyrophosphate and 0.1 M sodium hydroxide. The supernatant was acidified to pH 2 with HCl and let stand 24 h at room temperature. To separate fulvic acids (FA) from humic acids (HA), the solution was centrifuged and the precipitate containing HA dissolved with sodium hydroxide [13]. The TEC and the carbon of the humic extract (HAC) were determined by the method of Walkley and Black [12].

Available Fe, Mn, Cu, Zn, Cd, Co, Ni and Pb were extracted with a mixture of 0.005 M DTPA (diethylenetriaminepentaacetic acid), 0.01 M $CaCl_2$ and triethanolamin 0.1 M [14] and measured by ICP-OES.

Dehydrogenase (DH-ase) activity was measured following the method proposed by Trevors [15] and modified by García et al. [16], using INT (2-*p*-iodophenyl 3-*p*-nitrophenyl-5-phenyltetrazolium chloride) as electron acceptor. Phosphatase (Phosp) and ß-glucosidase (β-glu) activities were determined according to Tabatabai [17] using *p*-nitrophenylphosphatase and *p*-nitrophenyl-ß-D-glucopyranoside as substrates for Phosp and β-glu, respectively. Urease (Ure) and BAA-protease (BAA) activities were measured according to Nannipieri et al. [18] using urea and N-α-benzoyl-L-argininamide (BAA) as substrates for urease and BAA-protease, respectively.

2.4. Statistical methods

The results were analyzed by ANOVA, considering the treatment as the independent variable. The means were separated by Tukey's test, considering a significance level of $p < 0.05$ throughout the study. A correlation matrix was calculated with the values of organic carbon, heavy metals and enzyme activities. The significance levels of the correlation matrix ($p < 0.01$ and $p < 0.05$) are based on the Pearson coefficients.

3. RESULTS AND DISCUSSION

3.1. Soil organic carbon contents

Soil organic carbon contents (TOC, TEC and HAC) after 3 years of treatment are reported in Figure 1.

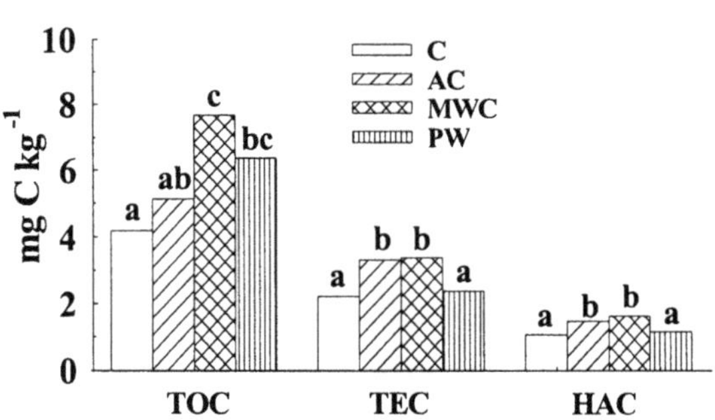

Figure 1. Organic carbon content (TOC, TEC and HAC) in the soils after 3 years of treatment. Values with the same letter are not significantly different ($p < 0.05$).

In general, the addition of the organic amendments had a noticeable effect on soil organic carbon contents. The TOC values in MWC- and PW-treated soils increased significantly when compared with the control soil. However, the dose of AC applied was not high enough to cause a significant increase of TOC. On the other hand, MWC and AC treatments increased significantly in soil humic fractions (TEC and HAC) when compared with the control soil (Figure 1). These materials (MWC and AC) are rich in humic and fulvic compounds because during the composting process part of their organic matter was humified. On the contrary, PW is a fresh material with a low level of humified compounds. Therefore, its application did not increase the content of humic substances in the soil.

3.2. Enzymatic activities

Enzymatic activities were evaluated through some specific enzymes: an oxidoreductase (dehydrogenase) and four hydrolases (phosphatase, ß-glucosidase, urease and BAA-protease).

Soil dehydrogenase activity reflects the total range of oxidative activity of soil microflora and consequently may be a good indicator of microbiological activity in the soil [19]. Dehydrogenase activity in the soil treated with both MWC and PW increased significantly when compared with the control soil (Figure 2), despite that dehydrogenase activity was low in those materials (Table 2). On the contrary, the relatively high value of dehydrogenase activity in AC did not increase the dehydrogenase activity in the soil amended with AC (Figure 2). These results suggest that the increase of the dehydrogenase activity in soils amended with organic residues is likely due to the stimulation of microbial activity rather than to the direct addition of the enzyme from the organic sources [2].

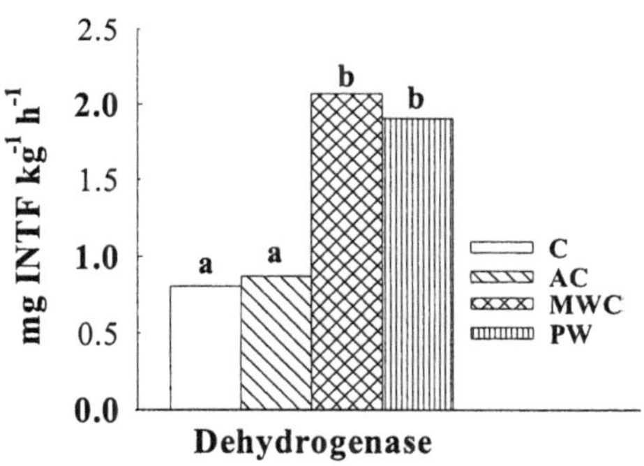

Figure 2. Dehydrogenase activity in the soils after 3 years of treatment. Values with the same letter are not significantly different ($p < 0.05$). (INTF: Iodonitrotetra zolium formazan).

Phosphatase and ß-glucosidase activities are involved in the P and C cycles, respectively, and play an important role in the hydrolytic processes that take place during organic matter breakdown [20]. Phosphatase activity increased in the soils treated with both MWC and PW, although differences between treatments were only significant for soil treated with MWC (Figure 3), which also showed a significant increase of ß-glucosidase activity.

Urease and BAA-protease activities, the other two hydrolytic enzymes studied, are involved in the last stages of organic N compound degradation by hydrolyzing C-N bonds of amide and urea, respectively. Urease activity in soil greatly increased upon the application of the organic residues. The effect was more noticeable in MWC and PW treatments (Figure 4). BAA-protease activity also increased in soils treated with organic residues, although the differences with the control soil were only statistically significant in the soil amended with MWC. Beside the positive effect caused by the application of organic matter, the ureic-N applied to these soils could also contribute to the increase of the urease activity.

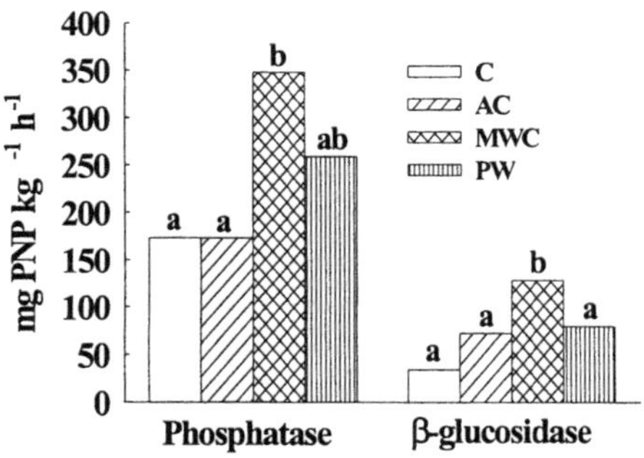

Figure 3. Phosphatase and β-glucosidase activities in the soils after 3 years of treatment. Values with the same letter are not significantly different ($p < 0.05$). (PNP: *p*-nitrophenol).

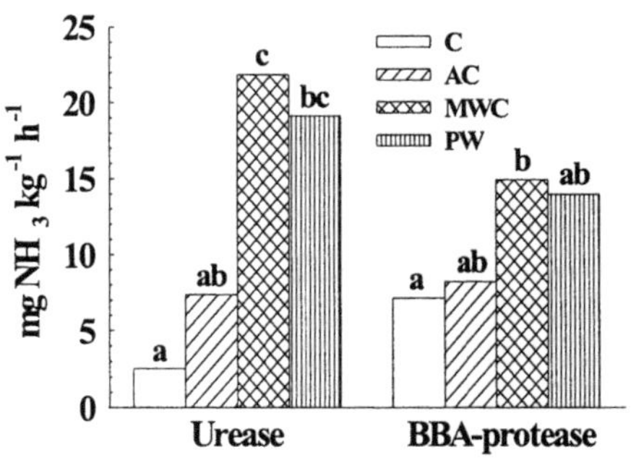

Figure 4. Urease and BAA-protease activities in the soils after 3 years of treatment. Values with the same letter are not significantly different ($p < 0.05$).

3.3. Correlation between organic matter and enzyme activities

Significant positive correlation between enzymatic activities and total organic carbon was found for all the treatments (Table 3), indicating the activation of soil microorganisms by the addition of the organic amendments [21, 22]. Organic residues added to soil promote microbial and soil enzyme activities [23]. Moreover, high correlations were found between the five enzyme activities themselves. These correlations also suggest that although each individual enzyme depends on specific substrates and takes part in specific reactions, the simultaneous measurement of different enzyme activities in soil might be a valid tool for estimating the overall soil microbiological activity and the response to management practices [24]. It is also interesting to note that β-glucosidase, which is related to the C cycle, was the only enzyme that was positively correlated with the content of soil humic substances.

Table 3
Correlation coefficients between enzyme activities and total organic carbon (TOC)

	TOC	TEC	HAC	DH-ase	Phosp	β-glu	Ure	BAA
TOC		0.400*	0.436*	0.864**	0.705**	0.602**	0.872**	0.681**
TEC			0.350*	0.236	0.293	0.531**	0.265	0.283
HAC				0.342	0.332	0.362*	0.320	0.187
DH-ase					0.658**	0.450**	0.872**	0.670**
Phosp						0.574**	0.620**	0.669**
β-glu							0.392*	0.472**
Ure								0.721**
BAA								

* Significant at the 0.05 level. ** Significant at the 0.01 level.

DH-ase: dehydrogenase activity; Phosp: phosphatase activity; β-glu: β-glucosidase; Ure: urease; BAA: BAA-protease.

Figure 5 shows the plot of the discriminant analysis useful to build a predictive model of each soil sample based on observed chemical and biochemical properties. The procedure generated, in this case, two canonical discriminant functions (F1 and F2) based on linear contributions of the variables (TOC, TEC, HAC, DH-ase, Phosp, β-glu, Ure, and BAA). These functions explain 98.6% of the variance between treatments (Table 4).

Through this analysis, it was observed that F1 is correlated with DH-ase, TOC and BAA (Table 4). This function clearly separates treatments C and AC from PW and MWC and underlines the effect of the higher doses of organic matter applied with MWC and PW during the three years of study. On the other hand, F2, which is correlated with humic substances (TEC and HAC) and β-glu (Table 4), separates treatments C and PW from AC and MWC. This function shows the difference of the effect of composted (AC and MWC) and fresh residues (PW) on soil properties.

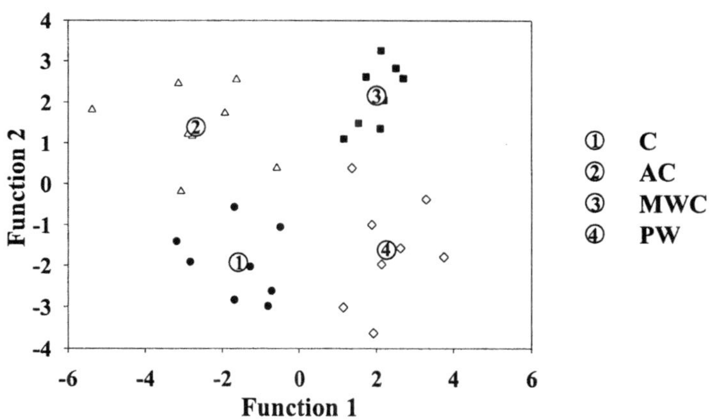

Figure 5. Discriminant analysis of some soil properties. Big circles are group centroids.

Table 4
Correlations between each variable and any discriminant function

	Function 1	Function 2
DH-ase	0.659*	0.205
TOC	0.440*	0.357
BBA	0.298*	0.111
TEC	-0.027	0.644*
HAC	0.110	0.436*
β-glu	0.291	0.407*
Phosp	0.326	0.207
Ure	0.335	0.165
% of variance	58.4	40.2
% of variance	58.4	98.6

* Largest absolute correlation.

The amount of organic matter added had a great influence on dehydrogenase, which is considered a good indicator of microbial activity in the soil. However, the nature of the organic matter added (composted or fresh material) had a clear influence on the quality of the soil organic matter and consequently on β-glucosidase, which is closely related to the C cycle.

3.4. Soil heavy metal contents

Generally, at the end of the experiment, available heavy metal contents were higher in soils treated with PW and MWC than in soils of the other treatments (Table 5). The highest mean values of available heavy metal were found in the soil treated with PW. However, high values of available soil heavy metals were not only due to the high contents of heavy metals in PW and MWC but also to the higher doses of these amendments applied to the soil.

Although several studies showed that heavy metals can inhibit soil enzymatic activities [25, 26], in the present study the increase of heavy metal availability did not affect negatively enzymatic activities. On the contrary, soils amended with the residues richest in heavy metals (MWC and PW) had the highest values of enzymatic activities. Therefore, the positive effect of the organic matter could mask the possible enzyme inhibition caused by the increase of heavy metals in soil. Similar results were found by Giusquiani et al. [27] in soils treated with composted municipal waste. Moreover, negative correlations were not found between enzymatic activities and available heavy metals in soils (data not show).

Table 5
Available (DTPA) heavy metal contents of the soil after 3 years of treatment

	Treatments							
Available-Metals	C		AC		MWC		PW	
Fe ($mg\ kg^{-1}$)	17.2	a	19.2	a	28.6	b	18.5	a
Mn ($mg\ kg^{-1}$)	10.5	a	9.88	a	11.5	a	18.3	b
Cu ($mg\ kg^{-1}$)	4.56	a	4.34	a	7.22	a	15.1	b
Zn ($mg\ kg^{-1}$)	8.36	a	8.88	a	15.2	b	24.2	c
Cd ($mg\ kg^{-1}$)	0.04	a	0.04	a	0.04	a	0.14	b
Co ($mg\ kg^{-1}$)	0.10	a	0.10	a	0.10	a	0.28	b
Ni ($mg\ kg^{-1}$)	0.26	a	0.34	ab	0.44	b	0.42	b
Pb ($mg\ kg^{-1}$)	1.32	a	1.70	a	3.22	b	1.82	a

Values followed by the same letter in the same row are not significantly different $p < 0.05$).

4. CONCLUSIONS

Repeated application of organic amendments to a sandy soil had a positive effect on soil organic matter content and soil enzymatic activities. In general, among treatments, municipal solid waste compost caused the higher positive effect on the soil properties. Moreover, a high positive correlation between soil organic carbon and soil enzymatic activities was found. The discriminant analysis underlined the effect of the higher doses of organic matter applied and the effect of composted and fresh residues on soil properties.

The relatively high concentration of heavy metals in municipal solid waste compost and in paper waste caused moderate increases of heavy metal contents in soil but no inhibition of soil enzymatic activities associated to heavy metals contamination was observed.

The use of these residues can contribute to enhance the level of organic matter and probably the fertility of the agricultural soils in SW Andalusia, which are particularly poor in organic matter. Additionally, this practice would be an effective way of recycling the residues and solving the problem of their disposal.

REFERENCES

1. Masciandaro, G., Ceccanti B., García, C., 1997. Soil agro-ecological management: fertirrigation and vermicompost treatments. Biores. Technol. 59, 199-206.
2. Martens, D.A., Johanson, J.B., Frankenberger, W.T., 1992. Production and persistence of soil enzyme with repeated addition of organic residues. Soil Sci. 153, 53-61.
3. Bellamy, K.L., Chong, C., 1995. Paper sludge utilization in agriculture and container nursery culture. J. Environ. Qual. 24, 1074-1083.
4. Giusquiani, P.L., Pagliai, M., Gigliotti, G., Businelli, D., Benetti, A., 1995. Urban waste composts: effects on physical, chemical and biochemical soil properties. J. Environ. Qual. 24, 175-182.
5. Barker, A.V., 1997. Composition and uses of compost. *In*: Rechling, J.E., Mackinnon, H.C. (Eds.), Agricultural Uses of By-products and Wastes. American Chemical Society, Washington DC, ACS Symposium Series Nº 668, Vol. 10, pp. 140-162.
6. Entry, J.A., Wood, B.H., Edwards, J.H., Wood, C.W., 1997. Influence of organic by-products and nitrogen source on chemical and microbiological status of an agricultural soil. Biol. Fertil. Soils. 24, 196-204.
7. Pascual, J.A., García, C., Hernández, T., Ayuso, M., 1997. Changes in the microbial activity of an arid soil amended with urban organic wastes. Biol. Fertil. Soils 24, 429-434.
8. Salam, A.K., Desvia, Y., Sutanto, E., Syam, T., Nugroho S.G., Kimura, M., 1999. Activities of soil enzymes in different land-use systems in middle terrace areas of Lampung province, South Sumatra, Indonesia. Soil Sci. Plant. Nutr. 45 (1), 89-99.
9. Ceccanti, B., García, C., 1994. Coupled chemical and biochemical methodologies to characterize a composting process and the humic substances. *In*: Senesi, N., Miano, T. (Eds.), Proceedings 6th Int. Meeting of Int. Humic Substances Society (IHSS), September 1993, Monopoli, Bari, Italy. Elsevier, Amsterdam, pp. 20-25.
10. Soil Survey Staff, 1996. Keys to Soil Taxonomy. US Department of Agriculture, Soil Conservation Service, Washington, DC.
11. Hesse, P.R., 1971. A Textbook of Soil Chemical Analysis. John Murray, London.
12. Walkley, A., Black, J.A., 1934. An examination of the Degtjareff method for determining soil organic matter and a proposed modification of the chromic acid titration method. Soil Sci. 37, 29-38.
13. Yeomans, J.C., Bremner, J.M., 1988. A rapid and precise method for routine determination of organic carbon in soil. Commun. Soil Sci. Plant Anal. 19, 1467-1476.
14. Lindsay, W.L., Norvell, W.A., 1978. Development of a DTPA soil test for zinc, iron, manganese and copper. Soil Sci. Soc. Am. J. 42, 421-428.
15. Trevors, J.T., 1984. Dehydrogenase activity in soil: a comparison between the INT and TTC assay. Soil Biol. Biochem. 16, 673-674.
16. García, C., Henández, T., Costa, F., Ceccanti, B., Masciandaro, G., 1993. The dehydrogenase activity of soils as an ecological marker in processes of perturbed system

regeneration. *In*: Gallardo, J.F., (Ed.), XI Int. Symp. of Environmental Biogeochemistry, 11th, Salamanca, Spain. September 1993. CSIC, Salamanca, Spain, pp 89-100.
17. Tabatabai, M.A., 1994. Soil enzymes. *In*: Weaver, R.W., Angle, J.S., Bottomley, P.S. (Eds.), Methods of Soil Analysis. Part 2. Microbiological and Biochemical Properties. Soil Sci. Soc. Am. Madison, WI, pp. 778–833.
18. Nannipieri, P., Ceccanti, B., Cervelli, S., Matarese, E., 1980. Extraction of phosphatase, urease, protease, organic carbon and nitrogen from soil. Soil Sci. Soc. Am. J. 44, 1011–1016.
19. Skujins, J., 1976. Enzymes in soil. *In*: Mc Laren, A.D., Peterson, G.H. (Eds.), Soil Biochemistry. Vol. 1, Marcel Dekker, New York, pp. 371-414.
20. Müller, U., Wegener, P., 1988. Interaction of humic substances with biota. *In*: Frimmel F.H., Christman, R.F. (Eds.), Humic Substances and their Role in the Environment. John Wiley & Sons, NY, pp. 179–192.
21. Dick, R.P., 1994. Soil enzyme activities as indicators of soil quality. Soil Sci. Soc. Am. J. 35, 107–124.
22. Masciandaro, G., Ceccanti, B., García, C., 1997. Changes in soil biochemical and cracking properties induced by "living mulch systems". Can. J. Soil Sci. 77, 579–587.
23. Nannipieri, P., Muccini, L., Ciardi, C., 1983. Microbial biomass and enzyme activities: production and persistence. Soil Biol. Biochem. 15, 679–685.
24. García, C., Hernández, T., Costa, F., Ceccanti, B., 1994. Biochemical parameters in soils regenerated by the addition of organic wastes. Waste Mngt. Res. 12, 457–466.
25. Frankenberger, W.T., Johanson, J.B., Nelson, C.O. 1983. Urease activity in sewage sludge-amended soils. Soil Biol. Biochem. 15, 543–549.
26. Reddy, G.B., Faza, A., Bennet, R., Jr., 1987. Activity of enzymes in rhizosphere and non-rhizosphere soils amended with sludge. Soil Biol. Biochem. 19, 203–205.
27. Giusquiani, P.L., Gigliotti, G., Businelli, D., 1994. Long-term effects of heavy metals from composted municipal waste on some enzyme activities in a cultivated soil. Biol. Fertil. Soils 17, 257–262.

Developments in Soil Science, Volume 28B
Editors: A. Violante, P.M. Huang, J.-M. Bollag and L. Gianfreda

OXYGENASE SYSTEMS IN AN OLIGOTROPHIC BACTERIAL COMMUNITY OF A SUBSURFACE WATER POLLUTED BY BTEX

L. Cavalca[a], E. Dell'Amico and V. Andreoni

Dipartimento di Scienze e Tecnologie Alimentari e Microbiologiche, Università degli Studi Via Celoria 2, 20133 Milano, Italy

[a] Corresponding author: Fax: +39.02.70630829. E-mail: lucia.cavalca@unimi.it

Molecular probes were designed to detect genes coding for various oxygenase systems. PCR amplifications were carried out either on DNA extracted directly from water samples or from isolates. In the DNA from water samples, genes for toluene-benzene monoxygenase (*tbmA*), for toluene dioxygenase (*todC1*), for catechol 2,3-dioxygenase (*xylE*), and for catechol 1,2-dioxygenase (*catA*) were present. Among 100 strains isolated to check the presence of hydrocarbon-degrading bacteria, 11 grew on aromatic substrates and most of them attacked aromatics through monoxygenase enzymatic systems. In strains able to grow on benzene, the nucleotide sequence of the 580 bp PCR fragment showed 78% homology to the *bmoB* gene for benzene monoxygenase of *Pseudomonas aeruginosa* JI104. In strains able to grow on *m*-xylene and *p*-xylene, the PCR fragment showed 66% homology to the *pheB* gene for phenol hydroxylase of *Ralstonia eutropha* JMP134. Degradation patterns of isolates monitored by HPLC analysis were in accord with the presence of these degradative enzymes. The probes developed in the present study enlarge the detection of degradative systems in the environment.

1. INTRODUCTION

The deliberate or accidental release of aromatic hydrocarbons leads to serious environmental and public health damage. One particular concern is the contamination of groundwater by benzene, toluene, ethylbenzene, and xylenes (BTEX); these are highly toxic and/or carcinogenic compounds. Microorganisms have evolved different pathways to utilize these compounds as carbon sources [1], and this capability can be used to accomplish bioremediation at contaminated sites.

The presence of one compound may affect the degradation of another. It has been reported that in a *Pseudomonas* sp. strain, the presence of toluene aided the degradation of benzene and xylene, whereas the presence of xylene retarded degradation of benzene and toluene [2]. In an aquifer, the biodegradation of an aromatic mixture can be even more complex. The concentration of substrates may change due to flowing groundwater, causing microbial population densities and distribution to be quite variable [3].

In the groundwater environment, the presence of a microbial population capable of degrading a mixture of xenobiotics even at low concentrations assures decontamination of oligotrophic habitats. More information is needed about microbial activities in the environment.

Optimization of biotreatment processes will be promoted when the most effective pollutant metabolizers are known and used under various conditions. Only cultivated microorganisms have been characterized for their degradative capabilities, but it is now recognized that they may not be the most effective microorganisms in the natural microbial community [4].

The objective of the present work was to check the presence of degradative systems in a BTEX-polluted subsurface water by detecting catabolic genes both directly in DNA water samples and in microbial degrading isolates.

2. MATERIALS AND METHODS

2.1. Water sampling

Water samples were taken from hydrocarbon-polluted groundwater underneath a paint factory located in the southern part of Milan. A pilot plant, consisting of an air sparging well and 8 monitoring wells, was set up to determine both the effectiveness of air sparging and the key design parameters; this knowledge could then be applied to the bioremediation of the aquifer using a full scale air sparging treatment.

Water was sampled from a monitoring well at 4.5 m depth at two successive times: the first one before starting of air sparging and the second one after 5 months of air sparging treatment. Water samples were put in sterile 1-liter bottles, brought to the laboratory and immediately used for microbiological analysis.

2.2. Strain isolation and growth conditions

Water samples were serially diluted in NaCl solution (9 g l^{-1}) and seeded on solid medium or inoculated in appropriate liquid medium. Counts of heterotrophic bacteria (CFU ml^{-1}) were carried out using tryptone yeast extract medium (TYEM) [5]; colonies were counted after 5 days of incubation at 30°C. To determine the number of toluene-, 1,2,4-trimethylbenzene- and *o*-, *m*-, and *p*-xylene-degrading bacteria, duplicates of each dilution were inoculated into 0.1x M9 mineral medium [6] to which 400 mg $l^{-1}$of each aromatic compound were separately added as a carbon source. After 14 days of incubation at 30°C, visible turbidity and microscopic observation indicated bacterial growth.

The number of bacteria was expressed as most probable number (MPN) ml^{-1} of sample. Strains were routinely cultured on TYEM and on 0.1x M9 liquid medium supplemented separately with various hydrocarbons. Fifty colonies were isolated randomly from plate count media obtained in the two subsequent samplings. Strains were characterized for Gram, catalase, and oxidase reactions. The growth characteristics of isolates on aromatic hydrocarbons were studied in 100 ml flasks containing 20 ml of 0.1x M9 medium with 250 mg l^{-1} of various aromatic substrates as the sole carbon and energy source. These flasks were incubated at 30°C in a jar containing the corresponding hydrocarbon in vapor phase. Growth and substrate utilization were registered as positive when cultures became turbid (optical density at 600 nm ($O.D._{600}$) > 0.5 in 3 days) in two subsequent transplants. Strains were maintained in 20% glycerol stock at –20°C.

2.3. Chemical analyses

The concentrations of hydrocarbons in water were determined (method number 6220 B and C) according to Standard Methods for Examination of Water and Wastewater [7].

The degradation of toluene and benzene by *Pseudomonas* sp. strain Z22 and of toluene and *m*-xylene by *Mycobacterium* sp. strain C3 was studied. Strains were grown as shaken liquid cultures in 100 ml vials each containing 15 ml of 0.1x M9 medium supplemented with 250 mg $l^{-1}$of the appropriate aromatic compound. Each vial was inoculated with 1 ml of cell suspension to an $O.D._{600}$ of 0.15 measured by a Beckman model DU 640 spectrophotometer, after growth on the same medium. Vials were then sealed with Teflon-coated grey butyl rubber stoppers and aluminium crimps and incubated at 30°C on a rotary shaker at 200 rpm. Three replicates were analyzed for the extent of degradation and the production of metabolites at successive incubation times. Samples were mixed with an equal volume of methanol, withdrawn from vials using a sterile syringe and stored at –20°C until HPLC analysis. This was carried out with an HPLC chromatograph (mod. LG 980/02, Jasco ternary gradient unit) equipped with an octyl silane column (Merck Lichrosorb RP-8, length 25 cm). The flow rate of eluent was 1.0 ml min^{-1}. Compounds were detected by their UV absorbance at 254 nm with a UV-975 Intelligent Jasco detector and identified by comparison of HPLC retention times to those of authentic standards. Concentrations of aromatic compounds in samples were calculated using standards with known concentrations.

2.4. Chemicals

Benzene, toluene, *o*-, *m*-, and *p*- isomers of xylene, and 1,2,4-trimethylbenzene (purity > 99%) were purchased from Merck (Darmstadt, Germany). Ethylbenzene, *o*-, *m*-, and *p*- isomers of cresol, benzoic acid, 3-methylcatechol, 4-methylcatechol, catechol, 3,4-dihydroxybenzoic acid, and phenol (purity > 99%) were purchased from Sigma-Aldrich (Steinheim, Germany).

2.5. DNA extraction

DNA extraction was carried out on cell suspension with an $O.D._{600}$ of 2.0 in 100 μl; cells were centrifuged at 13,000 rpm for 7 min and resuspended with 100 μl of sterile MilliQ water, 100 μl of 10 mM tris-HCl (pH 8.0), and 13 μl of proteinase K (1 mg ml^{-1}). The mix was incubated for 2 hours at 55°C then boiled for 10 min and centrifuged at 13,000 rpm for 5 min. The supernatant containing DNA was drawn off and put in sterile eppendorfs. DNA extraction from a 40-ml water sample was carried out as described above.

2.6. Polymerase chain reaction (PCR) amplifications

Amplifications of *xylE* were carried out according to Joshi and Walia [8]. Amplifications of *catA* were performed in a final volume of 50 μl containing: 5 μl of 10X buffer (Bioline, UK), 1.25 mM $MgCl_2$, 200 μM nucleotides (Pharmacia Applied Biotech, Sweden), 0.25 μM forward primer 5'-AAG GTC ACC TAC GAC GAA TAC AAC GGC CTC AAG-3' and 0.25 μM reverse primer 5'-AAG TAG AGC TGG GTG GTG AT-3' (Genenco, Florence, Italy), 1 U *Taq*-polymerase (Bioline, UK), and 1 μl of template DNA. Primers were placed in positions 761 and 1131 of the reference nucleotide of *catA* gene of strain *Rhodococcus erytropolis* ATCC4277 [9]. The thermal profile was: 94°C for 2 min, then *Taq* was added to PCR mixture and followed by 35 cycles (94°C for 40 s, 60°C for 50 s, 72°C for 1 min) and

72°C for 7 min. Amplification of *todC1* was carried out according to Whyte et al. [10] and of *xylA,M* according to Cavalca et al. [11].

On the basis of consensus of known sequences for toluene *o*-xylene monoxygenase [12,13,14], benzene monoxygenase *touA* [15], toluene 2-, 3-, and 4-monoxygenase *tbmA* [12, 13, 14], and benzene monoxygenase *bmoB* [16], degenerated primers were developed for PCR amplification of homologous *tbmA* genes in water isolates. Primers were placed in positions 376 and 1569 of the reference nucleotide of *touA* gene strain *P. stutzeri* OX1. Amplifications were performed in a total volume of 25 μl, containing: 2.5 μl of 10X buffer; 2.5 mM $MgCl_2$; 200 μM dNTPs; 0.25 μM forward primer 5'-AAG ACC TAT CCS GAR TAC GT-3' and 0.25 μM reverse primer 5'-GGC TGG ATC WGR CCT GCS AGG AA-3' (Pharmacia Applied Biotech, Sweden) and 1 U *Taq*-polymerase. To amplify different gene sequences, two different thermal profiles were set up. The first one started with 94°C for 3 min followed by the addition of 1 U of *Taq*-polymerase, then 35 cycles (94°C for 1 min, 55°C for 1 min, 72°C for 1 min), and a final extension step at 72°C for 10 min was completed. In the second profile, the DNA denaturation and *Taq* addition steps were the same as in the first one; then two series of 35 cycles were performed, one with the annealing temperature of 60°C and the second at 50°C.

16S rDNA genes were amplified as specified elsewhere [11], using eubacterial universal primers P27f and P1495r referred to *E. coli* nucleotide sequence of 16S rDNA gene. Digestions with endonucleases *HhaI*, *HinfI*, and *RsaI* (Pharmacia Applied Biotech, Sweden) were carried out as specified by the manufacturer's instructions.

2.7. Sequence analysis

The nucleotide sequences of 16S rDNA and of a 580 bp *tbmA* internal fragment were determined according to the Perkin Elmer Applied Biosystem protocol. The forward and reverse samples were run on an Applied Biosystems 373A sequencer. Primers used in the PCR reaction of sequencing products were the same as in normal PCR.

2.8. Statistical methods

Sizes of fragments from restriction profile gels were estimated by using a linear regression equation between the molecular mass of the DNA ladder and the log of the distance covered by fragments within the same gel run.

Similarity trees among catabolic gene sequences were built using the Jalview program (http://jura.ebi.ac.uk:6543/~michele/jalview/) on the basis of alignments obtained from the Clustalview program (http://searchlauncher.bcm.tmc.edu:9331/multi-align/).

2.9. Sequence accession numbers

Sequence accession numbers of 16S rDNA genes are: Y17229 *Microbacterium aurum;* U65012 *Pseudomonas stutzeri;* AF131104 *P. putida;* AJ131117 *Stenotrophomonas maltophilia;* AB021392 *P. pictorum*; Y08453 *Mycobacterium smegmatis;* U94745 *Mycobacterium duvalii;* Z35330 *Bradyrhizobium japonicum;* D86354 *Bradyrhizobium* sp.; AF064460 *P. veronii;* AF064457 *P. orientalis;* X74913 *Zooglea ramigera;* AF011344 *Azoarcus* sp.

Sequence accession numbers of toluene monoxygenase genes are: U04052 *tbuA1* of *P. pickettii* PKO1, AF065891 *pheB* of *Ralstonia eutropha* JMP134, D83068 *bmoB* of *P. aeruginosa* JI104, M65106 *tmoA* of *P. mendocina* KR1, AF001356 *tbhA,B* of *Burkholderia cepacia* AA1, AJ005663 *touA* of *P. stutzeri* OX1, L40033 *tbmA* of *Burkholderia* sp. JS150.

3. RESULTS AND DISCUSSION

3.1. Characterization of groundwater samples

The major contaminants in groundwater were benzene, toluene, ethylbenzene, and xylenes (Table 1).

Table 1
Characterization of groundwater samples

Sampling	Oxygen ($\mu g\ l^{-1}$)	Contaminant Concentrations ($\mu g\ l^{-1}$) Toluene	Xylenes[a]	Ethyl benzene	Microbial Titers A-H[b] (UFC ml^{-1})	T-O[c] (MPN ml^{-1})	X-O[d] (MPN ml^{-1})	TMB-O[e] (MPN ml^{-1})
10/31/1997	0.8	18,239	21,230	3,502	1.5×10^6	2.3×10^4	2.3×10^4	2.3×10^3
03/24/1998	2.8	88	160	139	1.5×10^5	6.2×10^3	2.3×10^3	2.3×10^3

[a]Xylene: equimolar concentration of *o*-, *m*- and *p*-xylenes; [b]A-H: aerobic heterotrophic bacteria; [c]T-O: toluene-oxidizing bacteria; [d]X-O: *o*-, *m*-, *p*-xylenes-oxidizing bacteria; [e]TMB-O: 1,2,4-trimethylbenzene-oxidizing bacteria.

Microbiological characterization of the groundwater samples showed the presence of aerobic hydrocarbon-degrading bacteria. The presence of potential degrading bacteria was checked also by molecular methods. PCR amplifications of catabolic genes on DNA extracted directly from water samples were positive for toluene-benzene monoxygenase (*tbmA*), toluene dioxygenase (*todC1*), catechol 2,3-dioxygenase (*xylE*) and catechol 1,2-dioxygenase (*catA*) (Figure 1).

3.2. Bacterial ARDRA fingerprints of aquifer community

To determine the diversity of the bacterial community, 60 strains are normally necessary to obtain a sufficient picture of the species richness, but in a polluted environment, the study of 40 strains was sufficient to obtain good results [17].

Restriction profiles of 16S rDNA genes amplified from isolates, gotten by *Hha*I, *Rsa*I, and *Hinf*I restriction enzymes, grouped the 100 isolates in different ARDRA groups (Figure 2). A high species diversity was shown in the isolates of the first sample (27 ARDRA groups). Groups P21 and P24 included 10% and 24%, respectively, of the community, probably representing the best suited strains for these environmental conditions. Changes in band pattern profiles were observed in the isolates of the second sample, reflecting changes in community composition. These microorganisms assembled in 15 ARDRA groups, with a more homogeneous distribution. These groups, P24, P1, and P4, were present in both samplings and the P24 group was the dominant group in both.

3.3. Growth characteristics of isolates

Among 100 isolates, 11 strains were capable of growing on aromatic hydrocarbons (Table 2). Eight strains isolated from the first sampling grew only on benzene and on mono-alkylated benzene; three strains isolated from the second sampling did not grow on benzene, but grew on different alkylated benzene. All 30 strains of the P24 ARDRA group grew on catechol, the central metabolite of various degradation pathways.

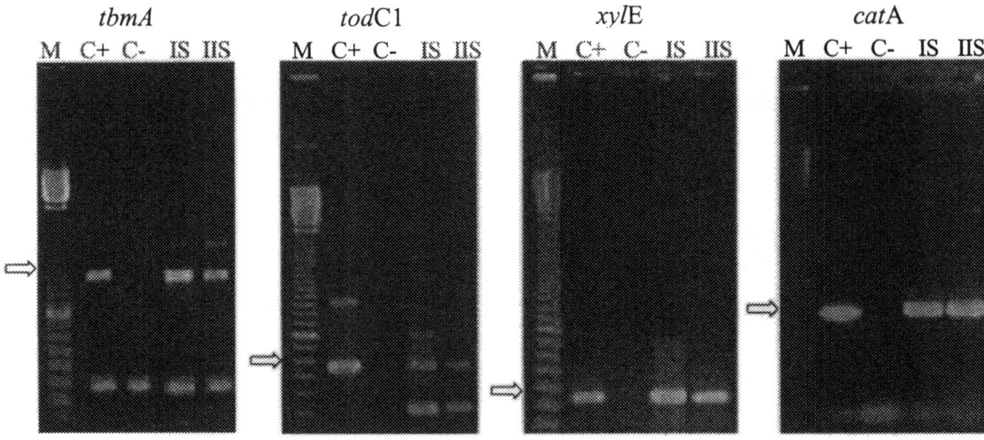

Figure 1. PCR amplifications of catabolic genes in DNA extracted from water at the two samplings. *TbmA*, toluene-benzene monoxygenase (1200 bp); *todC1*, toluene dioxygenase (560 bp); *xylE*, catechol 2,3-dioxygenase (350 bp); *catA*, catechol 1,2 dioxygenase (570 bp). M = 100 bp molecular mass ladder; C- = negative control; IS = first sampling; IIS = second sampling; *tbmA* C+ = *Mycobacterium* sp. C3; *todC1* C+ = *E. coli* JM109 pDTG601; *xylE* C+ = *E. coli* JM109 pGSH2960; *catA* C+ = *R. erytropholis* ATCC4277.

3.4. Phylogenetic analyses of bacteria

One representative strain of each ARDRA group of degrading bacteria and of the P24 group was identified on the basis of 550 bp nucleotide sequence of two strands of 16S rDNA (Figure 2). Strain Z22 was related to *P. stutzeri* Zobell (U65012) with a similarity of 88%, strain Z6 to *P. pictorum* ATCC23328T (AB021392) with a similarity of 95%, strain C27 to *P. veronii* (AF064460) with a similarity of 99%, strain H54 to *Azoarcus* sp. BH72 (AF011344) with a similarity of 93%, strain C9 to *Bradyrhizobium* sp. USDA4362 (D86354) with a similarity of 97%, strain C3 to *Mycobacterium smegmatis* (Y08453) with a similarity of 97%, and finally, strain H10 to *Microbacterium aurum* (Y17229) with a similarity of 93%. Isolates belong to the α, β, and γ subdivisions of proteobacteria and to the low G-C content Gram-positive group. The majority of the degrading bacteria were affiliated with the *Pseudomonas* genus.

Phylogenetic analyses of restriction pattern profiles of the 100 isolates from the two water samplings were performed by the neighbor joining method [18]. Comparison between ARDRA groups of water communities and sequence data of the degrading strains allows the inference that most of the isolates were closely related to the *Pseudomonas* sp. genus.

3.5. Characterization of oxygenase systems

PCR amplifications of catabolic genes were performed to detect the pathway present in degrading bacteria (Table 2). *Pseudomonas* sp. strains Z12, Z17, Z22, Z23, and Z24 gave amplifications with primers for toluene monoxygenase *tbmA*, with primers for catechol 1,2-dioxygenase *catA*, and for catechol 2,3-dioxygenase *xylE*. All 30 strains of the P24 group able to grow only on catechol showed the presence of *xylE*. *Mycobacterium* sp. C3 and *Bradyrhizobium* sp. C9 gave amplification with *tbmA* primers only. In *Microbacterium* sp.

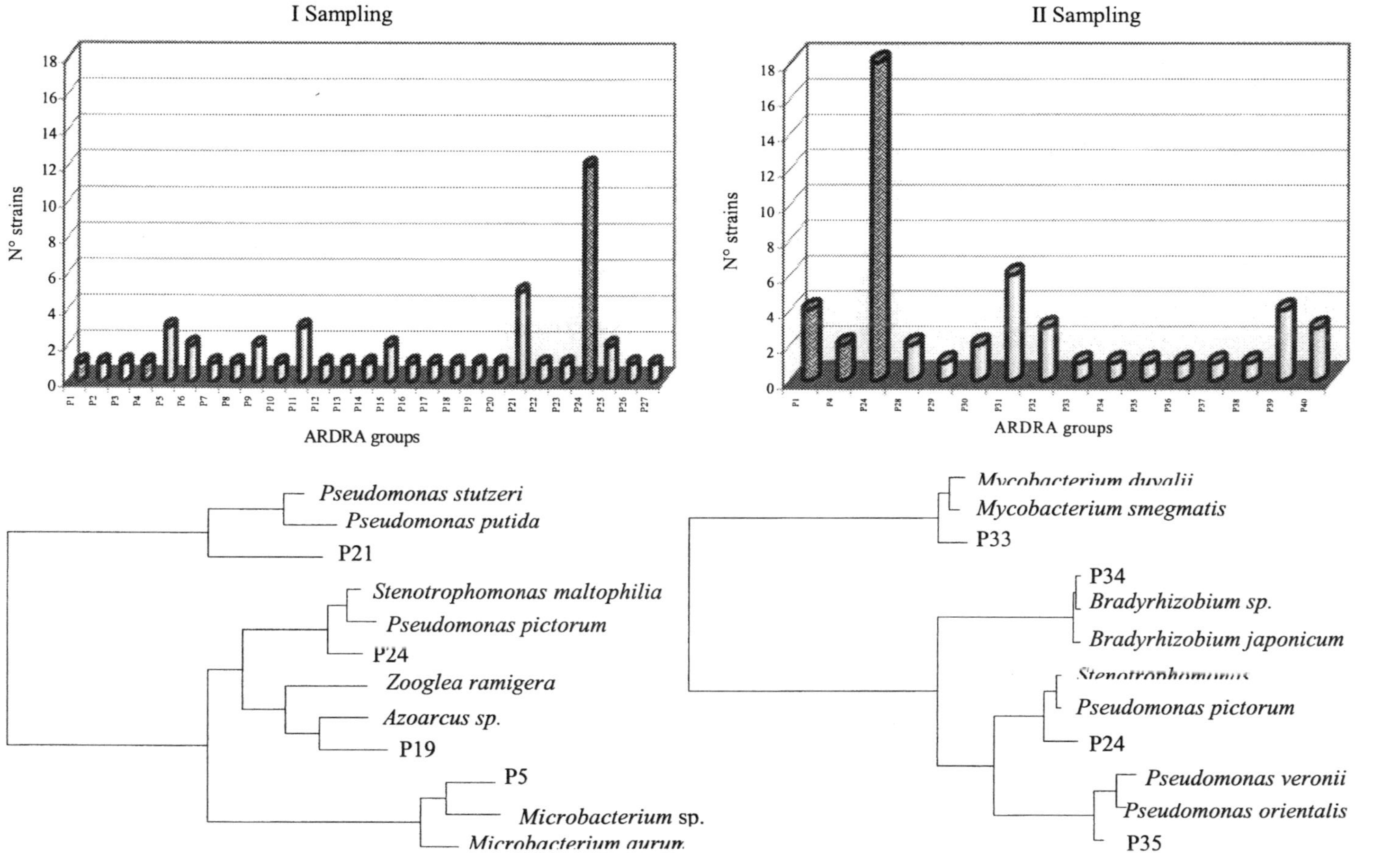

Figure 2. Distribution of isolates on the basis of ARDRA groups at the two samplings and phylogenetic distance trees [19] produced from partial 16S rDNA sequences showing affiliations of degrading bacteria with the most closely related 16S rDNA sequences present in databases.

Table 2
Growth characteristics of degrading strains and PCR amplifications of catabolic genes

	ARDRA groups	Strains	Identification	Aromatic compounds							PCR amplifications		
				T[a]	B[b]	EtB[c]	*m*-X[d]	*p*-X[e]	TMB[f]	C[g]	*tbmA*	*xylE*	*catA*
10/31/1997	P5	H10, H33	*Microbacterium* sp.	+	-	+	-	-	-	+	-	-	-
	P19	H54	*Azoarcus* sp.	+	-	+	-	-	-	+	-	-	-
	P21	Z12, Z17, Z22, Z23, Z24,	*Pseudomonas* sp.	+	+	+	-	-	-	+	+	+	+
	P24	Z1, Z2, Z3, Z4, Z5, Z6, Z7, Z8, Z9, Z13, Z14, Z18	*Pseudomonas* sp.	-	-	-	-	-	-	+	-	+	-
03/24/1998	P24	C4, C6, C7, C8, C13, C14, C25, C28, C32, C34, C36, V6, V10, V12, V16, V20, V32, V34	*Pseudomonas* sp.	-	-	-	-	-	-	+	-	+	-
	P34	C3	*Mycobacterium* sp.	+	-	-	+	+	+	+	+	-	-
	P33	C9	*Bradyrhizobium* sp.	+	-	-	+	+	+	+	+	-	-
	P35	C27	*P. veronii*	+	-	+	-	-	-	+	-	-	-

[a]T= toluene; [b]B= benzene; [c]EtB= ethylbenzene; [d]*m*-X= *m*-xylene; [e]*p*-X= *p*-xylene; [f]TMB= 1,2,4-trimethylbenzene; [g]C= catechol.

+ = growth after 3 day-incubation ($O.D._{600nm}>0.5$); - = absence of growth.

strains H10 and H33, *Azoarcus* sp. H54, and *P. veronii* C27, no catabolic genes were amplified. The absence of amplification could be due to the presence of gene sequences with a low homology with the sequences known in the literature.

These preliminary data seem to indicate that most of the subsurface water microorganisms utilize hydrocarbons through monoxygenase enzymatic systems. *Pseudomonas* strains of the P21 ARDRA group showed the presence of two lower pathways catechol 1,2-dioxygenase and catechol 2,3-dioxygenase. Because no isolated bacteria carried the *todC1* gene, but a positive signal was found in PCR amplification of total DNA from water samples, it is possible that toluene dioxygenase was present within the bacterial community of the aquifer in low concentration or in uncultivable strains. This result emphasizes the importance of molecular analyses applied to environmental samples within a multidisciplinary evaluation process to predict the applicability of biological treatments for the cleanup of contaminated sites.

Data obtained from restriction fragment analyses of *tbmA* genes present in the isolates revealed the presence of two different oxygenases. Nucleotide sequencing performed on the two different *tbmA* genes from representative strains *Pseudomonas* sp. Z22 and from *Mycobacterium* sp. C3 confirmed this diversity. *TbmA*-like gene of strain Z22 showed 78% homology to the benzene monoxygenase (*bmoB*) of *P. aeruginosa* JI104 (Figure 3); that of *Mycobacterium* C3 had 66% homology to phenol hydroxylase of *Ralstonia eutropha* JMP134 (*pheB*) and to toluene-3-monoxygenase of *P. pickettii* PK01 (*tbuA*).

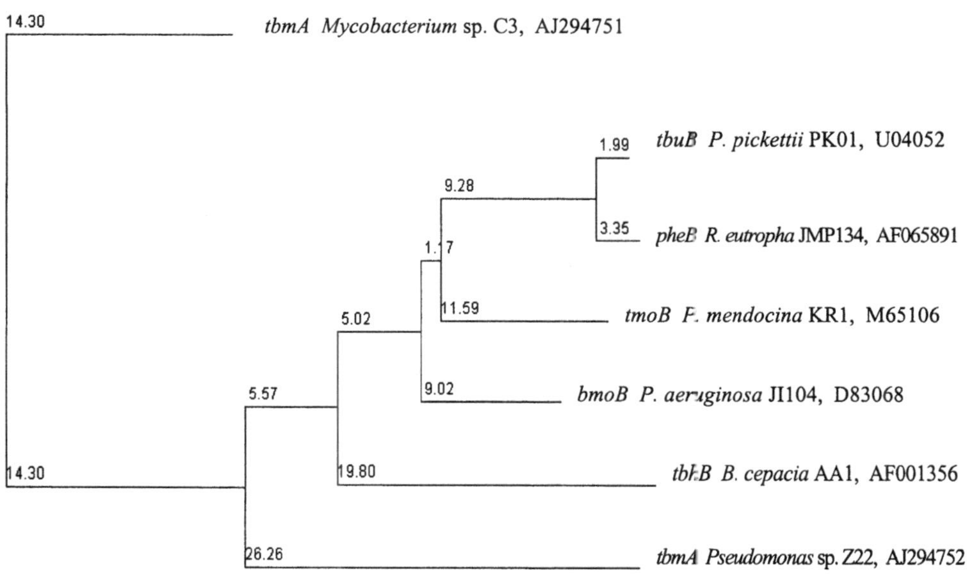

Figure 3. Toluene-benzene monoxygenase: phylogenetic tree generated from CLUSTALW 1.8 [19] alignment of translation products of *tbmA* fragments of *P. putida* Z22, *Mycobacterium* sp. C3 and published sequences by the neighbor joining method.

These low homologies suggest that monoxygenases of aquifer isolates were different from the ones reported in the literature, which were isolated from soil. To the best of our knowledge, the presence of putative monoxygenase genes in a Gram-positive bacterium has never been described.

3.6. Degradation pathway

To confirm amplification of a *tbmA*-like gene, kinetic degradation was performed with strains Z22 and C3.

Toluene was oxidized by Z22 cells to *p*-cresol, which was subsequently hydroxylated to 4-methylcatechol, as described in *Burkholderia* sp. JS150 [20] and to 3,4-dihydroxybenzoic acid as described in *P. mendocina* KR1 [14]. This strain degraded benzene via phenol and cathecol (Figure 4). This is in accordance with the amplifications of a benzene toluene-4-monoxygenase and the amplifications of two genes of lower pathways: *xylE* responsible for *meta* cleavage of catechol and methylcatechols and *catA* responsible for *ortho* cleavage of 3,4-dihydroxybenzoic acid.

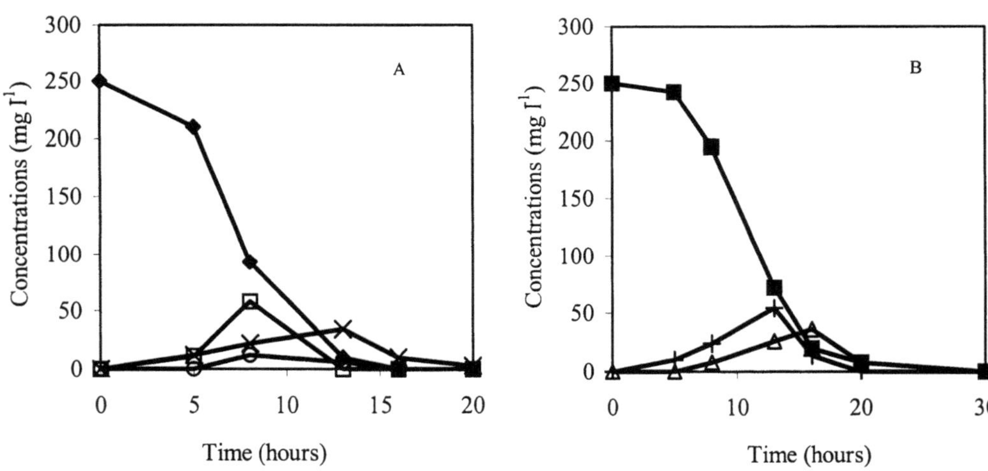

Figure 4. Time course of toluene (A) and benzene (B) degradation by growing cells of *Pseudomonas* sp. Z22: (♦), toluene; (□), *p*-cresol; (O), 4-methylcatechol; (✕), 3,4-dihydroxybenzoate; (■), benzene; (+), phenol; (△), catechol.

Toluene instead was differently degraded by C3 cells. This strain utilized completely 250 mg l^{-1} in 72 hours through the formation of catechol, of 3-methylcatechol, and of products with a retention time corresponding both to benzoic acid and to cresols. Such a finding has been recently documented in rhodococci strains for the degradation of *o*-xylene [21] and of toluene [22]. Finally, *m*-xylene was also degraded and 3-methylbenzoic acid and 3-methylcatechol were transiently formed in cultural broth (Figure 5). Thereby, this monoxygenase system was able to attack directly both the aromatic ring and the alkyl group.

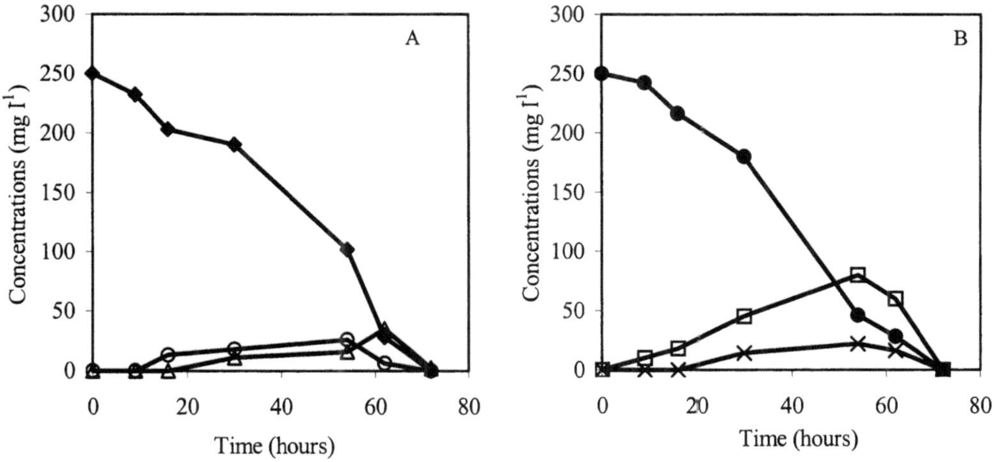

Figure 5. Time course of toluene (A) and *m*-xylene (B) degradation by growing cells of *Mycobacterium* sp. C3: (♦), toluene; (O), 3-methylcatechol; (△), catechol; (●), *m*-xylene; (□), 3-methylbenzoic acid; (✕), 3-methylsalicylic acid.

4. CONCLUSIONS

A high species diversity was encountered in the two communities of the polluted groundwater, even if species richness can be reduced by the presence of contaminants. On the seven 16S rDNA sequences analyzed, four gave good matches with already known sequences (97-99%). Lower matches were found for other strains, probably because they are members of less studied genera and species and fewer gene sequences are available. *Pseudomonas* genus was most represented and particularly the genotype P24 was always dominant, taking a feed advantage from the ability to grow on catechol.

Most BTEX-degrading bacteria harbored toluene-benzene monoxygenase systems. The low homologies encountered suggest that monoxygenases of aquifer isolates were different from those reported in literature, which were mostly isolated from soil. Biodiversity encountered in water isolates suggests a probable different evolution of degradative systems in bacteria retrieved in different environments.

ACKNOWLEDGMENT

This work was supported by a contribution of the Italian National Research Council (CNR).

REFERENCES

1. van der Meer, J.R., de Vos, W.M., Harayama, S. and Zehnder, A.J., 1992. Molecular mechanisms of genetic adaptation to xenobiotic compounds. Microbiol. Rev. 56, 677-694.

2. Alvarez, P.J.J., Vogel, T.M., 1991. Substrates interactions of benzene, toluene and *para*-xylene during microbial degradation by pure cultures and mixed culture aquifer slurries. Appl. Environ. Microbiol. 57, 2981-2985.
3. Mackay, D.M., Roberts, P.V., Cherry, J.A., 1985. Transport of organic contaminants in groundwater. Environ. Sci. Technol. 19, 384-392.
4. Nogales, B., Moore, E.R.B., Abraham, W.R., Timmis, K.N., 1999. Identification of the metabolically active members of a bacterial community in a polychlorinated biphenyl-polluted moorland soil. Environ. Microbiol. 1, 199-212.
5. Mikesell, M.D., Olsen, R.H., Kukor, J.J., 1991. Stratification of anoxic BTEX-degrading bacteria at three petroleum-contaminated sites. *In:* Hinchee, R.E., Olfenbuttel, R.F. (Eds.), In Situ Bioreclamation Application and Investigations for Hydrocarbon and Contaminated Site Remediation. Butterworth-Heinemann, Boston, pp. 351-361.
6. Kunz, D.A., Chapman, P.J., 1981. Catabolism of pseudocumene and 3-ethyltoluene by *Pseudomonas putida (arvilla)* mt-2: evidence for new function of the TOL (pWW0) plasmid. J. Bacteriol. 146, 179-191.
7. Greenberg, A.E., Eaton, A.D., Cleseri, L.S., (Eds.), 1997. Standard Methods for Examination of Water and Wastewater. American Technical Publishers Ltd, XVIII Ed., Washington.
8. Joshi, B., Walia, S., 1995. PCR amplification of catechol 2,3-dioxygenase gene sequences from naturally occurring hydrocarbon degrading bacteria isolated from petroleum hydrocarbon contaminated groundwater. FEMS–Microb. Ecol. 19, 5-15.
9. Murakami, S., Kodama, N., Shinke, R., Aoki, K., 1997. Classification of catechol 1,2-dioxygenase family: sequence analysis of a gene for the catechol 1,2-dioxygenase showing high specificity for methylcatechols from Gram+ aniline-assimilating *Rhodococcus erythropolis* AN-13. Gene 185, 49-54.
10. Whyte, L.G., Greer C.V., Innis, W.E., 1996. Assessment of the biodegradation potential of psychotrophic micro-organisms. Can. J. Microbiol. 42, 99-106.
11. Cavalca, L., Confalonieri, A., Larcher, S., Andreoni, V. 2000. Evolution of a degradative bacterial consortium able to degrade naphtha solvent J. Appl. Microbiol. 88, 1009-1018.
12. Shields, M.S., Reagin, M.J. Gerger, R.R., Campbell, R., Someerville, C., 1995. TOM, a new aromatic degradative plasmid from *Burkholderia cepacia* G4. Appl. Environ. Microbiol. 61, 1352-1356.
13. Kukor, J.J., Olsen, R.H., 1990. Molecular cloning, characterisation and regulation of a *Pseudomonas picketti* PK01 gene encoding phenol hydroxylase and expression of the gene in *Pseudomonas aeruginosa* PAO1. J. Bacteriol. 172, 4624-4630.
14. Whited, G.M., Gibson, D.T., 1991. Separation and partial characterization of the enzymes of the toluene 4-monoxygenase catabolic pathway in *Pseudomonas mendocina* KR1. J. Bacteriol. 173, 3017-3020.
15. Bertoni, G., Martino, M., Galli, E., Barbieri, P., 1998. Analysis of the gene cluster encoding toluene/*o*-xylene monoxygenase from *Pseudomonas stutzeri* OX1. Appl. Environ. Microbiol. 64, 3626-3632.
16. Kitayama, A., Suzuki, E., Kawakami, Y., Nagamune, T., 1996. Gene organization and low regiospecificity in aromatic-ring hydroxylation of a benzene monoxygenase of *Pseudomonas aeruginosa* JI104. J. Ferment. Bioeng. 82, 421-425.
17. Bianchi, M.A., Bianchi, A.J., 1981. Statistical sampling of bacterial strains and its use in bacterial diversity measurement. FEMS-Microbiol. Ecol. 8, 61-69.

18. Sneath, P.H., Sokal, R.R., 1973. Numerical Taxonomy. W.H. Freeman, San Francisco, CA.
19. Thompson, J.D., Higgins, D.G., Gibson, T.J., 1997. CLUSTALW: improving the sensitivity of progressive multiple sequence allignment through sequence weighting, position, specific gap penalties and weight matrix choice. Nucleid Acids Res. 22, 4673-4680.
20. Johnson, G.R., Olsen, R.H., 1997. Multiple pathways for toluene degradation in *Burkholderia* sp. JS150. Appl. Environ. Microbiol. 63, 4047-4052.
21. Bickerdike, S.R., Holt, R.A., Stephens, G.M., 1997. Evidence for metabolism of *o*-xylene by simultaneous ring and methyl group oxidation in a new soil isolate. Microbiology 143, 2321-2329.
22. Vanderberg, L.A., Krieger-Grumbine, R., Taylor, M.N., 2000. Evidence for diverse oxidations in the catabolism of toluene by *Rhodococcus rhodochrous* strain OFS. Appl. Microbiol. Biotechnol. 53, 447-452.

Developments in Soil Science, Volume 28B
Editors: A. Violante, P.M. Huang, J.-M. Bollag and L. Gianfreda

LICHEN-ROCK INTERACTIONS AND BIOFORMATION OF MINERALS

P. Adamo, S. Vingiani and P. Violante

Dipartimento di Scienze Chimico-Agrarie, Università di Napoli Federico II,
Via Università 100, 80055 Portici, Italy

The principal effects of lichen growth on mineral substrata are discussed with special reference to rock surface disintegration, mineral etching patterns and formation of oxalates, iron oxides and hydroxides, aluminosilicates and lichen acid-metal complexes. The release of organic molecules, such as oxalic acid and polyphenolic secondary products of lichen metabolism usually indicated as "lichen acids", have been proven to play a key role in lichen weathering and neogenesis of poorly and well crystalline biominerals. Both physical and chemical properties of the rock substrate and morphology of the thallus may strongly differentiate extent and assemblage of the lichen-substrate contact zone. Nevertheless, differences in the bioweathering capability seems to be more related to the physiology of the lichen species involved. Recent and further progress in the study of the interactions between lichens and their rock substrata appears to be mainly relying on the utilisation of more specialised analytical and instrumental techniques and on the close collaboration among scientists from different research fields. The significance of the lichen bioweathering in the colonisation of rock outcrops by more exigent plant species and in the process of soil formation is remarked.

1. INTRODUCTION

1.1. Basic types of lichens and their attachment to the substrate

Lichen symbiosis consists of algae and fungi living in a mutually beneficial relationship. The algae fix atmospheric nitrogen and photosynthesize, providing the organics necessary for the growth of the fungi. These, in turn, supply inorganic nutrients to the algae. Myco- and photobionts form a heterogeneous, stratified structure, the thallus, characterized by three basic growth forms: crustose, foliose, and fruticose [1, 2]. Crustose lichens cover rocks, trees, or soil as a thin, more or less well-developed crust. Foliose lichen thalli resemble the leaves of higher plants and adhere relatively firmly to the substrate. Fruticose species show upright, branching, brushlike forms. Generally, the fungus is in close contact with the substrate and therefore acts as the main weathering agent; on the contrary, the algae are buried in the upper layers of the thallus [3]. The attachment of the thallus to the substrate is ensured by the basal hyphae. In crustose species, the contact is so intimate that lichens are practically inseparable from the substrate. The thalli of foliose lichens are usually attached by rhizine or rhizoidal hyphae, which are composed of bundles of roughly parallel aligned, connected hyphae. Many fruticose thalli are attached to the substrate by a holdfast. The role of basal hyphae in the translocation of dissolved mineral elements or organic metabolites from the substrate to the thallus has not yet been clarified. In fact, it is still an open question whether lichens derive any nutriment from their substrate at all.

1.2. The weathering of rocks by lichens

The significance of lichens in rock weathering and in the formation of soil has been well documented [4, 5, 6, 7, 8, 9, 10]. Lichen-encrusted rocks provide an interesting environment for studying the biological weathering of minerals because, unlike the situation in soil, where many different types of minerals and microorganisms coexist, there are fewer interacting factors. Both physical and chemical processes are involved in the ability of these composite organisms to degrade their rock substrates [4]. Fungal hyphae penetration and thallus expansion and contraction are considered to be the most significant physical mechanisms of weathering. The mycobiont excretion of organic acids during growth represents the main process by which lichens decompose the rock-forming minerals. So far, the production of oxalic acid and of a variety of weak polyphenolic compounds called "lichen acids" has been proven important in lichen weathering. Chemical decomposition of the rocks proceeds at the same time as physical disintegration, which may even accelerate biogeochemical effects by increasing the substrate surface area.

As a result of both biogeophysical and biogeochemical weathering, the zone of contact between the lichen thallus and the rock substrate consists of an extremely complex and highly heterogeneous assemblage. Rock fragments and rock-forming mineral particles coexist with a range of biogenic minerals and organo-mineral complexes. Biomolecules, produced by lichens and active in the breakdown of primary minerals, slow down the crystallization processes, enriching the contact zone with the presence of noncrystalline and poorly ordered secondary products. Various microorganisms grow in association with the myco- and photobionts of lichen thalli.

Morphological characteristics of the lichen species may influence the extent and arrangement of the lichen-substrate contact zone. In crustose lichens, the entire lower surface of the thallus is in immediate contact with the substrate, whereas in foliose species, the access occurs through root-like structures, which may greatly extend the contact zone. Nonetheless, differences in thallus morphology do not necessarily imply differences in the lichen bioweathering potential, which appears to be more related to the physiology of the species involved [11].

In the last decades, the study of the interaction between lichens and mineral surfaces has been largely based on the application of highly specialized analytical and instrumental techniques and on the close collaboration among biologists, chemists, and geologists. The complementary use of optical and scanning and transmission electron microscopy (OM, SEM, TEM, HRTEM), energy and wavelength dispersive microprobes (EDS, WDS), X-ray diffractometry (XRD), and infra-red spectrometry (FT-IR) have permitted the study of the rock-lichen interface from macro to submicroscopic range. This allows instantaneous qualitative and quantitative in situ microanalysis of the inorganic components and provides new insights into the ultrastructural features of the biological components.

The aim of this review is to discuss the principal effects of lichen growth on mineral substrata. Attention is focused on rock surface disintegration, on mineral etching patterns and decomposition features, and on the formation of biogenic minerals.

2. ROCK SURFACE DISINTEGRATION

Disintegration of the rock surface immediately below the lichen thallus has been frequently reported. It is mainly considered the result of mechanical actions brought about by rhizine penetration and expansion and contraction of the thallus. However, a certain contribution from biogeochemical processes has in most cases to be taken into account.

The penetration of the adhesion organs into the rock on which the lichen grows is influenced by the physical and chemical properties of the lithic substrate. *Stereocaulon vesuvianum* is a fruticose lichen that grows on volcanic rock of Mt. Vesuvius (Italy). The basal ramification of its thallus, called the pseudopodetium, has been shown to penetrate down to 30 mm in the substrate [12] (Figure 1a). The occurrence of preexisting cavities and fissures into the volcanic rock facilitates this deep penetration. Crustose lichen species that grow on carbonate rocks are characterized by thalli partly or entirely buried in the substrate. In the weathering of dolomite rocks by *Lepraria* spp. lichens, the high solubility of calcium/magnesium carbonate likely favored the penetration of the lichen into the substrate, followed by detachment and incorporation of mineral fragments into the thallus [13] (Figure 1b). On siliceous schist or on rocks containing mica crystals, rhizine penetration has been shown to follow the lines of cleavage [14, 15, 16] (Figure 1c).

In some lichen species, the occurrence of significant morphological elaborations of the mycobiont, known as rhizomorphs (*sensu lato*), may greatly extend the lichen-substrate contact zone. Rhizomorphs are "mycelial cords" distinguishable from rhizine by their irregular form and indeterminate growth [17]. They are organs of exploration as well as attachment whose continued growth and proliferation produces an extensive substratic network with a potentially significant role in biodeterioration, as was found in a study of the physical interactions of rhizomorph-forming *Squamarina cartilaginea* and *Acarospora scotica* lichens with their rock substrates, a calcareous conglomerate and a siliceous schist, respectively [14]. For both species, interpenetration and separation of substrate particles by distal hyphae with incorporation of rock fragments into the developing rhizomorphs were observed (Figure 1d).

In addition to rhizine penetration, the thallus expansion and contraction is frequently indicated as a mechanism of physical weathering of rock substrate by lichens. The high proportion of gelatinous substances occurring in the lichen thallus accounts for the expansion and contraction effects that follow wetting and drying cycles. According to the work of Fry [18], when the thallus contracts on drying, marginal hyphae can be pulled from the substrate with consequent possible detachment of rock fragments. Although this physical mechanism of weathering is widely accepted and cited, it has not been closely examined and the actual strength of the forces involved is still unknown.

3. MINERAL ETCHING AND DECOMPOSITION FEATURES

Chemical decomposition processes, mainly by organic acids, occurring at the rock-lichen interface result in corrosion of the rock surfaces immediately below the lichen, as well of the substrate particles entrapped in the thallus. Differences in mineral structure and composition result in a large variety of observed dissolution patterns.

Because of their susceptibility to weathering, carbonate and ferromagnesian minerals can be found to be particularly badly corroded below lichens. Rounded edges and corners along with deep etch marks on dolomite rhombohedral crystals beneath *Lepraria sp.* are manifestations of rapid dissolution (Figure 2a) [13]. Calcite is commonly weathered in a succession of strata that leads to the individualization of forms close to scalenohedra (Figure 2b) [11, 19]. Large and deep perfectly formed pits and narrow trenches are the solubilization features of olivine and augite found, respectively, in the thallus of *Pertusaria corallina* on basalt and at the vesuvite rock-*Stereocaulon vesuvianum* interface [20, 21] (Figure 2c and d). The presence in the rock of minerals characterized by high solubility makes easier hyphae penetration in the substrate. Calcareous rocks have been found to be directly perforated by lichens through chemical dissolution [4, 14].

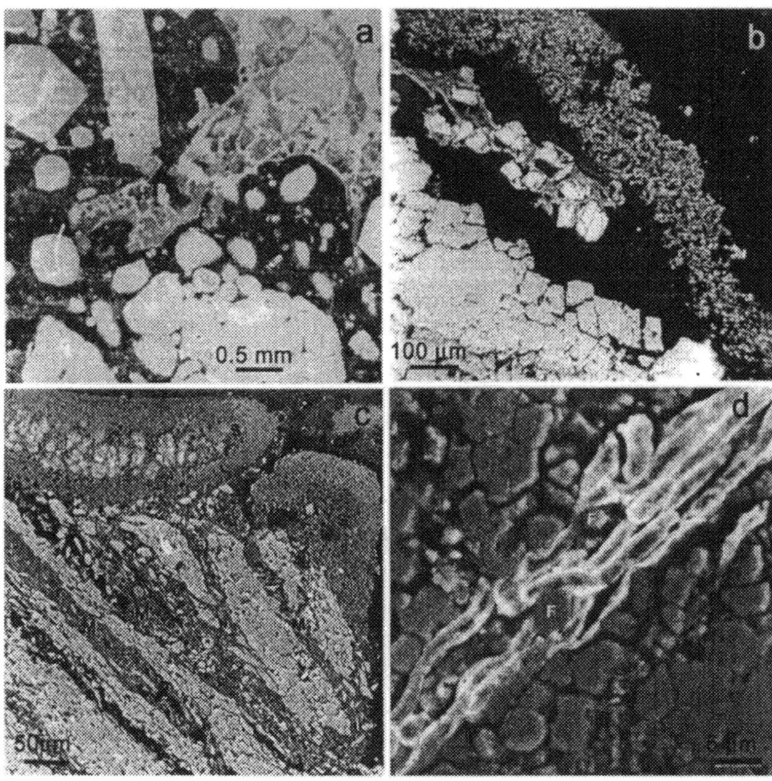

Figure 1. Thin section micrograph (plain polarised light) of (a) the interface between *Stereocaulon vesuvianum* and volcanic rock from Mt. Vesuvius showing the penetration of the lichen in the substrate. Scanning electron micrographs of: (b) detachment and incorporation of mineral fragments into the thallus of *Lepraria* genus lichen on dolomite rock from Mt. Faito, Campania; (c) siliceous schist laminae interpenetrated by *Acarospora scotica* mycobiontic tissue [14]; (d) interpenetration of substrate particles by distal hyphae, incorporating fragments (F) into developing rhizomorph of *Squamarina cartilaginea* on calcareous conglomerate rock [14].

Lichen weathering of feldspar proceeds by reactions at relatively fresh mineral surfaces and normally results in etching marks varying in shape according to the type of feldspar and the crystallographic face involved. On the (001) cleavage face of albite in lichen-encrusted granite at Aberchirder in north-east Scotland, diamond-shaped etch pits are the characteristic etch marks (Figure 2e), whereas smaller and differently shaped markings exist on other sodium feldspar faces (Figure 2f) [22]. Potassium feldspars tend to etch out preferentially in the form of narrow trenches [20]. In the case of labrodorite, the characteristic ridge and furrow pattern observed is considered to reflect the lamellar intergrowth between the two calcium sodium components [20].

The solubilizing effects of lichen compounds have even been detected on minerals extremely resistant to weathering such as quartz grains either isolated from the zone of contact between lichens and quartzites/quartzitic substrates or treated in laboratory with selected lichen extracts [22, 23, 24].

When the decomposing action leads to the complete extraction of all structural cations, no distinct etching pattern can be observed. In this case, residual siliceous relicts are typically recognized [3].

Dissolution processes take place preferentially at surface sites where structural dislocations in the mineral lattice occur. Dislocations can be considered as areas of the mineral structure characterized by lower stability due to the existence of active energy tensions. Their consideration can assist the analysis of the dissolution features observed on mineral particles beneath lichens.

Corrosion patterns caused by biological action on mineral surfaces have a certain ecological significance in the postfire recolonization of rock outcrops by lithobionts. Pits and holes can act as reservoirs for water, as well as sites where fine-grained material, organic matter, and fire ash can accumulate [25]. Furthermore, the release of constituent elements through weathering of primary minerals by lichens might contribute to increase the nutrient bioavailability in the rock-lichen interface.

4. PRECIPITATION AND FORMATION OF MINERALS

As a result of the biological activity, either direct or indirect, of lichens on mineral substrates, a range of secondary weathering products have been found in the zone of contact between lichen and rock and in the thallus itself. Insoluble crystalline oxalates, whose composition closely matches that of the substrate, iron oxides and hydroxides, aluminosilicates and lichen acid-metal complexes are the groups of biominerals more frequently observed.

4.1. Oxalates

Secondary crystalline oxalates originate from the reaction between the oxalic acid produced by the mycobiont and the minerals of the rock substrate. Calcium oxalate monohydrate [$CaC_2O_4 \cdot H_2O$], the mineral whewellite, and polyhydrate [$CaC_2O_4 \cdot (2+x)H_2O$], the mineral weddellite, are the most common forms found in lichens colonizing calcareous rocks and mineral substrates containing calcium [4, 11, 13, 15, 20, 26, 27, 28]. These oxalates can be distinguished by their morphology and X-ray reflections. Whewellite is commonly found in flat, platy crystals (Figure 3a) giving main X-ray peaks at 0.593, 0.365, and 0.297 nm; weddellite

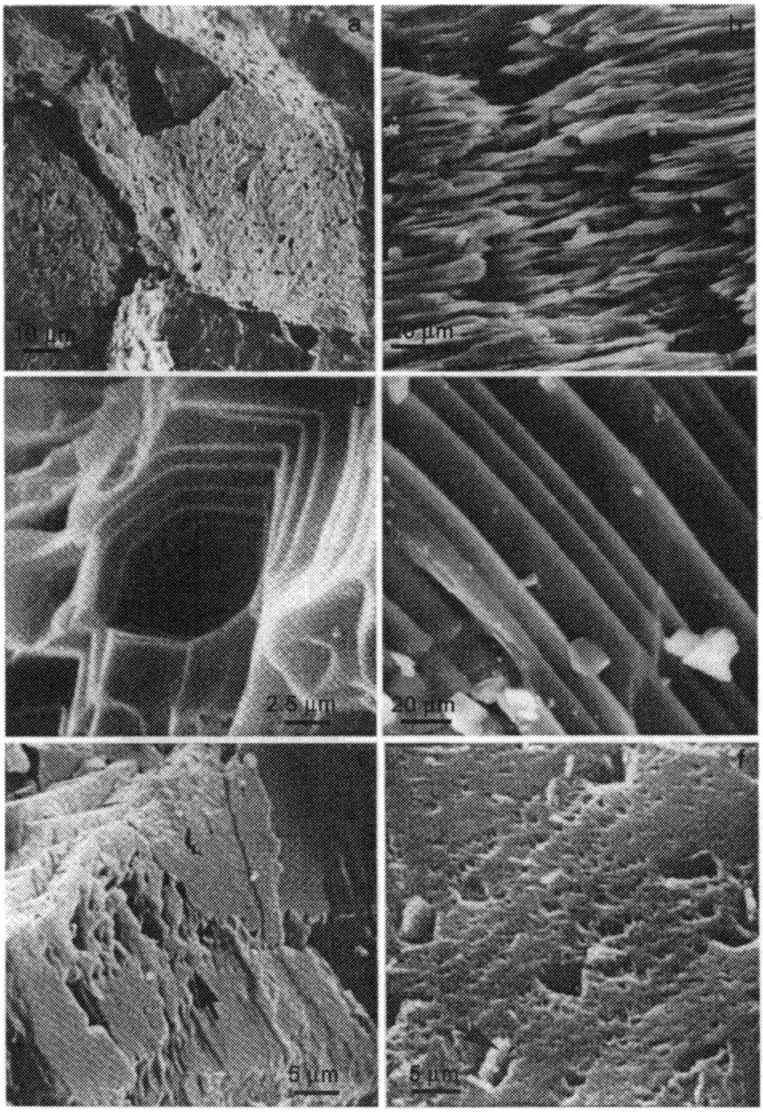

Figure 2. Decomposition features observed by Scanning Electron Microscope (SEM) beneath lichens. (a) Etched surface of dolomite rombohedric crystals beneath *Lepraria sp.* [13]; (b) scalenohedric dissolution pattern of calcite under *Xanthoria ectaneoides* on diabase [47]; (c) large pit on olivine in thallus of *Pertusaria corallina* on basalt [20]; (d) trenches on augite at the *Stereocaulon vesuvianum*-leucitic rock interface; (e) diamond shaped etch pits on (001) cleavage face and (f) on other faces of albite in lichen encrusted granite [20].

tends to exhibit a tetragonal pyramidal form (Figure 3b) with strong peaks at 0.618, 0.442, and 0.278 nm. On rocks that contain no or low amounts of calcium, the formation of other oxalates has been observed. On serpentinite, a magnesium-rich rock occurring in the Grampian Region (Scotland), the growth of the crustose lichen *Lecanora atra* brings about the precipitation and crystallization at the rock-lichen interface and into the thallus of magnesium oxalate, the mineral glushinskite [$MgC_2O_4 \cdot 2H_2O$] [29, 30]. Both magnesium and calcium oxalates have been found in a lichen growing on allivalite from Rhum, with direct association of the two forms with olivine and anorthitic plagioclase feldspar mineral components, respectively [20]. Oxalate crystals rich in manganese have been found in the lichen *Pertusaria corallina* colonizing a Mn ore, consisting of hard cryptomelane, KMn_8O_{16}, and powdery lithiophorite, $(Al,Li)MnO_2(OH)_2$ [31]. Copper oxalate hydrate, the mineral moolooite, has been revealed on substrates rich in copper as vivid blue thallus inclusions [32, 33].

In lichens, as in higher plants, oxalates seem to form intracellularly at early stages of growth with successive extracellular protrusion as their size increases [34]. Calcium oxalate formation has long been related to the necessity for the lichen to dispose of an excess of calcium, although the compounds have been found also in lichens growing on substrates not containing calcium [35]. Recently, Modenesi and coworkers have suggested that the production of heavy crystalline deposits of calcium oxalate at the surface of epiphytic lichens could be linked to exposure to oxidative stress in response to toxicity by SO_2 pollution, paraquat treatment, and high light intensity environments [36, 37]. In the case of high irradiance oxidative stress, the deposition of whitish layers of calcium oxalate would increase the surface albedo of the thallus [37]. An increased synthesis of the well known antioxidant ascorbic acid, along with glyoxylate, the major precursor of oxalic acid in fungi and plants, seems to constitute the relation between oxalate formation and oxidative stress [38]. Extracellular immobilization of heavy metals as oxalate salts in the lichen thallus seems to prevent element toxicity and enable an accumulation of metals far above the physiological needs of the organism [39, 40].

4.2. Iron oxides and hydroxides

With the exception of a study of Ascaso et al. [26], which gives evidence of the detection of ferric oxalate in the thallus of the lichen *Caloplaca callopisma* colonizing an Fe-bearing dolomite, the formation of iron oxalates as a result of the lichen weathering of rocks containing iron has never been reported. On the contrary, both amorphous and crystalline iron oxides have been revealed by several authors in the reddish weathering crust beneath various lichens. Ferrihydrite was identified, as a poorly crystallized form of Fe oxide, in the weathering crusts beneath *Sterocaulon vulcani* colonizing recent lava flow outcrops in Hawaii [41], as a hydrous Fe oxide with a formula close to $Fe_5HO_8.4H_2O$ in the thin ochreous layer at the interface between *Pertusaria corallina* and a weathered basalt substrate in western Scotland [20], and as the main component of the short-range ordered iron oxyhydroxide pool in the rusty ferruginous material located in the zone of contact between the thallus of *Stereocaulon vesuvianum* and the leucite-bearing rock of Mt. Vesuvius [12] (Figure 3c). Goethite (α-FeOOH) was detected by Ascaso et al. [23] at the interface between *Rhizocarpon geographicum* and granite, by Galvan et al. [42] associated with various lichens on metamorphic rocks, and by Adamo et al. [12] in association with hematite (α-Fe_2O_3) in the *Stereocaulon vesuvianum*-Vesuvius volcanic rock weathering interface (Figure 3d).

It seems likely that at the rock-lichen interface, iron, mobilized through weathering of iron silicates and/or through reduction of iron oxides, is primarily bound by biomolecules in organic complexes. Oxidation, possibly microbially, of organic ligands leads to the release of Fe with subsequent hydrolysis and precipitation of iron oxides on rock surfaces or within the lichen thalli [43]. Organics may completely inhibit iron-oxide formation, enriching the weathered material beneath lichens in poorly ordered iron phases or, if forming, organics may determine the nature of iron oxides formed. The treatments with oxalate [44] and dithionite [45] reagents coupled with differential X-ray diffraction (DXRD) analysis [46] have proven to be extremely useful in analyzing the different forms of Fe in selected particle-size fractions separated from the interface weathered material [12]. Distinctiveness in the rock-lichen interface of microsites each with separate pH, humidity, and redox potential conditions has been indicated as the cause leading to the neoformation of Fe crystalline phases of different nature [12]. Therefore, it has been suggested that the environmental conditions influencing the formation of hematite, through aggregation, dehydration, and internal structural rearrangement of the tiny ferrihydrite particles, are likely to be found in the more external zone of the rock-lichen interface where there are lower contents of water and organics. Conversely, goethite formation, via solution and through nucleation and crystal growth of monomeric monovalent Fe^{3+} ions supplied from any Fe source, is likely to occur closer to the lichen thallus, where humidity and organics are greater.

4.3. Aluminosilicates

Both poorly and well crystallized aluminosilicate forms have been detected in the weathered material that accumulates below lichens [20, 21, 23, 27, 47] (Figure 3e-f). As suggested for iron, aluminum might also be removed from the substrate minerals through complexation by lichen-derived organic compounds [3]. Oxidation of the ligands would liberate Al in a reactive form to combine with silica, leading to the formation of aluminosilicates. For a long time, many authors have claimed that due to the high level of biomolecules that characterize the rock-lichen interface, biochemical weathering typically results in secondary products of a poorly ordered nature [20, 30, 28, 41].

The question of whether clay minerals can be biogenically formed and transformed as a result of lichen activity at the lichen-rock interface has been recently addressed by Wierzchos and Ascaso [16, 48] by applying microscopical and analytical techniques (SEM, HRTEM, EDS). In these studies, distinct chemical changes in the granitic biotite mineral due to the strong action of *Parmelia conspersa* and *Aspicilia intermutans* thalli were demonstrated. Biogeneous mineralogical transformation of biotite from the *Parmelia conspersa* contact zone with formation of an interstratified biotite/vermiculite mineral phase was revealed. According to the authors, two parallel processes, physical exfoliation of trioctahedral mica and interlayer ionic exchange of potassium with subsequent vermiculite formation, are the mechanisms for biotite bioweathering induced by lichens.

Interpretation of the presence of clay minerals in the weathered material accumulated beneath the lichen thallus is confused by the possibility of external contamination by fine-grained particles as a result of aeolian deposition [4, 15]. Small amounts of phyllosilicates may occur in the fresh rock. The chemical treatments of the weathered material to remove any organic matter can alter the mineralogical characteristics of the sample, leading to the formation of false secondary products [49, 50, 51, 52].

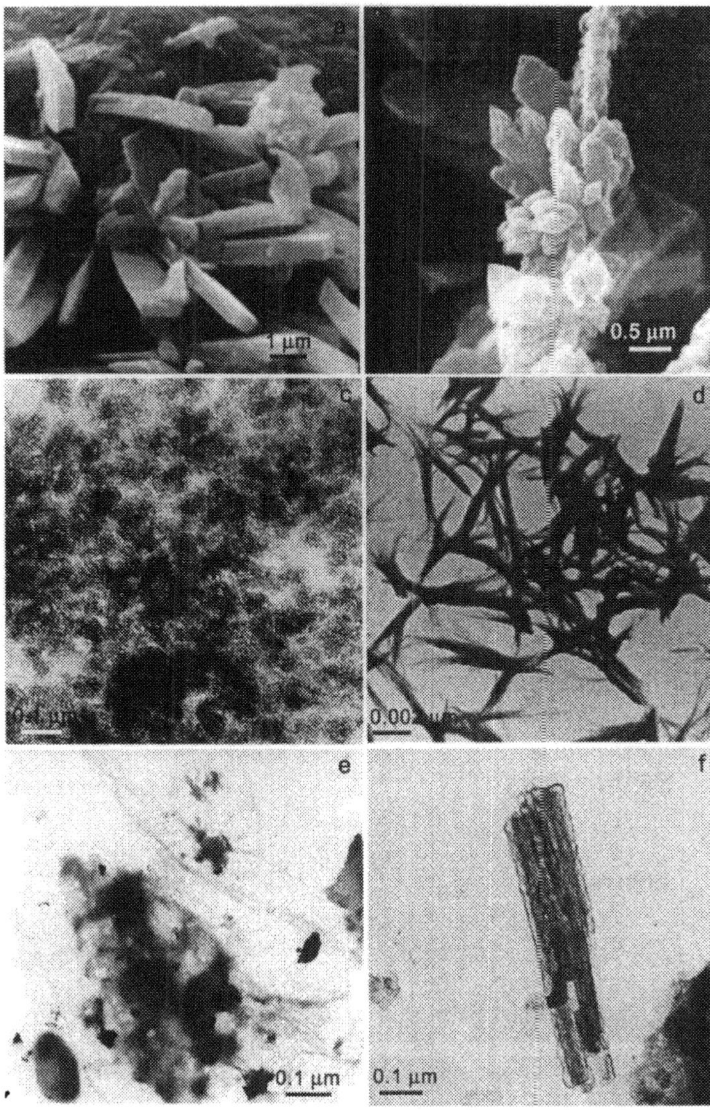

Figure 3. Scanning electron micrographs of: (a) play crystals of calcium oxalate monohydrate (whewellite) in the thallus of *Aspicilia radiosa* on gabbro from Impruneta, Toscana; (b) bypiramidal crystals of calcium oxalate dihydrate (weddellite) in the thallus of *Cladonia sp.* Lichen on dolomite from Mt. Faito, Campania. Transmission electron micrographs of: (c) ferrihydrite and (d) goethite from the fine clay fraction (< 0.5 µm) separated from the rusty interface material beneath *Stereocaulon vesuvianum* thallus on leucitic rock; (e) allophanes and imogolite fibres under the thallus of *Xanthoria elegans* on volcanic andesite [27]; (f) halloysite crystals (parallel association) at the interface between *Parmelia conspersa* and granite substrate [61].

4.4. Lichen acid-metal complexes

The term "lichen acids" usually indicates a class of secondary products of lichen metabolism that can account for up to 8% of the organism dry weight [4]. They are mainly esters and oxidative coupling products of variously substituted phenolic acids [53, 54]. Their water solubility and chelating capacity have been essentially related to the presence in their molecular structure of a hydroxyl (-OH) group ortho to a carboxyl (-COOH) or an aldehyde (-CHO) group on an aromatic ring [55, 56] (Figure 4).

Although the involvement of lichen acids in dissolving silicate minerals at the rock-lichen interface has been largely emphasized, only in recent years has the lichen acid-metal complexes formation been demonstrated under natural conditions [57, 58, 59]. Previous studies reported the formation of soluble colored complexes when lichen compounds or ground lichen thalli were mixed with water suspensions of minerals and rocks [23, 55, 56, 60]. Significant amounts of Ca, Mg, Fe, and Al were complexed by lichen acids, with prevalence of the divalent cationic species [56]. Morphological alterations were shown by granite, gneiss, and their primary minerals when treated with lichen compounds and, under favorable conditions, the formation of montmorillonite and halloysite was observed [61]. The weathering action of ground lichens was more effective than was that of solutions of single or combined selected lichen acids [23].

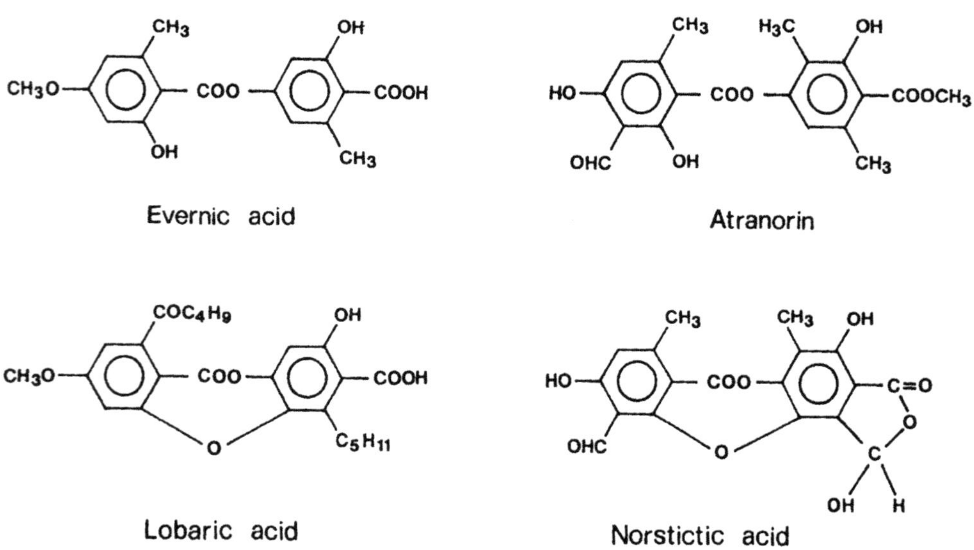

Figure 4. Structural formulae of some "lichen acids"

The formation of metal-lichen acid complexes has been conclusively demonstrated by Purvis and co-workers [62] in their investigations of lichens colonizing cupriferous substrata. Using in a complementary approach OM, SEM, EDS, and FT-IR, copper-norstictic and copper–psoromic complexes have been shown to occur in the cortex of *Acarospora smaragdula* and *Lecidea lactea*, in the apothecia of *Lecidella bullata,* and in the thallus of *Tephromela testaceoatra* growing on Cu-rich rocks [57, 58, 59] (Figure 5). Anomalous yellow-green color of the thallus and a high concentration of Cu reaching 5% by weight were observed and related to the incorporation of Cu from the rocks on which the lichens were growing [62].

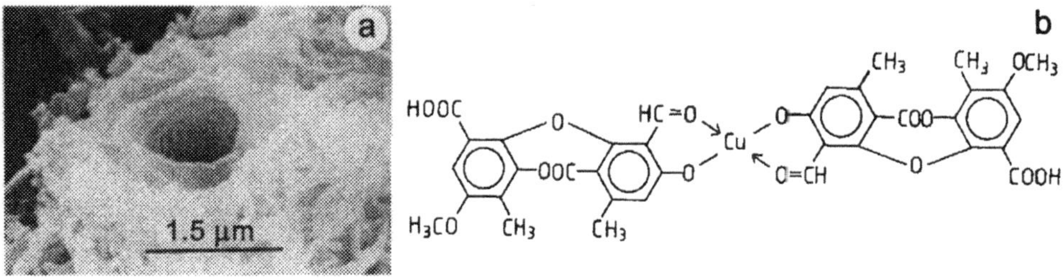

Figure 5. (a) Scanning electron micrograph of a section of a hypha within hypothecium of *Lecidella bullata* on Cu-rich substrata, encrusted in extracellular crystals (Cu-psoromic acid complex?); (b) proposed structure of synthetic Cu-psoromic acid complex [62].

5. CONCLUSIONS

Notable attention has been given in the last decades to study of the interactions between lichens and their rock substrata. The close collaboration among scientists from the fields of ecology, biology, chemistry, physiology, and mineralogy, along with the use of more and more specialized instrumental and analytical techniques has produced new, convincing, direct evidence of the physical and chemical actions that these pioneer organisms bring about on the colonized mineral substrate, and on these bases, further progress can be envisaged.

It might be predicted that in the future new biominerals will be described. On substrata of appropriate composition colonized by oxalic-acid producing lichens, it is likely that a range of previously unreported oxalates may exist. Similarly, new extracellular metal-lichen acids complexes can be found. It is likely that other simple low molecular weight organic acids excreted by lichen mycobionts, as well as organic acids produced during the decomposition of the lichen organic material, are also involved in the biogeochemical weathering of primary minerals.

It is clear that lichens show substrate preferences [63] and that the chemistry and the mineralogy of the substratum have a considerable influence on the lichen species assemblages occurring on rocks and derived soils [62]. Nevertheless, to what extent the mineral substratum determines the availability of elements to saxicolous lichens through weathering processes is still largely unknown. Whether the elements that define the chemical composition of these organisms derive from the substrate of growth or from atmospheric particulates trapped by thalli is a largely unresolved question [39, 64]. It would be worthwhile in the future to pay more attention to such aspects of the relation between lichens and their substrata as this might lead to a better understanding of lichen physiology as well as a more precise use of these organisms as biomonitors.

Particles of the underlying rock, more or less extensively altered by biogeochemical weathering processes, biominerals, biomolecules, organic matter produced by the decomposition of the thallus, organo-mineral complexes, atmospheric dust, and microorganisms, left on lichens or at the rock-lichen interface, all may facilitate, in certain situations, the accumulation of a thin layer of "lithomorphic" soil below lichens [4, 65]. The occurrence within this material of higher levels of bioavailable nutrients and the increased moisture holding capacity might make it possible that other more exigent plant species such as mosses may become established. Erosion processes of the primitive soil along with the slow growth rate and longevity of lichens make soil formation in such situations an extremely slow process.

REFERENCES

1. Ozenda, P., Clauzade, G., 1970. Les Lichens - Etude Biologique et Flore Illustrée, Masson, Paris.
2. Ahmadjan, V., Hale, M.E., 1973. The Lichens. Academic Press, London.
3. Wilson, M.J., Jones, D., 1983. Lichen weathering of minerals and implications for pedogenesis. *In:* Wilson, R.C.L., (Ed.), Residual Deposits: Surface Related Weathering Processes and Materials. Special Publication of the Geological Society, Blackwell, London, pp. 5-12.
4. Syers, J.K., Iskandar, I.K., 1973. Pedogenetic significance of lichens. *In:* Ahmadjian, V., Hall, M.E., (Eds.), The Lichens. Academic Press, London, pp. 225-248.
5. Jones, D., 1988. Lichens and pedogenesis. *In*: Galun, M. (Ed.), Handbook of Lichenology. CRC Press, Boca Raton, pp. 109-124.
6. Jones, D., Wilson, M.J., 1985. Chemical activity of lichens on mineral surfaces - A Review. Int.. Biodeterior. 21, 99-104
7. Gehrmann, C., Krumbein, W.E., Petersen, K., 1988. Lichen weathering activities on mineral and rock surfaces. Studia Geobot. 8, 33-45.
8. Ascaso, C., Wierzchos, J., 1995. Study of the biodeterioration zone between the lichen thallus and his substrate. Cryptogam. Bot. 5, 270-281.
9. Wilson, M.J., 1995. Interactions between lichens and rocks: a review. Cryptogam. Bot. 5, 299-305.
10. Adamo, P., Violante, P., 2000. Weathering of rocks and neogenesis of minerals associated with lichen activity. Appl. Clay Sci. 16, 229-256.
11. Adamo, P., Marchetiello, A., Violante, P., 1993. The weathering of mafic rocks by lichens. Lichenologist 25, 285-297.
12. Adamo, P., Colombo, C., Violante, P., 1997. Iron oxides and hydroxides in the weathering interface between *Stereocaulon vesuvianum* and volcanic rock. Clay Min. 32, 275-283.
13. Adamo, P., Violante, P., 1989. Bioalterazione di roccia dolomitica operata da una specie lichenica del genere *Lepraria*. Agr. Med. 119, 460-464.
14. Sanders, W.B., Ascaso, C., Wierzchos, J., 1994. Physical interactions of two rhizomorph-forming lichens with their rock substrate. Bot. Acta 107, 432-439.
15. Wierzchos, J., Ascaso, C., 1994. Application of back-scattered electron imaging to the study of the lichen-rock interface. J. Microsc. 175, 54-59.
16. Wierzchos, J., Ascaso, C., 1996. Morphological and chemical features of bioweathered granitic biotite induced by lichen activity. Clays Clay Miner. 44, 652-657.
17. Poelt, J., Baumgärtner, H., 1964. Über Rhizinenstränge bei placodialen Flechten. Öster. Bot. Z. 111, 1-18.

18. Fry, E.J., 1927. The mechanical action of crustaceous lichens on substrata of shale, schist, gneiss, limestone and obsidian. Ann. Bot. 41, 437-460.
19. Robert, M., Veneau, G., Berrier, J., 1980. Solubilisation comparée des silicates carbonates et hydroxydes en fonction del conditions du milieu. Bull. Miner. 103, 324-329.
20. Jones, D., Wilson, M.J., Tait, J.M., 1980. Weathering of a basalt by *Pertusaria corallina.* Lichenologist 12, 277-289.
21. Adamo, P., Violante, P., 1991. Weathering of volcanic rocks from Mt Vesuvius associated with the lichen *Stereocaulon vesuvianum.* Pedobiologia 35, 209-217.
22. Jones, D., Wilson, M.J., McHardy, W.J., 1981. Lichen weathering of rock-forming minerals: application of scanning electron microscopy and microprobe analysis. J. Microsc. 124, 95-104.
23. Ascaso, C., Galvan, J., Ortega, C., 1976. The pedogenetic action of *Parmelia conspersa, Rhizocarpon geographicum* and *Umbilicaria pustulata.* Lichenologist 8, 151-171.
24. Hallbauer, D.K., Jahns, H.M., 1977. Attack of lichens on quartzitic rock surfaces. Lichenologist 9, 119-122.
25. Garty, J., 1992. The postfire recovery of rock-inhabiting algae, microfungi and lichens. Can. J. Bot. 70, 301-312.
26. Ascaso, C., Galvan, J., Rodriguez-Pascual, C., 1982. The weathering of calcareous rocks by lichens. Pedobiologia 24, 219-229.
27. Ascaso, C., Sancho, L.G., Rodriguez-Pascal, C., 1990. The weathering action of saxicolous lichens in maritime Antarctica. Polar Biol. 11, 33-39.
28. Vidrich, V., Cecconi, C.A., Ristori, G.C., Fusi, P., 1982. Verwitterung Toskanischer Gesteine unter Mitwirkung von Flechten. Z. Phlanzenern. Bodenkunde 145, 384-389.
29. Wilson, M.J., Jones, D., Russell, J.D., 1980. Glushinskite, a naturally occurring magnesium oxalate. Miner. Mag. 43, 837-840.
30. Wilson, M.J., Jones, D., McHardy, W.J., 1981. The weathering of serpentinite by *Lecanora atra.* Lichenologist 13, 167-176.
31. Wilson, M.J., Jones, D., 1984. The occurrence and significance of manganese oxalate in *Pertusaria corallina.* Pedobiologia 26, 373-379.
32. Purvis, O.W., 1984. The occurrence of copper oxalate in lichens growing on copper sulphide-bearing rocks in Scandinavia. Lichenologist 16, 197-204.
33. Chisholm, J.E., Jones, G.C., Purvis, O.W., 1987. Hydrated copper oxalate, moolooite, in lichens. Miner. Mag. 51, 766-803.
34. Pinna, D., 1983. Fungal physiology and the formation of calcium oxalate films on stone monuments. Aerobiologia 9, 157-167.
35. Wadsten, T., Moberg, R., 1985. Calcium oxalate hydrates on the surface of lichens. Lichenologist 17, 239-245.
36. Modenesi, P., 1993. An SEM study of injury symptoms in *Parmotrema reticulatum* treated with paraquat or growing in sulphur dioxide-polluted air. Lichenologist 25, 423-433.
37. Modenesi, P., Canepa, R., Tafanelli A., 1998. The structural role of calcium oxalate and medullary architecture in *Menegazzia terebrata* and *Hypogymnia physodes.* In: Türk, R., Zorer R., (Eds.), Progress and Problems in Lichenology in the Nineties. IAL 3. Bibliotheca Lichenologica Vol. 68, J. Cramer in der Gebr. Verlagsbuchhandlung, Berlin-Stuttgart, Germany, pp. 101-110.
38. Caviglia, A.M., Modenesi, P. 1999. Oxidative stress and ascorbic acid contents in *Parmotrema reticulatum* and *Parmelia sulcata* thalli. Lichenologist 31, 105-110.

39. Bargagli, R., 1998. Trace Elements in Terrestrial Plants: an Ecophysiological Approach to Biomonitoring and Biorecovery. Springer-Verlag, Berlin, Germany.
40. Sarret, G., Manceau, A., Cuny, D., Van Haluwyn, C., Déruelle, S., Hazemann, J., Soldo, Y., Eybert-Berard, L., Menthonnex, J., 1998. Mechanisms of lichen resistance to metallic pollution. Environ. Sci. Technol. 32, 3325-3330.
41. Jackson, T.A., Keller, W.D., 1970. A comparative study of the role of lichens and "inorganic" processes in the chemical weathering of recent Hawaiian lava flows. Am. J. Sci. 269, 446-466.
42. Galvan, J., Rodriguez, C., Ascaso, C., 1981. The pedogenetic action of lichens on metamorphic rocks. Pedobiologia 21, 60-73.
43. Schwertmann, U., Taylor, R.M., 1989. Iron oxides. In: Dixon, J.B., Weed, S.B. (Eds.), Minerals in Soil Environments. 2nd ed. Soil Sci. Soc. Am., Madison, WI. pp. 379-438.
44. Schwertmann, U., 1964. Differenzierung der Eisenoxyde des Bodens durch Extraktion mit Ammoniumoxalat-Lösung. Z. Pflanzenern. Bodenkunde 105, 195-202.
45. Soil Survey Staff. 1996. Soil survey laboratory methods manual. Soil Survey Investigations Report No. 42. Version 3.0. U.S. Dept. of Agriculture, National Resources Conservation Service, National Soil Survey Center, Lincoln, NE.
46. Schulze, D.G., 1981. Identification of some iron oxide minerals by differential X-ray diffraction. Soil Sci. Soc. Am. J. 45, 437-440.
47. Adamo, P., 1996. Ruolo dell'attività dei licheni nell'alterazione di substrati rocciosi e nella neogenesi di entità mineralogiche. Atti XIII Convegno Nazionale SICA, pp. 13-26.
48. Wierzchos, J., Ascaso, C., 1998. Mineralogical transformation of bioweathered granitic biotite studied by HRTEM: evidence for a new pathway in lichen activity. Clays Clay Miner. 46 (3), 446-452.
49. Farmer, V.C., Mitchell, B.D., 1963. Occurrence of oxalates in soils clays following hydrogen peroxide treatment. Soil Sci. 96, 221-229.
50. Lakvulich, L.M., Wiens, J.H., 1970. Comparison of organic matter destruction by hydrogen peroxide and sodium hypochlorite and its effect on selected mineral constituents. Soil Sci. Soc. Am. Proc. 34, 755-758.
51. Douglas, L.A., Fiessinger, F., 1971. Degradation of clay minerals by H_2O_2 treatment to oxidise organic matter. Clays Clay Miner. 19, 67-68.
52. Jackson, M.L., 1975. Soil Chemical Analysis - Advanced Course. 2nd Edition, 10th printing. Published by the author, Madison, Wisconsin.
53. Sundholm, E.G., Huneck, S., 1980. ^{13}C NMR-spectra of lichen depsides, depsidones and depsones. Chem. Scripta 16, 197-200.
54. Huneck, S., Yoshimura, I., 1996. Identification of Lichen Substances. Springer Verlag, Berlin.
55. Schatz, A., 1963. Soil microorganisms and soil chelation. The pedogenic action of lichens and lichen acids. Agric. Food Chem. 11, 112-118.
56. Iskandar, I.K., Syers, J.K., 1972. Metal complex formation by lichen compounds. J. Soil Sci. 23, 255-265.
57. Purvis, O.W., Gilbert, O.L., James, P.W., 1985. The influence of copper mineralization on *Acarospora smaragdula*. Lichenologist 17, 111-116.
58. Purvis, O.W., Elix, J.A., Broomhead, J.A., Jones, G.C., 1987. The occurrence of copper-norstictic acid in lichens from cupriferous substrata. Lichenologist 19, 193-203.

59. Purvis, O.W., Elix, J.A., Gaul, L., 1990. The occurrence of copper-psoromic acid in lichens from cupriferous substrata. Lichenologist 22, 345-354.
60. Syers, I.K., 1969. Chelating ability of fumarprotocetraric acid and *Parmelia conspersa*. Plant Soil 31, 205-208.
61. Ascaso, C., Galvan, J., 1976. Studies on the pedogenic action of lichen acids. Pedobiologia 16, 321-331.
62. Purvis, O.W., Halls, C., 1996. A review of lichens in metal-enriched environments. Lichenologist 28, 571-601.
63. Brodo, I.M., 1973. Substrate Ecology. In: Ahmadjian, V., Hale, M.E., (Eds.), The Lichens. Academic Press, New York. pp. 401-441.
64. Bargagli, R.,1995. The elemental composition of vegetation and the possible incidence of soil contamination of samples. Sci. Tot. Environ. 176, 121-128
65. Adamo, P., Vingiani, S., Violante, P., 2001. Bioalterazione di tefriti fonolitiche dell'Etna (Sicilia) operata dai licheni *Sterocaulon vesuvianum* (Pers.) e *Lecidea fuscoatra* (L.) Ach. e dal muschio *Grimmia pulvinata*. CD-ROM Atti del Convegno Annuale della Società Italiana della Scienza del Suolo (SISS), Gressoney Saint Jean, Aosta, 22-25 giugno 1999.

Developments in Soil Science, Volume 28B
Editors: A. Violante, P.M. Huang, J.-M. Bollag and L. Gianfreda

SOIL HUMIC ACIDS FORMATION AND CHARACTERISTICS IN A XERIC MOLLISOL REFORESTED WITH TWO TREE SPECIES

M. T. Dell'Abate[a], F. Pinzari[a], A. Benedetti[a], B. M. Petronio[b] and C. Dazzi[c]

[a]Istituto Sperimentale per la Nutrizione delle Piante di Roma,
Via della Navicella 2/4, 00184 Roma, Italy

[b]Dipartimento di Chimica, Università di Roma "La Sapienza",
Piazzale Aldo Moro 5, 00185 Roma, Italy

[c]Dipartimento di Agronomia, Coltivazioni Erbacee e Pedologia, Università di Palermo,
Viale delle Scienze, 90128 Palermo, Italy

The present study deals with the influence on humic acids characteristics of different plant species [*Pinus halepensis* Miller and *Cedrus atlantica* (Endl) Carrière] used in the restoration of a Mollisol under xeric conditions. The area studied is in Sicily (Italy), and two stands were compared 40 years after planting. The objective was to contribute to the comprehension of the different levels at which the processes of pedogenesis acted in the study area. The differences between soil profile development in the two stands were investigated. Moreover, some chemical and spectroscopic characteristics of humic acids (HA) were analyzed. The results showed that the evolution of soils differed under the two tree stands, both in the formation of different humus forms and in different C storage in the forest floors. The HA chemical composition showed differences in the interactions between humic acids and inorganic matter in relation to the tree species and depth. Since the pedogenetic factors (mainly climate and parent material) were the same for the two pedons, and the HA extracting procedure adopted was the same for all samples, the data seem to suggest a different role for the microflora and plant coverage in the soil profile evolution of the two stands, which could address the pedogenesis process.

1. INTRODUCTION

The influence exerted by vegetation type on some soil properties due to both different organic matter dynamics and nature has been discussed and exploited by many authors [1, 2]. The study of the relationship between soil and vegetation and the natural succession of biocenosis in the reforested sites allows us to define the fitness of the stands for the specific environmental condition. In fact, the suitability of the tree species commonly used in reforestation for improving soil formation and avoiding soil degradation differs significantly, according to the characteristics of the areas, i.e., climate, soils physics, and chemistry. In particular, the formation of different humus form profiles, which are recognizable from a taxonomic point of view [3], is the consequence of both the rate and the type of organic carbon turnover, which depend on both the amount and the chemical characteristics of the litter. Since humic substances represent the final product of the organic matter degradation/building up processes that occurred in the soil following anthropogenic activity, such as reforestation practices, their amount and qualitative characteristics can yield some information on pedogenesis.

In the present study, the effects from humic acids characteristics of mineral-organic matter-biota interactions due to different plant species in the restoration of a Mollisol under xeric conditions were investigated to evaluate the effects of the tree species on site remediation and pedogenesis. The differences between soil profile development were investigated by chemical and biochemical methods in order to understand the different levels at which the processes of pedogenesis acted in the study area. In particular, some chemical and spectroscopic characteristics of humic acids (HA), separated from two selected pedons, were analyzed.

2. MATERIALS AND METHODS

2.1 Study area and sampling.

The study area (5000 m^2) is located in the Sicani range (western Sicily, Italy), within a wider area subjected to a reforestation program for soil conservation purposes from 1944 to 1960 (Figure 1). The climate, from data relative to years 1951-1990 recorded in Piano del Leone (slightly to the west of the test area), provides average monthly temperatures whose maximum values are found in July (21.7°C) and August (22°C) and minimum values in January (6°C) and February (6.5°C). Average rainfall is 800 mm, with maximum levels recorded in January and December (136 and 123 mm, respectively) and minimum in June and July (12 and 8 mm, respectively). The climate of the area was defined according to Bagnouls-Gaussen as meso-mediterranean, the udometric regime as xeric, and the thermometric regime as thermic.

Figure 1. Geographical location of the study area.

The testing area consists of two subunits, separated by a footpath, that were reforested in rows with *Pinus halepensis* Miller in one subunit and *Cedrus atlantica* (Endl) Carrière in the other during the winter of 1956. The study area is nearly uniform in terms of topography, lithology, slope, exposition, soil, climate, etc. Soils were classified as Lithic Haploxerolls [4]. Twenty pedons were selected, taking care to respect some similarities such as light exposition, pitch, etc. and the distance from the tree's trunk (1 meter) to avoid possible differences in soil characteristics due to stem flow [5]. Soils were described in the field in hand-dug pits; samples were collected from each organic horizon [3] and from each mineral horizon [6]. The pedological characteristics of the soil profiles are reported in Table 1. Each sample was air-dried and manually sieved (< 2 mm).

Table 1
Soil classification, description of soil-forming factors and morphological properties of twenty pedons in the testing area on the Sicani range, Sicily, Italy

Pedon N°	Site elevation (m)	Parent Material	Topography	Vegetation	Mean annual temp. (T °C)	Mean annual precip. (mm)	Classification (SSS, 1997)
1-10	1030	Dolomitic limestone	south-facing slope of 8%	*Pinus halepensis*	13.5	800	Lithic Haploxeroll, loamy, mixed, mesic
11-20	1030	Dolomitic limestone	south-facing slope of 8%	*Cedrus atlantica*	13.5	800	Lithic Haploxeroll, loamy, mixed, mesic

Pedon N°	Horizon	Depth* (cm)	Boundary*	Color* dry	Color* moist	Texture	Structure*	Consistence*	Roots*
1-10	Lv	6-3							
	Fm	3-0							
	Ah	0-8	gs	5YR3/3	5YR2/2	scl	3 fmcr	sh fr ss ps	2fm
	Bw	8-30	ai	7,5YR3/2	7,5YR2/2	cl	3 mabk	sh fr ss ps	1co 2fm
	R	30+							
11-20	Fa	4-0							
	Ah	0-7	gs	5YR3/3	5YR2/2	scl	1 fmcr	sh fr ss ps	2fm
	Bw	7-30	ai	7,5YR3/2	7,5YR2/2	cl	2 mabk	sh fr ss ps	1co 2fm
	R	30+							

Abbreviations: Boundary: a=abrupt; c=clear; g=gradual; d=diffuse; Ss=smooth; w=wavy; i=irregular. Texture: s=sand; l=loam; c=clay. Structure: 1=weak; 2=moderate; 3=strong; vf=very fine; f=fine; m=medium; c=coarse; cr=crumby; pr=prismatic; abk=angular blocky; sbk=subangular blocky. Consistence: (dry) so=soft; sh=slightly hard; vh=very hard; (moist) vfr=very friable; fr=friable; fi=firm; (wet) so=nonsticky; ss=slightly sticky; po=nonplastic; ps=slightly plastic. Root: 1=few; 2=common; f=fine; m=medium; co=coarse.
* average of prevalent.

2.2 Humic acids

Two pedons, under *Cedrus* and *Pinus*, respectively, were selected to carry out a comparative investigation on some chemical characteristics of humic acids. HA were extracted and purified according to Stevenson [7]. Thus, soil was shaken for 1 h with 0.1 M HCl (soil:solution ratio 1:10) to eliminate carbonates and alkaline-earth metals. After washing for the elimination of chloride, a solution of 0.1 M NaOH was added under N_2, to give a soil:extractant ratio of 1:10. The suspension was shaken at room temperature for 4 h and then allowed to stand for 12 h. After centrifugation, HA were precipitated from the supernatant by acidification with 6 M HCl to pH 1, then separated by centrifugation. Several purification steps with a solution of 0.1 M HCl and 0.3 M HF were included to reduce the amounts of inorganic impurities in the humic extracts. The HA obtained were then dialyzed by Spectrapor membrane (size exclusion limit, 1000 Dalton) to remove salts, and finally lyophilized.

2.3 Chemical analyses

Total organic carbon (C_{org}) was determined by mineralization with 2 N $K_2Cr_2O_7$ and H_2SO_4 solutions at 160°C according to Springer and Klee [8]. The extraction of the soil organic matter (SOM) was carried out according to Ciavatta et al. [9], by using a reagent containing 0.1 N NaOH and 0.1 N $Na_4P_2O_7$. Total extractable carbon (C_{ext}) and the combined humic and fulvic fractions carbon ($C_{(HA+FA)}$) were determined by dichromate oxidation method [9].

The humification rate (HR) and degree of humification (DH) were calculated according to Ciavatta et al. (1990), as follows:

$$HR\ (\%) = C_{(HA+FA)} \times 100 / C_{org} \qquad (1)$$

$$DH\ (\%) = C_{(HA+FA)} \times 100 / C_{ext} \qquad (2)$$

HA carbon, hydrogen, nitrogen and sulfur were determined by a Perkin-Elmer 240 B elemental analyzer.

2.4 FTIR analysis

FTIR spectra of HA were recorded on KBr pellets in the 4000 to 400 cm^{-1} wavenumber range using a Philips PU9800 FTIR spectrometer working in diffuse reflectance conditions. The results are given in Kubelka Munk units deriving from a mathematical formula applied to the diffuse reflectance spectra (Kortum, cited in [10]). The KBr pellets were prepared by pressing a mixture of 1 mg HA and 100 mg anhydrous KBr, spectrometry grade. The operative conditions for spectra acquisition were 100 scans with a 4 cm^{-1} resolution. Each analysis was replicated twice.

2.5 Thermal analysis

Differential scanning calorimetry (DSC) and thermogravimetry (TG) measurements of HA were carried out with a Netzsch Simultaneous Analyzer STA 409 equipped with a TG/DSC sample carrier supporting a type S thermocouple of PtRh10-Pt. The following conditions were used: heating rate of 10°C min^{-1} from 20°C to 900°C, static air atmosphere, alumina crucible, calcined kaolinite as reference, sample weight 5 mg. The thermo-balance was calibrated for buoyancy effects to obtain quantitative estimation of weight changes. Heat production in the "heat-flux" DSC was calibrated under the same measurement conditions by using a sapphire standard and subtracting a baseline obtained by an additional run for the empty crucibles. Netzsch applied software SW/cp/311.01 was used for data processing. Duplicate thermal runs were carried out to verify the measurement reproducibility.

3. RESULTS AND DISCUSSION

The two sites were previously compared on the basis of their contents of both total and exchangeable Ca, Mg, K, Na, N, and P; pH; humus forms; microbial biomass activity [11, 12]. Higher concentrations of the above mentioned mineral elements were found in the forest floor under *Cedrus* than under *Pinus,* with significant difference ($P < 0.01$) for total K contents (10.61 $cmolkg^{-1}$ and 7.83 $cmolkg^{-1}$, respectively). Exchangeable K was also higher in the A horizon under *Cedrus* (1.75 $cmolkg^{-1}$) than under *Pinus* (1.52 $cmolkg^{-1}$). In the B horizon, a significantly lower pH value under *Cedrus* (6.34) was found with respect to that under *Pinus* (7.24). Under the *Cedrus* stand, only two soil profiles showed an Lv horizon. The taxonomic classification of the humus forms [3] also revealed differences in the forest floors under the two tree stands: Hemimor under *Pinus* and Mormoder under *Cedrus* [4].

In Table 2, the contents of soil organic matter fractions in the two tree stands are reported. In the *Pinus* stand, a significantly higher amount of total organic carbon (C_{org}) was stored in both the Fm and A horizons than in the *Cedrus* horizons. In addition, the *Pinus* stand exhibited the highest contents of organic carbon extracted with a NaOH plus $Na_4P_2O_7$ 0.1 N solution (C_{ext}) and the highest humified fraction [$C_{(HA+FA)}$]. The organic matter from the *Pinus* Fm horizon showed a slightly lower chemical extractability, indicated by the C_{ext}/C_{org} ratio, which might be due to the mycological characteristics of this horizon that were recognized in the field observations.

Table 2
Average values of total organic carbon (C_{org}), total extractable carbon (C_{ext}), carbon of humic plus fulvic fraction ($C_{(HA+FA)}$), C_{org} to N ratio, C_{ext} to C_{org} ratio, humification rate (HR) and degree of humification (DH) along the profiles (average values of 10 profiles for each stand).

Ecosystem	C_{org} (g kg^{-1})	C_{org}/N	C_{ext} (g kg^{-1})	$C_{(HA+FA)}$ (g kg^{-1})	C_{ext}/C_{org}	HR (%)	DH (%)
Lv horizons							
*Cedrus**	343.2	32.1	218.7	133.9	0.637	39	61
Pinus	303.5	28.4	182.8	129.6	0.602	43	71
F horizons							
Cedrus	253.4	25.0	162.7	107.7	0.642	43	66
Pinus	351.6	27.5	202.3	143.3	0.575	41	71
Student t	3.18	0.75	3.15	3.73	1.29	0.52	1.69
Significance	P<0.010	NS	P<0.010	P<0.005	NS	NS	NS
A horizons							
Cedrus	65	20.2	42	29	0.657	45	69
Pinus	93	21.4	57	43	0.617	46	75
Student t	3.61	0.46	3.10	2.78	1.70	0.79	0.70
Significance	P<0.005	NS	P<0.010	P<0.025	NS	NS	NS
B horizons							
Cedrus	29	13.8	20	13	0.742	45	65
Pinus	31	14.2	21	14	0.669	46	67
Student t	1.41	0.47	0.59	1.06	1.75	0.43	0.88
Significance	NS	NS	NS	NS	NS	NS	NS

* only 2 samples. NS= values not significantly different (Student t-test).

According to the traditional approach, based on the different solubility at acidic conditions of extracted organic macromolecules, the calculation of the humification parameters [9] might give some indication of the extent to which the humification processes occur. From results reported in Table 2, it can be deduced that the HR and DH did not show statistically significant differences, although the small differences of DH could reflect a different quality of the extracted organic matter. Thus, the highest amount of humified fractions extracted under the *Pinus* stand did not correspond to a higher humification of the extracted organic matter on a gross scale.

Table 3 reports the results of correlation analysis for each A and B horizon between parameters related to the C cycle and the pH and K^+ values (ten values for each parameter and for each tree stand). The parameters related to the C cycle showed a significant correlation with the pH in A horizon under the *Cedrus* plantation. On the other hand, no correlation was found under *Pinus*. Since there were no significant pH differences in A horizons between the two tree stands, but since Pinzari et al. [12] reported significant differences in the microbial biomass activity and in the carbon mineralization rates between the A horizons, it was hypothesized that microbial biomass activity could play the main role in pedogenesis under *Cedrus*. In particular, a different litter quality, more favourable to microbial biomass activity and leading to better substrate utilization in the A horizons of the site with the *Cedrus* plantation, could determine different mineral-organic matter-biota interactions in the two stands [12].

Table 3
Correlation coefficients between C-parameters and the three chemical parameters that differed significantly between the two soils. The bold values correspond to a significant correlation ($0.05 > \alpha > 0.01$).

Parameters	pH				K^+			
	Cedrus A	*Cedrus* B	*Pinus* A	*Pinus* B	*Cedrus* A	*Cedrus* B	*Pinus* A	*Pinus* B
C_{org} (1)	0.75	0.59	0.19	0.20	0.03	0.46	0.17	-0.10
C_{ext} (2)	0.69	-0.60	-0.16	0.47	0.06	0.00	-0.31	0.35
$C_{(HA+FA)}$ (3)	0.72	0.51	0.04	-0.22	-0.09	0.51	-0.02	-0.23
C_{mic} (4)	0.77	0.53	0.09	-0.27	0.13	0.12	0.21	-0.39
$C_{28}{}^{th}$(5)	0.79	0.32	-0.05	0.05	0.11	0.50	-0.07	-0.02
C_o (6)	0.81	0.53	0.04	0.14	0.12	0.67	0.07	0.31
K (7)	-0.74	-0.44	0.14	-0.04	0.22	-0.17	-0.10	-0.11
C1 (8)	0.64	0.19	0.29	0.32	0.34	0.70	0.28	0.35

(1-2-3) see Table 3; (4) soil microbial biomass carbon; (5) microbial biomass basal respiration; (6) potentially mineralizable C; (7) rate constant of carbon mineralizable; (8) carbon respiration flush at the first day measurement. Data referred to (4) to (8) parameters are reported in Pinzari et al. [12].

The investigation carried out on HA from the two pedons selected under the two plant coverages allowed us to highlight some differences in their chemical composition. A different elemental composition of HA (Table 4) was found. In particular, the very high ash content of HA from the mineral horizons, despite the purification procedure adopted, demonstrates a strong linkage between humic acids and soil mineral components, more marked for HA formed under *Pinus* plants. This high HA ash content, different in both the profile and between the two studied pedons, makes the HA composition not directly comparable with respect to literature data, whereas the atomic ratio indicates the incorporation of N into the humic molecule (increase of the C/N ratio) and an increase in aliphatic (H/C) with depth.

Table 4
Compositions of humic acids (HA) and atomic ratio

	Elemental composition (%)					Atomic ratio	
Horizons	C	H	N	S	Ash*	H/C	N/C
			Cedrus atlantica				
Lv	52.49	5.63	3.46	0	2.87	1.289	0,057
Fa	52.40	5.25	3.10	0	4.50	1.202	0.051
A	30.81	2.99	1.93	0	36.62	1.165	0.054
Bw	17.14	2.78	1.35	0	56.63	1.946	0.068
			Pinus halepensis				
Lv	51.68	5.11	1.9	0	5.01	1.187	0.032
Fm	51.70	5.48	2.82	0	6.7	1.272	0.047
A	21.59	2.81	1.27	0	50.32	1.562	0.050
Bw	8.26	1.79	0.77	0	71.41	2.600	0.080

*Ash was measured by thermogravimetric analysis.

The FTIR spectra of the HAs studied are shown in Figures 2a and 2b. In all samples, the more intense bands were recorded in the following regions: at about 3400 cm^{-1} the -OH stretch, at 1730 cm^{-1} the carboxylic and/or carbonilic –C=O stretching, at 1630 cm^{-1} the conjugated and aromatic C=C vibrations, and at 1050 cm^{-1} the polysaccharide C-O stretch and/or the Si-O vibrations of associated clays. Less intense absorption was recorded at about 2900 cm^{-1} (aliphatic C-H stretching), 1520 cm^{-1} (aromatic C=C bonds), 1450 cm^{-1} (methylic and methylenic –C-H bending). The presence of a higher ash content in HA from the *Pinus* pedon was also confirmed. In fact, the increasing intensity of the absorption at 1050 cm^{-1} along the profile, accompanied by the decrease of the 3400 cm^{-1} signal, due to the aliphatic OH and/or NH groups, allowed us to assign the 1050 cm^{-1} band to the Si-O vibrations of associated clays. This behavior was not recorded for *Cedrus* HA. Moreover, HA under *Pinus* showed a more aromatic structure and a minor presence of alcoholic, carboxylic and carbonylic groups, in contrast to *Cedrus* HA, which were more hydrous and rich in both free and bound alcoholic groups. The minor presence of OH groups is considered indicative of a higher humification [13].

Differential scanning calorimetry and thermogravimetry measurements confirmed the observed differences between HA of the two pedons (Figures 3a and 3b). The oxidative thermal decomposition of HA from organic horizons (Lv and F) showed multisteps patterns that were reduced to two main reactions for HA from horizons A and B. In particular, the DSC curve of HA from *Pinus* Lv showed three well-resolved exothermic peaks at 331°, 456° and 531.7°C, each of which was associated with distinct weight losses registered on the TG curve. The correspondence between the heat flux rate on the DSC curve and the weight loss rate was checked by the calculation of the first derivative of the TG curve (DTG). The lower

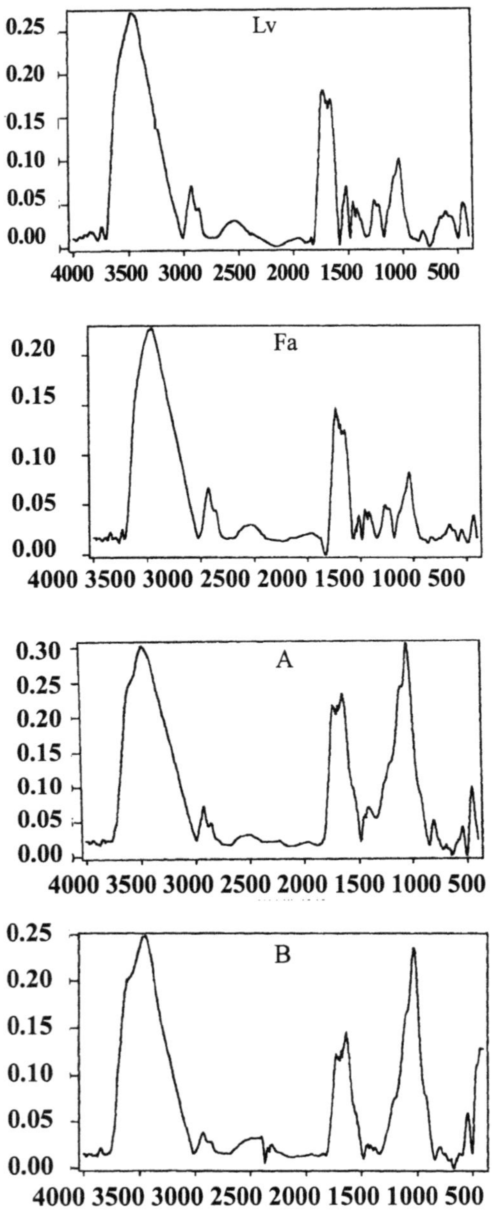

Figure 2a. FTIR spectra of HA separated from the horizons Lv, Fa, A and B of the *Cedrus* pedon. (Y axis: Kubelka Munk units; X axis: wavenumbers).

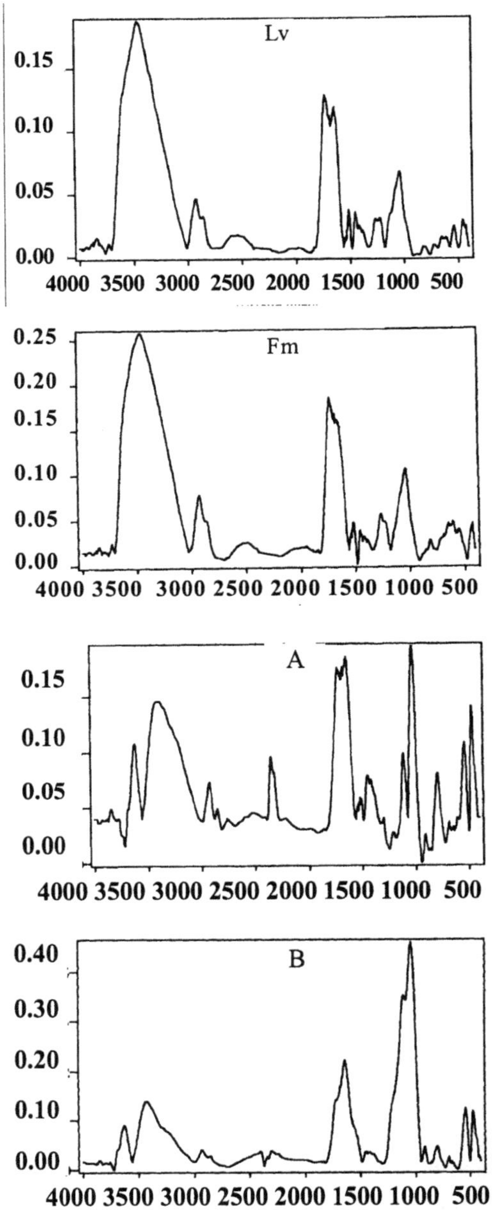

Figure 2b. FTIR spectra of HA separated from the horizons Lv, Fm, A and B of the *Pinus* pedon. (Y axis: Kubelka Munk units; X axis: wavenumbers

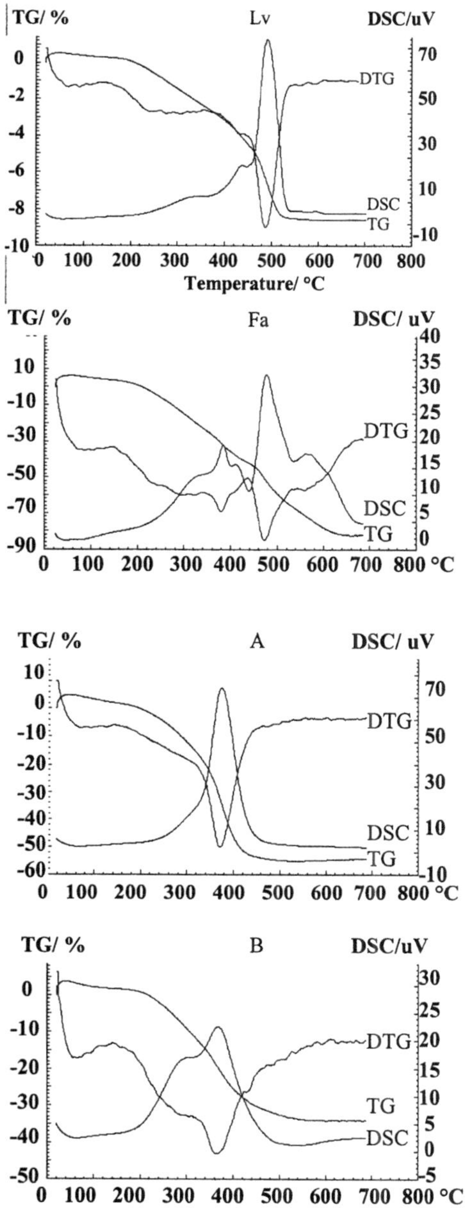

Figure 3a
Cedrus pedon: DSC, TG and DTG traces of HA from the horizons Lv, Fa, A and B.

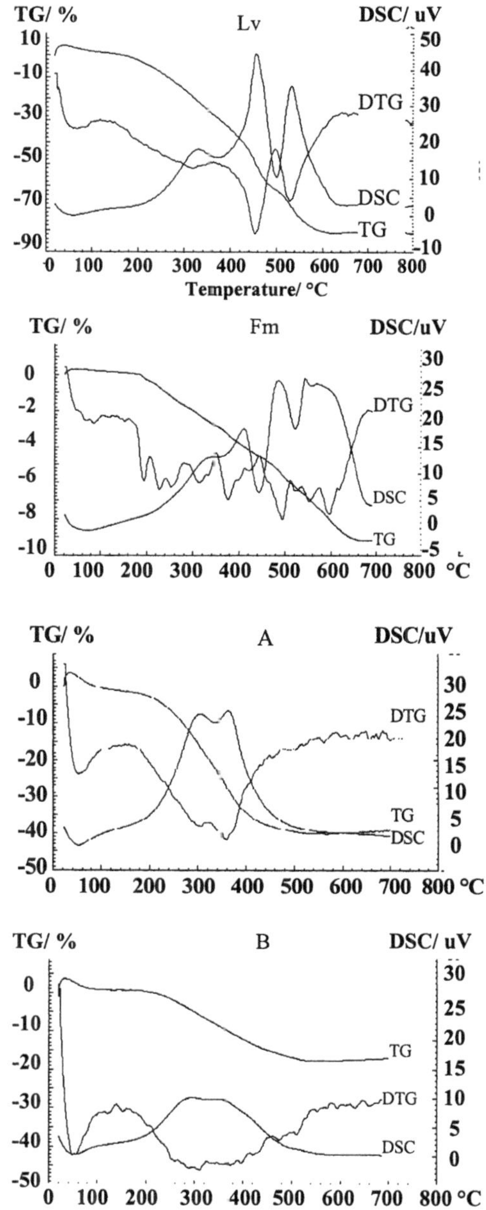

Figure 3b
Pinus pedon: DSC, TG and DTG traces of HA from the horizons Lv, Fm, A and B.

temperature exotherm was attributed to aliphatics and carbohydrates and the effects at higher temperature to thermal disassociation and the breakdown of aromatics [14].

HA from *Cedrus* Lv showed the most intense exotherm at 490.7°C and minor thermal effects at 435° and 350°C. Thermal behavior of HA from F horizons of both pedons was more complex in terms of number of reactions, which occurred over a wider temperature range. This could be the result of the organic resynthesis carried out by the microbial biomass acting in these horizons. Finally, thermograms of HA from *Pinus* mineral horizons showed two main reactions, which differed from those from *Cedrus* horizons.

The thermogravimetric data of weight losses that occurred during each thermally induced reaction allowed us to quantify the sizes of the different organic pools involved (Table 5). The results confirmed the difference in chemical composition between HA developed in the organic horizons under the two plantations and the strong linkage existing between humic acids and mineral components, as show by the high ash contents of HA; these data are reported in Table 4 and discussed above.

Table 5
Results of thermogravimetric measurements on HA: weight losses (%) related to the exothermic reactions; mean values of two replications.

Horizons	*Cedrus atlantica*		*Pinus halepensis*	
	Temperature range (°C)	TG weight loss (%)	Temperature range (°C)	TG weight loss (%)
Lv	149-366	31.55	146-365.1	30.2
	366-448.2	19.0	365.1- 497.5	34.94
	448.2-602.2	41.45	497.5- 640.4	20.57
F	144.9-344	28.2	152.4- 352	27.29
	344-437	18.24	352-371.8	3.17
	437-532.9	26.1	371.8-441.7	11.4
	532.9-678	17.8	441.7-518.0	15.07
A	145.2- 309.4	16.1	154.5-328.1	20.19
	309.4-558.7	42.6	322.5-583.8	20.7
B	157-300	12.4	143.8-601.4	22.03
	300-607.9	25.4		

4. CONCLUSIONS

The results obtained in this study showed a different evolution of soil profile under the two tree stands, in the formation of different humus forms and in different C storage in the forest floor and in some chemical parameters in the mineral horizons, such as pH. In particular, the parameters related to soil organic matter characteristics showed a significant correlation with pH in A horizons under the *Cedrus* plantation. It was hypothesized that microbial biomass activity could play a major role in pedogenesis under *Cedrus*, possibly due to the different litter quality more favorable to microbial biomass activity and leading to a better substrate utilization, as was found by Pinzari et al. [12]. The plantation difference influenced the amount of the humified organic fraction stored in the two stand soils: it was higher in both organic and A horizons under *Pinus*. However, the highest amount of humified fractions extracted under the *Pinus* stand did not correspond to a higher humification level of the extracted organic matter on a gross scale, as shown by the calculation of the humification parameters. Thus, it was not possible to define if either the quantity of organic matter stored in soil or its chemical characteristics were relevant for pedogenesis. On the other hand, data on plant biomass production were not available; thus this aspect was not investigated.

The investigation carried out on HA from the two pedons selected under the two plant stands allowed us to highlight some differences in their chemical composition. A different elemental composition of HA was shown. The atomic ratio indicated the incorporation of N into the humic molecule (increase of the C/N ratio) and an increase in aliphatic (H/C) with depth. Spectroscopic investigation carried out by FTIR showed a more aromatic structure and a minor presence of alcoholic, carboxylic and carbonylic groups for HA under *Pinus,* in contrast to *Cedrus* HA, which were more hydrous and rich in both free and bound alcoholic groups. Finally, HA thermal behavior was consistently different between HA from Lv horizons: in particular, under *Pinus*, more aromatic structures were revealed. Despite the HA purification procedure adopted, a very high ash content in the mineral horizons HA was found by thermogravimetry; this finding demonstrates a strong linkage between humic acids and soil mineral components, which was more marked for HA formed under *Pinus* plants.

Since the pedogenetic factors (mainly climate and parent material) were the same for the two pedons, and the HA extracting procedure adopted was the same for all samples, the formation of humic acids having different chemical composition could derive from different mineral-organic matter-biota interactions in the two tree plantations. In other words, a different role for the microflora and plant coverage in the soil profile evolution of the two stands could address the pedogenesis process. The results of investigations on energy use in the soil profile by soil microbial biomass can be found in Pinzari et al. [12].

REFERENCES

1. Fyles, J.W., Cote, B., 1994. Forest floor and soil nutrient status under Norway spruce and red pine in a plantation in southern Quebec. Can. J. Soil Sci. 74, 387-392.
2. Howard, P.J.A., Howard, D.M., Lowe, L.E., 1998. Effects of tree species and soil physico-chemical conditions on the nature of soil organic matter. Soil Biol. Biochem. 30, 285-297.
3. Green, R.N., Trowbridge, R.L., Klinka, K., 1993. Towards a Taxonomic Classification of Humus Forms. Supplement to Forest Science 29, 1-49.
4. Dazzi, C., Di Pisa, A., La Mela Veca, S., Vacante, G., 1996. Caratteristiche e distribuzione dei suoli forestali del bosco Ledera sui monti Sicani (Sicilia occidentale), Boll. Soc. Ital. Sc. Suolo, Palermo, 8, 295-307.
5. Pritchett, W.L., Fisher, R.F., 1985. Properties and Management of Forest Soils, 2nd Ed. Wiley & Sons, New York.
6. Soil Survey Staff, 1993. Soil Survey Manual, USDA Handbook, n. 18, Washington, D.C..
7. Stevenson, F. J., 1982. Humus Chemistry, John Wiley & Sons, New York.

8. Springer, U., Klee, J., 1954. Prüfung der Leistungsfähigkeit von einigen wichtigeren Verfahren zur Bestimmung des Kohlemstoffs mittels Chromschwefelsäure sowie Vorschlag einer neuen Schnellmethode, Z. Pflanzenernähr. Dang. Bodenk., 64, 1-8.
9. Ciavatta, C., Govi, M., Vittori Antisari, L., Sequi, P., 1990. Characterization of humified compounds by extraction and fractionation on solid polyvynilpyrrolidone. J. Chromat. 509, 141-146.
10. Painter, P., Starsinic, M., Coleman, M., 1985. Determination of functional groups in coal by Fourier transform interferometry. *In:* Ferraro, J.R., Basile, L.J. (Eds.), Fourier Transform Spectroscopy. Application to Chemical Systems. Vol. 4. Academic Press, Orlando, FL, pp.169-241.
11. Dazzi, C., 1996. Tassonomia del "Forest floor" e caratteristiche di fertilità di suoli forestali sotto pino d'Aleppo (*Pinus halepensis* Miller) e Cedro dell'Atlante [*Cedrus atlantica* (Endl.) Carriére], Boll. Soc. Ital. Sc. Suolo, Palermo, 8, 51-61.
12. Pinzari, F., Dell'Abate, M.T., Benedetti, A., Dazzi, C., 2002. Energy use in the A and B horizons of a soil under a pine and a cedar stand. *In:* Violante, A., Huang, P.M., Bollag, J.-M., Gianfreda, L. (Eds.), Proceedings of ISMOM Congress.
13. Schnitzer, M., Desjardin, J.G., 1965. Carboxyl and fenolic hydroxyl groups in some organic soils and their relation to the degree of humification. Can. J. Soil Sci. 45, 257-264.
14. Calace, N., Petronio, B.M., 1997. Characterization of high molecular weight organic compounds in landfill leachate: humic substances. J. Environ. Sci. Health, A32(8), 2229-2244.

Developments in Soil Science, Volume 28B
Editors: A. Violante, P.M. Huang, J.-M. Bollag and L. Gianfreda

ENERGY USE IN THE A AND B HORIZONS OF THE SOIL UNDER A PINE AND A CEDAR STAND

F. Pinzari[a], M.T. Dell'Abate[a], A. Benedetti[a] and C. Dazzi[b]

[a]Istituto Sperimentale per la Nutrizione delle Piante,
Via della Navicella, 2/4, 00184 Rome, Italy

[b]Dipartimento di Agronomia, Coltivazioni Erbacee e Pedologia, Università di Palermo,
Viale delle Scienze, 90128 Palermo, Italy

In this study, an estimate of the energy potentially available in soil profiles of two afforested sites in Sicily (Italy) was carried out. Our aim was to compare the relationship between organic carbon content, substrate availability, and metabolism by microbial biomass in two recently developed forest soils. In the two systems studied, the competition for available energy among species of bacteria, fungi, and other organisms was driven mainly by differences in the plant species that form the leaf litter. As in all terrestrial ecosystems, the carbon balance is highly dependent on the living microbial biomass, which is responsible for a substantial amount of the carbon flux from terrestrial to atmospheric systems.

1. INTRODUCTION

A proper understanding of energy dynamics, in terms of organic matter turnover and bioavailability for heterotrophic microbial communities in soil, is fundamental for remediation of terrestrial ecosystems. Carbon balance in terrestrial ecosystems is highly dependent on microbial populations, which are responsible for a substantial amount of the carbon flux from terrestrial to atmospheric systems [1]. The use of afforestation to promote the storage of organic carbon in eroded soils may lead to different outcomes according to the plant species employed. Substrate-dependent phenomena are, in fact, the basis of the differential functioning of saprophytic soil communities [2].

Microbial growth in soil depends on a combination of limiting factors interacting at the soil microsites where microbes are found. The structure of a soil microbial community is,therefore, determined by many variables but primarily by energy supply. In this study, an analysis of the energy theoretically available along the soil profiles of two afforested sites was performed.

Our aim was to compare organic carbon content, availability of substrates, and microbial metabolism in two forest soils developed under the same climatic and physical conditions but underlying different tree species. Given the key role of the soil microbiota in promoting energy exchanges and transformations in the soil profile, an extensive analysis of the microbial activity and of organic carbon pools was performed on twenty profiles (ten for each tree-stand). The parameters considered in this study were based on microbial activity (qCO_2 and C_{mic}:C_{org}), organic carbon fractions, and kinetics of organic C mineralization.

2. STUDY AREA

The study area (5000 m^2) is in the Sicani range in western Sicily, Italy, which is characterized by homogeneous environmental features of topography, lithology, slope, exposure, soil, climate, etc. Soils are classified as Lithic Haploxerolls [3]. Average monthly temperatures, from the meteorological data recorded between 1951 and 1990, indicated a maximum temperature in August of 22°C and a minimum temperature in January of 6°C. Average annual rainfall was 800 mm, with a maximum in January (136 mm) and a minimum in July (8 mm) [4]. The ombrothermic graph according to Bagnouls-Gaussen is characteristic of a meso-mediterranean climate. The udometric regime of the soil was defined as "xeric" and the thermometric regime as "thermic" [4].

3. MATERIALS AND METHODS

The stands were compared 40 years after planting. The afforestation was undertaken in the framework of a restoration aimed at preserving a pastured area prone to erosion. The tree species employed were *Pinus halepensis* Miller and *Cedrus atlantica* (Endl) Carrière. Despite the uniformity of lithology, climate, and age of the stands, the soils under pine and cedar turned out to be different with respect to their humus [4].

Twenty pedons (ten profiles for each tree stand) were studied to obtain a good representation of the spatial heterogeneity of soil development. Samples of each horizon of the soil profile were collected, although in this study, only the mineral horizons (A and B) were considered. During sampling, particular attention was paid to the observance of the same procedure in the excavation of each profile. A 50-cm^2 square-shaped mold was used to have a standard area from which to collect the soil samples. It was placed at a fixed distance from the tree trunks. All the profiles were chosen at comparable physical properties (light exposure, pitch, etc.). Each sample was air dried and manually sieved ($\leq$ 2 mm).

The physiological performances of the soil samples were tested under standard laboratory conditions (–33 kPa water tension and 30°C) to compare the features of the microbial community independent of temperature and soil moisture. The water retention of the soils was determined on sieved samples by a pressure cell apparatus [5]. Total nitrogen was determined by the Kjeldhal method [6]. Total organic carbon (%C_{org}] was determined after oxidation with 0.3 M $K_2Cr_2O_7$ at 160°C under acidic conditions [7].

The extraction of soil organic matter was carried out on 1 g of soil with 100 ml of a 0.1 N NaOH and 0.1 N $Na_4P_2O_7$ solution for 48 hours at 65°C under a N_2 atmosphere. The humic acids (HA) were precipitated by acidification of the extract with HCl to a pH value of 1 to 2; the fulvic acids (FA), which remained in solution, were purified on a polyvinyl-pyrrolidone column and then recombined with the humic portion [8]. Total extractable carbon (%C_{ext}) and the carbon content of the combined humic and fulvic fractions (%$C_{(HA+FA)}$) were determined by the dichromate oxidation method [8].

Microbial biomass carbon (C_{mic}) was measured by the fumigation-extraction method of

Vance et al. [9], with some slight modifications. The measurements were performed on air-dried soils, preconditioned by a 10-day incubation in open glass jars at –33 kPa water tension and 30°C. Incubation was used to restore the microbial activity of air-dried soils to that of soils in the field [10]. Four replicates of each soil sample were used. Average values are given in mg-C kg^{-1} of soil.

Soil respiration was measured in a closed system (1 L jar with a rubber ring), according the method of Isermeyer [11], with some slight modifications. The concentration and volume of the NaOH and HCl solutions for this study were chosen to ensure that less than 20% of the NaOH was neutralized by CO_2 during each measurement interval; indeed, low estimates of CO_2 can be accurately determined when large amounts of NaOH are neutralized [12]. Two beakers containing, respectively, the soil sample and the NaOH solution were placed in each jar; 4 ml of CO_2-free water was placed in the bottom of each jar to prevent soil moisture depletion by the NaOH [13]. Four replicates of each soil sample (25 g of each, oven dry-weight equivalent) were rewetted to–33 kPa water tension and incubated at 30°C. Carbon dioxide evolution was measured after 1, 2, 4, 7, 10, 14, 17, 21, and 28 days. Average values are given in mg CO_2-C kg^{-1} of soil, oven dry-weight equivalent. The 28-day measurements were taken as basal respiration values. Nonlinear least square regression analysis was used to calculate parameters affecting organic carbon mineralization from daily CO_2 evolution data (Stat Win 4.0 for Windows). The best fit was obtained with the exponential model of CO_2-C accumulation based on the negative exponential decay model: $C_m = C_o(1-e^{-kt})$, where C_m is the cumulative value of mineralized carbon during t days, k is a rate constant, and C_o is the potentially mineralizable carbon [14].

The metabolic quotient, $q(CO_2)$, defined as specific soil respiration of the microbial biomass, was calculated by considering the respiration values on an hourly basis and using the formula: $q(CO_2) = (mg\ C\text{-}CO_2) \cdot (mg\text{-}C_{mic} \cdot kg_{soil}^{-1})h^{-1}$ [15]. The ratio, $\%(C_{mic}:C_{org})$, was used as an index of the contribution of microbial biomass to soil organic carbon [16].

All results are expressed on the basis of oven-dry weight of soil (at 105°C), and they are the averages of measurements performed on ten field replicates. The significance of the differences between the data obtained for pine and cedar cover and between the data of the A and B horizons within the same vegetation was statistically assessed by the Student's t test for nonhomogeneous data. The probability of noncasual differences was considered $0.1 > \alpha > 0.00001$, corresponding to $90\% < P < 99.9\%$; for values of $\alpha > 0.1$, the difference was defined as not significant (NS) in the text.

4. RESULTS AND DISCUSSION

The long-term effects of the two types of trees on development of the soils are documented by the current differences of the relevant carbon pools, assuming that 40 years ago, the two soils were substantially the same [4]. Probably the processes that have led to these differences can also be inferred by means of microbial activity quotients. The soil profile in the site forested with *P. halepensis* appeared deeper and better developed than that forested with *C. atlantica* [4, 17].

The organic carbon content (C_{org}) is significantly higher in the A horizon of the soil under pine, and both the extractable and the humified fractions (C_{ext}, and $C_{(HA+FA)}$) are also higher in the soil under pine than in the soil under cedar (Table 1). The B horizons of the two soils do not differ in the amount of the stored organic carbon (Table 1). The averages of organic carbon to total nitrogen ratio values (C_{org}/N) did not differ significantly between the two sites (Table 2).

Table 1
Comparison between the average values of total organic carbon content, extractable carbon, humic carbon, and ratio of the extractable to total organic carbon of soils in the A and B horizons of 10 profiles for each stand. The Student's t values and the significance of the difference between the soils under pine and cedar are shown.

Plant species	Soil horizon	C_{org} (1)	C_{ext} (2)	$C_{(HA+FA)}$ (3)	C_{ext}/C_{org} (4)
Cedar	A	6.5	4.2	2.9	0.657
Pine	A	9.3	5.7	4.3	0.617
Student's t		3.61	3.10	2.78	1.74
Significance		S	S	S	S
		(α<0.005)	(α<0.01)	(α<0.025)	(α<0.1)
Cedar	B	2.9	2.0	1.3	0.742
Pine	B	3.1	2.1	1.4	0.669
Student's t		1.41	0.59	1.06	1.75
Significance		NS	NS	NS	S
					(α<0,1)

(1) total organic carbon, %; (2) carbon extractable with a $NaOH+Na_4P_2O_7$ solution, %; (3) carbon of the humic + fulvic acid fraction, %; (4) %C_{ext} to %C_{org} ratio. NS = values not significantly different (Student's t test).

Based on the microbial biomass (C_{mic}, Table 3), the soil under pine seems to sustain a smaller community; the C_{mic} values are higher in the cedar horizons despite the lower amount of available substrate (C_{org} and the correlated C-fractions, Table 1). Such a disproportion leads to a highly significant difference between the values of the C_{mic}:C_{org} ratios of the two soils (Table 2); the microbial biomass carbon normalized by the organic carbon values is related to the degradability of the substrate.

The lower value of the C_{mic} :C_{org} ratio in the soil under pine can be attributed to the higher complexity of the organic matter or to the effects of one or more limiting factors, such as the presence of terpenic compounds [18, 19]. Theoretically, microbial growth should be directly proportional to a given amount of available substrate [20]. But, if differences in growth exist, one or more of the following could apply: i) the characteristics of the substrates are not the same; ii) the microbial communities are not the same; or iii) other limiting factors are involved.

The comparison between the two soils on the basis of the activity of living microbial biomass was performed by means of metabolic quotients and substrate mineralization kinetics (Tables 2 and 4).

Table 2
Comparison between the average values of the metabolic quotient, the microbial biomass to total organic carbon ratio, and the total organic carbon to total nitrogen ratio of soils in the A and B horizons of 10 profiles for each stand. The Student's t values and the significance of the difference between the soils under pine and cedar are shown.

Plant species	Soil horizon	$q(CO_2)$ (1)	(C_{mic}/C_{org}) (2)	C_{org}/N (3)
Cedar	A	0.0024	1.12	20.2
Pine	A	0.0036	0.77	21.4
Student's t value		2.84	4.17	0.46
Significance		S ($\alpha<0.025$)	S ($\alpha<0.001$)	NS
Cedar	B	0.0022	0.71	13.8
Pine	B	0.0024	0.60	14.2
Student's t value		0.46	1.23	0.47
Significance		NS	NS	NS

(1) metabolic quotient (mg $C\text{-}CO_2$)· (mg-C_{mic}· kg_{soil}^{-1})h^{-1}; (2) percentage of microbial C (C_{mic}) to total organic carbon (C_{org}); (3) total organic carbon (C_{org}) to total nitrogen (N) ratio. NS = values not significantly different (Student's t test).

Table 3
Comparison between the average values of microbial biomass carbon and basal respiration of microbial biomass of soils in the A and B horizons of 10 profiles for each stand. The Student's t values and the significance of the difference (between the soils under pine and cedar) are shown.

Plant species	Soil horizon	C_{mic} (1)	$C_{28}{}^{th}$ (2)
Cedar	A	724	41.7
Pine	A	694	59.2
Student's t value		0.41	2.00
Significance		NS	S ($\alpha<0.1$)
Cedar	B	205	9.0
Pine	B	184	10.2
Student's t value		0.64	0.90
Significance		NS	NS

(1) microbial biomass carbon expressed in mg· kg^{-1} of soil dry weight (T= 105°C); (2) microbial biomass basal respiration, mg- $C\text{-}CO_2$· $hour^{-1}$· kg^{-1} of soil. NS = values not significantly different (Student's t test).

Table 4
Comparison between the average values of potentially mineralizable carbon, rate of carbon mineralization, ratio of potentially mineralizable carbon to effectively mineralized carbon, and ratio of potentially mineralizable carbon to total organic carbon of soils in the A and B horizons of 10 profiles for each stand. The Student's t values and the significance of the difference between the soils under pine and cedar are shown.

Plant species	Soil horizon	C_o (1)	K (2)	C_o/C_m (3)	C_o/C_{org} (4)
Cedar	A	2297.6	0.113	0.98	3.44
Pine	A	2952.7	0.088	1.04	3.15
Student's t		1.76	2.65	2.36	1.06
Significance		S ($\alpha<0.1$)	S ($\alpha<0.05$)	S ($\alpha<0.05$)	NS
Cedar	B	663.1	0.15	0.95	2.31
Pine	B	728.9	0.15	0.94	2.40
Student's t		0.72	0.23	0.80	0.29
Significance		NS	NS	NS	NS

(1) mg· kg^{-1} soil (dry weight at 105°C) of potentially mineralizable carbon; (2) rate constant of carbon mineralization; (3) mg· kg^{-1} soil (dry weight at 105°C); ratio of potentially mineralizable carbon to total carbon mineralized during the 28-day incubation; (4) mg· kg^{-1} soil (dry weight at 105°C); ratio of potentially mineralizable carbon to total organic carbon. NS = values not significantly different (Student's t test).

The analysis of the metabolic quotient (qCO_2) and the metabolic rate (k) is highly informative with regard to the difference in the energy balance between the two soils. In the soil under pine, the microbial biomass apparently uses more energy to obtain smaller results in terms of time required to mineralize the available substrate. The microbial biomass in the A horizon of soil under pine, in fact, used more energy (high respiration rate) but grew more slowly and degraded less organic carbon than the microbial biomass in the A horizon of soil under cedar. In the A horizons of the soil under cedar, the qCO_2 was significantly lower than in the soil under pine, the kinetic constant (k) was significantly higher, and the ratio of potentially mineralizable carbon (Co) to effectively mineralized carbon (Cm) was < 1, thus indicating a high efficiency of energy use (Tables 2 and 4). The lower value of k in the soil under pine (Table 4) indicates that it takes longer for the microbial biomass to use the organic matter in the soil, and the higher value of the qCO_2 (basal respiration normalized by the amount of microbial biomass respiring) (Table 2) indicates metabolic stress (i.e., consistently harsh conditions, which are likely when microorganisms find difficulties in using substrates: the higher metabolic activity in enzymes production brings microorganisms to increase their respiration rate) [21].

Further investigations are necessary to assess whether or not the activity of the microbial biomass is itself responsible for the higher storage of organic carbon in soil under pine. Alternatively, it might be that the substrate is so different that the composition of the microbial population or its activity is affected, given the similarity of the geological and climatic variables. Changes in the organic carbon content of soils may depend on climatic fluctuations and the amounts of organic input. In a given pedoclimatic condition, turnover of organic matter is controlled not only by the inputs but also by the outputs represented by the mineralization of organic matter. Microbial activity in soil is controlled by physical and chemical conditions, such as oxygen supply, pH, and ionic activity [22, 23]. Limiting factors for maximal microbial activity may, therefore, differ appreciably between superficially similar environments [24]. In forest soils, organic inputs are mainly the result of litterfall, and plant litter is, therefore, the major determinant of properties of the ecosystems (such as the availability of nutrients in soil, the soil organic matter quality, the water content of the top soil horizons, etc.). In the soils both under pine and cedar, the above- and below-ground systems are strongly dependent upon each other, and the soil-based biotic component is of great importance in determining the activity in the ecosystems, e.g., by affecting nutrient cycling, decomposition, and energy flow [1, 19, 21].

Differences between the microbial communities of the two soils were evident, even during a first field observation: the F horizon [4, 25], classified as "amphi" under cedar, and as "mycogenous" under pine, showed, in fact, macroscopic peculiarities corresponding to a different kind of saprophytic attack on the organic residues [4].

The F horizon under cedar trees was, in fact, characterized by partially comminuted and decomposed foliage with a friable structure, rich in mesofauna and, in particular, in arthropods [4]. The F horizon under pine trees appeared, instead, poorer in mesofauna but rich in fungal hyphae, which forms a dense reticulum that gives to the horizon a firm structure and a diagnostic aspect [4, 25].

The B horizons of both soils under pine and cedar were statistically similar according to most of the parameters studied, as only extractability of organic matter (C_{ext}/C_{org}) differed between the two soils at the B horizon level (Table 1). Based on these results, two interpretations can be suggested: i) 40 years of forestation on the sites studied were not sufficient to induce net changes in the chemistry of the B horizon; and ii) transformations did occur in the B horizons but resulted in the acquisition of similar characteristics. To analyze this aspect in greater detail, the energy balances in the two mineral horizons (A and B) within each soil were compared. Table 5 reports the significance of the differences between A and B horizons in the profiles under both cedar and pine (Student's t values).

High values of the Student's t correspond to a highly significant statistical difference between the upper and the lower horizons within the soil profile under each tree species. Low values, even when there was a significant difference, indicate a less sharp transition from one horizon to the next along the profile.

According to most of the parameters, both horizons in the profile under pine were distinct: the depth-dependent decrease in the C-linked values was highly significant and the Student's t values were higher than in the soil under cedar (Table 5, parameters 1–6 and 8, 9 and 11). The interface between the A and B horizons in terms of substrate availability and metabolic activity was less sharp in the soil under cedar than under pine. The metabolic quotient (qCO_2), for example, was significantly different between the A and B horizons of the soil under pine. In the soil under cedar, the metabolic activity, based on the differences between qCO_2 values, was the same in the A and B horizons (very low Student's t value, Table 5). It can be hypothesized that under cedar, the metabolic activity of the soil biomass was proportional to the availability of substrates in both horizons, which resulted in similar qCO_2 values.

Table 5
Significance of the difference between A and B horizons in the profiles under cedar and pine (Student's t values). High values correspond to a highly significant statistical difference between the upper and the lower horizons within each soil profile. Low values, even when addressing a significant difference, also indicate a less sharp transition from one horizon to the other along each profile.

Parameter	Cedar		Pine	
	Student's t	α	Student's t	α
C_{org} (1)	6.9	<0.001	9.0	<0.0001
C_{ext} (2)	7.6	<0.0001	9.0	<0.0001
$C_{(HA+FA)}$ (3)	6.9	<0.001	7.8	<0.0001
C_{ext}/C_{org} (4)	1.9	<0.10	2.4	<0.05
C_{mic} (5)	7.6	<0.0001	12.3	<0.00001
$q(CO_2)$ (6)	0.5	NS	2.7	<0.05
(C_{mic}/C_{org}) (7)	4.5	<0.001	2.0	<0.10
C_{28}^{th}(8)	5.9	<0.001	7.1	<0.0001
C_o (9)	5.9	<0.001	7.7	<0.0001
K (10)	2.8	<0.05	0.2	NS
C_{org}/N (11)	2.5	<0.05	4.1	<0.001
C_o/C_{org} (12)	4.6	<0.001	2.2	<0.05

(1) Organic carbon; (2) extractable carbon; (3) humic and fulvic acid carbon; (4) extractable to total carbon; (5) microbial biomass carbon; (6) metabolic quotient; (7) microbial to organic carbon ratio; (8) basal respiration (C-CO_2); (9) potentially mineralizable carbon; (10) carbon mineralization rate; (11) organic carbon to total nitrogen ratio; (12) mineralizable carbon to total organic carbon.

This suggests a greater influence of the topsoil on the deeper layers in the soil under cedar, perhaps
consisting of a different solubility of organic matter or different ion equilibrium.

A better understanding of the processes at work in the pedogenesis of the two soils might be derived from further analysis of the quality and stability of the organic matter. The chemical and physical characteristics of leaf litter are factors toward which additional studies should be directed, to obtain a better definition of the mechanisms that regulate energy use along a soil profile.

ACKNOWLEDGMENTS

We thank Dr. A. Zilli (Rome) and the journal referees for helpful comments and suggestions on the editing of the manuscript.

REFERENCES

1. Howard, P.J.A., Howard, D.M., Lowe, L.E., 1998. Effects of tree species and soil physico-chemical conditions on the nature of soil organic matter. Soil Biol. Biochem. 30, 285-297.
2. Priha, O., Smolander, A., 1997. Microbial biomass and activity in soil and litter under *Pinus sylvestris*, *Picea abies* and *Betula pendula* at originally similar afforestation sites. Biol. Fert. Soils 24, 45-51.
3. Dazzi, C., Di Pisa, A., La Mela Veca, S., Vacante, G., 1996. Caratteristiche e distribuzione dei suoli forestali del bosco Ledera sui monti Sicani (Sicilia occidentale). Boll. Soc. Ital. Sci. Suolo 8, 295-307.
4. Dazzi, C., 1996. Tassonomia del "Forest floor" e caratteristiche di fertilità di suoli forestali sotto pino d'Aleppo (*Pinus halepensis* Miller) e Cedro dell'Atlante [*Cedrus atlantica* (Endl.) Carriére]. Boll. Soc. Ital. Sci. Suolo 8, 51-61.
5. Richards, L.A., Fireman, M., 1943. Pressure plate apparatus for measuring moisture sorption and transmission by soils. Soil Sci. 56, 395-404.
6. Forster, J.C., 1995. Soil nitrogen. *In:* Alef, K., Nannipieri, P. (Eds.), Methods in Applied Soil Microbiology and Biochemistry. Academic Press, London, pp. 79-81.
7. Springer, U., Klee, J., 1954. Prüfung der Leistungsfähigkeit von einigen wichtigeren Verfahren zur Bestimmung des Kohlenstoffs mittels Chromschwefelsäure sowie Vorschlag einer neuen Schnellmethode. Z. Pflanzenernähr. Dang. Bodenk. 64, 1-8.
8. Ciavatta, C., Govi, M., Vittori Antisari, L., Sequi, P., 1990. Characterization of humified compounds by extraction and fractionation on solid polyvinyl-pyrrolidone. J. Chromat. 509, 141-146.
9. Vance, E.D., Brookes, P.C., Jenkinson, D.S., 1987. An extraction method for measuring microbial biomass C. Soil Biol. Biochem. 19, 703-707.
10. Stotzky, G., Goos, R.D., Timonin, M.I. 1962. Microbial changes occurring in soil as a result of storage. Plant Soil 16, 1-18.
11. Isermeyer, H., 1952. Eine einfache Methode sur Bestimmung der Bodenatmung und der Karbonate im Boden. Z. Pflanzanern. Bodenk. 56, 26-38.
12. Gupta, S.R., Singh, J.S., 1977. Effect of alkali concentration, volume and absorption area on the measurement of soil respiration in a tropical sward. Pedobiologia 17, 233-239.
13. Stotzky, G., 1965. Microbial respiration. *In:* Black, C.A. et al (Eds.), Methods of Soil Analysis. Part 2, Chemical and Microbiological Properties. American Society of Agronomy, Madison, WI, pp. 1550-1572.
14. Riffaldi, R., Saviozzi, A., Levi-Minzi, R., 1996. Carbon mineralization kinetics as influenced by soil properties. Biol. Fertil. Soils 22, 293-298.
15. Anderson, T.-H., Domsch, K.H., 1990. Application of eco-physiological quotients (qCO_2 and qD) on microbial biomass from soils of different cropping histories. Soil Biol. Biochem. 10, 251-255.
16. Anderson, T.-H., Domsch, K.H., 1989. Ratios of microbial biomass carbon and total organic carbon in arable soils. Soil Biol. Biochem. 21, 471-479.
17. Pinzari, F., Dell'Abate, M.T., Benedetti, A., Dazzi, C., 1998. Effects of reforestation made up by *Cedrus atlantica* and *Pinus halepensis* on soil chemistry and fertility. *In:*

Proceedings of the World Congress of Soil Science. Supplement to: Revue Etude et Gestion des Sols, Vol. 5, 1988 n.2, Montpellier, France (abstract).

18. Rice, E.L., 1979. Allelopathy-an update. Bot. Rev. 45, 15-109.
19. Rice, E.L., Pancholy, S.K., 1972. Inhibition of nitrification by climax vegetation. Am. J. Bot. 59, 1033-1040.
20. Odum , E.P., 1985. Trends expected in stressed ecosystems. Bioscience 35, 419-422.
21. Wardle, D.A., Ghani, A., 1995. A critique of the microbial metabolic quotient (qCO_2) as a bioindicator of disturbance and ecosystem development. Soil Biol. Biochem. 27, 1601-1610.
22. Stotzky, G., 1997. Soils as an environment for microbial life. *In:* van Elsas, J.D., Wellington, E.M.H., Trevors, J.T. (Eds.), Modern Soil Microbiology. Marcel Dekker, NY, pp. 1-20.
23. Stotzky, G., Norman, A.G., 1961. Factors limiting microbial activities in soil: I. The level of substrate, nitrogen, and phosphorus. Arch. Mikrobiol. 40, 341-369.
24. Pinzari, F., Trinchera, A., Benedetti, A., Sequi, P., 1999. Use of biochemical indexes in the Mediterranean environment: Comparison among soils under different forest vegetation. J. Microbiol. Met. 36, 21-28.
25. Green, R.N., Trowbridge, R.L., Klinka, K., 1993. Towards a taxonomic classification of humus forms. Forest Sci. 29, 1-49.

Developments in Soil Science, Volume 28B
Editors: A. Violante, P.M. Huang, J.-M. Bollag and L. Gianfreda

FUTURE PERSPECTIVES ON THE ENVIRONMENT AND HUMAN HEALTH INTERACTIONS

C. De Kimpe

Research Branch, Agriculture and Agri-Food Canada, 930 Carling Avenue, Ottawa, Ontario, Canada K1A 0C5. E-mail: dekimpec@em.agr.ca

At the meeting point between the atmosphere, the biosphere, the lithosphere and the hydrosphere, soils play a central role in supporting life on the earth. The health of the soil ecosystem is controlled by the equilibrium that exists between the mineral and organic, living and non-living soil components. The paper considers examines and discusses some new knowledge in the areas of the mineral-organic matter-microorganisms interactions in the soil, in particular the understanding of the soil ecosystem in space and time, and its functioning. Inorganic and organic soil components, due to specific surface properties are interacting with the heavy metals and xenobiotics. The dynamics, mobility, availability and fate of such toxic ions and molecules are largely controlled by the soil microorganisms. The management and transformation of soils for human activities, in particular food production, have major impacts on the mineral-organic interactions, and this is discussed from the points of view of soil physical condition and nutrient dynamics. As a consequence, it becomes increasingly important to understand how the soil functions that are supported by the living soil organisms, may contribute to soil resilience and bioremediation. Considering the growing pressure of the increasing world's population on the natural resources, soil, water, biodiversity, for food production and disposal of waste, research has to provide decision-makers with tools based on sound science and the full understanding of the soil processes and interactions in order to manage the natural resources in a sustainable way. Progress made recently in the domain of soil ecology, especially the dynamics, the role and activity of the soil organisms, opens new perspectives for maintaining and improving ecosystem health in the future, for the benefit of human health, for example by securing safe food and drinking water.

1. INTRODUCTION

An ecosystem is the complex of a community and its environment functioning as a unit in nature.

The concept of "Ecosystem health" includes ecological integrity as well as social, economic and political concerns: an ecosystem-based management considers and analyzes the landscapes with its physical and biological components, the processes, and the human activities in order to maintain the integrity of the ecosystem. Soils, at the meeting point between the atmosphere, the lithosphere and the hydrosphere, are structural and functional elements of the terrestrial ecosystems; they result from the interactions of geological, climatic and biotic factors, i.e. flora, fauna, microorganisms, and humans.

In the ecosystem, soils play a major role by providing a variety of habitats and ensuring their stability. The soil functions include *biochemical and geochemical cycling*, *partitioning of water*, *storage and release*, *buffering*, and *energy partitioning*. Therefore, land use that is sustainable for human and environmental health requires the understanding of how the soil fulfills its functions in the ecosystem. Humans affect the biotic and abiotic components of ecosystems in many ways; because some changes are irreversible, it is the responsibility of the research community to provide the decision-makers with the scientific support to promote the sustainability of the natural resources.

2. THE CHALLENGE AHEAD

World's population reached 6 billion in 1999, and according to the United Nations Population

Division [1], it is expected to increase to about 8.47 billion by the year 2025; a projection by FAO [2] suggests a population of about 10 billion by the year 2050. Urban population was less than 30% of the world's population in 1950, but it will exceed 60% by 2025 [3]. Population increase and urbanization are creating new challenges in terms of the management of the natural resources, e.g. providing food and fibre to present generations while ensuring the sustainability of the soil and water for future generations, and disposing of huge amounts of waste material generated by human activities. Although much of the organic wastes can be recycled on the agricultural lands, they present some risks for human and environmental health.

The world's land base used for agriculture is about 36% of the total land mass, including 11.3 % of cropland and 24.7% of pastures. It cannot be much increased because constraints related to the nature of soils, climate and topography. Concerns for human and environmental health arise therefore from:

- the increased use of fertilizers and pesticides for more intensive food production
- the industrial and municipal wastes applied to agricultural soils, and susceptible to move to the food chain and the water bodies.

The Symposium "Soil Mineral-Organic Matter-Microorganisms Interactions and Ecosystem

Health" addressed the above issues from several points of view, and contributed scientific information for understanding and maintaining the ecosystem integrity. Six topics were addressed during this Conference, and some findings will be briefly discussed before considering the perspectives for future research.

2.1. Ecological significance of interactions of metals and metalloids with soil colloids, enzymes and microbial biomass

In order to understand the complex soil system, and the ecological effects of contaminants on community interactions, Rutgers et al. [4] studied the impact of Cu contamination and pH stress on the microbial communities under field conditions; they measured the intensity of 31 metabolic activities, in order to characterize the community-level physiological profile: Biolog plates allowed to assess the decrease in sensitivity of microbial functions in the presence of pollutants. Dialysis was used for determining the ability of natural humic material to form complexes with Cu and Co [5]. Compared to other methods, dialysis can be used under conditions (pH, ionic strength) close to natural systems and is applicable to a larger number of metals than techniques based on ion selective electrodes or fluorescence spectrometry. Ecophysiological indices were applied to detect the influence of metals on the microflora under field conditions, and to evaluate the quality of soil microbial biomass [6]: cellulase activity was found as a suitable index. A positive correlation between heavy metals and organic carbon indicated an interference of heavy metals on the mineralization of organic substances, and explained the accumulation of organic carbon. Similar observations were made at a different scale [7]: microorganisms communities in the different microhabitats corresponding to particle size fractions, are affected in their structure and functioning by soil pollution.

2.2. Dynamics and mobility of nutrients and inorganic pollutants in the soil environment

At a microscopic scale, fine-grained particles and associated microbial communities constitute a variety of ecosystems acting as sinks and secondary sources of nutrients and toxic contaminants; they mediate chemical transformations of these substances, controlling their bio-accumulation and availability, and bio-geochemical cycling [8]. Energy dispersive X-ray microanalysis (EDXM) coupled with transmission electron microscopy (TEM) of ultrathin sections allowed to investigate the processes occurring in natural systems, the characteristics and *in situ* activities of microbial populations, and the interactions among microorganisms and with the soil colloids.

Sparks [9] used X-ray absorption fine structure spectroscopy and diffuse reflectance spectroscopy to study the kinetics of metal sorption and to assess the speciation, mobility, and bio-availability of metals in natural systems. Mixed metal-Al hydroxide precipitates (layered double hydroxides, LDH) form when metals such as Ni(II), Co(II), and Zn(II) are introduced in an environment with a source of Al. Metal-hydroxides, especially LDHs, stabilize the metal. The kinetics of metal sorption/release is necessary for predicting the metal fate and mobility, e.g. in soil remediation. For some metals (e.g., Co, Ni, Zn), the residence time affects the rate of metal release while for other ones such as Pb, the effect of the residence time is negligible. Emissions of F from an aluminum plant had caused humus accumulation in nearby soils [10]. Later on, the installation of a filtration unit resulted in a decay of the accumulated humus, and this influenced the behavior of major and minor chemical constituents of the soil; particle size decreased, partly because of weathering, and partly because of the destruction of clay-humus complexes. Fe and Al extractability was modified by the change in humus content as mineral Fe-, Al-, Mg- and Mn-phases developed following humus decomposition. Pb content was related to humus content and change in Fe content. Cd correlated with pH, (earth)alkali ions, and to a lesser extent with Fe and Al, but no correlation was found with soil organic matter.

The interdependence of cation exchange capacity (CEC), pH, ionic strength, and organic and mineral forms of Al was investigated in organic matter-rich forest soils [11]. CEC is largely due to pH-dependent charges, and organic matter content is the most important single factor affecting CEC. Al in organic complexes contributes to pH buffering in acid soils, and the amount of Al in the complexes depends on Al availability. Krishnamurti et al. [12] found that the Al_{13} polycation, with the structure $[AlO_4Al_{12}\bullet(OH)_{(24+n)}(H_2O)_{(12-n)}]^{(7-n)+}$, is the dominant species formed by hydrolysis of Al at an OH/Al molar ratio of 2.5, but its presence in the soil has not been confirmed. Several factors can perturb its formation and stability, e.g.

the phenols, common in the decomposition products of plant and animal materials and microbial metabolites in soils, prevent its crystallization, and lead to the formation of x-ray amorphous products.

Torrent and Barron [13] determined quantitatively the poorly crystalline and crystalline Fe oxides and other Fe-containing phases by selective dissolution methods, using citrate/ascorbate and oxalate extractions. The second derivative of the soil diffuse reflectance spectrum provided a clear identification of various Fe oxides, and also a quantitative determination of goethite and hematite in soils and mineral mixtures when no other Fe oxides were present. Lamy et al. [14] recovered, by ultracentrifugation, colloids from gravitational water in agricultural, forest, and grassland soils near an industrial site. They studied their role in metal transport and redistribution in soil profiles, using TEM and EDXM. The nature and morphology of the colloidal fraction in soil solutions differed in the ecosystems, with dominant amorphous organic colloidal material in the forest soil, a mixture of amorphous and crystallized, organic and mineral colloids in the agricultural soil, and dominantly well-crystallized, non-aggregated clay minerals in the soil under the grassland. Zn was always associated with colloids. In a study of 29 grassland soils from the UK, differing by their microbial and biochemical properties, P level, organic matter and clay content, the soil microbial biomass controlled the inorganic and organic P release from the soil: the phosphodiesterase activity determined the labile soil organic P [15]. There was no relation with phosphatase activity for the inorganic P, but the uptake of inorganic P by the biomass controlled the availability of inorganic P for water contamination.

Organic residues are commonly used as amendments in order to maintain fertility, especially in sandy soils; this practice contributes an ecological way of disposing of the residues. Because the content of heavy metals may be high, it is essential to determine the bio-availability to crops. The mobilization of the heavy metals in the soil during the turnover of soil organic matter was studied through the biodegradation of metal-polluted plant residues and the organic matter-heavy metal interactions in a soil amended with contaminated and non-contaminated plant leaves [16]. The long-term effects of municipal solid waste compost, MWC, agroforestry compost, F, and paper mill waste, PW, on strawberry yield in Spain [17]. Organic carbon, heavy metals and enzyme activity were measured. Organic carbon content generally increased. Enzyme activity was enhanced, more with MWC and PW, despite a low enzyme activity in these residues. A positive correlation was found between organic carbon and enzyme activity. Small increases of available metal content did not have evident negative effects on the biological activity of the soils, but negative impacts could be masked by the positive effect of organic matter. Heavy metal uptake and redistribution was measured in the shoots and roots of celery plants grown on a soil irrigated with wastewater containing Cr, Cu and Cd [18]. Metals accumulated in the roots, while increasing Cr concentrations caused a noticeable reduction of root and shoot growth. Sappin-Didier et al. []19] used isotope exchange kinetics (IEK) to investigate Cd availability to crops: Cd content in winter wheat was correlated with the variables of IEK. The exchangeable Cd, the capacity of the soil to supply Cd in the solution, and the capacity of the soil to release Cd during wheat growth were determined.

2.3. Dynamics and transformations of organic pollutants in soil and their impact on ecosystem health

Soils, similar by their mineralogy and exchangeable cations, may differ in the accumulation of organic matter (OM): differences in drainage conditions affecting the soil processes [20] and the degree of organic matter aromaticity [21] had a role in Mollisols from Germany. This stresses the need for characterizing better the quantity, quality and properties of soil organic matter.

Land use and cultivation have strong impact on the amount and quality of the soil organic matter. In a rice production system, the reduction of organic matter content and soil productivity was related to an increasing use of mineral fertilizer [22]. Eco-farming, i.e. the application of organic management practices, may contribute to restoring the soil organic matter level. Clay-OM complexes and aggregation contribute to the physical protection and stability of soil OM. The clay-OM complexes are destroyed upon cultivation [23], leading to OM decrease, less particle stability and clay dispersion. Particulate organic matter, POM, with an intermediate turnover time, is easily affected by changes in land management. Leifeld and Kögel-Knaber [24] used ultrasonic energy to disrupt aggregates and separate free and occluded POM in samples from adjacent arable and grassland sites. Although OM and N content of the bulk soils was higher in the grassland, structural stability was low and soil aggregation was not related to land-use. Low aggregate stability was possibly explained by the sandy texture of the soils, that facilitates the OM turnover. In sandy soils, textural effects may supersede the effects of land use on the amount and structure of OM fractions.

Long-term addition of manure to a fine textured soil results in a significant increase of macro- and microporosity [25]. A positive correlation was observed between total porosity and soil C concentration: a 1 to 2 % increase of soil C induced an increase in porosity from 52 to 56%, a 9% relative increment, while the largest increase (90%) was found in the very fine macropore fraction (60-600 mm). Kiem et al. [26] analyzed soil organic C composition in plots with high and low C-input: low input causes the depletion of labile compounds and a relative accumulation of refractory C components. Coarse fractions (> 63 μm) lost more C than the intermediate (63-6 μm) and fine fractions (< 6 μm). The fine clay (< 0.2 μm) had the smallest decrease in OC content. NMR spectroscopy showed an increase of the signal intensity of aromatic carbon in the fractions > 6 μm; the lignin content was much reduced, and is not a major form of refractory aromatic carbon.

Impacts of Cu, Cd and Zn on biological properties were investigated in soils from Sri Lanka [27]. Organic matter content, microbial biomass, fungi and actinomycetes, basal respiration and substrate induced respiration were significantly lowered, and this effect increased with the number of years of cultivation. Microbial biomass content, population of fungi and actinomycetes, and respiration were correlated positively with the OM content and negatively with HCl-extractable Cu, Cd and Zn. Heterotrophic bacteria did not decrease significantly following cultivation while NH_4^+-oxidizing bacteria increased significantly.

Allelochemicals, e.g. phenolics, terpenoids, alkaloids, interact with mineral and organic soil components. Many chemicals suspected to have allelopathic properties have features similar to pesticides and xenobiotics, and can be used as models for characterizing the fate of xenobiotics and their interactions in the ecosystem. Fixation of xenobiotics in the soil contributes to reduce their bio-availability. Organic matter and monomeric humic substances produced by lignin decomposition form strong covalent bonds, difficult to disrupt by microbial activity or chemical treatment. Chemical bonds may be formed through oxidative coupling reactions, catalyzed biologically by polyphenol oxidases and peroxidases, and abiotically by clays and metal oxides. Bollag et al. [28] evaluated the degree of binding using ^{13}C- or ^{15}N-labeled xenobiotics combined with ^{13}C- and ^{15}N-NMR spectroscopy. Signal delocalization indicated whether covalent binding occurred, and the type of covalent bond. Low molecular weight organic acids (LMWOAs) from root exudates, decomposing organic matter, and other sources are important ligands. Chen et al. [29] identified several substances by gas chromatography, and used NMR to determine the organic C functional groups. Organic functional groups in the rhizosphere and bulk soils followed the trend: alkyl-C > O-alkyl-C > N-alkyl-C > acetal-C > aromatic-C > carboxylic-C > phenolic-C. Abiotic soil reactions that govern the bio-availability of organic pollutants in the soil are poorly documented [30]. Hydrolysis, oxidation and polymerization reactions catalyzed by clay minerals and Mn and Fe oxides, dominate the abiotic transformation of xenobiotics in soil. Sorption of organic contaminants by colloids prevents the direct contact between microorganisms and contaminants, that is necessary for bio-degradation to occur. Prediction of bio-degradation rates of sorbed xenobiotics is more complex than for free molecules. Microorganisms are found preferentially at the particle surface [31]: clay and humate surfaces provide high concentrations of carbon and energy sources, and microbes on surfaces survive longer, grow faster and respire more rapidly than those located away from surfaces. Selection pressure in oligotrophic conditions favors soil microorganisms that migrate and adhere to surfaces and, in combination with other species, form structured biofilms. Stability, spatial integrity and degradative capacity of such biofilms were studied using nucleic acid probes and fluorescent stains. The community diversity was demonstrated by amplified ribosomal DNA restriction analysis and single strand conformation polymorphism.

A Microbiological Index of Soil Pollution (MISP) as an indicator of soil pollution by PCBs was proposed [32]. It is based on two concepts:

- some microorganisms can better develop in soils when unstable ones are depressed by PCBs,
- stable microorganisms can develop in the presence of different pollutants.

MISP is measured by the total content of soil anaerobic micro-organisms per ml of nutrient medium containing the contaminant. The greater the PCB content in the soil, where unstable microorganisms have been suppressed and used by stable ones as a source of easily available nutrient, the higher is the MISP. Significant correlations between PCB content and MISP value suggest that the index provides an adequate ecological information under high level of soil pollution.

Fate and impact of pesticides on soil biota and non-target organisms is a concern. In a study of the degradation of four pesticides [33], soil physical, chemical and biological properties were measured in order to relate pesticide concentration with soil water potential and soil respiration rate over time. A relation between soil respiration rate and residual pesticide concentration indicated an effect of microbial activity on pesticide degradation. Soils with higher OM and moisture had a greater rate of degradation.

2.4. Effects of soil colloids on microorganisms, immobilization and activity of biopolymers

Soil, plant, and microorganisms interact in the rhizosphere, where soil chemical, physical and biological properties differ from the bulk soil [34]. Low molecular organic acids, phenols, amino acids and reduced sugars from plant and microbial origin play a major role in the biogeochemistry of trace elements. Organic acids, through their ability to solubilize and complex trace elements and as oxidizing agents, influence soil weathering and plant uptake of essential and toxic elements. Organic acids, tannins and humic acids promote and hinder anion retention by the Al or Fe-rich materials and, by forming complexes with Al and Fe, prevent their crystallization. Violante and De Cristofaro [35] studied the adsorption of phosphate on variable charge minerals. Phosphate may substitute for the organic ligands that have weak affinity for Al and Fe oxides. On the contrary, organic (oxalate) and inorganic (sulfate) ligands co-precipitated with Al, and involved in the structural network of amorphous precipitates, are only partly removed by phosphate. pH, nature of the ligand, order of anion addition, and molar ratio had an effect on these reactions.

The role of microorganisms in the degradation and mineralization of organic matter, and the weathering of soil minerals was discussed [36]. As the bio-degradation and mineralization of soil organic matter by the microflora proceeds, solubilization of major elements takes place, mainly through anaerobic and aero-anaerobic (facultative anaerobic) bacteria, in particular Fe-reducing bacteria. Fe from ferric oxyhydroxydes is found mainly as ferrous iron and in organo-metallic complexes (ferric and ferrous). Gaillard et al. [37] investigated the bio-degradation of wheat straw residue at a microscale. They observed millimetric gradients of C, enzyme activity and change of the faunal population near the residues. Four compartments were identified, i.e. the straw, the adhering soil, the soil up to 4 mm from the residues, and the soil beyond 4 mm. Only 3% of the ^{13}C added with the straw was found in the soil up to 4 mm after 10 days. Soil cores sampled in these compartments were incubated, and the $^{13}C-CO_2$ evolved provided a measure of the contribution of each compartment to the straw C mineralization.

2.5. Bioavailability of metals, metalloids and xenobiotics adsorbed on mineral-organic soil components

Microorganisms are exposed in the soil to the toxic effects of pollutants, e.g. heavy metals. To follow the evolution of soil ecology, studies are needed on gene exchange (conjugation, transformation and transduction), a driver of bacterial dynamics and evolution, and on the persistence of genetic material (transforming DNA molecules, transducing virus particles) in natural soils as a result of their interaction with the soil components, mainly the clay particles. Genetic material can persist for a long time in the soil as a result of its reaction with clay particles [38]. Mechanisms by which surface-active particles, e.g. clays and humic substances, contribute to the persistence of biomolecules that would otherwise be quickly degraded or inactivated by the microbiota were examined [39]: insecticidal proteins of *Bacillus thuringiensis* and of transgenic plants containing the toxin genes, enzymes, bacterial transforming DNA, and viruses provide examples of resistance to bio-degradation and retention of biological activity when bound to soil colloids. Clay particles affect the activity, ecology, and population dynamics of microbes (including viruses), the impacts of anthropogenic pollution (heavy metals, acid precipitation, pesticides) on microbes and plants, and the transfer of genetic information among bacteria in the soil. All these factors affect the "health" of soil habitats.

The role of humic substances (HS) in soil aggregation was studied by the flocculation and dispersion of montmorillonite as a function of exchangeable cation, humic and fulvic acid concentration and pH [40]. Flocculation values, edge-face and edge-edge interactions at various pH, heteroflocculation and its response to pH, and concentration of HS were examined in an experiment using soil columns treated with wastewater. A sharp decrease in hydraulic conductivity resulted from a decrease in pore size that reflects two processes: retention of some dissolved organic matter during water percolation, and a change in pore-size distribution due to swelling and dispersion of clay particles. A "sensitive" pool in the soil, that responds more quickly than bulk organic matter to changes in soil management and climate change was identified [41] by using enzyme activity in different soil fractions, 2000-200, 200-63, 63-2, 2-0.1 and <0.1 μm to evaluate the effect of amendments on soil OM. Organic carbon (OC) increased in the bulk soil, ranging from 10.8 (fallow) to 32.0 $g.kg^{-1}$ (peat). Total recovery of OC in the size fractions amounted from 94 to 106% of the bulk soil content. Size fraction contribution to total OC differed between treatments. The silt fraction yielded the highest amount of OC and its contribution increased with increasing bulk OC. Similar trends were observed for the sand fractions. The contribution of the fine fractions decreased significantly with increasing bulk OC. Xylanase and invertase activity in the bulk soil increased after addition of amendments: xylanase activity was located mainly in the silt fraction, and invertase in the clay and silt fraction. Invertase response to amendments was observed in the sand fraction. Hot water extraction was used to separate the easily decomposable, labile fraction of the active soil organic matter (SOM) that contains some microbial biomass and simple organic compounds [42]. SOM pools were isolated according to particle size and specific density. Mass distribution of isolated fractions reflected the soil type and the management dependence. Organo-chemical uptake by plants is negatively correlated with the concentration in hot water extractable fraction. Correlations between hot water-extractable SOM and soil biological processes and parameters suggest that the content of organo-chemicals in this fraction may be adequate for predicting the bioavailability of pollutants.

Celis et al. [43] assessed herbicide-organoclay formulations to prevent soil and water contamination. Hexadecyltrimethylammonium-montmorillonite (HDTMA) was mixed with the highly mobile herbicide hexazinone and the complexes (HDTMA-Hex) were assayed as controlled-release (CR) formulations of the herbicide. Release of the active compound was followed in water and in water-soil suspension by HPLC. Release of hexazinone to water was greatly reduced in the formulations based on organoclays.

2.6. The role of minerals-organic matter-soil biota interactions in the restoration of perturbed ecosystems

Phytoremediation, the use of plants for the treatment of contaminated soils, is an emerging technology for effective and inexpensive cleanup of various hazardous wastes. Germida et al. [44] reviewed the key processes, and the promises and limitations of phytoremediation. This *ecological engineering* capitalizes on synergistic relationships among plants, microorganisms, and the environment. It takes advantage of the extensive root systems of the plants that explore large soil volumes, support large bacterial populations in the rhizosphere, and produce exudates which affect the activity of rhizobacterial populations. Rhizosecretion is a technology based on molecular farming designed to produce useful and valuable products (exudates) from plant roots. This plant-based biotechnology rests on the ability of plant roots to secrete various substances in their immediate environment. The rhizosphere can enhance the population and activity of microrganisms susceptible to degrade xenobiotics. An agricultural waste, the olive mill-waste water was found to favor the growth of a specific fungal population, and thus modified the microbial composition [45].

Churchman [46] discussed remediation from the angle of soil degradation. A loss of clays, as a result of, or leading to a loss of the associated organic matter, is conducive to a breakdown of soil structure, a loss of the capacity to hold and supply nutrients and water to plants, and to support useful microorganisms; soil degradation can also be chemical, as a result of the presence of inorganic and organic pollutants. In healthy soils, clay particles are embedding organic matter and other minerals to form microaggregates, and they protect organic matter from degradation. Microaggregates provide habitats for microorganisms. In degraded soils, clays occur frequently dissociated from other components, so that their removal by erosion is enhanced. Remediation involves practices that will restore soil quality through the balance among soil components.

Soil biological sensors can be used to detect ecosystem stress and environmental pollution, and help developing sustainable soil management [47]. Enzymatic properties are rather stable but may reflect soil management or detect environmental stress. Most soil enzymes have a significant percentage of their activity associated with abiontic enzymes that exist in two locations: adsorbed on internal or external clay surfaces, and complexed with humic colloids through adsorption, entrapment, or co-polymerization. Many enzymes are substrate-specific and can be selected from different functional groups to represent the potential of a soil to perform a range of reactions critical for the functioning of the ecosystem. Their absence may be indicative of a "stressed or degraded" soil.

3. AN OVERVIEW

Presentations at the Symposium provided information in several areas:

Understanding the soil ecosystem in space and time. Progress in the knowledge of the role of microbial communities in the ecosystem was achieved with new or improved methods for characterizing the nature and functioning of biota. Techniques allow to identify key organisms and to determine genetic and physiological profiles, in communities and single species, and at different scales.

Inorganic and organic soil components and surface properties. Synthesis of information has improved the understanding of colloid surface properties, their reactivity and ability to immobilize and release heavy metals and xenobiotics. Colloids, both mineral and organic, play an active role in the protection of soil microorganisms. The easily decomposed organic matter is a major factor affecting physical and chemical soil characteristics.

Interactions of heavy metals with the soil ecosystem. Information was provided on the fate of heavy metals and their effects on the soil components. Regulations on total metal loading in the soil are intended to ensure crop and water quality, but do not consider the effects on microoorganisms and microbial processes; they assume little or no mobility of the metals in the soil. Yet, colloids can facilitate the transport of metals towards surface and groundwater.

Microorganisms affect the mobility and availability of heavy metals through adsorption, uptake, chelation, storage, release of extracellular complexing products, and they are affected by the presence of heavy metals in their environment. In metal-contaminated soils, the organic matter turnover is reduced: this prevents the soil from performing some key functions in the ecosystem.

Impacts of soil management. Cultivation has a strong impact on soil organic matter: the content may increase or decrease according to management practices, and this affects properties such as tilth, microbial activity and productivity levels. Addition of fresh organic matter from crop residues stimulates microbial growth, extracellular excretion and formation of microbial decay products, that may combine in different ways with clay particles.

Xenobiotics in the environment. Diversity of chemicals and soil environments constitutes a challenge for interpreting the fate, bioavailability and toxic effects of xenobiotics in the soils. Low molecular weight organic acids from root exudates and decaying organic matter, and monomeric humic substances are important ligands: bonds may be formed through oxidative coupling reactions, catalyzed biologically by enzymes, or abiotically by clays and oxides. The fate of pesticides and their degradation products is a major issue. Rate of degradation is related to organic matter and microbial activity. Soil structure influences also bio-degradation through its effects on the reactions between organic matter and microorganisms.

Nutrient availability. Soil fertility is strongly related to microbial activity; mobility and availability of major and minor elements is influenced by microorganisms. Fe is made available to plants by organic acids that release Fe from minerals by chelation or reduction. Weathering of minerals and solubilization of elements is achieved in part by the natural microflora, e.g. the anaerobic or facultative anaerobic bacteria releasing soluble ferrous Fe. Phosphate may also form complexes with Al and does so in competition with some organic ligands.

Bioremediation. The faculty of soil microorganisms to synthesize antibiotics when competing for space and food may be used as a mechanism to control undesirable organisms. Phytoremediation, the use of plants for *in situ* treatment of contaminated soils, is an effective and inexpensive cleanup technology for a variety of wastes. Substrate-specific enzymes in diverse functional groups are necessary for the soil capability to perform some key functions, and are indicative of a "stressed or degraded" soil. Soil remediation can be also achieved through the restoration of the equilibrium among the soil components.

4. RESEARCH PERSPECTIVES

The objectives and activities of the International Union of Soil Sciences (IUSS) Working Group MO, i.e. to supplement information on the “Interactions of soil minerals with organic components and microorganisms”, are directly relevant to Division 2, “Soil Properties and Processes”. This Division is expected to provide the scientific background and understanding of the soil ecosystem, that involves the integration of the physical, chemical, biological and mineralogical properties of the soil, and the interactions among the mineral, organic and biological components in the biotic and abiotic control of the soil functions. Soils play a key role for human and environmental health [48]; in order to develop and convey the knowledge on soil functioning to non-soil scientists and other users of this information, the approach that will deal with the soil as an entity must be favored as much as possible. This statement can be clearly illustrated by an example: insecticidal protein activity of *Bacillus thurigiensis.* is maintained when these molecules are adsorbed by the soil colloids, whereas they would be otherwise rapidly degraded by microorganisms. This observation calls for information in several areas: among others, the study of surface properties of the organic and mineral soil components to interpret the adsorption, exchange and protection mechanisms, the identification and ecology of soil microorganisms susceptible to participate in the bio-degradation, the processes involved, and the environmental conditions under which the bio-degradation can be actually performed,... The same diversity of information is required to evaluate the long-term impacts on human and ecosystem health of a variety of chemicals and products used by agriculture and other industrial activities. It is also the responsibility of the IUSS to address emerging concerns, e.g. the fate and effects of endocrine disruptive substances and other toxic substances present in or brought to the soil as a result of human activities: land use for sewage sludge and organic waste disposal is a growing concern. The accumulated knowledge will then be used by the other Divisions, especially Division 4, "Role of Soils in Sustaining Society and the Environment": addressing issues such as human and environmental health, food security and safety, biodiversity requires soil scientists and the scientific community to provide decision- and policy-makers with a strong scientific support for the development of policies that will lead to the sustainable land use by humans. How do these considerations translate in the future activities of the Working Group?

The evolution of soil functioning in space and time will obviously require further attention. Understanding the complexity of the soil and its functioning requires a multidisciplinary and integrated approach. Studies consider often binary system interactions among the soil components or between soil components and metals or xenobiotics. Knowledge is based largely on global properties and effects, and on the properties and functions of selected components and organisms. Investigations on such systems are a step towards the understanding of the soil ecosystem, but studies that will consider all the elements in a comprehensive approach (Figure 1) will be increasingly required. Fragmentation of the soil research on individual or a limited number of components may lead to an incomplete if not misleading interpretation of the global processes taking place in the ecosystem. Is it sufficient to focus on "representative" organisms of a microbial community to describe the soil ecosystem? The validity of binary system studies to interpret the soil phenomena must be assessed; in other words, is the understanding of the soil ecosystem functioning reducible to the study of its components? In the integrated approach, there will be a need for non-destructive techniques that will allow to study the soil under natural or close to natural conditions, and on a repetitive basis at the same site. Techniques are still often a limiting factor in the study of biotic processes under natural conditions. More extensive studies of microhabitats will be needed to add to the knowledge and understanding of the behavior and activity of microbial communities and their interactions with the other soil components. Duration of many studies is short or medium term. Because microbial populations and activity fluctuate considerably over the year, studies would cover longer periods of time in order to assess the effects of such variations on the soil ecosystem.

The spatial and temporal variability of soil properties prompts for a risk assessment when extrapolating the research findings to other areas or conditions. To what extent can the laboratory models be extrapolated to field conditions, and the results of field studies generalized at the regional level? This must be addressed to ensure the maximum benefit from the soil research activities, and to assist IUSS Divisions 1 "Soil in Space ands Time" and 3 "Soil Use and Management" in fulfilling their mandate and be responsive to the society. Unless soil scientists assume their responsibilities and speak clearly on topics in their areas of specialization, other groups will try to fill this role.

Increasing concerns about the entry of heavy metals and organic contaminants into the food chain through fertilizers, sewage sludge and other organic soil amendments must be addressed in relation to human health. Food production systems would be reassessed in relation to their impacts on the environment and be increasingly integrated in the ecosystem, in order to maintain the sustainability of the production system, the natural resources and the environment health. The Green Revolution, based on new crop varieties, was followed by the chemical revolution, based on fertilizers and pesticides, that allowed to obtain the full potential of the improved varieties while protecting them from pests and diseases. The biotechnological revolution has numerous applications in the use of land by humans, for food production and other benefits. However, biotechnology applications prompt many concerns, and it is the role of soil scientists to take part in the debate. The Life Science revolution, i.e. the application of research outputs to meet human needs in a range of areas, e.g. the bio-medicals and bio-fuels streams, opens new research areas and more opportunities for soil science in the future. The Life Science revolution is based largely on products grown on and found in the soil.

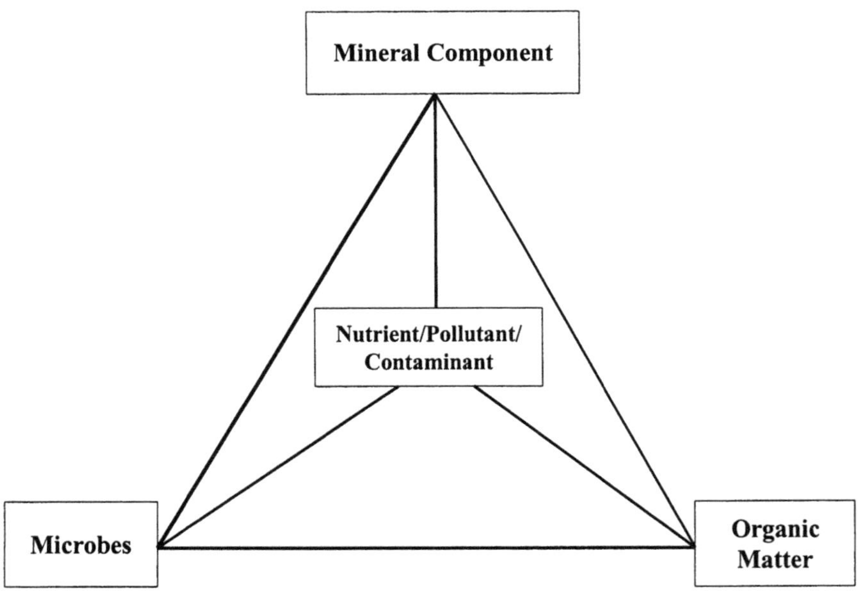

Figure 1. The interaction of the components in the soil system: an integrated approach.

In relation to the above observations, the cooperation between the Working Groups MO "Interactions of soil minerals with organic components and microorganisms" and SU "Soils of urban, industrial, traffic, and mining areas" would increase in the future in order to assess and understand the properties of the numerous "brownfield" sites around the world, and assist in developing guidelines for their use for food production and food safety. Urban and suburban soils provide indeed the food to a large percentage of the world's population, and many questions relevant to human and environmental health need to be addressed from the soil perspective.

Detailed rhizosphere studies will likely increase in the future. Several communications indicated various areas of interest: the rhizosphere hosts a larger variety of microorganisms than the bulk soil, and they combine with root exudates and decaying products to play an important role in many reactions taking place in the soil; this role has to be better understood. Rhizosecretion is a technology based on molecular farming designed to produce valuable or useful products (exudates) from plant roots. This biotechnology uses the ability of plant roots to absorb or release substances in their immediate environment. Taking advantage of the inherent soil biodiversity is a tool for pest control and a substitute for chemicals: promoting the growth of beneficial or neutral microorganisms will help controlling the growth of detrimental ones. There are also applications in soil remediation. Knowing the factors that ensure the survival of microorganisms in the ecosystems will allow to manipulate the diversity and numbers of beneficial and useful organisms, and to encourage the natural suppression of plant pathogens.

Soil degradation affects a large percentage of the agricultural lands: remediation can be approached from the soil degradation and restoration perspectives. In healthy soils, clays surround organic matter and other minerals in microaggregates, protecting the organic matter from degradation; aggregates provide habitats for microorganisms. In degraded soils, clays are in a dispersed state, and easily removed by erosion. A loss of clay particles can cause the loss of the associated organic matter, a breakdown of the soil structure, a loss of the capacity of soils to hold and supply nutrients and water, and to support microorganisms. Controlling soil degradation and facilitating the soil functions is a direct mechanism for soil remediation, and another avenue for ensuring the sustainability of the natural resources.

REFERENCES

1. United Nations, 1997. World population prospects. The 1996 revision. UN Population Division, New-York.
2. Wagner, W., 2000. Comparative analysis of AKS approaches in addressing problems related to food safety in OECD member countries. Paper AGR/CA/AKS(2000)%. Conference on Agricultural Knowledge Systems (AKS) addressing food safety and environmental issues, OECD, Paris, January 2000, pp. 16
3. World Resources Institute, 1996. The Urban Environment - World Resources. A guide to the Global Environment. Oxford University Press, New-York.
4. Rutgers, M., Wouterse, M.J., Bogte, J.J., Breure, A.M., 2000. Two ecological indicators for soil quality based on the analysis of microbial communities using Biolog plates. *In:*

Violante, A., Gianfreda, L. (Eds.), Soil Mineral-Organic Matter-Microorganism Interactions and Ecosystem Health. Abstracts, 3[d] Int. Symp. of the IUSS Working Group MO, Naples-Capri, Italy, p. 154.

5. Ciffroy, P., Charlatchka, R., 2000. Determination of conditional stability constants for the complexation of Cu and Co ions with natural fulvic acids. Development of a dialysis procedure. *In:* Violante, A., Gianfreda, L. (Eds.), Soil Mineral-Organic Matter-Microorganism Interactions and Ecosystem Health. Abstracts, 3[d] Int. Symp. of the IUSS Working Group MO, Naples-Capri, Italy, p. 57.
6. Gigliotti, C., Farini, A., 2002. Microbial biomass response to heavy metals in the field. *In:* Violante, A., Huang, P.M., Bollag, J.-M., Gianfreda, L. (Eds.), Soil Mineral-Organic Matter-Microorganism Interactions and Ecosystem Health. Ecological Significance of the Interactions among Clay Minerals, Organic Matter and Soil Biota. Elsevier, The Netherlands. This issue.
7. Kandeler, E., Tscherko, D., Bruce, K.D., Stemmer, M., Hobbs, P.J., Bardgett, R.D., Amelung, W., 2000. Structure and function of the soil microbial community in particle size fractions of a heavy metal polluted soil. *In*: Violante, A., Gianfreda, L. (Eds.), Soil Mineral-Organic Matter-Microorganism Interactions and Ecosystem Health. Abstracts, 3[d] Int. Symp. of the IUSS Working Group MO, Naples-Capri, Italy, p. 142.
8. Jackson, T.A., Leppard, G.G., 2000. Relations between heavy metals, microorganisms, and natural colloids in contaminated ecosystems: results of research employing energy dispersive X-ray microanalysis. *In*: Violante, A., Gianfreda, L. (Eds.), Soil Mineral-Organic Matter-Microorganism Interactions and Ecosystem Health. Abstracts, 3[d] Int. Symp. of the IUSS Working Group MO, Naples-Capri, Italy, p. 37.
9. Sparks, D.L., 2000. Advances in understanding the kinetics and mechanisms of metal sorption at the mineral/water interface. *In*: Violante, A., Gianfreda, L. (Eds.), Soil Mineral-Organic Matter-Microorganism Interactions and Ecosystem Health. Abstracts, 3[d] Int. Symp. of the IUSS Working Group MO, Naples-Capri, Italy, p. 34.
10. Egli, M., 2000. Short-term changes in heavy metal contents of forest soils near an aluminium industrial plant. *In*: Violante, A., Gianfreda, L. (Eds.), Soil Mineral-Organic Matter-Microorganism Interactions and Ecosystem Health. Abstracts, 3[d] Int. Symp. of the IUSS Working Group MO, Naples-Capri, Italy, p. 170.
11. Matschonat, G. 2002. Capacity of organically complexed aluminum, ionic strength, and pH to affect the CEC of organic samples. *In:* Violante, A., Huang, P.M., Bollag, J.-M., Gianfreda, L. (Eds.), Soil Mineral-Organic Matter-Microorganism Interactions and Ecosystem Health. Ecological Significance of the Interactions among Clay Minerals, Organic Matter and Soil Biota. Elsevier, The Netherlands. This book, Volume A.
12. Krishnamurti, G.S.R., Wang, M.K., Huang, P.M., 2000. Phenol-Al^{13} tridecamer interactions: implications in Al transformation and abiotic formation of humic substances. *In*: Violante, A., Gianfreda, L. (Eds.), Soil Mineral-Organic Matter-Microorganism Interactions and Ecosystem Health. Abstracts, 3[d] Int. Symp. of the IUSS Working Group MO, Naples-Capri, Italy, p. 111.
13. Torrent, J., Barron, V., 2000. Characterizing the reactivity of soil iron oxides: old and new methods. *In*: Violante, A., Gianfreda, L. (Eds.), Soil Mineral-Organic Matter-Microorganism Interactions and Ecosystem Health. Abstracts, 3[d] Int. Symp. of the IUSS Working Group MO, Naples-Capri, Italy, p. 45.
14. Lamy, I., Citeau, L., Denaix, L., Elsass, F., van Oort, F., 2000. Significance and fate of mineral and organic colloids in gravitational water of metal-polluted soils. *In*: Violante,

A., Gianfreda, L. (Eds.), Soil Mineral-Organic Matter-Microorganism Interactions and Ecosystem Health. Abstracts, 3[d] Int. Symp. of the IUSS Working Group MO, Naples-Capri, Italy, p. 60.

15. Turner, B.L., Haygarth, P.M., 2000. Microbiological and enzymatic controls on soil P release to water. *In*: Violante, A., Gianfreda, L. (Eds.), Soil Mineral-Organic Matter-Microorganism Interactions and Ecosystem Health. Abstracts, 3[d] Int. Symp. of the IUSS Working Group MO, Naples-Capri, Italy, p. 65.
16. Boucher, U., Balabane, M., 2000. Interactions between the biodegradation of metal contaminated plant debris and the fate of associated metals. *In*: Violante, A., Gianfreda, L. (Eds.), Soil Mineral-Organic Matter-Microorganism Interactions and Ecosystem Health. Abstracts, 3[d] Int. Symp. of the IUSS Working Group MO, Naples-Capri, Italy, p. 85.
17. Burgos, P., Madejon, E. and Cabrera, F., 2002. Changes in soil organic matter, enzymatic activities and heavy metal availability induced by application of organic residues. *In:* (Violante, A., Huang, P.M., Bollag, J.-M., Gianfreda, L. Eds.), Violante, A., Huang, P.M., Bollag, J.-M., Gianfreda, L. (Eds.), Soil Mineral-Organic Matter-Microorganism Interactions and Ecosystem Health. Ecological Significance of the Interactions among Clay Minerals, Organic Matter and Soil Biota. Elsevier, The Netherlands. This issue.
18. Mendoza, J., Garrido, T., Ahumada, I., 2000. Evaluation of Cr, Cu and Cd bioaccumulation by celery plants (Apium Graveolens) grown on contaminated substrate. *In*: Violante, A., Gianfreda, L. (Eds.), Soil Mineral-Organic Matter-Microorganism Interactions and Ecosystem Health. Abstracts, 3[d] Int. Symp. of the IUSS Working Group MO, Naples-Capri, Italy, p. 113.
19. Sappin-Didier, V., Brayette, S., Jade, C. Masson, P., Mench, M., 2000. Cd phyto-availability in sandy and calcareous soils assessed by the ^{109}Cd exchange kinetic technique. *In*: Violante, A., Gianfreda, L. (Eds.), Soil Mineral-Organic Matter-Microorganism Interactions and Ecosystem Health. Abstracts, 3[d] Int. Symp. of the IUSS Working Group MO, Naples-Capri, Italy, p. 122.
20. Jahn, R., Rössner, J., Kleber, M., Tanneberg, H., 2000. Clay mineralogy and carbon sequestration in loess-derived soils. I Mineralogical and pedological clues. *In*: Violante, A., Gianfreda, L. (Eds.), Soil Mineral-Organic Matter-Microorganism Interactions and Ecosystem Health. Abstracts, 3[d] Int. Symp. of the IUSS Working Group MO, Naples-Capri, Italy, p. 120.
21. Kleber, M., Rössner, J., Glaser, B., Knicker, H., Jahn, R., 2000. Clay mineralogy and carbon sequestration in loess-derived soils. II. Organic matter chemistry. *In*: Violante, A., Gianfreda, L. (Eds.), Soil Mineral-Organic Matter-Microorganism Interactions and Ecosystem Health. Abstracts, 3[d] Int. Symp. of the IUSS Working Group MO, Naples-Capri, Italy, p. 108.
22. Hossain, M.F., Elahi, S.F., White, S.K., Hossain, M, Sarker, S., Alam, Q.K., Gaunt, J.l., Rother, J., 2000. A comparison between ecological and conventional rice farming systems in Bangladesh. *In*: Violante, A., Gianfreda, L. (Eds.), Soil Mineral-Organic Matter-Microorganism Interactions and Ecosystem Health. Abstracts, 3[d] Int. Symp. of the IUSS Working Group MO, Naples-Capri, Italy, p. 181.
23. Chenu, C., Arias, M., Besnard, E., Arrouays, D., 2000. The influence of cultivation on the composition and properties of clay-organic matter associations in silty soils. *In*: Violante, A., Gianfreda, L. (Eds.), Soil Mineral-Organic Matter-Microorganism Interactions and Ecosystem Health. Abstracts, 3[d] Int. Symp. of the IUSS Working Group MO, Naples-

Capri, Italy, p. 91.
24. Leifeld, J., Kögel-Knaber, I., 2000. Particulate organic matter in agricultural soils as affected by aggregation and land management. *In*: Violante, A., Gianfreda, L. (Eds.), Soil Mineral-Organic Matter-Microorganism Interactions and Ecosystem Health. Abstracts, 3[d] Int. Symp. of the IUSS Working Group MO, Naples-Capri, Italy, p. 187.
25. Kirchmann, H., Gerzabek, M.H., 2002. Pore size changes in a long-term field experiment with organic amendments. *In:* Violante, A., Huang, P.M., Bollag, J.-M., Gianfreda, L. (Eds.), Soil Mineral-Organic Matter-Microorganism Interactions and Ecosystem Health. Ecological Significance of the Interactions among Clay Minerals, Organic Matter and Soil Biota. Elsevier, The Netherlands. This book, Volume A.
26. Kiem, R., Knicker, H., Körschens, M., Kögel-Knabner, I., 2000. Refractory soil organic carbon as studied in particle-size fractions of arable soils. *In*: Violante, A., Gianfreda, L. (Eds.), Soil Mineral-Organic Matter-Microorganism Interactions and Ecosystem Health. Abstracts, 3[d] Int. Symp. of the IUSS Working Group MO, Naples-Capri, Italy, p. 74.
27. Rajapaksha, R.M.C.P., Raufa, M.C.F., 2000. Diversity of microbial populations in cultivated lands contaminated with Cu, Cd and Zn. *In*: Violante, A., Gianfreda, L. (Eds.), Soil Mineral-Organic Matter-Microorganism Interactions and Ecosystem Health. Abstracts, 3[d] Int. Symp. of the IUSS Working Group MO, Naples-Capri, Italy, p. 198.
28. Bollag, J.-M., Strynar, M., Ahn, M.-Y., Dec, J., 2002. Characterization of enzymatic or abiotic immobilization of xenobiotics in soil. *In:* Violante, A., Huang, P.M., Bollag, J.-M., Gianfreda, L. (Eds.), Soil Mineral-Organic Matter-Microorganism Interactions and Ecosystem Health. Ecological Significance of the Interactions among Clay Minerals, Organic Matter and Soil Biota. Elsevier, The Netherlands. This issue.
29. Chen, M.C., Wang, M.K., Chiu, C.Y., 2000. Determination of low molecular weight dicarboxylic acids in rhizosphere and bulk soils of Tsuga and Yushania. *In*: Violante, A., Gianfreda, L. (Eds.), Soil Mineral-Organic Matter-Microorganism Interactions and Ecosystem Health. Abstracts, 3[d] Int. Symp. of the IUSS Working Group MO, Naples-Capri, Italy, p. 131.
30. Ruggiero, P., Pizzigallo, M.D.R., Crecchio, C., 2002. Effects of soil abiotic processes on the bioavailability of anthropogenic organic residues. *In:* Violante, A., Huang, P.M., Bollag, J.-M., Gianfreda, L. (Eds.), Soil Mineral-Organic Matter-Microorganism Interactions and Ecosystem Health. Ecological Significance of the Interactions among Clay Minerals, Organic Matter and Soil Biota. Elsevier, The Netherlands. This issue.
31. Burns, R.G., Stach, J.E.M., 2002. Microbial ecology of biofilms: activities, architecture and relevance to soil-microbe-xenobiotic interaction. *In:* Violante, A., Huang, P.M., Bollag, J.-M., Gianfreda, L. (Eds.), Soil Mineral-Organic Matter-Microorganism Interactions and Ecosystem Health. Ecological Significance of the Interactions among Clay Minerals, Organic Matter and Soil Biota. Elsevier, The Netherlands. This issue.
32. Motuzova, G., Skvorzova, I., Pavlenko, N., 2000. Microbiological diagnostics of soils polluted with polychlorbiphenyls. *In*: Violante, A., Gianfreda, L. (Eds.), Soil Mineral-Organic Matter-Microorganism Interactions and Ecosystem Health. Abstracts, 3[d] Int. Symp. of the IUSS Working Group MO, Naples-Capri, Italy, p. 192.
33. Webb, K.M., Aylmore, L.A.G., 2002. The role of soil organic matter and water potential in determining pesticide degradation. *In:* Violante, A., Huang, P.M., Bollag, J.-M., Gianfreda, L. (Eds.), Soil Mineral-Organic Matter-Microorganism Interactions and Ecosystem Health. Ecological Significance of the Interactions among Clay Minerals, Organic Matter and Soil Biota. Elsevier, The Netherlands. This book, Volume A.

34. Page, A.L., Chang, A.C., 2000. Biogeochemistry of trace elements in soils as influenced by soil-plant-microbial interactions. *In*: Violante, A., Gianfreda, L. (Eds.), Soil Mineral-Organic Matter-Microorganism Interactions and Ecosystem Health. Abstracts, 3[d] Int. Symp. of the IUSS Working Group MO, Naples-Capri, Italy, p. 32.
35. Violante, A., Pigna, M., Ricciardella, R., Gianfreda, L., 2002. Adsorption of phosphate on variable charge minerals and soils as affected by organic and inorganic ligands. *In:* Violante, A., Huang, P.M., Bollag, J.-M., Gianfreda, L. (Eds.), Soil Mineral-Organic Matter-Microorganism Interactions and Ecosystem Health. Ecological Significance of the Interactions among Clay Minerals, Organic Matter and Soil Biota. Elsevier, The Netherlands. This book, Volume A.
36. Stemmler, S., Berthelin, J., 2000. Relation between soil organic matter biodegradation and mobilization of mineral elements in humid tropical environment (watershed of Nsimi, Cameroon). *In*: Violante, A., Gianfreda, L. (Eds.), Soil Mineral-Organic Matter-Microorganism Interactions and Ecosystem Health. Abstracts, 3[d] Int. Symp. of the IUSS Working Group MO, Naples-Capri, Italy, p. 136.
37. Gaillard, V., Chenu, C., Recous, S., 2000. Fluxes of C in the wheat straw detritusphere. *In*: Violante, A., Gianfreda, L. (Eds.), Soil Mineral-Organic Matter-Microorganism Interactions and Ecosystem Health. Abstracts, 3[d] Int. Symp. of the IUSS Working Group MO, Naples-Capri, Italy, p. 175.
38. Franchi, M., Rimaldi, L., Romanelli, B., Gallori, E., 2000. Effect of heavy metals on bacterial gene transfer by transformation and transduction. *In*: Violante, A., Gianfreda, L. (Eds.), Soil Mineral-Organic Matter-Microorganism Interactions and Ecosystem Health. Abstracts, 3[d] Int. Symp. of the IUSS Working Group MO, Naples-Capri, Italy, p. 174.
39. Stotzky, G., 2002. Clays and humic acids affect persistence and biological activity of insecticidal proteins from *Bacillus Thuringiensis* in soil. *In:* Violante, A., Huang, P.M., Bollag, J.-M., Gianfreda, L. (Eds.), Soil Mineral-Organic Matter-Microorganism Interactions and Ecosystem Health. Ecological Significance of the Interactions among Clay Minerals, Organic Matter and Soil Biota. Elsevier, The Netherlands. This issue.
40. Chen, Y., Tarchitzky, J., 2000. Dissolved organic matter and pH effects on clay dispersion and hydraulic conductivity of soils. *In*: Violante, A., Gianfreda, L. (Eds.), Soil Mineral-Organic Matter-Microorganism Interactions and Ecosystem Health. Abstracts, 3[d] Int. Symp. of the IUSS Working Group MO, Naples-Capri, Italy, p. 216.
41. Gerzabek, M.H., Haberhauer, G., Kandeler, E. , Sessitsch, A., Kirchmann, H., 2002. Response of organic matter pools and enzyme activities in particle size fractions to organic amendments in a long-term field experiment. *In:* Violante, A., Huang, P.M., Bollag, J.-M., Gianfreda, L. (Eds.), Soil Mineral-Organic Matter-Microorganism Interactions and Ecosystem Health. Ecological Significance of the Interactions among Clay Minerals, Organic Matter and Soil Biota. Elsevier, The Netherlands. This issue.
42. Schulz, E., 2000. Bioavailability of lindane, and methoxychlor in two different soil types as influenced by their associations with mineral-organic complexes. *In*: Violante, A., Gianfreda, L. (Eds.), Soil Mineral-Organic Matter-Microorganism Interactions and Ecosystem Health. Abstracts, 3[d] Int. Symp. of the IUSS Working Group MO, Naples-Capri, Italy, p. 146.
43. Celis, R., Carrizosa, M.J., Hermosin, M.C., Escalona, M.L., Cornejo, J., 2000. Herbicide-organoclay formulations to prevent soil and water contamination. *In*: Violante, A., Gianfreda, L. (Eds.), Soil Mineral-Organic Matter-Microorganism Interactions and Ecosystem Health. Abstracts, 3[d] Int. Symp. of the IUSS Working Group MO, Naples-

Capri, Italy, p. 105.
44. Germida, J.J., Frick, C.M., Farrell, R.E., 2002. Phytoremediation of oil-contaminated soils. *In:* Violante, A., Huang, P.M., Bollag, J.-M., Gianfreda, L. (Eds.), Soil Mineral-Organic Matter-Microorganism Interactions and Ecosystem Health. Ecological Significance of the Interactions among Clay Minerals, Organic Matter and Soil Biota. Elsevier, The Netherlands. This issue.
45. Banné, E., Ragano-Caracciolo, M., Santori, F., 2000. Fungal rhizosphere modification under olive mill-waste water treatment. *In*: Violante, A., Gianfreda, L. (Eds.), Soil Mineral-Organic Matter-Microorganism Interactions and Ecosystem Health. Abstracts, 3[d] Int. Symp. of the IUSS Working Group MO, Naples-Capri, Italy, p. 197.
46. Churchman, G.J., 2002. The role of clays in the restoration of perturbed ecosystems. *In:* Violante, A., Huang, P.M., Bollag, J.-M., Gianfreda, L. (Eds.), Soil Mineral-Organic Matter-Microorganism Interactions and Ecosystem Health. Ecological Significance of the Interactions among Clay Minerals, Organic Matter and Soil Biota. Elsevier, The Netherlands. This book, Volume A.
47. Dick, R.P., 2000. Enzymes stabilized on soil colloids as ecosensors. *In*: Violante, A., Gianfreda, L. (Eds.), Soil Mineral-Organic Matter-Microorganism Interactions and Ecosystem Health. Abstracts, 3[d] Int. Symp. of the IUSS Working Group MO, Naples-Capri, Italy, p. 44.
48. Huang, P.M., 2000. Foreseeable impacts of soil mineral-organic component-microorganism interactions on society: ecosystem health. *In*: Violante, A., Gianfreda, L. (Eds.), Soil Mineral-Organic Matter-Microorganism Interactions and Ecosystem Health. Abstracts, 3[d] Int. Symp. of the IUSS Working Group MO, Naples-Capri, Italy, p. 47.

INDEX

A

Adsorption of
- proteins 1-16
- toxins 59-68, 62

Allochemicals 135-145
- extraction and retention 139-141
- sample handling 138
- sources 138
- transformation 141-142

Arylsulphatase activity 345-351

B

Bacillus subtilis
- transformation, transduction 257-265

Bacillus thuringiensis 1-16, 59-68, 69-78

Bacterial community 363-375

Bioaugmentation 29

Bioavailability 20, 95-133
- abiotic reactions 117
- concept of 98
- of 2,4-D 21
- of organic residues 95-133
- of simazine 21

Biogeochemistry 43-57
- of iron 48
- of phosphorous 48
- trace elements 50

Bioremediation 1, 25

Biostimulation 25

Biotic and abiotic catalysts 291-295

BTEX 363-375

Bti crystals 71, 74

Btt toxins 60,
- adsorption of 62

C

Clays 1, 59, 69

Culex pipiens 69

E

Energy use in soil horizons 405-414

Environment and human health, future perspectives 415-432

Enzyme activity 329-344
- changes by organic residues applications 353-362
- response to organic amendments 329-344

Enzymes in soil
- as detoxifying agents 317-322
- behaviour 301-327
- effect of natural and anthropogenic factors on 311-317
- potential applications 301-327
- properties 301-327

F

Fatty acid diversity 80

G

Gene transfer,
- effect of Cu2+ and Hg 2+ on 257-265

H

Heavy metals
- availability 353-362
- in polluted soils 187-193
- microbial influence on 187-193
- mobilization of 187-193
- response to microbial biomass 195-204

Humic acids 1
- formation of 393-404

I

IMS-PCR methods 239-246

Inoculants 30

Insecticidal proteins 1-16

L
Lychens
- effect on mineral bioformation 377-391
- interaction with rocks 377-391
- metal complexes 386-387

M
Metsulfuron-methyl 147
- environmental effect of 147-156
Microbes 104
Microbial biomass 147, 195, 247-255
- cadmium and zinc toxicity 267-273
- interaction with heavy metals 416
Microbial
- activity
  – cadmium and zinc toxicity 267-273
  – effect of fire on 205-215
  – impact of tillage 275-287
  – in a crop rotation 275-287
  – in Mediterranean area 229-237
- community 217-228
- diversity 79-95
- temporal variability 247-255
Mineral
- formation and precipitation 381-385

N
NAPL 101
Nitrate dynamics 275
N-Mineralization 147-156

O
Oil-contaminated soils 169-186
Oxidative coupling reactions 290-291
Oxygenase systems 363-375

P
Phytoremediation 169-186
- influence of environment on 179-180
- mechanisms of 170-172
- rhizosphere effect 172-175
Pollutants
- ageing 22, 102
- organic 97

Q
Quercus ilex 229-237

R
Rhizosphere
- effect 172-175
- effect of root exudates 44

S
Salmonella recovery 239-246
Soil biofilms 17-42, 19
Soil ecology 17-42
Soil ecosystems 17
- characterization 86
- methods 88
Soil organic matter
- changes by organic residues applications 353-362
- effect of fire on 205-212
- interaction with minerals and living organisms 135-145
- response of 329-344
Soil weathering 50
- effect of lichens 378

T
Toxins 59, 69
Transgenic plants 1, 5
Trinitrotoluene 2, 4, 6
- toxicity of 161-162
- transformation and binding in soil 157-168
Tropical soils 59-68

X
Xenobiotics 135-145
- abiotic immobilization of 289-299
- abiotic transformations of 109
- binding- NMR application 295-296
- enzymatic immobilization of 289-299
- extraction and retention 139-141
- organic 99
- reaction of 111, 115
- sample handling 138
- sorption-desorption processes 99
- sources 138
- transformation 141-142

www.ingramcontent.com/pod-product-compliance
Ingram Content Group UK Ltd.
Pitfield, Milton Keynes, MK11 3LW, UK
UKHW020228290926
15470UKWH00001B/415